Representation Theory and C*-algebras

Representation Theory and *C-algebras** is devoted to the representation theory of solvable Lie groups and the associated non-commutative harmonic analysis including the study of C*-algebras. It contains full proofs of long-standing problems in the theory, including several polynomial conjectures and primitive zero ideals descriptions. It provides an in-depth study of their structural properties, the classification of unitary representations using the orbit method, and the underlying algebraic and analytic frameworks.

The book is most suitable for doctoral students, postdoctoral fellows, and researchers specialising in Lie theory, non-commutative geometry, functional analysis, operator algebras, and theoretical physics.

Features

- Complete solutions to polynomial conjectures (Corwin–Greenleaf and Duflo) in both nilpotent and exponential Lie group settings
- Complete results to long-standing problems about intertwining operators and the structure of primitive ideals in the exponential setting
- Comprehensive analysis of Casimir elements based on a new approach and their role in several related problems
- Detailed study of C*-algebras of solvable Lie groups, via methods using the Fourier transform and the spectral analysis
- Rich examples and counterexamples, including Heisenberg, thread-like, and G_6 groups
- Bridge between classical and modern methods in representation theory, with applications to harmonic analysis and mathematical physics.

Representation Theory and C*-algebras

Ali Baklouti, Hidenori Fujiwara, and Jean Ludwig

CRC Press
Taylor & Francis Group
Boca Raton London New York

CRC Press is an imprint of the
Taylor & Francis Group, an **informa** business
A CHAPMAN & HALL BOOK

Designed Cover Image: Shutterstock ID: 2636489537

First edition published 2027
by CRC Press
2385 NW Executive Center Drive, Suite 320, Boca Raton, FL 33431

and by CRC Press
4 Park Square, Milton Park, Abingdon, Oxon, OX14 4RN

CRC Press is an imprint of Taylor & Francis Group, LLC

ISBN: 978-1-041-16223-0 (hbk)
ISBN: 978-1-041-16662-7 (pbk)
ISBN: 978-1-003-68573-9 (ebk)

DOI: 10.1201/9781003685739

Typeset in Latin Morden font
by KnowledgeWorks Global Ltd.

Publisher's Note: This book has been prepared from camera-ready copy provided by the author.

Contents

Preface

This work delves into harmonic analysis of solvable Lie groups and associated C^*-algebras, addressing both foundational and advanced topics. It aims to advance the understanding of representation theory and C^*-algebra structures in solvable Lie groups, with broader implications for fields engaging with algebraic and analytical methodologies, think for instance of number theory, algebraic geometry, differential geometry, operator algebras, PDEs, and also with farther fields like physics.

Though many problems in harmonic analysis on solvable homogeneous spaces remain difficult subjects to solve, the orbit method comes as a fundamental tool to link in an elegant way unitary duals to the space of coadjoint orbits in the exponential setting. In this framework, a central component of this study is to provide exhaustive proofs of the so-called polynomial conjectures, which assert that some algebras of differential operators on homogeneous spaces are polynomial rings. These conjectures in both cases of induction and restriction seek to deepen our understanding of the disintegration machine and the differential operators algebras structures, and feature many examples illustrating these phenomena.

Further, we tackle Duflo's polynomial conjecture in many settings, and study Casimir elements, which play a pivotal role in the problems cited above, with applications to both theoretical and practical aspects of representation theory. Moving to intertwining operators of irreducible representations, the study addresses definitely a solution how to concretely realise the transmutation procedure on the space of representations in question.

The first chapter offers an in-depth overview of the representation theory of solvable Lie groups, with a particular emphasis on the class of exponential solvable Lie groups, which play a central role in non-commutative harmonic analysis. It begins by introducing fundamental structural aspects of these groups, including the nature of minimal non-central ideals that are pivotal in understanding their Lie algebraic properties. The chapter then develops the theoretical framework for studying their unitary representations, covering key concepts such as modular functions, invariant measures on homogeneous spaces, and the construction of induced representations, which serve as a principal tool in describing the unitary dual of such groups. Through illustrative examples like the Heisenberg and thread-like Lie groups, the discussion provides concrete models that demonstrate the subtleties of the general theory. A detailed case study is devoted to the exponential Lie group G_6, highlighting the classification of its coadjoint orbits, the associated topology of its unitary dual, and the relevance of these orbits in the orbit method approach. The chapter concludes with an analysis of branching laws, describing how representations behave under restriction to subgroups, both in the context of induced representations and when dealing with analytic subgroups. This comprehensive treatment bridges structural theory and representation-theoretic techniques, offering valuable insight into the harmonic analysis on solvable Lie groups.

The second chapter explores polynomial conjectures in the context of nilpotent Lie groups, focusing on both inductions and restrictions. We introduce the algebras of differential operators and state the commutativity conjectures for induction and restriction, and Duflo's polynomial conjecture. Key results include proofs of the Corwin-Greenleaf polynomial conjecture and fundamental tempered solutions for differential operators. Our study delves into multiplicities, Penney distributions, and rationality, providing comprehensive proofs and examples. This work serves as a foundational study of polynomial behaviour in the structure of nilpotent and exponential Lie groups, highlighting connections between algebraic and geometric aspects of representation theory.

The third chapter introduces the notion of Casimir elements, focusing on their emergence from minimal non-central ideals within the universal enveloping algebra of a Lie algebra. These elements serve as pivotal central operators that commute with all elements of the algebra, thereby acting as invariants in representation theory. Their connection with the algebra's centre highlights their significance in analysing representations and symmetries.

The existence of Casimir elements is rigorously established, leveraging a layering of Lie algebras. A detailed proof of the existence theorem is provided, elucidating the conditions under which Casimir elements arise. This result underscores their universality in enveloping algebras and their pivotal role in the determination of generation families of primitive ideals. We also shift focus to the Poisson centre of the symmetric algebra associated with a Lie algebra. It establishes the relationship between Casimir elements and the Poisson centre, providing a bridge between classical mechanics and representation theory. This connection is crucial for understanding the quantisation process.

Primitive ideals are intimately connected to the geometry of coadjoint orbits. This chapter also analyzes the generation of primitive ideals in cases where the orbits are flat, and introduced by utilising non-saturation ideals. We then generalise the results to encompass the general case, where we establish a unified framework for primitive ideals in arbitrary settings, linking their structure to geometric and algebraic invariants. This comprehensive treatment solidifies their place in the broader theory of Lie algebras.

Frobenius exponential Lie algebras are characterised via primitive ideals using previous results. This characterisation highlights the unique algebraic and geometric properties of these algebras, such as their connection to non-degenerate invariant bilinear forms. The discussion underscores their importance in both pure and applied mathematics, particularly in areas like differential geometry and theoretical physics. Chapter 4 explores polynomial conjectures in the exponential setting, focusing on monomial and restricted representations of discrete type, as well as their implications for Duflo's conjecture.

The monomial setting begins with an analysis of branching rules and Plancherel formulas, which describe how representations decompose when restricted to sub-groups. It introduces an algebra of differential operators relevant to monomial repre-sentations, followed by a study of discrete-type monomial representations and their connection to Penney distributions, a special class of probability measures. The main result is then presented, alongside concrete examples and a counterexample to Duflo's problem, which challenges a key assumption in representation theory.

The second part examines restricted representations of discrete type, including conditions under which representation restrictions remain discrete. It explores trace formulae, extends Penney-type distributions to this setting, and studies commutativity properties of the related algebraic structures. The extension of the polynomial conjecture is also considered, with supporting examples illustrating key phenomena.

Finally, the chapter directly addresses Duflo's polynomial conjecture in the exponential setting, first in the case of monomial representations, then in the context of discrete-type restrictions. The discussion highlights how Duflo's problem behaves in these settings, emphasising significant structural differences between exponential solvable Lie groups and classical nilpotent cases.

Let $G = \exp \mathfrak{g}$ be an exponential solvable Lie group with Lie algebra $\mathfrak{g}$ and $\mathfrak{g}^*$ the dual vector space of $\mathfrak{g}$. We take two (real) polarisations $\mathfrak{h}_1, \mathfrak{h}_2$ of $\mathfrak{g}$ at $f \in \mathfrak{g}^*$ which satisfy the Pukanszky condition and define a unitary character χ_f of $H_j = \exp(\mathfrak{h}_j)$ $(j = 1, 2)$ by the formula $\chi_f(\exp X) = e^{if(X)}(X \in \mathfrak{h}_j)$. Then, it is well known that the two monomial representations $\pi_j = \mathrm{ind}^G \chi_f (j = 1, 2)$ are mutually equivalent, and we even have a natural candidate of the intertwining operator between them. In order to verify that it is a true intertwining operator, the principal obstacle is the convergence of the integral which intervenes the realisations of $\pi_j (j = 1, 2)$. A property which assures this convergence, is the closedness of the simple product set $H_2 H_1$ in G. In Chapter 5, we establish this property, generalising then a previous proof in the particular case when one of the polarisations is of Vergne type (Amer. J. Math. **118** (1996), 839 878).

The last chapter provides a comprehensive study of C^*-algebras associated with solvable and nilpotent Lie groups, with a particular focus on their structure, Fourier analysis, and continuity properties. It begins with foundational definitions and explores specific types of C^*-algebras, including special solvable C^*-algebras and those with norm-controlled dual limits. A key theme is the continuity of operator fields and the behaviour of Fourier transforms on these algebras. The chapter presents detailed treatments of C^*-algebras for important examples such as the Heisenberg group, thread-like Lie groups, and the six-dimensional exponential Lie group G_6, including their topological and spectral properties. The chapter concludes with an extensive examination of the C^*-algebras of $(ax + b)$-like groups, analysing their dual spaces, coadjoint orbits, and Fourier transforms. Altogether, this chapter lays a rich analytic and algebraic foundation for understanding the operator-algebraic aspects of Lie group representation theory.

Authors

Ali Baklouti is a Full Professor of Mathematics at the University of Sfax, Tunisia. He earned his Ph.D. in Mathematics from the University of Metz, France, in 1995. He served as Vice-President of the University of Sfax from December 2020 to July 2024 and held the position of President of the Tunisian Mathematical Society for two consecutive terms (April 2016–March 2019 and April 2019–March 2022). Since January 2012, he has been the Deputy Director of the Mediterranean Institute of Mathematical Sciences, an institution he co-founded that same year. In December 2016, he was elected as a permanent member of the Tunisian Academy of Sciences, Letters, and Arts. Professor Baklouti has received numerous prestigious awards, including the AMU-PaCOM 2022 Award and Medal, Category A in Mathematics, and the Royal Society Africa Prize 2024. In the same year, he was also honoured with the Order of Merit for Education and Teaching by the President of Tunisia. He currently serves as Co-Editor-in-Chief of the Tunisian Journal of Mathematics (published by MSP, USA) and as Editor-in-Chief of Advances in Pure and Applied Mathematics (published by ISTE, UK). Additionally, he is a member of the editorial boards of other several journals, including the Graduate Journal of Mathematics (MIMS), and the Arabian Journal of Mathematics (Springer).

Hidenori Fujiwara is an Emeritus Professor of Kinki University, Japan. He had been a Full Professor of Kinki University for 26 years and retired in 2013. He received his Ph.D. in Mathematics from Tokyo University (Japan) in 1977. He studies the unitary representations of solvable Lie groups and the harmonic analysis on solvable homogeneous spaces. He published over 40 papers in peer-reviewed international journals, as well as two books.

Jean Ludwig is currently an Emeritus Professor at the University of Lorraine, France. He received his Ph.D. in Mathematics from the University of Bielefeld (Germany) in 1976 and his habilitation in 1979. He was a professor of Metz University (France) from 1990 to 2014. He had 13 PHD students, and he published over 100 papers in peer-reviewed international journals and proceedings and acted as a co-editor of the Journal of Lie Theory. He had been Directeur du Labaoratoire LMAM for 3 years and had many administrative duties at the Department and UFR level at the University of Bielefeld and Metz.

CHAPTER 1

Representation theory of solvable Lie groups

The material on solvable Lie groups and Lie algebras is quite standard. The readers could consult the references [23] and [31] and many references therein. The goal of the chapter is to record some main background on nilpotent, solvable, and exponential solvable Lie groups. We first recall the definitions of real solvable and nilpotent Lie algebras and define their associated Lie groups. We also go through some details with famous and important examples, such as Heisenberg groups, thread-like groups, and the affine group of the real line.

We also record basic results on representation theory of solvable and exponential Lie groups, and harmonic analysis on homogeneous spaces as will be dealt with throughout the book.

1.1 SOLVABLE LIE GROUPS

1.1.1 Solvable and exponential solvable Lie groups

Let $\mathfrak{g}_{\mathbb{C}}$ denote the complexified Lie algebra of a given Lie algebra $\mathfrak{g}$, where $\mathbb{R} \subset \mathbb{C}$ denote the fields of complex and real numbers. Let $ad_X, X \in \mathfrak{g}$ be the adjoint endomorphism defined by

$$ad_X(Y) = [X, Y],$$

where $[,]$ denotes the Lie bracket. When $[X, Y] = 0$ for any $X, Y \in \mathfrak{g}$, we say that $\mathfrak{g}$ is **commutative** or **Abelian**. A linear subspace $\mathfrak{a}$ of $\mathfrak{g}$ is an **ideal** if $[X, \mathfrak{a}] \subset \mathfrak{a}$ for any $X \in \mathfrak{g}$. Then, the quotient space $\mathfrak{g}/\mathfrak{a}$ endowed with the Lie bracket induced from $\mathfrak{g}$ becomes a Lie algebra. Let

$$D\mathfrak{g} = [\mathfrak{g}, \mathfrak{g}] = \mathbb{R}\text{-span}\{[X, Y]; X, Y \in \mathfrak{g}\},$$

then obviously $D\mathfrak{g}$ is an ideal of $\mathfrak{g}$ and the quotient Lie algebra $\mathfrak{g}/D\mathfrak{g}$ is commutative. We define the following sequences of vector spaces:

$$\mathfrak{g} = D^0\mathfrak{g} \supset D^1\mathfrak{g} \supset D^2\mathfrak{g} \supset ... \qquad (1.1.1)$$

DOI: 10.1201/9781003685739-1

and

$$\mathfrak{g} = C^0\mathfrak{g} \supset C^1\mathfrak{g} \supset C^2\mathfrak{g} \supset ..., \tag{1.1.2}$$

where

$$D^0\mathfrak{g} = \mathfrak{g}, D^k\mathfrak{g} = D(D^{k-1}\mathfrak{g})(k = 1, 2, ...),$$

and

$$C^0\mathfrak{g} = \mathfrak{g}, C^k\mathfrak{g} = [\mathfrak{g}, C^{k-1}\mathfrak{g}](k = 1, 2, ...).$$

In both cases, the decreasing sequences (1.1.1) and (1.1.2) turn out to be respectively flags of ideals of $\mathfrak{g}$.

The Lie algebra $\mathfrak{g}$ is said to be **solvable** if there exists k such that $D^k\mathfrak{g} = \{0\}$ and **nilpotent** if there exists k such that $C^k\mathfrak{g} = \{0\}$. On the other hand, $\mathfrak{g}$ is said to be m-step nilpotent (for some integer m), if $C^m\mathfrak{g} \neq \{0\}$ and $C^{m+1}\mathfrak{g} = \{0\}$.

Besides, a connected Lie group is said to be **solvable** (resp. nilpotent, m-step nilpotent) if its Lie algebra is solvable (resp. nilpotent, m-step nilpotent).

Comparing (1.1.1) and (1.1.2), nilpotent Lie groups are solvable Lie groups, but the gap between these two classes is quite wide. Here, we define some classes of solvable non-necessarily nilpotent Lie groups. Let

$$\exp = \exp_G : \mathfrak{g} \to G \tag{1.1.3}$$

be the exponential mapping of G.

Definition 1.1.1 *Let G be a solvable Lie group and $\mathfrak{g}$ its Lie algebra. Then G said to be* **exponential solvable** *if the exponential mapping (1.1.3) is a C^∞-diffeomorphism from $\mathfrak{g}$ onto G. In this case, let $\log = \log_G$ designate the inverse map of $\exp_G$.*

Any exponential solvable Lie group is therefore connected and simply connected. The following result provides a characterisation of such a class:

Theorem 1.1.2 *([70]) Let G be a connected, simply connected, and solvable Lie group, and let $\mathfrak{g}$ be its Lie algebra. Then the following are equivalent:*

1. *G is exponential solvable.*
2. *$\exp_G$ is an injective mapping.*
3. *$\exp_G$ is a surjective mapping.*
4. *No endomorphism $ad_X, X \in \mathfrak{g}$ has purely imaginary eigenvalues.*

Definition 1.1.3 *Let $\mathfrak{g}$ be a Lie algebra such that dim $\mathfrak{g} = n$. When there exists a sequence of ideals*

$$\{0\} = \mathfrak{g}_0 \subset \mathfrak{g}_1 \subset \cdots \subset \mathfrak{g}_{n-1} \subset \mathfrak{g}_n = \mathfrak{g}, \ \dim \mathfrak{g}_j = j \ (0 \leq j \leq n),$$

we say that $\mathfrak{g}$ is **completely solvable**. *Any completely solvable Lie algebra is an exponential solvable Lie algebra for which any endomorphism $ad_X, X \in \mathfrak{g}$ has only real eigenvalues. A connected Lie group is said to be completely solvable, if its Lie algebra is completely solvable.*

The following notion of co-exponential bases to closed subgroups of connected Lie groups (cf. [70]) plays a capital role in the sequel:

Definition 1.1.4 *Let G be a connected and simply connected Lie group and let H be a closed connected subgroup of G. Let $\mathfrak{g}, \mathfrak{h}$ be the Lie algebras of G and H, respectively. A basis $\{X_1, \dots, X_p\}$, $p = \dim(\mathfrak{g}/\mathfrak{h})$, is said to be co-exponential to $\mathfrak{h}$ in $\mathfrak{g}$ if the map:*

$$\begin{array}{cccc} \varphi_{\mathfrak{g},\mathfrak{h}}: & \mathbb{R}^p \times H & \to & G \\ & ((t_1, \dots, t_p), h) & \mapsto & \exp(t_p X_p) \cdots \exp(t_1 X_1) \cdot h \end{array}$$

is a diffeomorphism.

We have the following:

Theorem 1.1.5 *([70], Proposition 2) Let G be a connected, simply connected, and solvable Lie group. Then every connected closed subgroup of G admits a co-exponential basis.*

Remark 1.1.6 *A constructive proof of Theorem 1.1.5 is based on the three following assertions:*

(*i*) *If $\mathfrak{h}$ is one co-dimensional ideal of $\mathfrak{g}$, then any vector in $\mathfrak{g} \setminus \mathfrak{h}$ is a co-exponential basis.*

(*ii*) *If $\mathfrak{g} \supset \mathfrak{h}' \supset \mathfrak{h}$ and $\{b_1, \dots, b_d\}$ respectively $\{c_1, \dots, c_r\}$ is a co-exponential basis to $\mathfrak{h}'$ in $\mathfrak{g}$ respectively to $\mathfrak{h}$ in $\mathfrak{h}'$, then $\{c_1, \dots, c_r, b_1, \dots, b_d\}$ is a co-exponential basis to $\mathfrak{h}$ in $\mathfrak{g}$.*

(*iii*) *If $\mathfrak{h}$ is a maximal subalgebra of $\mathfrak{g}$, which is not an ideal of $\mathfrak{g}$, then any co-exponential basis to $\mathfrak{h} \cap [\mathfrak{g}, \mathfrak{g}]$ in $[\mathfrak{g}, \mathfrak{g}]$ is also a co-exponential basis to $\mathfrak{h}$ in $\mathfrak{g}$.*

As a direct consequence from Remark 1.1.6, we get the following lemmas (cf. [70],[6]).

Lemma 1.1.7 *Let $\mathfrak{h}$ be a subspace of $\mathfrak{g}$ containing $[\mathfrak{g}, \mathfrak{g}]$ and V a linear subspace of $\mathfrak{g}$ such that $\mathfrak{g} = \mathfrak{h} \oplus V$. Then any basis of V is a co-exponential basis to $\mathfrak{h}$ in $\mathfrak{g}$.*

Lemma 1.1.8 *Let $\mathfrak{h}$ be a subalgebra of $\mathfrak{g}$ such that $\mathfrak{h} + [\mathfrak{g}, \mathfrak{g}] = \mathfrak{g}$. Then there exists a co-exponential basis $\{u_1, \dots, u_r\}$ to $\mathfrak{h}$ in $\mathfrak{g}$ such that $[\mathfrak{g}, \mathfrak{g}] = W \oplus (\mathfrak{h} \cap [\mathfrak{g}, \mathfrak{g}])$, where W is the linear span of the family $\{u_1, \dots, u_r\}$.*

Definition 1.1.9 *1) With the notations above, a strong Malcev (or Jordan-Hölder) basis $\{Z_1, \dots, Z_m\}$ of $\mathfrak{g}$ is a basis of $\mathfrak{g}$ such that $\mathfrak{g}_i = \mathbb{R}$-span$\{Z_1, \dots, Z_i\}$ is an ideal of $\mathfrak{g}$ for every $i \in \{1, \dots, m\}$. The obtained flags of ideals*

$$\{0\} = \mathfrak{g}_0 \subset \mathfrak{g}_1 \subset \dots \subset \mathfrak{g}_m = \mathfrak{g}$$

is called strong Malcev (or Jordan-Hölder) sequence of $\mathfrak{g}$. It is well known that such basis exists if $\mathfrak{g}$ is nilpotent.

2) *A family of vectors* $\{X_1, \ldots, X_p\}$, $p = \dim(\mathfrak{g}/\mathfrak{h})$ *is said to be a Malcev basis of* $\mathfrak{g}$ *relative to* $\mathfrak{h}$ *if* $\mathfrak{g} = span\{X_1, \ldots, X_p\} \oplus \mathfrak{h}$ *and for all* $i \in \{1, \ldots, p\}$, *the vector space* $\mathbb{R}$-*span*$\{X_1, \ldots, X_i\} \oplus \mathfrak{h}$ *is a subalgebra of* $\mathfrak{g}$.

Remark 1.1.10 1. It is well known that nilpotent Lie groups admit Jordan-Hölder bases.

2. Let $\mathfrak{g}$ be a nilpotent Lie algebra and let $\mathscr{B} = \{Z_1, \ldots, Z_m\}$ be a strong Malcev basis of $\mathfrak{g}$ and

$$\mathscr{S}_{\mathfrak{g}} : \{0\} = \mathfrak{g}^0 \subset \mathfrak{g}^1 \subset \ldots \subset \mathfrak{g}^{m-1} \subset \mathfrak{g}^m = \mathfrak{g} \tag{1.1.4}$$

the associated Jordan-Hölder sequence of $\mathfrak{g}$ associated with the basis $\mathscr{B}$. Here $\mathfrak{g}^i = \mathbb{R}$-span$\{Z_1, \ldots, Z_i\}$, $i = 1, \ldots, m$. Such a notion of basis is very important and will be crucially used to adapt our setup with convenient coordinates which provide an adequate model for the adjoint action. Given a subalgebra $\mathfrak{h}$ of $\mathfrak{g}$, we denote by $\mathscr{I}_{\mathfrak{g}}^{\mathfrak{h}} = \{i_1 < \cdots < i_p\}$, ($p = \dim \mathfrak{h}$) the set of indices i ($1 \leq i \leq m$) such that $\mathfrak{h} + \mathfrak{g}^{i-1} = \mathfrak{h} + \mathfrak{g}^i$. We note for all $i_s \in \mathscr{I}_{\mathfrak{g}}^{\mathfrak{h}}$,

$$\tilde{Z}_s = Z_{i_s} + \sum_{r<i_s} \alpha_{r,s} Z_r \in \mathfrak{h}.$$

We get a basis $\mathscr{B}_{\mathfrak{h}} = \{\tilde{Z}_1, \ldots, \tilde{Z}_p\}$ which is a strong Malcev basis of $\mathfrak{h}$ singled out from $\mathscr{B}$. We define the matrix of $\mathfrak{h}$ written in the basis $\mathscr{B}$, denoted by $M_{\mathfrak{h},\mathscr{B}} \in M_{m,p}(\mathbb{R})$ and will be used later.

3. When $\mathfrak{g}$ is nilpotent, every subalgebra $\mathfrak{h}$ of $\mathfrak{g}$ admits such a Malcev basis, which is obviously a co-exponential basis.

1.1.2 Minimal non-central ideals of exponential Lie algebras

We here develop some structure details on minimal non-central ideals of exponential solvable Lie algebras. Let $\ell \in \mathfrak{g}^*$ be a linear form of an exponential solvable Lie algebra $\mathfrak{g}$, which does not vanish on any non-trivial ideal of $\mathfrak{g}$. This means that $\dim \mathfrak{z} \leq 1$, where $\mathfrak{z}$ denotes the centre of $\mathfrak{g}$, and that $\ell|_{\mathfrak{z}} \neq 0$ if $\mathfrak{z} \neq \{0\}$.

Take a minimal non-central ideal $\mathfrak{a}$ of $\mathfrak{g}$, which is abelian and of dimension ≤ 3 (cf. [70], Page 34). Let $\mathfrak{m}$ be the centraliser of $\mathfrak{a}$, which is clearly an ideal of $\mathfrak{g}$. Let

$$\mathfrak{a}^{\ell} = \{X \in \mathfrak{g}; \ell([X, \mathfrak{a}]) = \{0\}\} \tag{1.1.5}$$

and

$$\mathfrak{g}_0 = \{X \in \mathfrak{g}; [X, \mathfrak{a}] \subset \mathfrak{z}\}. \tag{1.1.6}$$

We shall encounter throughout the book chapters the following situations:

Assume first that $\mathfrak{z} = \{0\}$, then $\dim \mathfrak{a} = 1$ or 2 in this case.

(i) When $\dim \mathfrak{a} = 1$, we write $\mathfrak{a} = \mathbb{R}Y$ with $\ell(Y) = 1$. Then there exists a non-zero vector $X \in \mathfrak{g}$ such that

$$\mathfrak{g} = \mathfrak{m} \oplus \mathbb{R}X \text{ and } [X, Y] = Y. \tag{1.1.7}$$

(ii) Assume now that $\dim \mathfrak{a} = 2$ and write $\mathfrak{a} = \mathbb{R}Y_1 \oplus \mathbb{R}Y_2$. Then as in the previous case, one can write

$$\mathfrak{g} = \mathfrak{m} \oplus \mathbb{R}X \text{ with } [X, Y_1] = Y_1 - \alpha Y_2 \text{ and } [X, Y_2] = \alpha Y_1 + Y_2, \tag{1.1.8}$$

for some non-zero real number α. In this case, $(\ell(Y_1), \ell(Y_2)) \neq (0, 0)$.

We now assume that $\dim \mathfrak{z} = 1$. Let first $\dim \mathfrak{a} = 2$. We write $\mathfrak{z} = \mathbb{R}Z$ with $\ell(Z) = 1$, and $\mathfrak{a} = \mathbb{R}Z \oplus \mathbb{R}Y$ with $\ell(Y) = 0$.

(iii) If $[\mathfrak{g}, \mathfrak{a}] = \mathfrak{z}$,

$$\mathfrak{g} = \mathfrak{m} \oplus \mathbb{R}X \text{ with } [X, Y] = Z. \tag{1.1.9}$$

(iv) Assume now that $[\mathfrak{g}, \mathfrak{a}] = \mathfrak{a}$. There exist non-zero vectors $X, T \in \mathfrak{g}$ such that $\mathfrak{g} = \mathfrak{m} \oplus \mathbb{R}X \oplus \mathbb{R}T$ with the commutators

$$[X, Y] = Z \text{ and } [T, Y] = Y. \tag{1.1.10}$$

In this case, using Jacobi's identity, one gets:

$$[T, X] = -X \bmod(\mathfrak{m}). \tag{1.1.11}$$

Further, we have $\mathfrak{g} = \mathbb{R}T \oplus (\mathbb{R}X \oplus \mathfrak{m})$. In particular,

$$\mathfrak{a}^\ell = \mathbb{R}T \oplus \mathfrak{m}, \text{ and } \mathfrak{g}_0 = \mathbb{R}X \oplus \mathfrak{m}. \tag{1.1.12}$$

(v) We finally look at the case where $\dim \mathfrak{a} = 3$. Then there exist non-zero vectors X_1, X_2, T, Y_1, Y_2 such that $\mathfrak{a} = \mathbb{R}Z \oplus \mathbb{R}Y_1 \oplus \mathbb{R}Y_2$ with $\ell(Y_1) = \ell(Y_2) = 0$, and

$$\mathfrak{g} = \mathfrak{m} \oplus \mathbb{R}X_1 \oplus \mathbb{R}X_2 \oplus \mathbb{R}T \tag{1.1.13}$$

with the commutators

$$[X_i, Y_j] = \delta_{i,j} Z (1 \leq i, j \leq 2), \quad [T, Y_2] = Y_2 + \alpha Y_1 \text{ and } [T, Y_1] = Y_1 - \alpha Y_2, \tag{1.1.14}$$

for some non-zero $\alpha \in \mathbb{R}$. In this case, we have

$$\mathfrak{a}^\ell = \mathbb{R}T \oplus \mathfrak{m}, \text{ and } \mathfrak{g}_0 = \mathbb{R}X_2 \oplus \mathbb{R}X_1 \oplus \mathfrak{m}. \tag{1.1.15}$$

Further, we have by Jacobi's identity

$$[T, X_1] = -X_1 - \alpha X_2 \bmod(\mathfrak{m}). \tag{1.1.16}$$

Similarly, we get

$$[T, X_2] = \alpha X_1 - X_2 \bmod(\mathfrak{m}). \tag{1.1.17}$$

1.2 REPRESENTATION THEORY

We begin with preliminaries to set up necessary mathematical definitions and structures, then proceed with a detailed exploration of the disintegration of unitary representations.

Let $G = \exp \mathfrak{g}$ be an exponential solvable Lie group with Lie algebra $\mathfrak{g}$. It follows that G is type I and monomial, namely every irreducible unitary representation of G is equivalent to one induced from a unitary character of a connected closed subgroup of G. Since the orbit method was launched by Kirillov [65], many studies are done in this framework, and we now know many interesting results. Here we focus our attention to two operations which are in some sense reciprocal each other: induction and restriction. Let H, K be analytic subgroups of G, χ a unitary character of H, and π an irreducible unitary representation of G. Two main objects of our study are the induced representation $\tau = \mathrm{ind}_H^G \chi$ of G and the restriction $\pi|_K$ of π to K. First, we are interested in their canonical central decompositions. How to compute Borel measures and multiplicities for these decompositions. How to construct explicitly intertwining operators between these representations and their decompositions.

1.2.1 Modular functions and invariant measures on homogeneous spaces

Let $G = \exp \mathfrak{g}$ be an exponential solvable Lie group with Lie algebra $\mathfrak{g}$. Let dg be a left Haar measure on G and Δ_G the modular function of G so that we have

$$\int_G \varphi(gx^{-1})dg = \Delta_G(x) \int_G \varphi(g)dg \quad (x \in G)$$

for all functions φ belonging to the space $K(G)$ of the continuous functions on G with compact support. We have

$$\Delta_G(x) = |\det(\mathrm{Ad}x)|^{-1} \quad (x \in G).$$

A closed subgroup H with Lie algebra $\mathfrak{h}$ being given, we denote by $\Delta_{H,G}$ the character of H with values in $\mathbb{R}_+$ defined by

$$\Delta_{H,G}(h) = \frac{\Delta_H(h)}{\Delta_G(h)} \quad (h \in H).$$

It follows that, for $X \in \mathfrak{h}$,

$$\Delta_{H,G}(\exp X) = \exp(\mathrm{Tr}\ \mathrm{ad}_{\mathfrak{g}/\mathfrak{h}} X).$$

Let $K(G, H)$ be the space of the numerical continuous functions φ on G with compact support modulo H and which verify $\varphi(gh) = \Delta_{H,G}(h)\varphi(g)$ for all $g \in G$ and $h \in H$. G acting on $K(G, H)$ by left translations, we know that there exists on this space, up to a scalar multiplication, one and only one G-invariant positive linear form. We denote it by $\nu_{G,H}$ or more simply ν and write it in the form of an integral

$$\nu_{G,H}(\varphi) = \oint_{G/H} \varphi(g)d\nu(g).$$

If $\Delta_H = \Delta_G$ on H, $\nu_{G,H}$ is nothing but a G-invariant measure on the homogeneous space G/H. The next lemma will play an important role in the sequel.

Lemma 1.2.1 (Transitivity Lemma (cf.[23], chap. V)) *Let $K \subset H$ be two closed subgroups of G and let φ be a $\nu_{G,K}$-integrable function. Then the set of all the g's in G, for which the function*

$$h \mapsto \varphi(gh)\Delta_{H,G}^{-1}(h)$$

on H is not $\nu_{H,K}$-integrable is $\nu_{G,H}$ negligible. The function defined on G by

$$g \mapsto \oint_{H/K} \varphi(gh)\Delta_{H,G}^{-1}(h)d\nu_{H,K}(h)$$

is $\nu_{G,H}$-integrable and, up to a normalisation, we have the formula

$$\oint_{G/K} \varphi(x)d\nu_{G,K}(x) = \oint_{G/H} d\nu_{G,H}(g) \oint_{H/K} \varphi(gh)\Delta_{H,G}^{-1}(h)d\nu_{H,K}(h). \tag{1.2.1}$$

1.2.2 Induced representations

Let for more generality G be a Lie group. Let H be a closed subgroup of G. We denote by Δ_G and Δ_H the modular functions of G and H, respectively.

Let π be a unitary representation of H in the Hilbert space $\mathcal{H}_\pi$, $\mathcal{K}(G,H,\pi)$ the space of continuous functions ψ on G with values in $\mathcal{H}_\pi$, having the compact support modulo H and verifying the covariance condition

$$\psi(gh) = \Delta_{H,G}^{1/2}(h)\pi(h)^{-1}(\psi(g)) \quad (g \in G, \quad h \in H). \tag{1.2.2}$$

We define the norm $\|\psi\|$ in $\mathcal{K}(G,H,\pi)$ by

$$\|\psi\| = \left(\oint_{G/H} \|\psi(g)\|_{\mathcal{H}_\pi}{}^2 d\nu_{G,H}(g)\right)^{1/2}.$$

Then G acts isometrically on the space $\mathcal{K}(G,H,\pi)$ by left translations, and we obtain the induced unitary representation $\tau = \mathrm{ind}_H^G\pi$ of G in the completion $\mathcal{H}_\tau$ of $\mathcal{K}(G,H,\pi)$. In the case where $\pi = \chi$ is a unitary character of H, τ is the induced monomial representation $\mathrm{ind}_H^G\chi$.

1.2.3 The unitary dual of an exponential Lie group

Let $\mathrm{ad}_\mathfrak{g}$ denote the adjoint endomorphism:

$$\mathrm{ad}_\mathfrak{g}(X)(Y) = \mathrm{ad}(X)(Y) = [X,Y], \ X, \ Y \in \mathfrak{g}.$$

The group G acts on $\mathfrak{g}$ by the adjoint representation Ad_G, i.e.

$$\mathrm{Ad}_G(g)(Y) = \mathrm{Ad}(g)(Y) = e^{\mathrm{ad}X}Y, \ g = \exp X \in G, \text{ for } X \text{ and } Y \in \mathfrak{g}$$

and on $\mathfrak{g}^*$, the dual vector space of $\mathfrak{g}$, by the coadjoint representation Ad_G^*, i.e.

$$\mathrm{Ad}_G^*(g)\ell(X) = \mathrm{Ad}^*(g)\ell(X) = g \cdot \ell(X) = \ell(\mathrm{Ad}(g^{-1})(X)), \ g \in G, \ \ell \in \mathfrak{g}^*, \ X \in \mathfrak{g}.$$

The coadjoint orbit of ℓ is the set $\Omega_\ell = \{g \cdot \ell,\ g \in G\}$. The space of coadjoint orbits is noted by $\mathfrak{g}^*/G$.

We designate by $\widehat{G}$ the unitary dual of G, namely the set of all equivalence classes of irreducible unitary representations of G. The Kirillov-Bernat-Vergne orbit method makes it possible to parameterise $\widehat{G}$ by the space of coadjoint orbits of G in $\mathfrak{g}^*$. For $\ell \in \mathfrak{g}^*$, let B_ℓ be the bilinear form on $\mathfrak{g}$ defined by $B_\ell(X,Y) = \ell([X,Y])$ and $\mathfrak{g}(\ell)$ the radical of B_ℓ. Then, $\mathfrak{g}(\ell)$ is the Lie algebra of the stabiliser $G(\ell)$ of G at ℓ with respect to the coadjoint action. Moreover, $G(\ell)$ is connected [23], i.e. $G(\ell) = \exp(\mathfrak{g}(\ell))$.

We take a polarisation $\mathfrak{b} \subset \mathfrak{g}$ at ℓ (a totally isotropic subalgebra of $\mathfrak{g}$ (or Lagrangian) with respect to B_ℓ of maximal dimension equal to $\frac{1}{2}(\dim \mathfrak{g} + \dim \mathfrak{g}(\ell))$), satisfying Pukanszky condition $B \cdot \ell = \ell + \mathfrak{b}^\perp$, where $B := \exp \mathfrak{b}$. We define the unitary character $\chi_\ell : B \to U(1)$ associated to ℓ by

$$\chi_\ell(\exp X) := e^{i\ell(X)} \quad (X \in \mathfrak{b}), \tag{1.2.3}$$

and denote by $\pi_{\ell,\mathfrak{b}}$ the induced unitary representation $\mathrm{ind}_B^G \chi_\ell$ of G. Then $\pi_{\ell,\mathfrak{b}}$ is an irreducible representation of G and its equivalence class $[\pi_{\ell,\mathfrak{b}}]$ only depends upon the coadjoint orbit Ω_ℓ through ℓ. Every unitary and irreducible representation π of G is equivalent to the representation $\pi_{\ell,\mathfrak{b}}$ induced from a character χ_ℓ as in (1.2.3), and a Pukanszky polarisation $\mathfrak{b}$ at ℓ. Let θ_G be the Kirillov-Bernat-Vergne mapping

$$\begin{array}{rccc} \theta_G : & \mathfrak{g}^* & \longrightarrow & \widehat{G} \\ & \ell & \longmapsto & \pi_{\ell,\mathfrak{b}}. \end{array} \tag{1.2.4}$$

Then the mapping (1.2.4) factors through the quotient to

$$\begin{array}{rccc} \bar{\theta}_G : & \mathfrak{g}^*/G & \longrightarrow & \widehat{G} \\ & \Omega_\ell & \longmapsto & [\pi_{\ell,\mathfrak{b}}] =: \pi_{\Omega_\ell} \end{array} \tag{1.2.5}$$

which is homeomorphism, when the relative spaces are endowed with adequate topologies (quotient topology and Fell topology, respectively). For more details, see [70].

In what follows, we shall often identify an equivalence class of a unitary representation with its representative and designate the equivalence relation by $\simeq$ or even by the equality. Through the homeomorphism (1.2.5), we denote by $\Omega(\pi) = \Omega_\ell$ for a given $\pi \in \widehat{G}$, the coadjoint orbit of G associated to $\pi \simeq \pi_\ell$.

1.2.4 Saturation with respect to a normal subgroup of codimension 1

Let $H = \exp \mathfrak{h}$ be a normal subgroup of codimension 1 in G. With respect to the restriction $p : \mathfrak{g}^* \to \mathfrak{h}^*$, a coadjoint orbit Ω of G is either saturated or non-saturated with respect to $\mathfrak{h}$, depending whether $\mathfrak{g}(\ell) \subset \mathfrak{h}$ ($\ell \in \Omega$) or not, and it occurs generically in the same eventuality of this alternative (cf. [18]). If Ω is saturated, $p(\Omega)$ is composed by a one-parameter family $\{\omega_t\}_{t\in\mathbb{R}}$ of H-orbits with $\dim \Omega = \dim \omega_t + 2$, and $\bar{\theta}_G(\Omega) \simeq \mathrm{ind}_H^G \bar{\theta}_H(\omega_t)$ for all $t \in \mathbb{R}$. In this case,

$$\bar{\theta}_G(\Omega)|_H \simeq \int_{\mathbb{R}}^{\oplus} \bar{\theta}_H(\omega_t)dt. \tag{1.2.6}$$

On the other hand, if Ω is non-saturated, $\omega = p(\Omega)$ is an H-orbit, and p gives a diffeomorphism from Ω onto ω. In this case, $p^{-1}(\omega)$ is composed of a one-parameter family $\{\Omega_t\}_{t\in\mathbb{R}}$ of G-orbits and the following hold: $\bar{\theta}_G(\Omega)|_H \simeq \bar{\theta}_H(\omega)$, and

$$\mathrm{ind}_H^G \bar{\theta}_H(\omega) \simeq \int_{\mathbb{R}}^{\oplus} \bar{\theta}_G(\Omega_t)dt.$$

1.3 FUNDAMENTAL EXAMPLES

1.3.1 Heisenberg Lie groups

These are different manners to define the Heisenberg group. Towards a real realisation, one can define the Heisenberg group H_{2n+1} (or sometimes merely H_n) as the connected Lie group associated to the 2-step real nilpotent Lie algebra $\mathfrak{g} := \mathfrak{h}_{2n+1}$ of dimension $2n+1$ admitting a basis $\{Z, X_1, Y_1, \ldots, X_n, Y_n\}$ with the unique non-vanishing brackets:

$$[X_l, Y_l] = Z, \qquad \text{for all } l = 1, \ldots, n. \tag{1.3.1}$$

Here, n is a positive integer. The underlying variety is the vector space $\mathbb{R}^n \times \mathbb{R}^n \times \mathbb{R}$ and on which the multiplication is given by

$$(x, y, t)(x', y', t') = (x + x', y + y', t + t' + \frac{1}{2}(x \cdot y' - x' \cdot y)),$$

where $x \cdot y = x_1 y_1 + \cdots + x_n y_n$ denotes the Euclidean scalar product on $\mathbb{R}^n$.

The centre of H_n is the subgroup $\mathcal{Z} := \{0_n\} \times \{0_n\} \times \mathbb{R}$, and the commutator subgroup $[H_n, H_n]$ of H_n is given by $[H_n, H_n] = \mathcal{Z}$. The Lie algebra $\mathfrak{g}$ of H_n has the basis

$$\mathcal{B} := \{X_j, Y_j, j = 1, ..., n, Z = (0_n, 0_n, 1)\},$$

where $X_j = (e_j, 0_n, 0), Y_j = (0_n, e_j, 0), j = 1, \ldots, n$ and e_j is the jth canonical basis vector of $\mathbb{R}^n$, with the non-trivial brackets

$$[X_i, Y_j] = \delta_{i,j} Z.$$

1.3.1.1 *The unitary dual of H_n*

The unitary dual $\widehat{H}_n$ of H_n can be described as follows.

1.3.1.2 *The infinite-dimensional irreducible representations*

For every $\lambda \in \mathbb{R}^*$, there exists a unitary representation π_λ of H_n on the Hilbert space $L^2(\mathbb{R}^n)$, which is given by the formula

$$\pi_\lambda(x, y, t)\xi(s) := e^{-2\pi i\lambda t - 2\pi i\frac{\lambda}{2}x\cdot y + 2\pi i\lambda s\cdot y}\xi(s - x), \;\; s \in \mathbb{R}^n, \xi \in L^2(\mathbb{R}^n), (x, y, t) \in H_n.$$

It is easily seen that π_λ is in fact irreducible and that π_λ is equivalent to π_ν if and only if $\lambda = \nu$. The representation π_λ is equivalent to the induced representation $\tau_\lambda := \mathrm{ind}_P^{H_n}\chi_\lambda$, where $P = \{0_n\} \times \mathbb{R}^n \times \mathbb{R}$ is a polarisation at the linear functional $\ell_\lambda((x,y,t)) := \lambda t, (x,y,t) \in \mathfrak{g}$ and where χ_λ is the character of P defined by $\chi_\lambda(0_n, y, t) = e^{-2\pi i \lambda t}$.

The theorem of Stone-Von Neumann tells us that every infinite-dimensional unitary representation of H_n is equivalent to one of the π_λ's (see [31]).

1.3.1.3 *The finite-dimensional irreducible representations*

Since H_n is nilpotent, every irreducible finite-dimensional representation of H_n is one-dimensional, by Lie's theorem.

Any one-dimensional representation is a unitary character $\chi_{a,b}$, $(a,b) \in \mathbb{R}^n \times \mathbb{R}^n$, of H_n, which is given by

$$\chi_{a,b}(x,y,t) = e^{-2\pi i(a\cdot x + b\cdot y)}, (x,y,t) \in H_n.$$

For $f \in L^1(H_n)$, let

$$\hat{f}(a,b) := \chi_{a,b}(f) = \int_{H_n} f(x,y,t) e^{-2\pi i(x\cdot a + y\cdot b)} dx dy dt, \;\; a, b \in \mathbb{R}^n,$$

and

$$\|f\|_{\infty,0} := \sup_{a,b\in\mathbb{R}^n} |\chi_{a,b}(f)| = \|\hat{f}\|_\infty.$$

1.3.2 Thread-like Lie groups

For $N \geq 3$, let $\mathfrak{g}_N$ be the N-dimensional real nilpotent Lie algebra with basis $X_1, \ldots, X_N$ and non-trivial Lie brackets

$$[X_N, X_{N-1}] = X_{N-2}, \ldots, [X_N, X_2] = X_1.$$

The Lie algebra $\mathfrak{g}_N$ is $(N-1)$-step nilpotent and is a semi-direct product of $\mathbb{R}X_N$ with the abelian ideal

$$\mathfrak{b} := \sum_{j=1}^{N-1} \mathbb{R}X_j. \tag{1.3.2}$$

Let $G_N := \exp(\mathfrak{g}_N)$ be the associated connected, simply connected Lie group, and

$$\mathfrak{b}_j := \mathrm{span}\{X_i, i = 1, ..., j\}, 1 \leq j \leq N-1.$$

Let also $B_j := \exp(\mathfrak{b}_j)$ and $B := \exp(\mathfrak{b})$. Then for $3 \leq M \leq N$ we have $G_M \simeq G_N/B_{N-M}$.

Note that $\mathfrak{g}_3$ is the three dimensional Heisenberg Lie algebra.

1.3.2.1 The unitary dual of G_N

In this section, we describe the unitary irreducible representations of G_N up to a unitary equivalence. For $\xi = \sum_{j=1}^{N-1} \xi_j X_j^* \in \mathfrak{g}_N^*$, the coadjoint action is given by

$$\mathrm{Ad}^*(\exp(-tX_N))\xi = \sum_{j=1}^{N-1} p_j(\xi, t) X_j^*,$$

where, for $1 \leq j \leq N-1$, $p_j(\xi, t)$ is a polynomial in t defined by

$$p_j(\xi, t) = \sum_{k=0}^{j-1} \frac{t^k}{k!} \xi_{j-k}.$$

Moreover, if $\xi_j \neq 0$ for at least one $1 \leq j \leq N-2$, then $\mathrm{Ad}^*(G_N)\xi$ is of dimension two, and $\mathrm{Ad}^*(G_N)\xi = \{\mathrm{Ad}^*(\exp(tX_N))\xi + \mathbb{R}X_N^*, t \in \mathbb{R}\}$. We shall always identify $\mathfrak{g}_N^*$ with $\mathbb{R}^N$ via the mapping $(\xi_N, \ldots, \xi_1) \to \sum_{j=1}^N \xi_j X_j^*$ and the subspace $V = \{\xi \in \mathfrak{g}_N^* : \xi_N = 0\}$ with the dual space of $\mathfrak{b}$. For $\xi \in V$ and $t \in \mathbb{R}$, let

$$\begin{aligned} t \cdot \xi &= \mathrm{Ad}^*(\exp(tX_N))\xi \\ &= \Big(0, \xi_{N-1} - t\xi_{N-2} + \ldots + \frac{1}{(N-2)!}(-t)^{N-2}\xi_1, \ldots, \xi_2 - t\xi_1, \xi_1\Big). \end{aligned} \tag{1.3.3}$$

We define the function $\widehat{\xi}$ on $\mathbb{R}$ by

$$\widehat{\xi}(t) := (t \cdot \xi)_{N-1} = \xi_{N-1} - t\xi_{N-2} + \ldots + \frac{1}{(N-2)!}(-t)^{N-2}\xi_1. \tag{1.3.4}$$

Then the mapping $\xi \to \widehat{\xi}$ is a linear isomorphism of V onto P_{N-2}, the space of real polynomials of degree at most $N-2$. In particular, $\xi_k \to \xi$ coordinate-wise in V as $k \to \infty$ if and only if $\widehat{\xi_k}(t) \to \widehat{\xi}(t)$ for all $t \in \mathbb{R}$. Also, the mapping $\xi \to \widehat{\xi}$ intertwines the Ad^*-action and translation in the following way:

$$\begin{aligned} \widehat{t \cdot \xi}(s) &= (s \cdot (t \cdot \xi))_{N-1} \\ &= ((s+t) \cdot \xi)_{N-1} = \widehat{\xi}(s+t) \end{aligned}$$

for $\xi \in V$ and $s, t \in \mathbb{R}$.

By Kirillov's orbit picture of the dual space of a nilpotent Lie group, we can describe the irreducible unitary representations of G_N in the following way (see [31] for details). For any non-constant polynomial $p = \hat{\ell} \in P_{N-2}$ we consider the induced representation $\pi_\ell = \mathrm{ind}_B^G \chi_\ell$, where χ_ℓ denotes the unitary character of the abelian group B defined by:

$$\chi_\ell(\exp U) = e^{-2\pi i \ell(U)}, U \in \mathfrak{b}.$$

Since $\mathfrak{b}$ is abelian of codimension 1, it is a polarisation at ℓ and so π_ℓ is irreducible. Every infinite-dimensional irreducible unitary representation of G_N arises in this manner up to equivalence.

Let us describe the representation $\pi_\ell, \ell \in \mathfrak{b}^*$, explicitly. The Hilbert space $\mathcal{H}_\ell$ of the representation π_ℓ is the space $L^2(G_N/B, \chi_\ell)$ consisting of all measurable functions $\tilde{\xi} : G_N \to \mathbb{C}$, such that $\tilde{\xi}(gb) = \chi_\ell(b^{-1})\tilde{\xi}(g)$ for all $b \in B$ and all $g \in G$ outside some set of Lebesgue measure 0 and such that the function $|\tilde{\xi}|$ is contained in $L^2(G_N/B)$. We can identify the space $L^2(G_N/B, \chi_\ell)$ in an obvious way with $L^2(\mathbb{R})$ via the isomorphism $U : \xi \mapsto \tilde{\xi}$ where $\tilde{\xi}(\exp(sX_N)b) := \chi_\ell(b^{-1})\xi(s)$, $s \in \mathbb{R}$, $b \in B$. Hence for $g = \exp(tX_N)b$ and $\xi \in L^2(\mathbb{R})$ we have an explicit expression for the operator $\pi_\ell(g)$:

$$\begin{aligned}
\pi_\ell(g)\xi(s) &= \tilde{\xi}(g^{-1}\exp(sX_N)) \\
&= \tilde{\xi}(b^{-1}\exp((s-t)X_N)) \\
&= \tilde{\xi}(\exp((s-t)X_N)(\exp((t-s)X_N)b^{-1}\exp((s-t)X_N))) \\
&= \chi_\ell(\exp((t-s)X_N)b\exp((s-t)X_N))\xi(s-t) \\
&= e^{-2\pi i\ell(\mathrm{Ad}(\exp((t-s)X_N)\log(b)))}\xi(s-t), s \in \mathbb{R}.
\end{aligned}$$

We can parametrise the orbit space $\mathfrak{g}_N^*/G_N$ in the following way. First, we have a decomposition

$$\mathfrak{g}_N^*/G_N = \bigcup_{j=1}^{N-2} \mathfrak{g}_N^{*j}/G_N \bigcup X^*,$$

where

$$\mathfrak{g}_N^{*j} := \{\ell \in \mathfrak{g}_N^*, \ell(X_i) = 0, i = 1, \cdots, j-1, \ell(X_j) \neq 0\}$$

and where

$$X^* := \{\ell \in \mathfrak{g}_N^*, \ell(X_j) = 0, j = 1, \cdots, N-2\}$$

denotes the characters of G_N. A character of the group G_N can be written as $\chi_{a,b}, a, b \in \mathbb{R}$, where

$$\chi_{a,b}(x_N, x_{N-1},, \cdots, x_1) := e^{-2\pi i a x_N - 2\pi i b x_{N-1}}, \ (x_N, \cdots, x_1) \in G_N.$$

For any $\ell \in \mathfrak{g}_N^{*j}, N-2 \geq j \geq 1$ there exists exactly one element ℓ_0 in the G_N-orbit of ℓ, which satisfies the conditions

$$\ell_0(X_j) \neq 0, \ell_0(X_{j+1}) = 0, \ell_0(X_N) = 0.$$

We can thus parametrise the orbit space $\mathfrak{g}_N^*/G_N$, and hence also the dual space $\widehat{G_N}$, with the sets

$$S_N := \bigcup_{j=1}^{N-2} S_N^j \bigcup X^*,$$

where $S_N^j := S_N \cap \mathfrak{g}^{*j} = \{\ell \in \mathfrak{g}_N^{*j}, \ell(X_k) = 0, k = 1, \cdots, j-1, j+1, \ell(X_j) \neq 0\}$. Let

$$S_N^{gen} := \bigcup_{j=1}^{N-2} S_N^j$$

be the family of points in S_N, whose G_N-orbits are of dimension 2.

1.4 THE EXPONENTIAL LIE GROUP G_6

Definition: Let $\mathfrak{g} = \mathfrak{g}_6$ be the Lie algebra spanned by the vectors A, B, P, Q, R, S and equipped with the Lie brackets:

$$[P,Q] = R, \qquad [P,R] = S, \qquad [A,P] = \frac{1}{2}P, \qquad [B,P] = -\frac{1}{2}P,$$

$$[B,Q] = Q, \qquad [A,R] = \frac{1}{2}R, \qquad [B,R] = \frac{1}{2}R, \qquad [A,S] = S,$$

We introduce the following closed subgroups and coordinate functions on $G = \exp \mathfrak{g}$, write

$$\mathfrak{g}_0 := \operatorname{span}(A,B,P), \ G_0 := \exp(\mathfrak{g}_0),$$

$$\mathfrak{h} := \operatorname{span}(Q,R,S), \ H := \exp \mathfrak{h}.$$

Then the subset G_0 is a closed subgroup of G, and the subset H is a closed normal subgroup. We have $G = G_0 \cdot H$ as a topological product. For $(a,b,p,q,r,s) \in \mathbb{R}^6$, let

$$g = E(a,b,p,q,r,s) := \exp(aA)\exp(bB)\exp(pP)\exp(qQ)\exp(rR)\exp(sS).$$

1.4.1 A parametrisation of the set of coadjoint orbits

We give in this section a system of representatives of the set of coadjoint orbits $\mathfrak{g}^*/\mathrm{Ad}^*G$.

Let $\{A^*, B^*, P^*, Q^*, R^*, S^*\}$ be the dual basis of $\{A, B, P, Q, R, S\}$, and $f = a^*A^* + b^*B^* + p^*P^* + q^*Q^* + r^*R^* + s^*S \in \mathfrak{g}^*$. Then

$$\begin{aligned} \mathrm{Ad}^*(g)(f) = {} & \left(a^* + \frac{p}{2}p^* + \frac{1}{2}(pq+r)r^* + \left(s + \frac{pr}{2}\right)s^*\right)A^* \\ & + \left(b^* - \frac{p}{2}p^* + qq^* + \frac{1}{2}(r-pq)r^* - \frac{pr}{2}s^*\right)B^* \\ & + e^{\frac{-a+b}{2}}(p^* + qr^* + rs^*)P^* \\ & + e^{-b}\left(q^* - pr^* + \frac{1}{2}p^2s^*\right)Q^* + e^{\frac{-a-b}{2}}(r^* - ps^*)R^* + e^{-a}s^*S^*. \end{aligned} \tag{1.4.1}$$

In the sequel, we write $\mathbb{R}^+ := \{x \in \mathbb{R};\ x > 0\}$ and $\mathbb{R}^{+,0} := \mathbb{R}^+ \cup \{0\}$. We note that the determinant of the matrix $(f([V,W]))_{V,W\in\{A,B,P,Q,R,S\}}$ is $\frac{1}{4}s^{*2}(2q^*s^* - r^{*2})^2$, and there are open orbits. The following list gives a parametrisation of the coadjoint orbits $\mathfrak{g}^*/\mathrm{Ad}^*G$.

0-dimensional orbits $a^*A^* + b^*B^*$, where $a^*, b^* \in \mathbb{R}$.

2-dimensional orbits

(2d-1) $\alpha^*(\frac{A^*+B^*}{2}) + \varepsilon P^*$, where $\alpha^* \in \mathbb{R}$, $\varepsilon = \pm 1$,

$$\Omega_{\alpha^*(\frac{A^*+B^*}{2})+\varepsilon P^*} = \{(\frac{\alpha^*}{2} + \frac{p}{2}\varepsilon)A^* + (\frac{\alpha^*}{2} - \frac{p}{2}\varepsilon)B^* + e^{-\alpha}\varepsilon P^*, \alpha, p \in \mathbb{R}\},$$

[cf. (1.4.11), (1.4.12) below].

(2d-2) $a^*A^* + \nu Q^*$, where $a^* \in \mathbb{R}$, $\nu = \pm 1$,

$$\Omega_{a^*A^*+\nu Q^*} = \{a^*A^* + q\nu B^* + e^{-b}\nu Q^*, b, q \in \mathbb{R}\},$$

[cf. (1.4.10)].

4-dimensional orbits

(4d-1) $\varepsilon P^* + \nu Q^*$, where $\varepsilon = \pm 1, \nu = \pm 1$,

$$\Omega_{\varepsilon P^*+\nu Q^*} = \{pA^* + qB^* + e^{\frac{-a+b}{2}}(\varepsilon)P^* + e^{-b}\nu Q^*, p, q, a, b \in \mathbb{R}\},$$

[cf. (1.4.9)].

(4d-2) . $b^*B^* + \varepsilon R^*$, where $b^* \in \mathbb{R}$, $\varepsilon = \pm 1$,

$$\Omega_{b^*B^*+\varepsilon R^*} = \{rA^* + (b^* + r - pq\varepsilon)B^* + qP^*$$

$$+(-pe^{-a})Q^* + e^{-a}\varepsilon R^*, a, p, q, r \in \mathbb{R}\},$$

[cf. (1.4.7), (1.4.8)].

(4d-3) $b^*B^* + \varepsilon S^*$, where $b^* \in \mathbb{R}$, $\varepsilon = \pm 1$,

$$\begin{aligned}\Omega_{b^*B^*+\varepsilon S^*} & \\ = & \{sA^* + \left(b^* - \frac{pr}{2}\varepsilon\right)B^* + e^{\frac{-a}{2}}(r\varepsilon)P^* \\ & + \left(\frac{1}{2}p^2\varepsilon\right)Q^* + e^{\frac{-a}{2}}(-p\varepsilon)R^* + e^{-a}\varepsilon S^*, \\ & s, p, r, a \in \mathbb{R}\}.\end{aligned}$$

[cf. (1.4.4), (1.4.5)].

6-dimensional orbits (6d) $\nu Q^* + \varepsilon S^*$, where $\nu = \pm 1$, $\varepsilon = \pm 1$,

$$\Omega_{\nu Q^*+\varepsilon S^*} = \{sA^*+rB^*+qP^*+e^{-b}(\nu+\frac{p^2\varepsilon}{2})Q^*+(-e^{-\frac{a+b}{2}}p\varepsilon)R^*+(e^{-a}\varepsilon)S^*,$$

$$s, r, q, a, b, p \in \mathbb{R}\}.$$

[cf.(1.4.2),(1.4.3)].

Proof and description of each orbit. Let Ω be a coadjoint orbit and $f \in \Omega$.

Case I) Suppose $f(S) \neq 0$. Then by (1.4.1), there exists $g = E(a, 0, p, 0, r, s)$ such that $g \cdot f = b^*B^* + q^*Q^* + \varepsilon S^*$, where $\varepsilon = \pm 1$ and $b^*, q^* \in \mathbb{R}$. Thus let $f = b^*B^* + q^*Q^* + \varepsilon S^*$.

I-1) If $q^* \neq 0$, then there exists $g = E(0, b, 0, q, 0, 0)$ such that $\mathrm{Ad}^*g(f) = (b^* + qq^*)B^* + e^{-b}q^*Q^* + \varepsilon S^* = \nu Q^* + \varepsilon S^*$, where $\nu = \pm 1$.

$$\begin{aligned} f : &= \nu Q^* + \varepsilon S^*, \qquad (1.4.2)\\ \mathrm{Ad}^*E(a, b, p, q, r, s)(f) &= \bar{a}A^* + \bar{b}B^* + \bar{p}P^* + \bar{q}Q^* + \bar{r}R^* + \bar{s}S^*,\end{aligned}$$

where

$$\begin{aligned}\bar{a} &= \left(s+\frac{1}{2}pr\right)\varepsilon, \quad \bar{b}=\left(q\nu-\frac{1}{2}pr\varepsilon\right), \quad \bar{p}=e^{\frac{-a+b}{2}}r\varepsilon,\\ \bar{q} &= e^{-b}\left(\nu+\frac{1}{2}p^2\varepsilon\right), \quad \bar{r}=-e^{\frac{-a-b}{2}}p\varepsilon, \quad \bar{s}=e^{-a}\varepsilon.\end{aligned}$$

Let $\Phi(l) := 2l(Q)l(S) - l(R)^2$ for $l \in \mathfrak{g}^*$. Then $\Phi(\bar{f}) = 2\bar{q}\bar{s} - \bar{r}^2 = 2e^{-a-b}\nu\varepsilon \neq 0$, and we have

$$\Omega_f = \{l;\ \Phi(l) \neq 0,\ l(S) \neq 0\}. \tag{1.4.3}$$

I-2) If $q^* = 0$, then $f = b^*B^* + \varepsilon S^*$ and $\mathfrak{g}(f) = \mathbb{R}\text{-span}\{B, Q\}$.

$$\mathrm{Ad}^*E(a,0,p,0,r,s)(f) = \bar{a}A^* + \bar{b}B^* + \bar{p}P^* + \bar{q}Q^* + \bar{r}R^* + \bar{s}S^*, \tag{1.4.4}$$

where

$$\begin{aligned}\bar{a} &= \left(s+\frac{pr}{2}\right)\varepsilon, \quad \bar{b}=\left(b^*-\frac{pr}{2}\varepsilon\right), \quad \bar{p}=re^{-\frac{a}{2}}\varepsilon,\\ \bar{q} &= \frac{1}{2}p^2\varepsilon, \quad \bar{r}=-pe^{-\frac{a}{2}}\varepsilon, \quad \bar{s}=e^{-a}\varepsilon.\end{aligned}$$

Let Φ, Φ_2 be functions on $\mathfrak{g}^*$ defined by

$$\Phi(l) = 2l(Q)l(S) - l(R)^2, \quad \Phi_2(l) = l(R)l(P) - 2(l(B) - b^*)l(S).$$

Then we have

$$\begin{aligned}\Omega_f = \Phi^{-1}(0) \cap \Phi_2^{-1}(0) \cap (&\mathbb{R}A^* + \mathbb{R}B^* + \mathbb{R}P^* + (\mathbb{R}^{+,0})\varepsilon Q^*\\ &+ \mathbb{R}R^* + \mathbb{R}^+\varepsilon S^*).\end{aligned} \tag{1.4.5}$$

In fact, let $\bar{f} := (\bar{a}, \bar{b}, \bar{p}, \bar{q}, \bar{r}, \bar{s})$ be an element of the right hand side. We take

$$a := -\log(\varepsilon\bar{s}), \quad p := -\bar{r}e^{\frac{a}{2}}\varepsilon, \quad r := \bar{p}e^{-\frac{a}{2}}\varepsilon, \quad s := \bar{a} - \frac{pr}{2}.$$

Then we have $\mathrm{Ad}^*E(a,0,p,0,r,s)f = \bar{f}$, this is, $\bar{f} \in \Omega_f$. The reverse inclusion is easily verified.

Case II) Suppose $f(S) = 0$. Then $\mathrm{Ad}^*G(f)(S) = \{0\}$. Then (1.4.1) is reduced to

$$\begin{aligned}\mathrm{Ad}^*g(f) \;=\;& \left(a^* + \frac{p}{2}p^* + \frac{1}{2}(pq+r)r^*\right)A^*\\ &+ \left(b^* - \frac{p}{2}p^* + qq^* + \frac{1}{2}(r-pq)r^*\right)B^*\\ &+ e^{\frac{-a+b}{2}}(p^*+qr^*)P^* + e^{-b}(q^*-pr^*)Q^* + e^{\frac{-a-b}{2}}r^*R^*.\end{aligned} \tag{1.4.6}$$

II-1) Suppose $f(R) = r^* \neq 0$. Then by (1.4.6), there exists $g = E(\alpha, \alpha, p, q, r, 0)$ such that $\mathrm{Ad}^*g(f) = b^*B^* + \varepsilon R^*$, where $\varepsilon = \pm 1$, $b^* \in \mathbb{R}$. Let $f = b^*B^* + \varepsilon R^*$. Then $\mathfrak{g}(f) = \mathbb{R}\text{-span}\{S, A - B\}$. Let $X = A + B$, $Y = A - B$, $X^* = \frac{1}{2}(A^* + B^*)$, $Y^* = \frac{1}{2}(A^* - B^*)$. Then $\{X^*, Y^*, P^*, Q^*, R^*, S^*\}$ is the dual basis of $\{X, Y, P, Q, R, S\}$.

$$\begin{aligned} f: &= b^*B^* + \varepsilon R^*, \\ \mathrm{Ad}^*E(\alpha, \alpha, p, q, r, 0)(f) &= \frac{1}{2}(pq + r)\varepsilon A^* + \left(b^* + \frac{1}{2}(r - pq)\varepsilon\right) B^* \\ &\quad + q\varepsilon P^* - e^{-\alpha}p\varepsilon Q^* + e^{-\alpha}\varepsilon R^* \\ &= (b^* + r\varepsilon)X^* + (-b^* + pq\varepsilon)Y^* \\ &\quad + q\varepsilon P^* - e^{-\alpha}p\varepsilon Q^* + e^{-\alpha}\varepsilon R^*, \end{aligned} \tag{1.4.7}$$

$$\Omega_f = \{\bar{x}X^* + \bar{y}Y^* + \bar{p}P^* + \bar{q}Q^* + \bar{r}R^*;\ \bar{r} \in \varepsilon\mathbb{R}^+, \bar{y} = -b^* - (\bar{p}\bar{q}/\bar{r})\} \tag{1.4.8}$$

II-2) Suppose $f(R) = 0$. Then $G \cdot f(R) = \{0\}$, and we have

$$\mathrm{Ad}^*g(f) = \left(a^* + \frac{p}{2}p^*\right) A^* + \left(b^* - \frac{p}{2}p^* + qq^*\right) B^* + e^{\frac{-a+b}{2}}p^*P^* + e^{-b}q^*Q^*.$$

II-2-i) Suppose $q^* \neq 0$ and $p^* \neq 0$. Then there exists $g = E(a, b, p, q, 0, 0)$ such that $\mathrm{Ad}^*g(f) = \varepsilon P^* + \nu Q^*$, where $\varepsilon = \pm 1, \nu = \pm 1$. We have $\mathfrak{g}(\varepsilon P^* + \nu Q^*) = \mathbb{R}\text{-span}\{S, R\}$, and

$$\begin{aligned} f &:= \varepsilon P^* + \nu Q^*, \\ \mathrm{Ad}^*E(a, b, p, q, 0, 0)(f) &= \frac{p}{2}\varepsilon A^* + \left(-\frac{p}{2}\varepsilon + q\nu\right) B^* + e^{\frac{-a+b}{2}}\varepsilon P^* \\ &\quad + e^{-b}\nu Q^*, \end{aligned}$$

$$\Omega_f = \mathbb{R}A^* + \mathbb{R}B^* + \mathbb{R}^+\varepsilon P^* + \mathbb{R}^+\nu Q^*. \tag{1.4.9}$$

II-2-ii) Suppose $q^* \neq 0$ and $p^* = 0$. Then there exists $g = E(0, b, 0, q, 0, 0)$ such that $g \cdot f = a^*A^* + \nu Q^*$. We have $\mathfrak{g}(a^*A^* + \nu Q^*) = \mathbb{R}\text{-span}\{A, P, R, S\}$,

$$\begin{aligned} f &:= a^*A^* + \nu Q^*, \\ \mathrm{Ad}^*E(0, b, 0, q, 0, 0)(f) &= a^*A^* + q\nu B^* + e^{-b}\nu Q^*, \end{aligned}$$

$$\Omega_f = a^*A^* + \mathbb{R}B^* + \mathbb{R}^+\nu Q^*. \tag{1.4.10}$$

II-2-iii) Suppose $q^* = 0$ and $p^* \neq 0$. Then there exists $g = \exp(\alpha(A - B))\exp(pP)$ such that $g \cdot f = \alpha^*(\frac{A^*+B^*}{2}) + \varepsilon P^*$, where $\alpha^* \in \mathbb{R}$, $\varepsilon = \pm 1$. Let $f = \alpha^*(\frac{A^*+B^*}{2}) + \varepsilon P^*$. Then we have $\mathfrak{g}(f) = \mathbb{R}\text{-span}\{A + B, R, S, Q\}$.
Letting $X = A + B$, $Y = A - B$, $X^* = \frac{A^*+B^*}{2}$, $Y^* = \frac{A^*-B^*}{2}$, we have

$$f := \alpha^*(\frac{A^* + B^*}{2}) + \varepsilon P^* = \alpha^*X^* + \varepsilon P^*, \tag{1.4.11}$$

$$\begin{aligned}\mathrm{Ad}^*E(\alpha,-\alpha,p,0,0,0)(f) &= \left(\frac{\alpha^*}{2}+\frac{p}{2}\varepsilon\right)A^*+\left(\frac{\alpha^*}{2}-\frac{p}{2}\varepsilon\right)B^* \\ &+ \ +e^{-\alpha}\varepsilon P^* \\ &= \alpha^*X^*+p\varepsilon Y^*+e^{-\alpha}\varepsilon P^*,\end{aligned}$$

$$\Omega_f=\alpha^*X^*+\mathbb{R}Y^*+\mathbb{R}^+\varepsilon P^*. \tag{1.4.12}$$

II-2-iv) Suppose $q^*=p^*=0$. Then $G\cdot f=\{a^*A^*+b^*B^*\}$ which is a point. □

1.4.2 The topology of $\widehat{G}$

1.4.2.1 The open orbits of $\Omega_{\varepsilon S^*\pm\varepsilon Q^*}, \varepsilon=\pm 1$

Let $f=\varepsilon S^*+\nu Q^*$, where $\nu=\pm\varepsilon$. Then $\mathrm{Ad}^*G(f)$ is an open orbit, and we have

$$\Omega_f=\{l;\ \Phi(l)\neq 0,\ l(S)\neq 0\},$$

where $\Phi(l):=2l(Q)l(S)-l(R)^2$ for $l\in\mathfrak{g}^*$. Noting that $\Phi(l)\leq 0$ if $l(S)=0$, we have

$$\overline{\Omega_f}\setminus\Omega_f=\begin{cases}\{l;\Phi(l)=0,l(S)\geq 0\} & \varepsilon=1,\nu=1,\\ \{l;\Phi(l)\leq 0,l(S)=0\}\cup\{l;\Phi(l)=0,l(S)\geq 0\} & \varepsilon=1,\nu=-1,\\ \{l;\Phi(l)\leq 0,l(S)=0\}\cup\{l;\Phi(l)=0,l(S)\leq 0\} & \varepsilon=-1,\nu=1,\\ \{l;\Phi(l)=0,l(S)\leq 0\} & \varepsilon=-1,\nu=-1.\end{cases}$$

Considering the natural projection $\mathfrak{g}^*\to\mathfrak{h}^*$ defined by restriction and denoting $G\cdot l:=\mathrm{Ad}^*G(l)|_{\mathfrak{h}}$ for $l\in\mathfrak{g}^*$, we have

$$(\overline{\Omega_f}|_{\mathfrak{h}})\setminus(\Omega_f|_{\mathfrak{h}})=\begin{cases}\{G\cdot S^*,G\cdot Q^*,0\} & \varepsilon=1,\nu=1,\\ \{G\cdot S^*,\pm G\cdot Q^*,\pm G\cdot R^*,0\} & \varepsilon=1,\nu=-1,\\ \{-G\cdot S^*,\pm G\cdot Q^*,\pm G\cdot R^*,0\} & \varepsilon=-1,\nu=1,\\ \{-G\cdot S^*,-G\cdot Q^*,0\} & \varepsilon=-1,\nu=-1.\end{cases}$$

Realisation of $\pi_f\in\widehat{G}$:

We realise π_f by taking the Pukanszky polarisation $\mathfrak{h}=\mathbb{R}\text{-span}\{Q,R,S\}$ on $L^2(\exp(\mathbb{R}A+\mathbb{R}B)\exp(\mathbb{R}P))$. We note that $\mathfrak{g}_0:=\mathbb{R}\text{-span}\{A,B,P\}$ is a subalgebra, and $G=G_0H$, where $G_0:=\exp(\mathfrak{g}_0)$. Writing $g_0=g_0(a,b,p):=\exp(aA+bB)\exp(pP)$, $g_1=g_1(q,r,s)=\exp(qQ+rR+sS)$, $h=h(t,u,v)=\exp(tA+uB)\exp(vP)$, we have

$$\pi_f(g_0g_1)\xi(h)=\chi_f(h^{-1}g_0g_1g_0^{-1}h)\xi(g_0^{-1}h)=\chi_{\mathrm{Ad}^*(g_0^{-1}h)(f)}(g_1)\xi(g_0^{-1}h).$$

We take the left Haar measure dg_0, dg_1, and dg on G_0, H, and G, respectively, which are defined by transferring Lebesgue measures by $(a,b,p)\in\mathbb{R}^3\mapsto g_0(a,b,p)\in G_0$, $(q,r,s)\in\mathbb{R}^3\mapsto g_1(q,r,s)\in H$, and $dg=dg_0dg_1$. Let $F\in L^1(G)$. For $l\in\mathfrak{h}^*$ we denote

$$\widehat{F}^{\mathfrak{h}}(a,b,p,l):=\widehat{F}^{\mathfrak{h}}(g_0(a,b,p))(l):=\int_{\mathfrak{h}}F(g_0g_1)\chi_l(g_1)dg_1,\quad a,b,p\in\mathbb{R}.$$

Then we have

$$\begin{aligned}\pi_f(F)\xi(h) &= \int_{G_0H} \pi_f(g_0g_1)\xi(h)F(g_0g_1)dg_0dg_1\\ &:= \int_{\mathbb{R}^6} F(g_0(a,b,p)g_1(q,r,s))\pi_f(g_0(a,b,p)g_1(q,r,s))\xi(h)dadbdpdqdrds\\ &= \int_{G_0H} F(g_0g_1)\chi_{\mathrm{Ad}^*(g_0^{-1}h)(f)}(g_1)\xi(g_0^{-1}h)dg_0dg_1\\ &= \int_{G_0H} F(hg_0g_1)\chi_{\mathrm{Ad}^*(g_0^{-1})(f)}(g_1)\xi(g_0^{-1})dg_0dg_1\\ &= \int_{G_0H} F(hg_0^{-1}g_1)\chi_{\mathrm{Ad}^*g_0(f)}(g_1)\xi(g_0)\Delta_{G_0}^{-1}(g_0)dg_0dg_1\\ &= \int_{G_0} \widehat{F^{\mathfrak{h}}}(hg_0^{-1})(\mathrm{Ad}^*g_0(f|_{\mathfrak{h}}))\Delta_{G_0}^{-1}(g_0)\xi(g_0)dg_0\\ &=: \int_{G_0} \mathcal{K}_F(h,g_0)\xi(g_0)dg_0,\end{aligned}$$

where Δ_{G_0} is the modular function of G_0 and

$$\begin{aligned}\mathcal{K}_F(h,g_0) &= \mathcal{K}_F(h(t,u,v),g_0(a,b,p))\\ &= \widehat{F^{\mathfrak{h}}}(hg_0^{-1})(\mathrm{Ad}^*g_0(f|_{\mathfrak{h}}))\Delta_{G_0}^{-1}(g_0)\\ &= \widehat{F^{\mathfrak{h}}}(g_0(t-a,u-b,e^{\frac{a-b}{2}}(v-p))(\mathrm{Ad}^*g_0(a,b,p)(f|_{\mathfrak{h}}))e^{\frac{a-b}{2}}\\ &= \widehat{F}^{\mathfrak{h}}(t-a,u-b,e^{\frac{a-b}{2}}(v-p),\varepsilon((e^{-b}(\pm1+\frac{p^2}{2})Q^*\\ &\quad +(-e^{-\frac{a+b}{2}}p)R^*+e^{-a}S^*))e^{\frac{1}{2}(a-b)}.\end{aligned} \tag{1.4.13}$$

1.4.2.2 The orbits of $\Omega_{b^*B^*+\varepsilon S^*}, \varepsilon=\pm1, b^*\in\mathbb{R}$

Let $f=b^*B^*+\varepsilon S^*$. Then $\mathfrak{g}(f)=\mathbb{R}\text{-span}\{B,Q\}$. We have

$$\Omega_f=\Phi^{-1}(0)\cap\Phi_2^{-1}(0)\cap(\mathbb{R}A^*+\mathbb{R}B^*+\mathbb{R}P^*+(\varepsilon\mathbb{R}^{+,0})Q^*+\mathbb{R}R^*+\varepsilon\mathbb{R}^+S^*),$$

where

$$\Phi(l)=2l(Q)l(S)-l(R)^2,\quad \Phi_2(l)=l(R)l(P)-2(l(B)-b^*)l(S).$$

Then

$$\begin{aligned}\overline{\Omega_f}\setminus\Omega_f &= (\mathbb{R}A^*+\mathbb{R}B^*+\varepsilon\mathbb{R}^+Q^*)\cup(\mathbb{R}A^*+\mathbb{R}B^*+\mathbb{R}P^*)\\ &= \bigcup_{a^*\in\mathbb{R}}\mathrm{Ad}^*G(a^*A^*+\varepsilon Q^*)\cup\bigcup_{x^*\in\mathbb{R},\nu=\pm1}\mathrm{Ad}^*G(x^*X^*+\nu P^*)\\ &\qquad \bigcup_{a^*,\lambda\in\mathbb{R}}(a^*A^*+\lambda B^*).\end{aligned} \tag{1.4.14}$$

Proof of (1.4.14)**:** Since

$$\overline{\Omega_f}\subset\Phi^{-1}(0)\cap\Phi_2^{-1}(0)\cap(\mathbb{R}A^*+\mathbb{R}B^*+\mathbb{R}P^*+\mathbb{R}^{+,0}\varepsilon Q^*+\mathbb{R}R^*+\mathbb{R}^{+,0}\varepsilon S^*),$$

we first note that

$$\overline{\Omega_f} \setminus \Omega_f \subset \Phi^{-1}(0) \cap \Phi_2^{-1}(0) \cap (\mathbb{R}A^* + \mathbb{R}B^* + \mathbb{R}P^* + \mathbb{R}^{+,0}\varepsilon Q^* + \mathbb{R}R^*).$$

In fact, let $l \in \overline{\Omega_f}$, and suppose $l(S) \neq 0$. Then there exists $g \in G$ such that $\mathrm{Ad}^*g(l) = b'B^* + q'Q^* + \varepsilon S^*$. Since $\overline{\Omega_f}$ is Ad^*G-invariant, we have $\Phi(\mathrm{Ad}^*g(l)) = 2q'\varepsilon = 0$ and $\Phi_2(\mathrm{Ad}^*g(l)) = -2(b' - b^*)\varepsilon = 0$, thus we have $\mathrm{Ad}^*g(l) = f$, this is, $l \in \Omega_f$.

Let $f_\infty = \bar{a}A^* + \bar{b}B^* + \bar{p}P^* + \bar{q}Q^* + \bar{r}R^* \in \overline{\Omega_f} \setminus \Omega_f$. Then $\bar{r} = 0$, since $\Phi(f_\infty) = 0$. Suppose $\bar{f} = \lim_{k\to\infty} \mathrm{Ad}^*E(a_k, 0, p_k, 0, r_k, s_k)(f)$. Since $\bar{f}(S) = \lim_k e^{-a_k}\varepsilon = 0$, we have $\lim_k a_k = \infty$.

Suppose $\bar{q} = \lim_k \frac{1}{2}p_k^2\varepsilon \neq 0$, since $\bar{b} = \lim_k (b^* - \frac{p_k r_k}{2}\varepsilon)$, we have that the sequence $\{r_k\}$ is bounded and $\bar{p} = \lim_k r_k e^{-\frac{a_k}{2}}\varepsilon = 0$. It follows that $\overline{\Omega_f} \setminus \Omega_f \subset (\mathbb{R}A^* + \mathbb{R}B^* + \varepsilon\mathbb{R}^+Q^*) \cup (\mathbb{R}A^* + \mathbb{R}B^* + \mathbb{R}P^*)$.

Conversely, let $\bar{f} = \bar{a}A^* + \bar{b}B^* + \bar{q}Q^*$ with $\bar{q} \in \varepsilon\mathbb{R}^+$. When $a \to \infty$, $p \to \sqrt{2\bar{q}\varepsilon}$, $r \to 2\varepsilon(b^* - \bar{b})\sqrt{2\bar{q}\varepsilon}^{-1}$ and $s \to \varepsilon(\bar{a} + \bar{b} - b^*)$, we have $\mathrm{Ad}^*E(a, 0, p, 0, r, s)f \to \bar{f}$. Suppose $\bar{f} = \bar{a}A^* + \bar{b}B^* + \bar{p}P^*$ with $\bar{p} \neq 0$, and let $p(a) := 2(b^* - \bar{b})\bar{p}^{-1}e^{-\frac{a}{2}}$, $r(a) := \varepsilon\bar{p}e^{\frac{a}{2}}$, and $s(a) := \varepsilon\bar{a} - \frac{p(a)r(a)}{2}$. Then $\mathrm{Ad}^*E(a, 0, p(a), 0, r(a), s(a))f \to \bar{f}$ as $a \to \infty$. Suppose $\bar{f} = \bar{a}A^* + \bar{b}B^*$. Let $p(a) := 2(b^* - \bar{b})\varepsilon e^{-\frac{a}{4}}$, $r(a) := e^{\frac{a}{4}}$, $s(a) := \varepsilon\bar{a} - \frac{p(a)r(a)}{2}$. Then $\mathrm{Ad}^*E(a, 0, p(a), 0, r(a), s(a))f \to \bar{f}$ as $a \to \infty$. □

Realisation of $\pi_f \in \widehat{G}$: Taking the Pukanszky polarisation $\mathfrak{b} = \mathbb{R}\text{-span}\{B, Q, R, S\}$ and $F \in L^1(G)$, we realise π_f and $\pi_f(F)$ on $L^2(\exp(\mathbb{R}X + \mathbb{R}P))$, where $X = A + B$, as follows: Let

$$E_0(x,p) := \exp(xX + pP), h = h(b,q,r,s) := \exp(bB)\exp(qQ + rR + sS) \in \exp(\mathfrak{b}),$$

$$E_0(x,p)_b := \exp(bB)E_0(x,p)\exp(-bB) \in \exp(\mathbb{R}X + \mathbb{R}P).$$

For $\xi \in L^2(\exp(\mathbb{R}X + \mathbb{R}P))$, we have

$$\begin{aligned}
&\pi_f(E_0(x,p)h(b,q,r,s))\xi(E_0(t,u)) \\
= \ &\chi(F)(E_0(t-x,u-p)^{-1}_{(-b)}h(0,q,r,s)E_0(t-x,u-p)_{(-b)})\chi_f(\exp(bB)) \\
&\left(\frac{\Delta_{\exp\mathfrak{b}}}{\Delta_G}(\exp(-bB))\right)^{\frac{1}{2}} \xi(E_0(t-x,u-p)_{(-b)}) \\
= \ &\chi_{\mathrm{Ad}^*E_0(t-x,u-p)_{(-b)}(f)}(h(0,q,r,s))\chi_f(\exp(bB)) \times \\
&\left(\frac{\Delta_{\exp\mathfrak{b}}}{\Delta_G}(\exp(-bB))\right)^{\frac{1}{2}} \xi(E_0(t-x,u-p)_{(-b)}) \\
= \ &\exp(\Big(i\varepsilon e^(- t + x)(\frac{1}{2}q(u-p)^2e^b - r(u-p)e^{\frac{b}{2}} + s) + ib^*b + \frac{b}{4}\Big)) \\
&\xi(t-x, e^{\frac{b}{2}}(u-p)).
\end{aligned}$$

We denote $\widehat{F}^{\mathfrak{h}}(g)(l) = \int_{\mathfrak{h}} \chi_l(\exp(qQ + rR + sS))F(g\exp(qQ + rR + sS))dqdrds$ for $l \in \mathfrak{h}^*$ satisfying $l([\mathfrak{h}, \mathfrak{h}]) = \{0\}$, and for integrable functions F as before. We take the left Haar measure dg on G defined by

$$\int_G F(g)dg := \int_{\mathbb{R}^6} F(E_0(x,p)h(b,q,r,s))e^{\frac{b}{2}}dxdpdbdqdrds.$$

Then we have

$$
\begin{aligned}
& \pi_f(F)\xi(E_0(t,u)) \\
= & \int_{\mathbb{R}^6} F(E_0(x,p)h(b,q,r,s))\pi_f(E_0(x,p)h(b,q,r,s))\xi(E_0(t,u))e^{\frac{b}{2}}dxdpdbdqdrds \\
= & \int_{\mathbb{R}^6} F(E_0(x,p)h(b,q,r,s))\chi_{\mathrm{Ad}^*E_0(t-x,u-p)_{(-b)}(f)}(h(0,q,r,s))\chi_f(\exp(bB)) \\
& \left(\frac{\Delta_{\exp\mathfrak{b}}}{\Delta_G}(\exp(-bB))\right)^{\frac{1}{2}}\xi(E_0(t-x,u-p)_{(-b)})e^{\frac{b}{2}}dxdpdbdqdrds \\
= & \int_{\mathbb{R}^3} \widehat{F^{\mathfrak{h}}}(E_0(t-x,u-p)\exp(bB))\left(\mathrm{Ad}^*E_0(x,p)_{(-b)}(f|_{\mathfrak{h}})\right)\chi_f(\exp(bB)) \\
& \left(\frac{\Delta_{\exp\mathfrak{b}}}{\Delta_G}(\exp(-bB))\right)^{\frac{1}{2}}\xi(E_0(x,p)_{(-b)})e^{\frac{b}{2}}dxdpdb \\
= & \int_{\mathbb{R}^3} \widehat{F^{\mathfrak{h}}}(E_0(t-x,u-e^{-\frac{b}{2}}p)\exp(bB))\,(\mathrm{Ad}^*E_0(x,p)(f|_{\mathfrak{h}}))\,\chi_f(\exp(bB)) \\
& \xi(E_0(x,p))e^{\frac{b}{4}}dxdpdb \\
= & \int_{\mathbb{R}^3} \widehat{F^{\mathfrak{h}}}(E_0(t-x,u-e^{-\frac{b}{2}}p)\exp(bB))\left(\varepsilon e^{-x}(\frac{1}{2}p^2Q^*-pR^*+S^*)\right)e^{ib^*b}e^{\frac{b}{4}} \\
& \xi(E_0(x,p))dxdpdb \\
=: & \int_{\mathbb{R}^2} \mathcal{K}_F(t,u;x,p)\xi(E_0(x,p))dxdp,
\end{aligned}
$$

where

$$
\begin{aligned}
& \mathcal{K}_F(t,u;x,p) && (1.4.15)\\
:= & \int_{\mathbb{R}} \widehat{F^{\mathfrak{h}}}(E_0(t-x,u-e^{-\frac{b}{2}}p)\exp(bB))\left(\varepsilon e^{-x}(\frac{1}{2}p^2Q^*-pR^*+S^*)\right)e^{ib^*b}e^{\frac{b}{4}}db.
\end{aligned}
$$

1.4.2.3 The orbits of $b^*B^*+\varepsilon R^*, b^*\in\mathbb{R}, \varepsilon=\pm1$

Let $f=b^*B^*+\varepsilon R^*$. Then $\mathfrak{g}(f)=\mathbb{R}\text{-span}\{S,A-B\}$. Writing $X=A+B, Y=A-B$, $X^*=\frac{1}{2}(A^*+B^*)$, $Y^*=\frac{1}{2}(A^*-B^*)$, we have

$$
\Omega_f=\{\bar{x}X^*+\bar{y}Y^*+\bar{p}P^*+\bar{q}Q^*+\bar{r}R^*;\ \bar{r}\in\varepsilon\mathbb{R}^+,\bar{y}=-b^*-(\bar{p}\bar{q}/\bar{r})\},
$$

$$
\begin{aligned}
\overline{\Omega_f}\setminus\Omega_f = & \ \{\bar{x}X^*+\bar{y}Y^*+\bar{p}P^*;\ \bar{x},\bar{y},\bar{p}\in\mathbb{R}\}\cup\{\bar{x}X^*+\bar{y}Y^*+\bar{q}Q^*;\ \bar{x},\bar{y},\bar{q}\in\mathbb{R}\} \\
= & \ \bigcup_{\bar{x}\in\mathbb{R},\varepsilon=\pm1}\mathrm{Ad}^*G(\bar{x}X^*+\varepsilon P^*)\bigcup_{a^*\in\mathbb{R},\nu=\pm1}\mathrm{Ad}^*G(a^*A^*+\nu Q^*) \\
& \bigcup_{a^*,\lambda\in\mathbb{R}}\{a^*A^*+\lambda B^*\}. \qquad (1.4.16)
\end{aligned}
$$

Proof of (1.4.16)**:**
Recall that

$$
\mathrm{Ad}^*E(\alpha,\alpha,p,q,r,0)\cdot f=(b^*+r\varepsilon)X^*+(-b^*+pq\varepsilon)Y^*+q\varepsilon P^*-e^{-\alpha}p\varepsilon Q^*+e^{-\alpha}\varepsilon R^*.
$$

Let Ψ be a function on $\mathbb{R}$-span$\{X^*, Y^*, P^*, Q^*, R^*\}$ defined by

$$\Psi(x,y,p,q,r) := r(y+b^*) + pq \quad \text{for } (x,y,p,q,r) \in \mathbb{R}\text{-span}\{X^*, Y^*, P^*, Q^*, R^*\}.$$

Then $\overline{\Omega_f} \subset \Psi^{-1}(0) \cap (\mathbb{R}^4 \times (\varepsilon\mathbb{R}^{+,0}))$ and $\overline{\Omega_f} \setminus \Omega_f \subset \Psi^{-1}(0) \cap (\mathbb{R}^4 \times \{0\}) = \{(x,y,p,0,0); x,y,p \in \mathbb{R}\} \cup \{(x,y,0,q,0); x,y,q \in \mathbb{R}\}$.

Conversely, let $\bar{x}, \bar{y}, \bar{p} \in \mathbb{R}$ and suppose $\bar{p} \neq 0$, then

$$\begin{aligned} \mathrm{Ad}^* E(\alpha, \alpha, \tfrac{\bar{y}+b^*}{q}, q\varepsilon, r\varepsilon, 0) f = \; & (b^*+r)X^* + \bar{y}Y^* + qP^* - e^{-\alpha}\tfrac{\varepsilon(\bar{y}+b^*)}{q} Q^* + e^{-\alpha}\varepsilon R^* \\ \to \; & \bar{x}X^* + \bar{y}Y^* + \bar{p}P^* \;\text{ as } \alpha \to \infty, q \to \bar{p}, r \to -b^* + \bar{x}. \end{aligned}$$

If $\bar{p} = 0$, we take

$$\begin{aligned} & \mathrm{Ad}^* E(\alpha, \alpha, \varepsilon(\bar{y}+b^*)e^{\frac{\alpha}{2}}, e^{-\frac{\alpha}{2}}, r\varepsilon, 0) f \\ = \; & (b^*+r)X^* + \bar{y}Y^* + e^{-\frac{\alpha}{2}}\varepsilon P^* - (\bar{y}+b^*)e^{-\frac{\alpha}{2}} Q^* + e^{-\alpha}\varepsilon R^* \\ \to \; & \bar{x}X^* + \bar{y}Y^* + 0P^* \;\text{ as } \alpha \to \infty, r \to -b^* + \bar{x}. \end{aligned}$$

Thus we have $\bar{x}X^* + \bar{y}Y^* + \bar{p}P^* \in \overline{\Omega_f}$. We can similarly verify $\bar{x}X^* + \bar{y}Y^* + \bar{q}Q^* \in \overline{\Omega_f}$. □

Realisation of $\pi_f \in \widehat{G}$:

We take the Pukanszky polarisation $\mathfrak{q} = \mathbb{R}$-span$\{Y, Q, R, S\}$ at f to realise π_f. Noting that $[X, P] = 0$, let

$$E_0(t,u) := \exp(tX + uP) \text{ and } h = h(y,q,r,s) := \exp(yY)\exp(qQ + rR + sS).$$

We realise $\pi_f = \mathrm{ind}_{\exp \mathfrak{q}}^G \chi_f$ on $L^2(\exp(\mathbb{R}X + \mathbb{R}P))$ as follows: Noting that $[Y, \mathbb{R}X + \mathbb{R}P] \subset \mathbb{R}P$, we write $E_0(t,u)_y := \exp(yY)E_0(t,u)\exp(-yY)$. Let $\xi \in L^2(\exp(\mathbb{R}X + \mathbb{R}P))$. Then

$$\begin{aligned} & \pi_f(E_0(x,p)h(y,q,r,s))\xi(E_0(t,u)) \\ & = \chi_f(E_0(t-x,u-p)_{(-y)}^{-1} h(0,q,r,s) E_0(t-x,u-p)_{(-y)})\chi_f(\exp(yY)) \\ & \quad \left(\frac{\Delta_{\exp \mathfrak{q}}}{\Delta_G}(\exp(-yY))\right)^{\frac{1}{2}} \xi(E_0(t-x,u-p)_{-y}) \\ & = \chi_{\mathrm{Ad}^* E_0(t-x,u-p)_{(-y)}(f)}(h(0,q,r,s))\chi_f(\exp(yY)) \\ & \quad \left(\frac{\Delta_{\exp \mathfrak{q}}}{\Delta_G}(\exp(-yY))\right)^{\frac{1}{2}} \xi(E_0(t-x,u-p)_{(-y)}) \\ & = \exp\left(-i\varepsilon(e^{-y}q(u-p) - r)e^{x-t} + ib^*y - \frac{y}{2}\right)\xi(t-x, e^{-y}(u-p)). \end{aligned}$$

We take the left Haar measure dg on G defined by

$$\int_G F(g)dg := \int_{\mathbb{R}^6} F(E_0(x,p)h(y,q,r,s))e^{-y} dx dp dy dq dr ds$$

for integrable functions $F \in L^1(G)$, and define

$$\widehat{F^{\mathfrak{h}}}(g)(l) := \int_{\mathfrak{h}} F(g\exp(qQ + rR + sS))\chi_l(\exp(qQ + rR + sS)) dq dr ds,$$

for $l \in \mathfrak{h}^*$ satisfying $l([\mathfrak{h},\mathfrak{h}]) = \{0\}$. Then

$$\begin{aligned}
\pi_f(F)\xi(E_0(t,u)) &= \int_{\mathbb{R}^6} F(E_0(x,p)h)\pi_f(E_0(x,p)h)\xi(E_0(t,u))e^{-y}dxdpdydqdrds \\
&= \int_{\mathbb{R}^6} F(E_0(x,p)h)\chi_{\mathrm{Ad}^* E_0(t-x,u-p)_{(-y)}(f)}(h(0,q,r,s))\chi_f(\exp(yY)) \\
&\quad \left(\frac{\Delta_{\exp \mathfrak{q}}}{\Delta_G}(\exp(-yY))\right)^{\frac{1}{2}} \xi(E_0(t-x,u-p)_{(-y)})e^{-y}dxdpdydqdrds \\
&= \int_{\mathbb{R}^6} F(E_0(t-x,u-p)h)\chi_{\mathrm{Ad}^* E_0(x,p)_{(-y)}(f)}(h(0,q,r,s))\chi_f(\exp(yY)) \\
&\quad \left(\frac{\Delta_{\exp \mathfrak{q}}}{\Delta_G}(\exp(-yY))\right)^{\frac{1}{2}} \xi(E_0(x,p)_{(-y)})e^{-y}dxdpdydqdrds \\
&= \int_{\mathbb{R}^6} F(E_0(t-x,u-e^y p)h(y,q,r,s))\chi_{\mathrm{Ad}^* E_0(x,p)(f)}(h(0,q,r,s))\chi_f(\exp(yY)) \\
&\quad e^{-\frac{y}{2}}\xi(E_0(x,p))dxdpdydqdrds \\
&= \int_{\mathbb{R}^3} \widehat{F^{\mathfrak{h}}}(E_0(t-x,u-e^y p)\exp(yY))(\mathrm{Ad}^* E_0(x,p)(f|_{\mathfrak{h}}))e^{ib^* y}e^{-\frac{y}{2}}\xi(E_0(x,p))dxdpdy \\
&= \int_{\mathbb{R}^3} \widehat{F^{\mathfrak{h}}}(E_0(t-x,u-e^y p)\exp(yY))(-\varepsilon p e^{-x}Q^* + \varepsilon e^{-x}R^*)e^{ib^* y}e^{-\frac{y}{2}} \\
&\quad \xi(E_0(x,p))dxdpdy \\
&=: \int_{\mathbb{R}^2} \mathcal{K}_f(t,u;x,p)\xi(E_0(x,p))dxdp,
\end{aligned}$$

where

$$\mathcal{K}_f(t,u;x,p) := \int_{\mathbb{R}} e^{ib^* y}e^{-\frac{y}{2}}\widehat{F^{\mathfrak{h}}}(E(t-x,u-e^y p)\exp(yY))$$
$$(-\varepsilon p e^{-x}Q^* + \varepsilon e^{-x}R^*)dy. \tag{1.4.17}$$

1.4.2.4 *The orbits of* $\nu Q^* + \varepsilon P^*, \nu, \varepsilon = \pm 1$

Let $f = \varepsilon P^* + \nu Q^*$ for $\varepsilon, \nu = \pm 1$. Then we have $\mathfrak{g}(f) = \mathbb{R}\text{-span}\{S, R\}$. Let $X^* = \frac{1}{2}(A^* + B^*)$, $Y^* = \frac{1}{2}(A^* - B^*)$.

$$\Omega_f = \mathbb{R}A^* + \mathbb{R}B^* + \mathbb{R}^+\varepsilon P^* + \mathbb{R}^+\nu Q^*,$$

$$\begin{aligned}
\overline{\Omega_f} \setminus \Omega_f = &\ (\mathbb{R}A^* + \mathbb{R}B^* + \mathbb{R}^+\varepsilon P^*) \cup (\mathbb{R}A^* + \mathbb{R}B^* + \mathbb{R}^+\nu Q^*) \cup (\mathbb{R}A^* + \mathbb{R}B^*) \\
= &\ \bigcup_{\alpha^* \in \mathbb{R}} (\alpha^* X^* + \mathbb{R}Y^* + \mathbb{R}^+\varepsilon P^*) \bigcup_{a^* \in \mathbb{R}} (a^* A^* + \mathbb{R}B^* + \mathbb{R}^+\nu Q^*) \\
&\ \bigcup_{a^*, b^* \in \mathbb{R}} \{a^* A^* + b^* B^*\} \\
= &\ \bigcup_{\alpha^* \in \mathbb{R}} \mathrm{Ad}^*(G)(\alpha^* X^* + \varepsilon P^*) \bigcup_{y^* \in \mathbb{R}} \mathrm{Ad}^*(G)(y^* Y^* + \nu Q^*) \\
&\ \bigcup_{a^*, b^* \in \mathbb{R}} \{a^* A^* + b^* B^*\}.
\end{aligned} \tag{1.4.18}$$

We realise π_f in $\widehat{G}$ by taking a Pukanszky polarisation $\mathfrak{l} = \mathbb{R}\text{-span}\{P, Q, R, S\}$ and $\pi_f = \operatorname{ind}_{\exp \mathfrak{l}}^G \chi_f$. Writing $g_0 = E_0(a,b) := \exp(aA + bB)$, $n = n(p,q,r,s) := \exp(pP)\exp(qQ + rR + sS)$ and $t_0 = E_0(t,u) := \exp(tA + uB)$, we have

$$\begin{aligned} \pi_f(g_0 n)\xi(t_0) &= \chi_f((g_0^{-1}t_0)^{-1} n g_0^{-1} t_0)\xi(g_0^{-1}t_0)) = \chi_{\mathrm{Ad}^*(g_0^{-1}t_0)f}(n)\xi(g_0^{-1}t_0)) \\ &= \chi_{\mathrm{Ad}^* E_0(t-a,u-b)(f)}(n)\xi(E_0(t-a,u-b)) \\ &= \exp i(\varepsilon p e^{\frac{-t+a+u-b}{2}} + \nu q e^{-u+b})\xi(E_0(t-a,u-b)). \end{aligned}$$

Let $F \in L^1(G)$ and $l \in \mathfrak{l}^*$ such that $l([\mathfrak{l},\mathfrak{l}]) = \{0\}$. We define

$$\widehat{F}^{\mathfrak{l}}(g_0)(l) := \int_{\mathfrak{l}} F(g_0 n(p,q,r,s)) e^{il(pP+qQ+rR+sS)} dp dq dr ds.$$

Let $dg_0 dn$ be the left Haar measure transferred the Lebesgue measure $dadbdpdqdrds$ on $\mathfrak{g}$ by the mapping $(a,b,p,q,r,s) \mapsto g_0(a,b)n(p,q,r,s)$.

$$\begin{aligned} \pi_f(F)\xi(E_0(t,u)) &= \int_G \pi_f(g_0 n)\xi(t_0)F(g_0 n) dg_0 dn \\ &= \int_{\mathbb{R}^6} \chi_{\mathrm{Ad}^* E_0(t-a,u-b)(f)}(n)\xi(E_0(t-a,u-b))F(E_0(a,b)n) \\ &\quad dadbdpdqdrds \\ &= \int_{\mathbb{R}^6} \chi_{\mathrm{Ad}^* E_0(a,b)(f)}(n)\xi(E_0(a,b))F(E_0(t-a,u-b)n) \\ &\quad dadbdpdqdrds \\ &=: \int_{\mathfrak{g}/\mathfrak{l}} \widehat{F}^{\mathfrak{l}}(E_0(t-a,u-b))(\mathrm{Ad}^* E_0(a,b)(f|_{\mathfrak{h}}))\xi(E_0(a,b)) dadb \\ &= \int_{\mathbb{R}^2} \widehat{F}^{\mathfrak{l}}(E_0(t-a,u-b))(\varepsilon e^{\frac{-a-b}{2}} P^* + \nu e^{-b} Q^*)\xi(E_0(a,b)) dadb \\ &=: \int_{\mathbb{R}^2} \mathcal{K}_F(t,u;a,b)\xi(E_0(a,b)) dadb, \end{aligned}$$

where

$$\begin{aligned} \mathcal{K}_F(t,u;a,b) &:= \widehat{F}^{\mathfrak{l}}(E_0(t-a,u-b))(\mathrm{Ad}^* E_0(a,b)(f|_{\mathfrak{l}})) \\ &= \widehat{F}^{\mathfrak{l}}(E_0(t-a,u-b))(\varepsilon e^{\frac{-a-b}{2}} P^* + \nu e^{-b} Q^*). \end{aligned} \tag{1.4.19}$$

1.4.2.5 The orbits of $a^*A^* + \varepsilon Q^*, a^* \in \mathbb{R}, \varepsilon = \pm 1$

Let $f := a^*A^* + \varepsilon Q^*$. Then $\mathfrak{g}(a^*A^* + \varepsilon Q^*) = \mathbb{R}\text{-span}\{A, P, R, S\}$,

$$\Omega_f = a^*A^* + \mathbb{R}B^* + \mathbb{R}^+ \varepsilon Q^*,$$

$$\overline{\Omega_f} \setminus \Omega_f = a^*A^* + \mathbb{R}B^* + 0 \cdot Q^* = \cup_{\lambda \in \mathbb{R}}(a^*A^* + \lambda B^*). \tag{1.4.20}$$

We realise π_f in $\widehat{G}$ as $\pi_f = \mathrm{ind}_{\exp \mathfrak{l}_A}^G \chi_f$, $\mathfrak{l}_A = \mathbb{R}\text{-span}\{A, P, Q, R, S\}$, on $L^2(G/\exp \mathfrak{l}_A, \chi_f) \simeq L^2(\exp(\mathbb{R}B))$. Write

$$E_0(b) := \exp(bB), \quad h = h(a,p,q,r,s) := \exp(aA)\exp(pP)\exp(qQ + rR + sS)$$

for $b \in \mathbb{R}$, $(a,p,q,r,s) \in \mathbb{R}^5$. We have

$$\begin{aligned}\pi_f(E_0(b)h(a,p,q,r,s))\xi(E_0(t)) &= \chi_{\mathrm{Ad}^* E_0(t-b)(f)}(h)\xi(E_0(t-b)) \\ &= \exp i(aa^* + q\varepsilon e^{b-t})\xi(E_0(t-b)),\end{aligned}$$

$$\begin{aligned}\pi_f(F)\xi(E_0(t)) &= \int_{\mathbb{R}^6} F(E_0(b)h)\chi_{\mathrm{Ad}^* E_0(t-b)(f)}(h)\xi(E_0(t-b))dbdadpdqdrds \\ &= \int_{\mathbb{R}^6} F(E_0(t-b)h)\chi_{\mathrm{Ad}^* E_0(b)(f)}(h)\xi(E_0(b))dbdadpdqdrds \\ &= \int_{\mathbb{R}} \widehat{F^{\mathfrak{l}_A}}(E_0(t-b))(\mathrm{Ad}^* E_0(b)(f|_{\mathfrak{l}_A})\xi(E_0(b))db \\ &= \int_{\mathbb{R}} \mathcal{K}_F(t,b)\xi(E_0(b))db,\end{aligned}$$

where

$$\widehat{F^{\mathfrak{l}_A}}(g)(l) := \int_{\mathfrak{l}_A} F(gh(a,p,q,r,s))\chi_l(h(a,p,q,r,s))dadpdqdrds$$

for $l \in \mathfrak{g}^*$ such that $l([\mathfrak{l}_A, \mathfrak{l}_A]) = \{0\}$, and

$$\mathcal{K}_F(t,b) := \widehat{F^{\mathfrak{l}_A}}(E_0(t-b))(\mathrm{Ad}^* E_0(b)(f|_{\mathfrak{l}_A}) = \widehat{F^{\mathfrak{l}_A}}(E_0(t-b))(a^*A^* + \varepsilon e^{-b}Q^*).$$

1.4.2.6 The orbits of $x^*X^* + \varepsilon P^*, x^* \in \mathbb{R}, \varepsilon = \pm 1$

Let $f = \alpha^*(\frac{A^*+B^*}{2}) + \varepsilon P^* = \alpha^* X^* + \varepsilon P^*$, $X = A + B$, $Y = A - B$, $X^* = \frac{A^*+B^*}{2}$, $Y^* = \frac{A^*-B^*}{2}$. Then $\mathfrak{g}(f) = \mathbb{R}\text{-span}\{X, Q, R, S\}$. By (1.4.11) and (1.4.12), we have

$$\Omega_f = \alpha^* X^* + \mathbb{R}Y^* + \mathbb{R}^+ \varepsilon P^*,$$

$$\overline{\Omega_f} \setminus \Omega_f = \alpha^* X^* + \mathbb{R}Y^* = \bigcup_{\lambda \in \mathbb{R}} \{\alpha^* X^* + \lambda Y^*\}. \tag{1.4.21}$$

We realise π_f in $\widehat{G}$ as $\pi_f = \mathrm{ind}_{\exp \mathfrak{l}_X}^G \chi_f$, where $\mathfrak{l}_X = \mathbb{R}\text{-span}\{X, P, Q, R, S\}$, on $L^2(G/\exp \mathfrak{l}_X, \chi_f)$
$\simeq L^2(\exp(\mathbb{R}Y))$. We write

$$E_0(y) := \exp(yY), \quad h = h(x,p,q,r,s) := \exp(xX)\exp(pP)\exp(qQ + rR + sS)$$

for $y \in \mathbb{R}$ and $(x,p,q,r,s) \in \mathbb{R}^5$, and transfer a Lebesgue measure on $\mathbb{R}^6$ to a left Haar measure on G by $(x,y,p,q,r,s) \mapsto E_0(y)h(x,p,q,r,s)$. For $\xi \in L^2(G/\exp \mathfrak{l}_X, \chi_f)$ we have

$$\begin{aligned}\pi_f(E_0(y)h(x,p,q,r,s))\xi(E_0(t)) &= \chi_{\mathrm{Ad}^* E_0(t-y)(f)}(h)\xi(E_0(t-y)) \\ &= \exp(ix\alpha^* + ip\varepsilon e^{-t+y})\xi(E_0(t-y)),\end{aligned}$$

and for an integrable function F, we have

$$\begin{aligned}\pi_f(F)\xi(E_0(t)) &= \int_{\mathbb{R}^6} F(E_0(y)h(x,p,q,r,s))\pi_f(E_0(y)h(x,p,q,r,s))\xi(E_0(t)) \\ &\quad dydxdpdqdrds \\ &= \int_{\mathbb{R}^6} F(E_0(y)h)\chi_{\mathrm{Ad}^*E_0(t-y)(f)}(h)\xi(E_0(t-y))dydxdpdqdrds \\ &= \int_{\mathbb{R}^6} F(E_0(t-y))h)\chi_{\mathrm{Ad}^*E_0(y)(f)}(h)\xi(E_0(y))dydxdpdqdrds \\ &=: \int_{\mathbb{R}} \widehat{F^{\mathfrak{l}_X}}(E_0(t-y))(\mathrm{Ad}^*E_0(y)(f|_{\mathfrak{h}}))\xi(E_0(y))dy \\ &= \int_{\mathbb{R}} \widehat{F^{\mathfrak{l}_X}}(E_0(t-y))(\alpha^*X^* + \varepsilon e^{-y}P^*)\xi(E_0(y))dy.\end{aligned}$$

For $l \in \mathfrak{g}^*$ satisfying $l([\mathfrak{l}_X, \mathfrak{l}_X]) = \{0\}$, we write

$$\widehat{F^{\mathfrak{l}_X}}(g)(l) := \int_{\mathfrak{l}_X} e^{il(Z)}F(g\exp Z)dZ,\ g \in G,$$

with some fixed Lebesgue measure dZ on $\mathfrak{h}$. And $\pi_f(F)$ is described by the operator $\pi_f(F)\xi(E_0(y)) = \int_{\mathbb{R}} \mathcal{K}_f(t,y)\xi(E_0(y))dy$, where

$$\mathcal{K}_f(t,y) := \widehat{F^{\mathfrak{l}_X}}(E_0(t-y))(\mathrm{Ad}^*E_0(y)(f|_{\mathfrak{l}_X})) = \widehat{F^{\mathfrak{l}_X}}(E_0(t-y))(\alpha^*X^* + \varepsilon e^{-y}P^*). \tag{1.4.22}$$

1.5 BRANCHING LAWS

1.5.1 Branching laws for induced representations

We consider here the setting monomial representations $\tau = \mathrm{ind}_H^G\chi_f$. Let $S(f,\mathfrak{g})$ be the set of subalgebras $\mathfrak{h}$ of $\mathfrak{g}$ which are isotropic for the bilinear form B_f, i.e. $B_f(\mathfrak{h},\mathfrak{h}) = \{0\}$. Let $M(f,\mathfrak{g}) \subset S(f,\mathfrak{g})$ be the set of subalgebras $\mathfrak{h}$ which are maximally isotropic for B_f, i.e. $\dim\mathfrak{h} = \frac{1}{2}(\dim\mathfrak{g} + \dim\mathfrak{g}(f))$. The element of $M(f,\mathfrak{g})$ is called (real) polarisation of $\mathfrak{g}$ at f. Finally, let $I(f,\mathfrak{g}) \subset M(f,\mathfrak{g})$ be the set of subalgebras such that $\tau = \mathrm{ind}_H^G\chi_f$ is irreducible. Then, we know $I(f,\mathfrak{g}) \neq \emptyset$ and:

Theorem 1.5.1 ([98]) *The map which brings $\mathfrak{h} \in I(f,\mathfrak{g})$ to the class of $\tau = ind_H^G\chi_f$ in $\widehat{G}$ is surjective.*

This theorem means that the Kirillov-Bernat map $\theta : \mathfrak{g}^* \to \widehat{G}$ induces a bijection, denoted also by θ, of the orbit space $\mathfrak{g}^*/G$ onto $\widehat{G}$. The elements of $I(f,\mathfrak{g})$ are characterised by the following Pukanszky condition.

Theorem 1.5.2 ([23]) *Let $\mathfrak{h} \in S(f,\mathfrak{g})$. Then, $\mathfrak{h}$ belongs to $I(f,\mathfrak{g})$ if and only if we have $H\cdot f = f + \mathfrak{h}^\perp$.*

Here we give a standard construction of an element of $I(f, \mathfrak{g})$ due to Vergne [23]. A sequence of subalgebras

$$\{0\} = \mathfrak{g}_0 \subset \mathfrak{g}_1 \subset \mathfrak{g}_2 \subset \cdots \subset \mathfrak{g}_n = \mathfrak{g}, \ \dim(\mathfrak{g}_j) = j(1 \leq j \leq n)$$

is called a good sequence if it satisfies the following conditions. If $\mathfrak{g}_j$ is not an ideal of $\mathfrak{g}$, then $\mathfrak{g}_{j\pm 1}$ are ideals of $\mathfrak{g}$ and the action of $\mathfrak{g}$ on $\mathfrak{g}_{j+1}/\mathfrak{g}_{j-1}$ is irreducible.

Theorem 1.5.3 *A good sequence of subalgebras $\{\mathfrak{g}_j\}_{j=0}^n$ being given, put $f_j = f|_{\mathfrak{g}_j} \in (\mathfrak{g}_j)^*$ for $1 \leq j \leq n$. Then the sum $\mathfrak{p} = \sum_{j=1}^n \mathfrak{g}_j(f_j)$ is an element of $I(f, \mathfrak{g})$.*

Let $\tilde{\mu}$ be a finite measure equivalent to the Lebesgue measure on $\Gamma_\tau = f + \mathfrak{h}^\perp$. We regard $\tilde{\mu}$ as a measure on $\mathfrak{g}^*$ and put $\mu = \theta_*(\tilde{\mu})$. Then, we have

Theorem 1.5.4 ([48]) *The canonical central decomposition of τ is described as*

$$\tau \simeq \int_{\widehat{G}}^{\oplus} m(\pi)\pi d\mu(\pi), \tag{1.5.1}$$

where the multiplicity $m(\pi)$ is given by the number of the H-orbits contained in $\Gamma_\tau \cap \Omega(\pi)$.

Since the algebraic structure of solvable Lie algebras is not so rigid that the proofs of many assertions are made by induction. First, we prepare very useful lemmas.

Definition 1.5.5 Let $\mathfrak{g}$ be a real solvable Lie algebra and V a $\mathfrak{g}$-module. We say that V is exponential type if any weight of $\mathfrak{g}$ in V is written as $X \mapsto \lambda(X)(1 + \sqrt{-1}\alpha)$ with certain $\lambda \in \mathfrak{g}^*$ and $\alpha \in \mathbb{R}$.

Then, it is easy to see (cf. [56]):

Lemma 1.5.6 *Let V be a finite-dimensional real vector space equipped with a G-action of exponential type. Let $v \in V$ be a nonzero G-invariant vector, i.e. $g{\cdot}v = v$ for all $g \in G$. We arbitrarily fix $x \in V$ and consider the line $L_x = x + \mathbb{R}v$. We then have the following two possibilities:*

$$L_x \cap G{\cdot}x = \{x\} \text{ or } L_x \cap G{\cdot}x = L_x.$$

In other words, the line passing x and having the direction of the invariant vector v either meets the orbit $G{\cdot}x$ at only one point or is completely contained there.

Definition 1.5.7 In the situation of this lemma, if $L_x \subset G{\cdot}x$, the orbit $G{\cdot}x$ is said to be saturated in the v-direction and otherwise $G{\cdot}x$ is said to be non-saturated.

When there is an ideal $\mathfrak{g}_0$ of $\mathfrak{g}$ such that $\dim(\mathfrak{g}/\mathfrak{g}_0) = 1$, a linear form $\ell \in \mathfrak{g}^*$ vanishing on $\mathfrak{g}_0$ is an invariant vector relative to the coadjoint representation of G. Write $\mathfrak{g} = \mathbb{R}X + \mathfrak{g}_0$ and $\ell_0 = \ell|_{\mathfrak{g}_0}$ for $\ell \in \mathfrak{g}^*$. Let $G_0 = \exp(\mathfrak{g}_0)$ and $p : \mathfrak{g}^* \to (\mathfrak{g}_0)^*$ the restriction map so that $\ell_0 = p(\ell)$ for $\ell \in \mathfrak{g}^*$. We denote by θ_0 the Kirillov-Bernat map for G_0.

Lemma 1.5.8 ([58]) *Let $\ell \in \mathfrak{g}^*$ and $\Omega = G{\cdot}\ell$ corresponding to $\sigma \in \widehat{G}$.*
(1) If the orbit Ω is saturated with respect to $\mathfrak{g}_0$, there is a one-parameter family $\{\omega_s\}_{s\in\mathbb{R}}$ of G_0-orbits in $\mathfrak{g}_0^*$ such that $p(\Omega) = \cup_{s\in\mathbb{R}}\omega_s$ and $\exp(tX)\cdot\omega_s = \omega_{s+t}$. Accordingly,

$$\sigma|_{G_0} \simeq \int_{\mathbb{R}}^{\oplus} \theta_0(\omega_s)ds. \tag{1.5.2}$$

(2) If the orbit Ω is non-saturated with respect to $\mathfrak{g}_0$, then $p(\Omega) = G_0{\cdot}\ell_0$. Accordingly, $\sigma|_{G_0}$ is irreducible and corresponds to $G_0 \cdot \ell_0$.

Lemma 1.5.9 ([58]) *Let $\pi_0 \in \widehat{G_0}$ correspond to the orbit $G_0 \cdot \ell_0$ with $\ell_0 \in (\mathfrak{g}_0)^*$. Let $\ell \in \mathfrak{g}^*$ be an extension of ℓ_0 and put $\Omega = G \cdot \ell$.*
(1) If Ω is saturated with respect to $\mathfrak{g}_0$, we have $ind_{G_0}^G\pi_0 \simeq \theta(\Omega)$.
(2) If Ω is non-saturated with respect to $\mathfrak{g}_0$, we have

$$ind_{G_0}^G\pi_0 \simeq \int_{\mathbb{R}}^{\oplus} \theta(\Omega_t)dt. \tag{1.5.3}$$

Here, X being a fixed element in $\mathfrak{g}(\ell)$ outside of $\mathfrak{g}_0$, $\Omega_t = G \cdot \ell_t$ with $\ell_t \in \mathfrak{g}^*$ defined by $\ell_t|_{\mathfrak{g}_0} = \ell_0$ and $\ell_t(X) = t$.

To prove a branching law for τ, we proceed by induction on $\dim G + \dim G/H$. We first suppose $\mathfrak{g} \neq \mathfrak{h} + [\mathfrak{g}, \mathfrak{g}]$. Then there is an ideal $\mathfrak{g}_0$ of codimension 1 in $\mathfrak{g}$ containing $\mathfrak{h}$. We can apply the induction hypothesis to $G_0 = \exp(\mathfrak{g}_0)$ and make use of the last lemma. Now suppose $\mathfrak{g} = \mathfrak{h} + [\mathfrak{g}, \mathfrak{g}]$. We may assume that $\mathfrak{h}$ contains the centre $\mathfrak{z}$ of $\mathfrak{g}$. Moreover, if $\mathfrak{h} \cap \ker f$ contains a non-trivial ideal $\mathfrak{a}$ of $\mathfrak{g}$, we may apply the induction hypothesis to the quotient group $G/\exp\mathfrak{a}$. Thus we may assume $\dim\mathfrak{z} \leq 1$. Let $\mathfrak{a}$ be a minimal non-central ideal of $\mathfrak{g}$ (cf. Subsection 1.1.2). When the adjoint action of $\mathfrak{g}$ on $\mathfrak{a}/(\mathfrak{a} \cap \mathfrak{z})$ produces a root $0 \neq \lambda \in \mathfrak{g}^*$, there is by our assumption $T \in \mathfrak{h}$ verifying $\lambda(T) = 1$, while $\mathfrak{g}_0 = \ker\lambda$ is an ideal of $\mathfrak{g}$ so that $\mathfrak{g} = \mathbb{R}T \oplus \mathfrak{g}_0$. If $\mathfrak{a} \subset \mathfrak{h}$, set $\mathfrak{a}^f = \{X \in \mathfrak{g}; B_f(X, \mathfrak{a}) = \{0\}\}$ and $A^f = \exp(\mathfrak{a}^f)$. We apply the induction hypothesis to A^f and use the relation $\tau \simeq \mathrm{ind}_{A^f}^G(\mathrm{ind}_H^{A^f}\chi_f)$. If $\mathfrak{a} \not\subset \mathfrak{h}$, we put $\mathfrak{m} = \mathfrak{h} + \mathfrak{a}$ and $M = \exp\mathfrak{m}$ and use the relation $\tau = \mathrm{ind}_M^G(\mathrm{ind}_H^M\chi_f)$. This time we apply the induction hypothesis to the process from M to G and study explicitly the disintegration of $\mathrm{ind}_H^M\chi_f$.

We generalise this theorem a little bit. Starting from a subgroup $K = \exp\mathfrak{k}$ and $\sigma \in \widehat{K}$, we examine the induced representation $\mathrm{ind}_K^G\sigma$. We denote by $\omega(\sigma) \subset \mathfrak{k}^*$ the coadjoint orbit of K corresponding to σ and by $p : \mathfrak{g}^* \to \mathfrak{k}^*$ the restriction mapping. A K-invariant measure on $\omega(\sigma)$ and a Lebesgue measure on $\mathfrak{k}^\perp$ provide the measure $\hat{\mu}$ on the submanifold $p^{-1}(\omega(\sigma))$ of $\mathfrak{g}^*$. Let $\tilde{\mu}$ be a finite measure on $\mathfrak{g}^*$ equivalent to $\hat{\mu}$, and put $\mu = \theta_*(\tilde{\mu})$. Moreover, for $\pi \in \widehat{G}$, we denote by $n_\pi(\sigma)$ the number of the K-orbits contained in $\Gamma(\pi,\sigma) = \Omega(\pi) \cap p^{-1}(\omega(\sigma))$. Using the fact that σ is a monomial representation of K, we get:

Theorem 1.5.10 *We have that*

$$ind_K^G \sigma \simeq \int_{\widehat{G}}^{\oplus} n_\pi(\sigma)\pi d\mu(\pi).$$

Example 1.5.11 Let $\mathfrak{g} = \langle T, X, Y, Z\rangle_{\mathbb{R}} : [T, X] = X, [T, Y] = -Y, [X, Y] = Z$ (completely solvable oscillator). Put $O(\alpha, \beta) = G{\cdot}f_{\alpha,\beta}(\beta \neq 0)$, where $f_{\alpha,\beta} = \alpha T^* + \beta Z^*$ relative to the dual basis of $\mathfrak{g}^*$. A simple computation shows

$$O(\alpha, \beta) = tT^* + xX^* + yY^* + \beta Z^*$$

with the coefficients satisfying $xy = \beta(\alpha - t)$.

Take first $\mathfrak{h} = \mathbb{R}T \oplus \mathbb{R}X$. We have $f_{\alpha_0,\beta_0} + \mathfrak{h}^\perp = \alpha_0 T^* + \mathbb{R}Y^* + \mathbb{R}Z^*$ and consequently, for nonzero β, $(f_{\alpha_0,\beta_0} + \mathfrak{h}^\perp) \cap O(\alpha, \beta)$ is a H-orbit $H{\cdot}f_{\alpha,\beta} = \alpha_0 T^* + \mathbb{R}Y^* + \beta Z^*$ if $\alpha = \alpha_0$ or empty if $\alpha \neq \alpha_0$. Let $\mathfrak{k} = \langle T, X, Z\rangle_{\mathbb{R}}, K = \exp \mathfrak{k}$ and $\pi(\alpha, \beta) \in \widehat{K}$ associated to the K-orbit passing $\alpha T^* + \beta Z^* \in \mathfrak{k}^*$. Then,

$$\begin{aligned}
\text{ind}_H^G \chi_{f_{\alpha_0,\beta_0}} &\simeq \text{ind}_K^G(\text{ind}_H^K \chi_{f_{\alpha_0,\beta_0}}) \simeq \text{ind}_K^G\left(\int_{\mathbb{R}} \pi(\alpha_0, \beta) d\beta\right) \\
&\simeq \int_{\mathbb{R}} \theta(O(\alpha_0, \beta)) d\beta.
\end{aligned}$$

Now, let $\mathfrak{h} = \mathbb{R}T \oplus \mathbb{R}Z$. For $\beta \neq 0$, $(f_{\alpha_0,\beta_0} + \mathfrak{h}^\perp) \cap O(\alpha, \beta)$ is empty if $\beta \neq \beta_0$ or, if $\beta = \beta_0$, equal to $\alpha_0 T^* + xX^* + yY^* + \beta_0 Z^*$ with $xy = \beta_0(\alpha - \alpha_0)$ which is composed by two H-orbits if $\alpha \neq \alpha_0$.

By the way, let $\mathfrak{g}_2$ be the $ax+b$ algebra: $\mathfrak{g}_2 = \mathbb{R}e_1 \oplus \mathbb{R}e_2$ with $[e_1, e_2] = e_2$. It is well known that $G_2 = \exp(\mathfrak{g}_2)$ has two irreducible unitary representations $\pi_\pm$ of infinite dimension, which correspond to two open orbits $\pm G{\cdot}e_2^*$. Let $f = \alpha e_1^* \in \mathfrak{g}_2^*, \mathfrak{h}_0 = \mathbb{R}e_1$ and $H_0 = \exp(\mathfrak{h}_0)$. For any $\ell \in f + \mathfrak{h}_0^\perp \subset \mathfrak{g}_2^*$, we have $\text{ind}_{H_0}^{G_2} \chi_f = \text{ind}_{H_0}^{G_2} \chi_\ell \simeq \pi_+ \oplus \pi_-$.

Returning to our example, put $\mathfrak{n} = \langle X, Y, Z\rangle_{\mathbb{R}}$ and $N = \exp \mathfrak{n}$. K being as above, K is isomorphic to the direct product $G_2 \times \mathbb{R}$. Let χ_{β_0} be the character of $\mathbb{R}$ defined by $\chi_{\beta_0}(s) = e^{is\beta_0}$ and $\times$ the outer Kronecker product. Then we have

$$\begin{aligned}
\text{ind}_H^G \chi_{f_{\alpha_0,\beta_0}} &\simeq \text{ind}_K^G(\text{ind}_H^K \chi_{f_{\alpha_0,\beta_0}}) \simeq \text{ind}_K^G((\pi_+ \times \chi_{\beta_0}) \oplus (\pi_- \times \chi_{\beta_0})) \\
&\simeq \text{ind}_{H'}^G(\chi_{f_1} \oplus \chi_{f_2}) \simeq \text{ind}_N^G(\text{ind}_{H'}^N(\chi_{f_1} \oplus \chi_{f_2})) \\
&= 2\text{ind}_N^G \pi_{\beta_0} \simeq 2\int_{\mathbb{R}}^{\oplus} \theta(O(\alpha, \beta_0)) d\alpha,
\end{aligned}$$

where $\mathfrak{h}' = \mathbb{R}X \oplus \mathbb{R}Z, H' = \exp(\mathfrak{h}'), f_j = \gamma_j X^* + \beta_0 Z^*(j = 1, 2)$ with $\gamma_1 > 0, \gamma_2 < 0$ and $\pi_{\beta_0} \in \widehat{N}$ associated to the N-orbit passing $\beta_0 Z^* \in \mathfrak{n}^*$. In this way, we see again that the multiplicities are given by the number of H-orbits.

1.5.2 Branching laws for restrictions to analytic subgroups

We proceed to the disintegration of the restriction $\pi|_K$ of $\pi \in \widehat{G}$ to an analytic subgroup $K = \exp \mathfrak{k}$. Let $p : \mathfrak{g}^* \to \mathfrak{k}^*$ be the restriction map and $\theta_K : \mathfrak{k}^* \to \widehat{K}$ the Kirillov-Bernat map for K. We take in $\mathfrak{g}^*$ a finite measure $\tilde{\nu}$ equivalent to the G-invariant measure on the orbit $\Omega(\pi)$ and set $\nu = (\theta_K \circ p)_*(\tilde{\nu})$. Using this measure ν on $\widehat{K}$ and the same multiplicity function $n_\pi(\sigma)$ as in the above theorem, we have:

Theorem 1.5.12 *We have the following*(cf. [49])*:*

$$\pi|_K \simeq \int_{\widehat{K}}^{\oplus} n_\pi(\sigma)\sigma d\nu(\sigma).$$

To get this result we proceed as usual by induction on $\dim G$. When $\mathfrak{k}$ is contained in a proper ideal of $\mathfrak{g}$. We introduce an ideal $\mathfrak{g}_0$ of codimension 1 in $\mathfrak{g}$ containing $\mathfrak{k}$. Setting $G_0 = \exp(\mathfrak{g}_0)$, we have $\pi|_K = (\pi|_{G_0})|_K$. We use the previous lemmas to go down to G_0 and apply the induction hypothesis for the process from G_0 to K.

Now suppose that $\mathfrak{k}$ is not contained in any proper ideal of $\mathfrak{g}$. It is easy to see that we may assume that $\mathfrak{k}$ contains the centre $\mathfrak{z}$ of $\mathfrak{g}$. Using a polarisation $\mathfrak{b}$ at $f \in \Omega(\pi)$, we realise π as $\pi \simeq \mathrm{ind}_B^G \chi_f$ with $B = \exp \mathfrak{b}$. We further assume that $\ker f$ does not contain any non-trivial ideal of $\mathfrak{g}$, otherwise we may pass to the quotient Lie algebra by such an ideal and apply the induction hypothesis. Consequently, $\dim \mathfrak{z} \leq 1$ and $f|_{\mathfrak{z}} \neq 0$ if $\dim \mathfrak{z} = 1$. Take a minimal non-central ideal $\mathfrak{a}$ of $\mathfrak{g}$ (cf. Subsection 1.1.2). Then, $\mathfrak{a}^f$ is a proper subalgebra of $\mathfrak{g}$ and we can take $\mathfrak{b}$ in $\mathfrak{a}^f$. Hence, $\pi \simeq \mathrm{ind}_{A^f}^G (\mathrm{ind}_B^{A^f} \chi_f)$. We know that it is possible to choose a linear subspace $\mathfrak{t}$ complementary to $\mathfrak{a}^f$ in $\mathfrak{g}$ in such a fashion that $G = TA^f$ with $T = \exp \mathfrak{t}$. Then the proof goes through a case-by-case study.

The last two theorems say:

Corollary 1.5.13 *The Frobenius reciprocity holds in our situations. Namely, the multiplicity of $\pi \in \widehat{G}$ in $\mathrm{ind}_H^G \sigma$ equals that of $\sigma \in \widehat{H}$ in $\pi|_H$ almost everywhere.*

Let π_j $(j = 1, 2)$ be two irreducible unitary representations of G. The outer Kronecker product $\pi_1 \times \pi_2$ of π_1 and π_2 corresponds to the orbit

$$\Omega_{G_1 \times G_2}(\pi_1 \times \pi_2) = (\Omega(\pi_1), \Omega(\pi_2)) \subset \mathfrak{g}^* \oplus \mathfrak{g}^*.$$

We identify G with the subgroup of $G \times G$ consisting of all diagonal elements. Since the (inner) Kronecker product $\pi_1 \otimes \pi_2$ of π_1 and π_2 is by definition the restriction of $\pi_1 \times \pi_2$ to G:

Corollary 1.5.14 ([49]) *Let $p : \mathfrak{g}^* \oplus \mathfrak{g}^* \to \mathfrak{g}^*$ be the restriction map. Then*

$$\pi_1 \otimes \pi_2 \simeq \int_{\widehat{G}}^{\oplus} m(\pi)\pi d\nu(\pi),$$

Here $\nu = (\theta_G \circ p)_*(\tilde{\nu}_{\pi_1 \times \pi_2})$ and the multiplicity $m(\pi)$ is obtained as the number of all G-orbits contained in $(\Omega(\pi_1), \Omega(\pi_2)) \cap p^{-1}(\Omega_G(\pi))$.

Remark 1.5.15 When G is nilpotent, the multiplicities $n_\pi(\sigma)$ are either uniformly bounded or equal to the infinity almost everywhere. According to these two possibilities, we say τ or $\pi|_K$ has finite or infinite multiplicities.

CHAPTER 2

The polynomial conjectures for nil-manifolds

Two main objects of our study are the induced representation $\tau = \mathrm{ind}_H^G \chi$ of G and the restriction $\pi|_K$ of π to a subgroup K. First, we are interested in their canonical central decompositions, and we saw in the last chapter how to compute Borel measures and multiplicities for these decompositions. Then, according to these two situations, we define certain algebras $D_\tau(G/H)$ and $D_\pi(G)^K$ of invariant differential operators related to the representations τ and $\pi|_K$. What can we say about the structures of these algebras? The polynomial conjectures assert that under the conditions that τ and $\pi|_K$ are of finite multiplicities, the algebras $D_\tau(G/H)$ and $D_\pi(G)^K$ are isomorphic to polynomial rings.

We are concerned in this chapter with the setting where G is nilpotent, where we shall provide complete proofs to both the conjectures. We also extend our study to prove the so-called Duflo's polynomial conjectures in this context.

2.1 THE POLYNOMIAL CONJECTURES FOR INDUCTION

2.1.1 An algebra of differential operators

We start our study here with an exponential solvable Lie group G, and let $\mathfrak{g}$ be its Lie algebra. Let $\mathfrak{h}$ be the Lie algebra of a closed connected Lie subgroup H of G, and χ a unitary character of H. There exists $f \in \mathfrak{g}^*$ satisfying $\chi = \chi_f$, where χ_f is defined by $\chi_f(\exp X) = e^{if(X)}, X \in \mathfrak{h}, i = \sqrt{-1}$. Put $\mathfrak{h}^\perp = \{\ell \in \mathfrak{g}^*; \ell|_\mathfrak{h} = 0\}$ and denote as earlier $\Gamma_\tau = f + \mathfrak{h}^\perp$.

We begin with the monomial representation $\tau = \mathrm{ind}_H^G \chi_f$. We denote by $\mathcal{U}(\mathfrak{g})$ the enveloping algebra of $\mathfrak{g}_\mathbb{C}$. We consider the algebra $\mathrm{Diff}(G)$ of the C^∞- differential operators on G which leave $C^\infty(G,\tau) = C^\infty(G) \cap \mathcal{K}(G,\chi)$ stable. Our algebra $D_\tau(G/H)$ is the set of the restrictions $D|_{C^\infty(G,\tau)}$ of $D \in \mathrm{Diff}(G)$ which commute with the left translations of G on $C^\infty(G,\tau)$. Let us describe $D_\tau(G/H)$ in terms of $\mathcal{U}(\mathfrak{g})$ (cf. [32] and [40]). For $A \in \mathcal{U}(\mathfrak{g})$, we denote $R(A)$ its right action. Let $\mathcal{U}(\mathfrak{g},\tau)$ be the set of $A \in \mathcal{U}(\mathfrak{g})$ such that $R(A)$ leaves $C^\infty(G,\tau)$ stable, $\mathfrak{a}_\tau$ the vector subspace spanned by $Y + if(Y) - \frac{1}{2}\mathrm{tr}\, \mathrm{ad}_{\mathfrak{g}/\mathfrak{h}} Y, Y \in \mathfrak{h}$, and $\mathcal{U}(\mathfrak{g})\mathfrak{a}_\tau$ the left ideal of $\mathcal{U}(\mathfrak{g})$ generated by $\mathfrak{a}_\tau$.

DOI: 10.1201/9781003685739-2

Then,

$$\mathcal{U}(\mathfrak{g},\tau) = \{A \in \mathcal{U}(\mathfrak{g}); [A,\mathfrak{h}] \subset \mathcal{U}(\mathfrak{g})\mathfrak{a}_\tau\}$$

and the mapping $A \mapsto R(A)|_{C^\infty(G,\tau)}$ gives us a homomorphism of $\mathcal{U}(\mathfrak{g},\tau)$ onto $D_\tau(G/H)$ with the kernel $\mathcal{U}(\mathfrak{g})\mathfrak{a}_\tau$. In sum,

$$D_\tau(G/H) \cong \mathcal{U}(\mathfrak{g},\tau)/\mathcal{U}(\mathfrak{g})\mathfrak{a}_\tau = (\mathcal{U}(\mathfrak{g})/\mathcal{U}(\mathfrak{g})\mathfrak{a}_\tau)^H,$$

where the last member represents the algebra of the H-invariant elements of $\mathcal{U}(\mathfrak{g})/\mathcal{U}(\mathfrak{g})\mathfrak{a}_\tau$.

We focus in this chapter on the setting of connected and simply connected nilpotent Lie groups. In this case, we have some more tools in hand than the exponential case. For instance, the space of C^∞-vectors of an irreducible unitary representation is realised as a Schwartz space of rapidly decreasing functions and this makes easier to discuss the convergence of some integrals or operators. Let us start with the so called commutativity conjecture for induction, which concerns the representation τ.

2.1.2 The commutativity conjecture for induction

Corwin-Greenleaf [32] proved that the finite multiplicities appearing in the disintegration (1.5.1) of τ, implies the commutativity of $D_\tau(G/H)$, and they substantiated the following:

Conjecture 2.1.1 (Commutativity Conjecture for induction) *Let G be a connected and simply connected nilpotent Lie group, H an analytic subgroup of G and χ a unitary character of H. Then the algebra $D_\tau(G/H)$ is commutative if and only if τ has finite multiplicities.*

Conjecture 2.1.1 is now settled. We have the following:

Theorem 2.1.1 *(cf. [54], [10]) Conjecture 2.1.1 holds for an arbitrary nilpotent data (G,H,χ).*

Towards the proof of Theorem 2.1.1, we write the following steps. We first have the:

Theorem 2.1.2 ([55]) *Assume that τ has infinite multiplicities. Let $\mathfrak{g}_0$ be an ideal of codimension 1 of $\mathfrak{g}$ containing $\mathfrak{h}$ and $G_0 = \exp(\mathfrak{g}_0)$. Suppose that $\tau_0 = \mathrm{ind}_H^{G_0}\chi_f$ has finite multiplicities. If there exists $W \in \mathcal{U}(\mathfrak{g},\tau)$ such that $W \notin \mathcal{U}(\mathfrak{g}_0) + \mathcal{U}(\mathfrak{g})\mathfrak{a}_\tau$, then there is $T \in \mathcal{U}(\mathfrak{g}_0,\tau_0)$ such that $[W,T] \notin \mathcal{U}(\mathfrak{g})\mathfrak{a}_\tau$.*

In the situation of this theorem, τ has infinite multiplicities because in Γ_τ the generic H-orbits are non-saturated with respect to $\mathfrak{g}_0$ but the section of generic G-orbits are saturated. In this way, the conjecture will be a by-product of the following theorem.

Theorem 2.1.3 ([54]) *The following properties are equivalent:*
(1) $\mathcal{U}(\mathfrak{g},\tau) \subset \mathcal{U}(\mathfrak{g}_0) + \mathcal{U}(\mathfrak{g})\mathfrak{a}_\tau$.
(2) The generic H-orbits in Γ_τ are saturated with respect to $\mathfrak{g}_0$.

In fact, if this theorem is true, we get the conjecture as follows. We know the equivalences for generic $\ell \in \Gamma_\tau$: τ has finite multiplicities $\Leftrightarrow \mathfrak{h} + \mathfrak{g}(\ell)$ is a Lagrangian subspace for B_ℓ generically in $\Gamma_\tau \Leftrightarrow \dim(H{\cdot}\ell) = \frac{1}{2}\dim(G{\cdot}\ell) \Leftrightarrow 2(\dim\mathfrak{h} - \dim(\mathfrak{h}(\ell))) = \dim\mathfrak{g} - \dim(\mathfrak{g}(\ell))$. Therefore, it suffices to show that $2(\dim\mathfrak{h} - \dim(\mathfrak{h}(\ell))) < \dim\mathfrak{g} - \dim(\mathfrak{g}(\ell))$ generically in Γ_τ implies the non-commutativity of $D_\tau(G/H)$. Take $\mathfrak{g}_0$ as above. If already $D_{\tau_0}(G_0/H) \subset D_\tau(G/H)$ is non-commutative, obviously $D_\tau(G/H)$ too. So we may assume that $D_{\tau_0}(G_0/H)$ is commutative or equivalently, using the induction argument, that $2(\dim\mathfrak{h} - \dim(\mathfrak{h}(\ell_0))) = \dim(\mathfrak{g}_0) - \dim(\mathfrak{g}_0(\ell_0))$, where $\ell_0 = \ell|_{\mathfrak{g}_0}$. Subtracting both relations, we have

$$2(\dim(\mathfrak{h}(\ell_0)) - \dim(\mathfrak{h}(\ell))) < 1 + \dim(\mathfrak{g}_0(\ell_0)) - \dim(\mathfrak{g}(\ell)) \leq 2$$

which implies $\mathfrak{h}(\ell_0) = \mathfrak{h}(\ell)$. Then the theorem asserts that there exists an element $W \in \mathcal{U}(\mathfrak{g},\tau)$ such that $W \notin \mathcal{U}(\mathfrak{g}_0) + \mathcal{U}(\mathfrak{g})\mathfrak{a}_\tau$. Finally, using Theorem 2.1.2, we obtain an element $T \in \mathcal{U}(\mathfrak{g}_0,\tau_0)$ such that $[W,T] \notin \mathcal{U}(\mathfrak{g})\mathfrak{a}_\tau$ as expected.

2.1.3 Layering the dual of an exponential Lie algebra

Let $\mathfrak{g}$ be a nilpotent Lie algebra and take a composition series of $\mathfrak{g}$:

$$\mathcal{S} : \{0\} = \mathfrak{g}_0 \subset \mathfrak{g}_1 \subset \cdots \subset \mathfrak{g}_n = \mathfrak{g}$$

Let $\{X_j\}_{1\leq j\leq n}$ be a Malcev basis of $\mathfrak{g}$ according to this composition series, i.e. $X_j \in \mathfrak{g}_j \backslash \mathfrak{g}_{j-1}$ $(1 \leq j \leq n)$ and $\{X_j^*\}_{1\leq j\leq n}$ its dual basis in $\mathfrak{g}^*$. We denote the coordinates of $\ell \in \mathfrak{g}^*$ by $(\ell_1,\dots,\ell_n), \ell_j = \ell(X_j)$. Then $\mathfrak{g}_j^\perp = \langle X_{j+1}^*,\dots,X_n^*\rangle_\mathbb{R} \subset \mathfrak{g}^*, \mathfrak{g}_j^* \cong \mathfrak{g}^*/\mathfrak{g}_j^\perp$ and the projection $p_j : \mathfrak{g}^* \to \mathfrak{g}_j^*$ intertwines the actions of G on $\mathfrak{g}^*$ and $\mathfrak{g}_j^*$. For $\ell \in \mathfrak{g}^*$, we define $e_j(\ell) = \dim\left(G{\cdot}p_j(\ell)\right), e(\ell) = (e_1(\ell),\dots,e_n(\ell))$ and set $\mathcal{E} = \{e(\ell); \ell \in \mathfrak{g}^*\}$. We also recognise $e_j(\ell) = \dim(G_j{\cdot}\ell)$ with $G_j = \exp(\mathfrak{g}_j)$. In fact,

$$\begin{aligned}\dim\left(G{\cdot}p_j(\ell)\right) &= \dim\left(\mathfrak{g}/\mathfrak{g}_j^\ell\right) = \dim(\mathfrak{g}/\mathfrak{g}(\ell)) - \dim\left(\mathfrak{g}_j^\ell/\mathfrak{g}(\ell)\right)\\ &= \dim(\mathfrak{g}/\mathfrak{g}(\ell)) - \left(\dim(\mathfrak{g}/\mathfrak{g}(\ell)) - \dim\left(\mathfrak{g}_j/\mathfrak{g}_j(\ell)\right)\right) = \dim(G_j{\cdot}\ell).\end{aligned}$$

For $e \in \mathcal{E}$ we define the G-invariant layer $U_e = \{\ell \in \mathfrak{g}^*; e(\ell) = e\}$. With $e_0 = 0$ we define the set of jump indices $S(e) = \{1 \leq j \leq n; e_j = e_{j-1} + 1\}$ and that of non-jump indices $T(e) = \{1 \leq j \leq n; e_j = e_{j-1}\}$. We introduce among $e(\ell), \ell \in \mathfrak{g}^*$, the lexicographic order such that

$$\begin{aligned}e(\ell_1) \leq e(\ell_2) \Longleftrightarrow e_k(\ell_1) \leq e_k(\ell_2), e_{k+1}(\ell_1) = e_{k+1}(\ell_2),\\ \dots, e_m(\ell_1) = e_m(\ell_2).\end{aligned}$$

and let $e_{\max} \in \mathcal{E}$ be the maximal element. Then, $U_{e_{\max}}$ is a G-invariant Zariski open set of $\mathfrak{g}^*$. More generally, take any $e \in \mathcal{E}$. We then see that $\{\ell \in \mathfrak{g}^*; e \leq e(\ell)\}$ is a Zariski open set of $\mathfrak{g}^*$.

2.1.4 Corwin-Greenleaf e-central elements

Let G be a nilpotent Lie group, $\mathfrak{g}$ its Lie algebra, and $\mathcal{U}(\mathfrak{g})$ the enveloping algebra of $\mathfrak{g}_\mathbb{C}$. We recall here the notion of e-central elements defined by Corwin and Greenleaf.

Fix $e \in \mathcal{E}$ as in the previous section. An element $A \in \mathcal{U}(\mathfrak{g})$ is called e-central element if, with $\pi_\ell = \theta(\ell)$, $\pi_\ell(A) = d\pi_\ell(A)$ is a scalar operator for any $\ell \in U_e$.

Now we mention a fundamental result of Corwin-Greenleaf. There exists a Zariski open set $\mathcal{Z}$ of $\mathfrak{g}^*$ which satisfies the following. $\mathcal{Z} \cap U_e$ is a non-empty G-invariant set and for each $j \in T(e)$ there is $A_j \in \mathcal{U}(\mathfrak{g}_j)$ with the following properties.

a. Each A_j is e-central on $\mathcal{Z} \cap U_e$, i.e. $\pi_\ell(A_j)$ is a scalar operator for $\ell \in \mathcal{Z} \cap U_e$. Furthermore, A_j has a form $A_j = P_j X_j + Q_j$ such that:

   1. P_j is a polynomial of A_k $(k \in T(e), k < j)$, in particular $P_j \in \mathcal{U}(\mathfrak{g}_{j-1})$;
   2. P_j is e-central on $\mathcal{Z} \cap U_e$;
   3. $Q_j \in \mathcal{U}(\mathfrak{g}_{j-1})$, in particular $P_1, Q_1 \in \mathbb{C}1$.

b. $\pi_\ell(P_j) \neq 0$ at $\forall \ell \in \mathcal{Z} \cap U_e$.

c. $\pi_\ell(A_j) = \varphi_j(\ell) Id$, where we can write $\varphi_j(\ell) = \tilde{p_j}(\ell')\ell_j + \tilde{q_j}(\ell')$ with two rational functions $\tilde{p_j}, \tilde{q_j}$ on $\mathcal{Z} \cap U_e$, which depend only on $\ell' = (\ell_1, \dots, \ell_{j-1})$.

d. $\tilde{p_j}(\ell')$ is G-invariant and $\tilde{p_j}(\ell') \neq 0$ for $\forall \ell \in \mathcal{Z} \cap U_e$.

The functions $(\varphi_j(\ell))_j$ called Corwin-Greenleaf functions corresponding to $(A_j)_j$.

Remark 2.1.4 Let U_e be the first layer which intersects Γ_τ. Having taken the Zariski open set $\Gamma_\tau \cap U_e$ of Γ_τ, having given the construction of A_j, Corwin-Greenleaf [15] remarked the following: Dropping out the Zariski open set $Z \cap U_e$ from U_e, we notice that, $U_e \backslash Z$ being G-invariant and semi-algebraic, the parametrisation of the orbits in U_e is carried out and retains all its properties on this sub-layer in U_e. We are able to repeat the whole process starting from $U_e \backslash Z$. Since U_e is semi-algebraic, the ascendant chain condition for the ideals in $\mathbb{C}[\mathfrak{g}^*]$ assures that the process terminates after a finite number of steps. So, patching the pieces together, we may suppose that $Z \cap U_e = U_e$.

Example 2.1.5 Let $\mathfrak{g} = \langle X_1, X_2, X_3, X_4, X_5 \rangle_{\mathbb{R}} : [X_5, X_k] = X_{k-1} (4 \leq k \leq 2)$. Let $f = \alpha X_3^* + \beta X_4^*$ with $\alpha, \beta \in \mathbb{R}$ and $\mathfrak{h} = \mathbb{R}X_3 \oplus \mathbb{R}X_4$. As $\mathfrak{h}$ is commutative, we construct $\tau = \mathrm{ind}_H^G \chi_f$ and introduce

$$\Gamma_\tau = \{\ell = \ell_1 X_1^* + \ell_2 X_2^* + \alpha X_3^* + \beta X_4^* + \ell_5 X_5^*; (\ell_1, \ell_2, \ell_5) \in \mathbb{R}^3\} \simeq \mathbb{R}^3.$$

We put $\mathfrak{b} = \oplus_{j=1}^4 \mathbb{R}X_j$, so that $\mathfrak{g} = \mathbb{R}X_5 \ltimes \mathfrak{b}$. We choose a flag of ideals

$$\{0\} \subset \mathfrak{g}_1 = \mathbb{R}X_1 \subset \mathfrak{g}_2 = \sum_{j=1}^{2} \mathbb{R}X_j \subset \mathfrak{g}_3 = \sum_{j=1}^{3} \mathbb{R}X_j$$
$$\subset \mathfrak{g}_4 = \sum_{j=1}^{4} \mathbb{R}X_j \subset \mathfrak{g}_5 = \mathfrak{g}.$$

Consider the Zariski open set

$$\mathcal{O} = \{\ell \in \Gamma_\tau; \ell_1 \neq 0, \ell_2 \neq 0\}$$

of Γ_τ. Immediately, $T(e) = \{1,3,4\}$ and $S(e) = \{2,5\}$. We find e-central elements

$$A_1 = X_1, A_3 = 2X_1X_3 - X_2^2, A_4 = X_1^2X_4 - X_1X_2X_3 + \frac{{X_2}^3}{3}.$$

To make explicit the corresponding functions $\{\varphi_j\}_{j\in\{1,3,4\}}$ of Corwin-Greenleaf, we first remark that, for all $\ell \in \mathcal{O}$, $\mathfrak{b}$ is a polarisation of $\mathfrak{g}$ at ℓ and $\{X_5\}$ is an adapted supplementary basis of $\mathfrak{b}$ in $\mathfrak{g}$ so that we obtain a realisation of representations π_ℓ of $\mathcal{U}(\mathfrak{g})$ in $\mathcal{S}(\mathbb{R})$ and $\mathcal{S}^*(\mathbb{R})$. Let $\Theta : \mathcal{S}(\mathbb{R}) \to \mathcal{H}_{\pi_\ell}^\infty$ be the isomorphism given by $\psi(x) = \Theta(\psi)(\exp(xX_5))(x \in \mathbb{R})$. Since $A_j (j = 1,3,4)$ are e-central, we have $\pi_\ell(A_j) = \Theta^{-1}R({}^tA_j)\Theta$, where ${}^tA_j$ denotes the image of A_j under the principal anti-automorphism of $\mathcal{U}(\mathfrak{g})$. We remark $R(X) = -i\ell(X)\mathrm{Id}$ on $\mathcal{H}_{\pi_\ell}^\infty$ for $X \in \mathfrak{b}$. Here, putting $\ell(X_j) = \ell_j$ for $j = 1,2,5$ and $\ell(X_3) = \alpha, \ell(X_4) = \beta$, it follows:

$$-i\pi_\ell(A_1) = \ell_1\mathrm{Id}, -\pi_\ell(A_3) = (2\alpha\ell_1 - \ell_2^2)\mathrm{Id},$$

$$i\pi_\ell(A_4) = (\beta\ell_1^2 - \alpha\ell_1\ell_2 + \frac{\ell_2^3}{3})\mathrm{Id}.$$

It follows that

$$\begin{aligned} -i\varphi_1(\ell) &= \ell_1, \\ -\varphi_3(\ell) &= 2\alpha\ell_1 - \ell_2^2, \\ i\varphi_4(\ell) &= \beta\ell_1^2 - \alpha\ell_1\ell_2 + \frac{\ell_2^3}{3}. \end{aligned}$$

Besides, we see for $\psi \in \mathcal{S}(\mathbb{R})$:

$$\begin{aligned} -i(\pi_\ell(X_1)\psi)(x) &= \ell_1\psi(x), \\ -i(\pi_\ell(X_2)\psi)(x) &= (\ell_2 - \ell_1x)\psi(x), \\ -i(\pi_\ell(X_3)\psi)(x) &= (\alpha - \ell_2x + \ell_1\frac{x^2}{2})\psi(x), \\ -i(\pi_\ell(X_4)\psi)(x) &= (\beta - \alpha x + \ell_2\frac{x^2}{2} - \ell_1\frac{x^3}{3!})\psi(x). \end{aligned}$$

Let us verify, for example, the last formula. We have

$$\begin{aligned} &(\pi_\ell(X_4)\psi) = (L(X_4)\Theta\psi)(\exp(xX_5)) \\ =&\frac{d}{dt}(\Theta\psi)(\exp(xX_5)\cdot \mathrm{Ad}(\exp(-xX_5))(\exp(-tX_4)))|_{t=0} \\ =&\frac{d}{dt}(\Theta\psi)(\exp(xX_5)\cdot \exp(t[-X_4 + xX_3 - \frac{x^2}{2}X_2 + \frac{x^3}{3!}X_1]))|_{t=0} \\ =&(\Theta\psi)(\exp(xX_5))\frac{d}{dt}e^{it\left(\beta-\alpha x+\ell_2\frac{x^2}{2}-\ell_1\frac{x^3}{3!}\right)}|_{t=0}, \end{aligned}$$

from which comes the result.

There uniquely exists $e \in \mathcal{E}$ such that $\Gamma_\tau \cap U_e$ is a non-empty Zariski open set of Γ_τ.

Theorem 2.1.6 ([53]) *Let $j \in T(e)$ and take the e-central element A_j. Then there exists a polynomial function $\varphi_j(\ell)$ on Γ_τ such that $\pi_\ell(A_j) = \varphi_j(\ell) Id$ at all $\ell \in \Gamma_\tau$.*

2.1.5 Penney distributions and rational eigenvalues

For a unitary representation ρ of G, we denote by $\mathcal{H}_\rho$ (respectively $\mathcal{H}_\rho^\infty$ and $\mathcal{H}_\rho^{-\infty}$) the Hilbert space of ρ (respectively the space of the C^∞-vectors of ρ and the anti-dual of $\mathcal{H}_\rho^\infty$). For $a \in \mathcal{H}_\rho^{\pm\infty}$ and $b \in \mathcal{H}_\rho^{\mp\infty}$, we denote by $\langle a, b\rangle$ the image of b by a, therefore $\langle a, b\rangle = \overline{\langle b, a\rangle}$.

Let

$$\mathcal{S} : \{0\} = \mathfrak{g}_0 \subset \mathfrak{g}_1 \subset \cdots \subset \mathfrak{g}_n = \mathfrak{g} \tag{2.1.1}$$

be a Jordan-HÖlder sequence of $\mathfrak{g}$, from which we extract a Jordan-Hölder basis $\mathcal{Z} = \{Z_j; j = 1, \cdots, n\}$ of $\mathfrak{g}$. Let

$$\mathcal{I} = \{i_1 < \cdots < i_d\} = \{1 \leq j \leq n; \dim(\mathfrak{h} \cap \mathfrak{g}_j) \neq \dim(\mathfrak{h} \cap \mathfrak{g}_{j-1})\},$$

where $d = \dim \mathfrak{h}$,

$$\mathcal{J} = \{j_1 < \cdots < j_q\} = \{1, 2, \ldots, n\} \setminus \mathcal{I},$$

where $q = \dim(\mathfrak{g}/\mathfrak{h})$ and $\mathfrak{h}_s = \mathfrak{h} \cap \mathfrak{g}_{i_s}$ for $1 \leq s \leq d$. Then,

$$\{0\} = \mathfrak{h}_0 \subset \mathfrak{h}_1 \subset \cdots \subset \mathfrak{h}_d = \mathfrak{h} \tag{2.1.2}$$

is a Jordan-Hölder sequence of $\mathfrak{h}$. For $1 \leq r \leq q$, put $\mathfrak{h}_{d+r} = \mathfrak{h} + \mathfrak{g}_{j_r}$. Then,

$$\mathfrak{h} = \mathfrak{h}_d \subset \mathfrak{h}_{d+1} \subset \cdots \subset \mathfrak{h}_n = \mathfrak{g} \tag{2.1.3}$$

is a sequence of subalgebras of $\mathfrak{g}$. The flags (2.1.2) and (2.1.3) give us a sequence of subalgebras $\{\mathfrak{h}_j\}_{j=0}^n$ of $\mathfrak{g}$ and we extract an adapted basis $\{X_j\}_{j=1}^n$ of $\mathfrak{g}$.

For $\ell \in \Gamma_\tau$, let $\mathfrak{b}[\ell]$ be the Vergne polarisation of $\mathfrak{g}$ at ℓ constructed from the flag $\mathcal{S}$, $B[\ell] = \exp(\mathfrak{b}[\ell])$ and $\pi_\ell = \mathrm{ind}_{B[\ell]}^G \chi_\ell$. We define a generalised vector $a_\ell := a_\ell^H \in \mathcal{H}_{\pi_\ell}^{-\infty}$ by

$$\langle a_\ell^H, \varphi\rangle = \int_{H/(H\cap B[\ell])} \overline{\varphi(h)\chi_\ell(h)} d\dot{h}. \tag{2.1.4}$$

Remark first the following important fact:

Proposition 2.1.2 *For $W \in \mathcal{U}(\mathfrak{g})$, the relation $\overline{W} \cdot a_\ell = 0$ almost everywhere on Γ_τ is equivalent to $W \in \mathcal{U}(\mathfrak{g})\mathfrak{a}_\tau$.*

Proof 2.1 *If $W \in \mathcal{U}(\mathfrak{g})\mathfrak{a}_\tau$ then it is clear that $\overline{W} \cdot a_\ell = 0$ almost everywhere on Γ_τ. We prove the converse by induction on* $\dim \mathfrak{g}$. *If* $\dim \mathfrak{g} = 1$ *or more generally if $\mathfrak{g}$ is abelian, then*

$$\overline{W} \cdot a_\ell^H = \widehat{W}(\ell) a_\ell^H, \ell \in \mathfrak{g}^*.$$

for some complex number $\widehat{W}(\ell)$*. On the other hand, we have that*

$$\mathcal{U}(\mathfrak{g})\mathfrak{a}_\tau \quad = \quad \{W \in \mathcal{U}(\mathfrak{g}); \widehat{W}(\ell) = 0, \ell \in \Gamma_\tau\}.$$

We suppose now that our fact is true for any $\mathfrak{g}$ *with* $n-1 = \dim \mathfrak{g} > 1$*. Take* W *such that* $\overline{W}\cdot a_\ell = 0$ *almost everywhere on* Γ_τ*.*

Let $\tilde{\mathfrak{g}} = \mathfrak{h}_{n-1}$ *and* $\tilde{G} = \exp \tilde{\mathfrak{g}}$*. For* $\ell \in \Gamma_\tau$*, put* $\tilde{\ell} = \ell|_{\tilde{\mathfrak{g}}} \in (\tilde{\mathfrak{g}})^*$*. Let* $X_n \in \mathfrak{g} \setminus \tilde{\mathfrak{g}}$*. If the generic* G*-orbits in* Γ_τ *are saturated with respect to* $\tilde{\mathfrak{g}}$*, we have that* $\pi_\ell = \mathrm{ind}_{\tilde{G}}^{G} \pi_{\tilde{\ell}}$ *and as in subsection 2.1.6,* $W \in \mathcal{U}(\tilde{\mathfrak{g}})$ *modulo* $\mathcal{U}(\mathfrak{g})\mathfrak{a}_\tau$*. As* a_ℓ *can now be identified with* $a_{\tilde{\ell}}$*, our claim follows from the induction hypothesis.*

Suppose next that the generic G*-orbits in* Γ_τ *are not saturated with respect to* $\tilde{\mathfrak{g}}$*. Then* $\mathfrak{g} = \mathfrak{g}(\ell) + \tilde{\mathfrak{g}}$ *(*ℓ *being generic in* Γ_τ*). There exists also an e-central element* $\sigma_n = a_n X_n + b_n$ *for* Γ_τ *with* $a_n, b_n \in \mathcal{U}(\tilde{\mathfrak{g}})$ *and* $a_n \neq 0$ *is e-central too. Write* $W = \sum_{j=0}^{r} w_j X_n^j, w_j \in \mathcal{U}(\tilde{\mathfrak{g}})(0 \leq j \leq r)$ *modulo* $\mathcal{U}(\mathfrak{g})\mathfrak{a}_\tau$*. Then* $w_r, \overline{\sigma_n}$ *commute with each other modulo* $\mathcal{U}(\mathfrak{g})\mathfrak{a}_\tau$*.*

We make now a further induction on the degree r *of* W *in* X_n*. If* $r = 0$*, our* W *is contained in* $\mathcal{U}(\tilde{\mathfrak{g}})$*. We see that the degree of* $w_r\overline{\sigma_n}^r - \overline{a_n}^r W$ *relative to* X_n *is smaller than or equal to* $r-1$ *modulo* $\mathcal{U}(\mathfrak{g})\mathfrak{a}_\tau$*. Let* $\varphi_n(\ell)$ *be the function of Corwin-Greenleaf corresponding to* σ_n*. Hence, if* $\overline{W}\cdot a_\ell = 0$ *almost everywhere, then*

$$\overline{P_{w_r}(\ell)\varphi_n^r(\ell)a_\ell^H} \quad = \quad (\overline{w_r}\sigma_n^r - a_n^r\overline{W}) \cdot a_\ell^H$$

and therefore $\varphi_n^r(\ell)\overline{w_r}\cdot a_\ell$ *must be of degree smaller than or equal to* $r-1$ *relative to* $\ell(X_n)$*. This means that* $\overline{w_r}\cdot a_\ell = 0$ *almost everywhere and consequently by the induction hypothesis* w_r *belongs to* $\mathcal{U}(\mathfrak{g})\mathfrak{a}_\tau$ *and also that* $w_r\overline{\sigma_n}^r - \overline{a_n}^r W \in \mathcal{U}(\mathfrak{g})\mathfrak{a}_\tau$*. This implies that* $\overline{a_n}^r W \in \mathcal{U}(\mathfrak{g})\mathfrak{a}_\tau$ *and then also* $W \in \mathcal{U}(\mathfrak{g})\mathfrak{a}_\tau$ *since* a_n *is e-central and not identically* 0.

2.1.6 Differential operators and saturation with respect to a one codimensional ideal

Let $\tilde{\mathfrak{g}}$ be an ideal of codimension 1 in $\mathfrak{g}$ containing $\mathfrak{h}$. Then $\mathcal{U}(\mathfrak{g},\tau) \subset \mathcal{U}(\tilde{\mathfrak{g}}) + \mathcal{U}(\mathfrak{g})\mathfrak{a}_\tau$ if and only if generic H-orbits in Γ_τ are saturated with respect to $\tilde{\mathfrak{g}}$ (cf. Theorem 2.1.3). On the other hand, if the generic G-orbits in Γ_τ are saturated with respect to $\tilde{\mathfrak{g}}$, then any $W \in \mathcal{U}(\mathfrak{g},\tau)$ satisfying relation (2.1.6) belongs to $\mathcal{U}(\tilde{\mathfrak{g}})$. Otherwise, $\pi(\overline{W})$ contains an outer partial derivative with respect to $\tilde{G} = \exp \tilde{\mathfrak{g}}$, which contradicts equation (2.1.6). This fact also happens in the case where τ is of finite multiplicities.

2.1.7 Corwin-Greenleaf polynomial Conjecture

The polynomial conjecture for induction in the nilpotent setting was posed by L. Corwin and F. Greenleaf in [32] and it states the following:

Conjecture 2.1.3 (Corwin-Greenleaf) *Let* G *be a connected and simply connected nilpotent Lie group,* H *an analytic subgroup of* G *and* χ *a unitary character of* H*. When* τ *has finite multiplicities, the algebra* $D_\tau(G/H)$ *is isomorphic to the algebra* $\mathbb{C}[\Gamma_\tau]^H$ *of the* H*-invariant polynomial functions on the affine space* Γ_τ*.*

Developing widely some arguments based on e-central elements (cf. Subsection 2.1.4), L. Corwin and F. Greenleaf proved that Conjecture 2.1.3 holds under the very restrictive assumption that there exists a common polarisation $\mathfrak{b}$ for μ-almost all $\ell \in \Gamma_\tau$ such that $\mathfrak{h}$ normalises $\mathfrak{b}$ (cf. [33]).

A. Baklouti [5] proposed an alternative proof of the last upshot (together with other different settings) as an application of his results on concrete intertwining operators between τ and its disintegration. Fujiwara generated in [51] a different method to study Conjectures 2.1.1 and 2.1.3. His arguments were based on the Plancherel formula for τ and the second part of the thesis of Benoist (cf. [22]). Many of the ideas developed therein and several related results allowed to establish Conjecture 2.1.1 in this restrictive case.

Conjecture 2.1.3 was proved by Fujiwara (cf. [50]) in the context where one of the following assumptions holds:

(1) there exists a common polarisation $\mathfrak{b}$ for μ-almost all $\ell \in \Gamma_\tau$;
(2) $\dim G \cdot \ell \leq 2$ for generic $\ell \in \Gamma_\tau$;
(3) $\dim \mathfrak{g} + \dim \mathfrak{g}(\ell) - 2 \leq 2 \dim \mathfrak{h}$ for generic $\ell \in \Gamma_\tau$.

The case (1), valid in particular when $\mathfrak{h}$ is an ideal of $\mathfrak{g}$, was also established by Baklouti and Ludwig [21] by different arguments.

2.1.8 Proof of Corwin-Greenleaf's Conjecture

Conjecture 2.1.3 has been established recently (cf. [19], [18]) and we provide here a proof in its full generality. Our tools here are Penney's distributions which turn out to be eigenvectors for the elements of our algebras. Considering the eigenvalue at almost every point, we associate some function to the element of our algebra. Our strategy consists in constructing for almost all $\ell \in \Gamma_\tau$ the Penney distribution a_ℓ (2.1.4), which diagonalises the action of $D_\tau(G/H)$, i.e.:

$$\overline{D} \cdot a_\ell = \overline{P_D(\ell)} a_\ell, \ D \in D_\tau(G/H) \tag{2.1.5}$$

with a scalar $P_D(\ell)$ almost everywhere on Γ_τ, which extends to a rational non-singular function on Γ_τ. Then the key point of the proof is to show that $\ell \mapsto P_D(\ell)$ turns out to be polynomial, generically on Γ_τ. It is worth-noting that the assumption of the finiteness of the multiplicities of τ turns out to be irrelevant at this stage and involves some complications at some parts of the induction procedure, we will be dealing with. Note also that an important step towards the polynomial conjecture is to see that the e-central functions of Corwin-Greenleaf are polynomial functions.

Suppose for a while that τ has finite multiplicities. Then $\mathfrak{h} + \mathfrak{g}(\ell)$ is a Lagrangian subspace for B_ℓ at almost every point $\ell \in \Gamma_\tau$. At such a ℓ, our distribution a_ℓ does essentially not depend on the choice of the (real) polarisation $\mathfrak{b}[\ell]$ used to construct π_ℓ. Namely, let $\mathfrak{b}_1, \mathfrak{b}_2$ be two polarisations at ℓ and T_{21} an intertwining operator from $\pi_\ell^1 = \mathrm{ind}_{B_1}^G \chi_\ell$ to $\pi_\ell^2 = \mathrm{ind}_{B_2}^G \chi_\ell$ with $B_j = \exp(\mathfrak{b}_j)(j = 1, 2)$. Let a_ℓ^j be our distributions a_ℓ relative to π_ℓ^j for $j = 1, 2$. Then $T_{21}(a_\ell^1) = c a_\ell^2$ with a certain $0 \neq c \in \mathbb{C}$ (cf. [47]).

Hence, if a_ℓ^1 is an eigenvector of $\pi_\ell^1(V), V \in \mathcal{U}(\mathfrak{g})$, a_ℓ^2 is that of $\pi_\ell^2(V)$ and their eigenvalues are the same.

It descends from Proposition 2.1.2 that the homomorphism

$$D_\tau(G/H) \cong \mathcal{U}(\mathfrak{g},\tau)/\mathcal{U}(\mathfrak{g})\mathfrak{a}_\tau \ni [W] \mapsto P_W,$$

where $[W]$ denotes the canonical image of W, is injective. To establish therefore Conjecture 2.1.3, we have to address following two questions.

Question 2.1.7 *Does the function P_W extend to a polynomial function on Γ_τ?*

Question 2.1.8 *Is the homomorphism*

$$\mathcal{U}(\mathfrak{g},\tau)/\mathcal{U}(\mathfrak{g})\mathfrak{a}_\tau \ni [W] \mapsto P_W \in \mathbb{C}[\Gamma_\tau]^H$$

surjective?

The following fact will be of frequent and important use in the sequel: We shall say that a property is generically true (on an algebraic variety) if it is verified on a non-empty Zariski open set (of this variety). An element of such a Zariski open set will be called generic.

We first prepare some results, which will be needed in the proof. Let $W \in \mathcal{U}(\mathfrak{g},\tau)$. Let

$$\mathcal{K} = \{k_1 < k_2 < \cdots < k_m\}$$

be the set of indices $k \in \{d+1,\ldots,n\}$ such that the generic orbits of $H_k = \exp(\mathfrak{h}_k)$ in $\Gamma_\tau|_{\mathfrak{h}_k}$ are non-saturated with respect to $\mathfrak{h}_{k-1}$. Then for such a position k, the orbit theory gives us an element

$$\sigma_k = a_k X_k + b_k,\ a_k, b_k \in \mathcal{U}(\mathfrak{h}_{k-1}), a_k \neq 0,$$

which is e-central for $\Gamma_\tau|_{\mathfrak{h}_k}$.

We start by proving the following:

Lemma 2.1.4 *Let $W \in \mathcal{U}(\mathfrak{g},\tau)$ satisfying relation (2.1.6). There exist two polynomials A, B of $\sigma_{k_1},\ldots,\sigma_{k_m}$ such that*

$$AW \equiv B \ mod\, \mathcal{U}(\mathfrak{g})\mathfrak{a}_\tau.$$

Proof 2.2 *We proceed by induction on the dimension of $\mathfrak{g}/\mathfrak{h}$. If $\mathfrak{h} = \mathfrak{g}$, then f is a character of $\mathfrak{g}$ and the assertion is trivial. We can therefore assume that $\mathfrak{g} \neq \mathfrak{h}$.*

Let $\tilde{\mathfrak{g}} = \mathfrak{h}_{n-1}$ be an ideal of codimension 1 in $\mathfrak{g}$ containing $\mathfrak{h}$. Then, if $W \in \mathcal{U}(\mathfrak{h}_{n-1})$, the result immediately follows from the induction hypothesis applied to $\mathfrak{h}_{n-1}$. This is the case for example if the generic G-orbits which encounter Γ_τ are saturated with respect to $\mathfrak{h}_{n-1}$ as in subsection 2.1.6.

Suppose $W \notin \mathcal{U}(\mathfrak{h}_{n-1})$. This means that the generic G-orbits which encounter Γ_τ are non-saturated with respect to $\tilde{\mathfrak{g}}$. In other words, $n \in \mathcal{K}$ and there exists an element

$$\sigma_n = a_n X_n + b_n,\ a_n, b_n \in \mathcal{U}(\mathfrak{h}_{n-1}),$$

which is e-central for Γ_τ and so is a_n for Γ_τ as in subsection 2.1.4. Write

$$W = \sum_{j=0}^{r} w_j X_n^j,\ w_j \in \mathcal{U}(\mathfrak{h}_{n-1})\ (0 \leq j \leq r),$$

where $r \geq 1$ and $w_r \notin \mathcal{U}(\mathfrak{g})\mathfrak{a}_\tau$. We use a new induction on r. Since $\overline{a_n}$ and w_r commute each other modulo $\mathcal{U}(\mathfrak{g})\mathfrak{a}_\tau$, $\widetilde{W} = \overline{a_n}^r W - w_r \overline{\sigma_n}^r$ is of degree less than or equal to $r-1$ with respect to X_n. Applying the new induction hypothesis to $\widetilde{W}$, we obtain the result for W.

Theorem 2.1.9 ([51]) *Suppose that τ has finite multiplicities. Let Γ_τ^0 be the set of $\ell \in \Gamma_\tau$ where $\mathfrak{h} + \mathfrak{g}(\ell)$ is a Lagrangian subspace for B_ℓ and such that $\dim(G{\cdot}\ell)$ is maximal when ℓ runs over Γ_τ. Let $\ell \in \Gamma_\tau^0$. For $A \in \mathcal{U}(\mathfrak{g}, \tau)$, $\pi_\ell(\overline{A})a_\ell$ is a multiple of a_ℓ: i.e. there exists a scalar $P_A(\ell) \in \mathbb{C}$ such that $\pi_\ell(\overline{A})a_\ell = \overline{P_A(\ell)}a_\ell$.*

Proof 2.3 *Remark first $\pi_\ell(\overline{A})a_\ell = 0$ for all $A \in \mathcal{U}(\mathfrak{g})\mathfrak{a}_\tau$. We can suppose that $\mathfrak{h}$ contains the centre $\mathfrak{z}$ of $\mathfrak{g}$. In fact, Put $\mathfrak{k} = \mathfrak{h} + \mathfrak{z}, K = \exp\mathfrak{k}, \sigma = ind_K^G \chi_\ell$. It follows that $\ell \in \Gamma_\sigma^0, \pi_\ell(k)a_\ell = \chi_\ell(k)a_\ell(k \in K)$ and $\mathcal{U}(\mathfrak{g},\tau) \subset \mathcal{U}(\mathfrak{g},\sigma)$.*

We now proceed by induction on $\dim G$. Passing to the quotient by $\exp(\mathfrak{z}\cap\ker f)$ if necessary, we are led to the case where $\mathfrak{z}\cap\ker f = \{0\}$. Let $\mathfrak{z} = \mathbb{R}Z$ with $f(Z) = 1$. It is well known [65] that there exist two elements X, Y in $\mathfrak{g}\cap\ker f$ such that $[X,Y] = Z$ and $\mathfrak{g} = \mathbb{R}X + \mathfrak{g}_0$ with the centraliser $\mathfrak{g}_0$ of Y in $\mathfrak{g}$. In the flag $\mathcal{S}$, $\mathfrak{g}_1 = \mathfrak{z}$ and take Y in $\mathfrak{g}_2 \setminus \mathfrak{g}_1$. Then, the polarisation $\mathfrak{b} = \mathfrak{b}[\ell]$ is in $\mathfrak{g}_0$. Put $G_0 = \exp(\mathfrak{g}_0), \ell_0 = \ell|_{\mathfrak{g}_0} \in (\mathfrak{g}_0)^*$ and $\pi_0 = ind_B^{G_0}\chi_{\ell_0}$.

We separate two cases. First let $\mathfrak{h} \subset \mathfrak{g}_0$. The theorem 5.4, (b) in [32] says $\mathcal{U}(\mathfrak{g},\tau) = \mathcal{U}(\mathfrak{g}_0,\tau_0) + \mathcal{U}(\mathfrak{g})\mathfrak{a}_\tau$ with $\tau_0 = ind_H^{G_0}\chi$. Because $\ell_0 \in \Gamma_{\tau_0}^0$ and a_ℓ is regarded as $a_{\ell_0} \in \mathcal{H}_{\pi_0}^{-\infty}$, it suffices to apply the induction hypothesis to G_0.

Next let $\mathfrak{h} \not\subset \mathfrak{g}_0$. We take X in $\mathfrak{h}\cap\ker f$ and put $\mathfrak{h}_0 = \mathfrak{h}\cap\mathfrak{g}_0, H_0 = \exp(\mathfrak{h}_0)$ and $\tau_0 = ind_{H_0}^{G_0}(\chi|_{H_0})$. Immediately, $\ell_0 \in \Gamma_{\tau_0}^0$. Each element $A \in \mathcal{U}(\mathfrak{g})$ is written as $A = \sum_{j=0}^r A_j X^j$ with some $A_j \in \mathcal{U}(\mathfrak{g}_0), 0 \leq j \leq r$. Hence $A \equiv A_0$ modulo $\mathcal{U}(\mathfrak{g})\mathfrak{a}_\tau$ and $\pi_\ell(\overline{A})a_\ell = \pi_\ell(\overline{A_0})a_\ell$. Besides, if $A \in \mathcal{U}(\mathfrak{g},\tau)$, $[A_0, W] \in \mathcal{U}(\mathfrak{g}_0)\mathfrak{a}_{\tau_0}$ for any $W \in \mathfrak{h}$. Recall $\mathfrak{b} \subset \mathfrak{g}_0$, which means that $\mathfrak{b}$ is a polarisation at $\ell \in \mathfrak{g}^*$ and at $\ell_0 \in (\mathfrak{g}_0)^*$ too. Let us calculate $\langle\pi_\ell(\overline{A_0})a_\ell, \psi\rangle$ for $\psi \in \mathcal{H}_{\pi_\ell}^\infty$ having the form $\psi(\exp(xX){\cdot}g_0) = \zeta(x)\xi(g_0)(x \in \mathbb{R}, g_0 \in G_0)$ with $\zeta \in \mathcal{S}(\mathbb{R}), \xi \in \mathcal{H}_{\pi_0}^\infty$. For $V \in \mathcal{U}(\mathfrak{g})$, we denote by $g{\cdot}V(\forall g \in G)$ the image of V under the adjoint action of g. Put $h(x) = \exp(xX) \in H$ for $x \in \mathbb{R}$. We then have

$$\begin{aligned}
\langle\pi_\ell(\overline{A_0})a_\ell, \psi\rangle &= \langle a_\ell, \pi_\ell({}^t(A_0))\psi\rangle \\
&= \int_{\mathbb{R}} \overline{\zeta(x)} \int_{H_0/(H_0\cap B)} \overline{(\pi_0({}^t(h(-x){\cdot}A_0))\xi)(h_0)\chi(h_0)}d\dot{h_0}dx \\
&= \int_{\mathbb{R}} \overline{\zeta(x)}\langle\pi_0(\overline{h(-x){\cdot}A_0})a_{\ell_0}, \xi\rangle dx \\
&= (\int_{\mathbb{R}} \overline{\zeta(x)}dx)\langle\pi_0(\overline{A_0})a_{\ell_0}, \xi\rangle,
\end{aligned}$$

where $a_{\ell_0} \in \mathcal{H}_{\pi_0}^{-\infty}$ is given by

$$\langle a_{\ell_0}, \xi \rangle = \int_{H_0/(H_0 \cap B)} \overline{\xi(h_0)\chi(h_0)} d\dot{h}_0$$

with $d\dot{h} = dxd\dot{h}_0$ for $h = \exp(xX)\cdot h_0$. By induction hypothesis, $\pi_0(\overline{A_0})a_{\ell_0} = \overline{P_{A_0}(\ell_0)}a_{\ell_0}$ and hence $\langle \pi(\overline{A})a_\ell, \psi\rangle = \overline{P_{A_0}(\ell_0)}\langle a_\ell, \psi\rangle$ for $\psi = \zeta \otimes \xi$. Since $\mathcal{S}(\mathbb{R})\widehat{\otimes}\mathcal{H}_{\pi_0}^\infty \cong \mathcal{H}_\pi^\infty$, we deduce $\pi(\overline{A})a_\ell = \overline{P_{A_0}(\ell_0)}a_\ell$.

Remark 2.1.10 Suppose that τ has finite multiplicities. Let $W \in \mathcal{U}(\mathfrak{g}, \tau)$. Then, the function $P_W(\ell)$ extends to a rational function on Γ_τ. In fact, let $\mathfrak{g}' = \mathfrak{h}_{n-1}$ as above. If $W \in \mathcal{U}(\mathfrak{g}')$, it suffices to apply the induction hypothesis. Suppose $W \notin \mathcal{U}(\mathfrak{g}')$ and write $W = \sum_{j=0}^r w_j X_n^j$, $w_j \in \mathcal{U}(\mathfrak{g}')(0 \leq j \leq r)$ modulo $\mathcal{U}(\mathfrak{g})\mathfrak{a}_\tau$, where we suppose $w_r \notin \mathcal{U}(\mathfrak{g})\mathfrak{a}_\tau$. As we already saw above, there exists an e-central element $\sigma = aX_n + b$, where a, b belong to $\mathcal{U}(\mathfrak{g}')$, $w_r \in \mathcal{U}(\mathfrak{g}, \tau)$ and the degree of $w_r\overline{\sigma}^r - \overline{a}^r W$ relative to X_n is smaller than or equal to $r - 1$ modulo $\mathcal{U}(\mathfrak{g})\mathfrak{a}_\tau$. Now we use a new induction on r to get our conclusion.

Well, we have the following potential result:

Theorem 2.1.11 *(cf. [19]) Assume that τ has finite multiplicities. Let $W \in \mathcal{U}(\mathfrak{g}, \tau)$. Then at μ-almost every point $\ell \in \Gamma_\tau$, a_ℓ^H is a H-semi-invariant tempered eigendistribution for all operators $\pi(\overline{W})$, $W \in \mathcal{U}(\mathfrak{g}, \tau)$. That is, for any $W \in \mathcal{U}(\mathfrak{g}, \tau)$,*

$$\overline{W}a_\ell^H = \pi(\overline{W})a_\ell^H = \overline{P_W(\ell)}a_\ell^H \tag{2.1.6}$$

for some complex number $P_W(\ell)$ Further, $\ell \mapsto P_W(\ell)$ extends to a H-invariant rational function on Γ_τ.

Remark 2.1.12 *Define using the Penney distribution (2.1.4) for $\varphi_\ell \in \mathcal{H}_{\pi_\ell}^\infty$, the function $\varphi_\ell^H : G \to \mathbb{C}$ by*

$$\varphi_\ell^H(g) := \int_{H/(H\cap B[\ell])} \varphi_\ell(gh)\chi_\ell(h)d\dot{h}, g \in G.$$

Then for $A \in \mathcal{U}(\mathfrak{g}, \tau)$ we have that

$$\begin{aligned} \langle \bar{A} \cdot a_\ell^H, \varphi_\ell\rangle &= \langle a_\ell^H, R(A^t)\cdot\varphi_\ell\rangle \\ &= \overline{R(A^t)(\overline{\varphi_\ell^H})(e)}. \end{aligned}$$

The answer to Question 2.1.7 above, is subject of the following:

Proposition 2.1.5 *Retain the same notations as above. Let Φ_τ be an affine H-invariant subspace of Γ_τ. Let $W \in \mathcal{U}(\mathfrak{g}, \tau)$ be such that $\overline{W}\cdot a_\ell = \overline{P_W(\ell)}a_\ell$ with a scalar $P_W(\ell)$ almost everywhere on Φ_τ, which extends to a rational function on Φ_τ. Then, the function P_W extends to a polynomial function on Φ_τ.*

Proof 2.4 *If* $\dim \mathfrak{g} = 1$, *there is nothing to prove. We can therefore assume that* $\dim \mathfrak{g} > 1$. *If* $\mathfrak{h} = \mathfrak{g}$, *then* f *is a character of* $\mathfrak{g}$ *and* $\Gamma_\tau = \{f\}$, *there is also nothing to prove. Hence, we can assume that* $\dim \mathfrak{h} < \dim \mathfrak{g} = n$ *and we can take any ideal* $\tilde{\mathfrak{g}}$ *of* $\mathfrak{g}$ *of codimension 1 containing* $\mathfrak{h}$.

Put $\tilde{G} = \exp \tilde{\mathfrak{g}}$, X_n *(as above) an element of* $\mathfrak{g} \setminus \tilde{\mathfrak{g}}$ *and write*

$$W \quad = \quad w_r X_n^r + w_{r-1} X_n^{r-1} + \cdots + w_0,$$

where $w_r, \cdots, w_0 \in \mathcal{U}(\tilde{\mathfrak{g}})$.

Let us observe that in the case where the generic G*-orbits in* Φ_τ *are not saturated with respect to* $\tilde{\mathfrak{g}}$, *there exists an e-central element* $\sigma_n = a_n X_n + b_n$ *for* Φ_τ *with* $a_n \neq 0$. *Then*

$$\sigma_n \cdot a_\ell^H \quad = \quad P_{\sigma_n}(\ell) a_\ell^H, \ell \in \Phi_\tau \tag{2.1.7}$$

with a polynomial function P_{σ_n} *on* Φ_τ.

To prove Proposition 2.1.5, we proceed now by induction on

$$d(\mathfrak{g}) := \dim \mathfrak{g} + \dim \mathfrak{h} + r.$$

For $\ell \in \Gamma_\tau$ *and* $\tilde{\ell} := \ell|_{\tilde{\mathfrak{g}}}$, *let* $\tilde{\tau} := \operatorname{ind}_H^{\tilde{G}} \chi_{\tilde{\ell}}$ *and let* $a_{\tilde{\ell}} = a_{\tilde{\ell}}^H \in \mathcal{H}_{\pi_{\tilde{\ell}}}^{-\infty}$ *be defined as before by*

$$\langle a_{\tilde{\ell}}, \tilde{\varphi} \rangle \quad = \quad \int_{H/(B[\tilde{\ell}] \cap H)} \overline{\tilde{\varphi}(h) \chi_{\tilde{\ell}}(h)} d\dot{h}, \tilde{\varphi} \in \mathcal{H}_{\pi_{\tilde{\ell}}}^\infty.$$

We have that $B[\ell] \cap \tilde{G} = B[\tilde{\ell}]$ *for any* $\ell \in \Phi_\tau$ *and so we have for every* $\varphi \in \mathcal{H}_{\pi_\ell}^\infty, \tilde{\varphi} := \varphi|_{\tilde{G}}$ *the relation*

$$\langle a_\ell^H, \varphi \rangle \quad = \quad \langle a_{\tilde{\ell}}^H, \tilde{\varphi} \rangle.$$

If now $A \in \mathcal{U}(\tilde{\mathfrak{g}})$, *it follows that*

$$\langle A \cdot a_\ell^H, \varphi \rangle \quad = \quad \langle A \cdot a_{\tilde{\ell}}^H, \tilde{\varphi} \rangle,$$

and so if $A \in \mathcal{U}(\mathfrak{g}, \tau) \cap \mathcal{U}(\tilde{\mathfrak{g}})$ *satisfies the relation*

$$\overline{A} \cdot a_\ell^H \quad = \quad \overline{P_A(\ell)} a_\ell^H, \ell \in \Phi_\tau,$$

then we see that

$$\overline{A} \cdot a_{\tilde{\ell}} \quad = \quad \overline{P_A(\ell)} a_{\tilde{\ell}}, \ell \in \Phi_\tau.$$

Therefore, if $W \in \mathcal{U}(\tilde{\mathfrak{g}})$, *i.e. if* $r = 0$, *then the induction hypothesis for the triplet* $(\tilde{\mathfrak{g}}, \Phi_\tau|_{\tilde{\mathfrak{g}}} \subset \Gamma_{\tilde{\tau}}, H)$ *immediately ensures the result. This is the case if the generic* G*-orbits which meet* Φ_τ *are saturated with respect to* $\tilde{\mathfrak{g}}$ *as in subsection 2.1.6.*

This is also the case if $\dim \mathfrak{h} = 0$. *Indeed, in this situation we have that* $\Gamma_\tau = \mathfrak{g}^*$, $\mathcal{U}(\mathfrak{g}, \tau) = \mathcal{U}(\mathfrak{g})$ *and the Penney's distribution* a_ℓ $(\ell \in \Phi_\tau)$ *is nothing but* $\overline{\delta_\ell}$, *where* δ_ℓ *denotes the Dirac distribution for* π_ℓ.

Our condition on W *then says that*

$$R(\overline{W^t})\overline{\varphi_\ell}(e) \quad = \quad \overline{P_W(\ell)}\overline{\varphi_\ell}(e), \varphi \in \mathcal{H}^\infty_{\pi_\ell}.$$

Therefore W *must now be an element of* $\mathcal{U}(B[\ell])$ *for any* $\ell \in \Phi_\tau$. *Since we can assume that generically no* $\ell \in \Phi_\tau$ *is a character of* $\mathfrak{g}$, *we see that* W *is contained in the enveloping algebra of a proper subalgebra of* $\mathfrak{g}$ *and therefore* $W \in \mathcal{U}(\tilde{\mathfrak{g}})$ *for some one co-dimensional ideal* $\tilde{\mathfrak{g}}$ *of* $\mathfrak{g}$ *and we are done.*

We can now suppose $\dim \mathfrak{h} \geq 1$ *and that* $W \notin \mathcal{U}(\mathfrak{h}_{n-1})$ *and so the generic* G-*orbits which encounter* Φ_τ *are non-saturated with respect to* $\tilde{\mathfrak{g}} = \mathfrak{h}_{n-1}$. *Therefore we have the e-central element* $\sigma_n = a_n X_n + b_n, a_n, b_n \in \mathcal{U}(\mathfrak{h}_{n-1})$ *for* Φ_τ. *We now start a new induction on* $d(\mathfrak{g})$ *of the following fact: Let* α *be the denominator of the rational function* P_W *on* Φ_τ *and suppose that* α *is non-trivial. Then, there exists in* $\mathcal{U}(\mathfrak{g}, \tau)$ *an element*

$$\nu = aX_n + b, \ a, b \in \mathcal{U}(\mathfrak{h}_{n-1}),$$

such that P_a *is not trivial on* Φ_τ, *it has no common divisors with* α *and such that* $\overline{\nu}\cdot a_\ell = \overline{P_\nu(\ell)} a_\ell$ *almost everywhere on* Γ_τ *with a function* $P_\nu(\ell)$ *which extends to a polynomial function on* Γ_τ. *Remark first that nothing happens for small* $d(\mathfrak{g})$. *Let next* $\widetilde{\mathfrak{h}} = \mathfrak{h}_{d-1}, \widetilde{H} = \exp \widetilde{\mathfrak{h}}$. *Denote by* $\widetilde{a}_\ell = a_\ell^{\widetilde{H}}$ *the Penney's distribution for* $\widetilde{\mathfrak{h}}$. *Let also* $\widetilde{\tau} = ind_{\widetilde{H}}^G \chi_f$. *We now show that*

$$\overline{W}\cdot\widetilde{a}_\ell = \overline{P_W(\ell)}\widetilde{a}_\ell \tag{2.1.8}$$

for almost all $\ell \in \Phi_\tau$. *Since* $\sigma_n = a_n X_n + b_n$ *is e-central for* Φ_τ,

$$\sigma_n\cdot\widetilde{a}_\ell = \varphi_n(\ell)\widetilde{a}_\ell, \tag{2.1.9}$$

where $\varphi_n(\ell)$ *denotes the Corwin-Greenleaf function corresponding to* σ_n, *for almost all* $\ell \in \Phi_\tau$. *Similarly for* a_n *which is also e-central for* Φ_τ:

$$a_n\cdot\widetilde{a}_\ell = \psi_n(\ell)\cdot\widetilde{a}_\ell, \tag{2.1.10}$$

where $\psi_n(\ell)$ *is the Corwin-Greenleaf function corresponding to* a_n, *for almost all* $\ell \in \Phi_\tau$. *Now the degree of*

$$\widetilde{W} = w_r\overline{\sigma_n}^r - \overline{a_n}^r W, \tag{2.1.11}$$

where w_r *is the dominant coefficient of* W *with respect to* X_n *is less than or equal to* $r-1$ *on* $a_\ell, \ell \in \Phi_\tau$, *because* $\overline{w_r} a_n \cdot a_\ell^H = a_n \overline{w_r} \cdot a_\ell^H$ *on* Φ_τ. *From the induction hypothesis it follows that*

$$\overline{\widetilde{W}}\cdot\widetilde{a}_\ell = \overline{P_{\widetilde{W}}(\ell)}\widetilde{a}_\ell \tag{2.1.12}$$

and

$$\overline{w_r}\cdot\widetilde{a}_\ell = \overline{P_{w_r}(\ell)}\widetilde{a}_\ell \tag{2.1.13}$$

for almost all $\ell \in \Phi_\tau$. *According to the relations (2.1.9), and (2.1.13), we see the desired claim (2.1.8). Hence, by the induction hypothesis applied to* $\widetilde{\mathfrak{h}}$, Φ_τ *and* W, *there exists in* $\mathcal{U}(\mathfrak{g}, \widetilde{\tau})$ *an element*

$$\widetilde{\nu} = \widetilde{a}X_n + \widetilde{b}, \ \widetilde{a}, \widetilde{b} \in \mathcal{U}(\mathfrak{h}_{n-1}),$$

such that $P_{\widetilde{a}}$ and α have no common divisor and such that $\overline{\widetilde{\nu}} \cdot \widetilde{a}_\ell = \overline{\widetilde{P_{\widetilde{\nu}}}(\ell)}\widetilde{a}_\ell$ almost everywhere on $\Gamma_{\widetilde{\tau}}$ with a function $\widetilde{P_{\widetilde{\nu}}}(\ell)$ which extends to a polynomial function on $\Gamma_{\widetilde{\tau}}$ and such that $P_{\widetilde{a}}$ is not 0 on Φ_τ.

We need to prove now that $\widetilde{\nu} \in \mathcal{U}(\mathfrak{g},\tau)$. To do so, we distinguish two cases. For $\ell \in \mathfrak{g}^*$, put $\Omega_\ell = G\cdot\ell$. First, suppose that the generic H-orbits contained in $\Omega_\ell|_\mathfrak{h}$ are non-saturated with respect to $\mathfrak{h}_{d-1} = \widetilde{\mathfrak{h}}$ for almost all $\ell \in \Gamma_{\widetilde{\tau}}$. Then, there exists an element

$$\zeta = \gamma X_d + \eta,\ \ \gamma,\eta \in \mathcal{U}(\widetilde{\mathfrak{h}}), \gamma \neq 0,$$

which is e-central for $\cup_{\ell\in\Gamma_{\widetilde{\tau}}}(\Omega_\ell|_\mathfrak{h})$. Then γ is e-central too. Let φ be the function of Corwin-Greenleaf corresponding to ζ. For $\ell \in \Gamma_{\widetilde{\tau}}$, let

$$\pi_\ell|_H \simeq \int_{\widehat{H}}^{\oplus} m(\rho)\rho d\mu(\rho) \tag{2.1.14}$$

be the canonical central decomposition of $\pi_\ell|_H$ as in subsection 1.5.2 and

$$\widetilde{a}_\ell = \int_{\widehat{H}}^{\oplus} \widetilde{a}_\rho d\mu(\rho),\ \ \widetilde{a}_\rho \in \mathcal{H}_{m(\rho)\rho}^{-\infty} \tag{2.1.15}$$

the corresponding disintegration of $\widetilde{a}_\ell$. Then,

$$\zeta\cdot\widetilde{a}_\ell = \int_{\widehat{H}}^{\oplus} \varphi(\rho)\widetilde{a}_\rho d\mu(\rho).$$

We apply now Lemma 2.1.4 to $\widetilde{\nu}$. We have

$$A\widetilde{\nu} \equiv B \ mod\, \mathcal{U}(\mathfrak{g})\mathfrak{a}_{\widetilde{\tau}}$$

with two polynomials A, B of $\overline{\sigma_{k_1}}, ..., \overline{\sigma_{k_m}}$. Applying this relation to $\widetilde{a}_\ell$, we find

$$\overline{A\widetilde{\nu}}\cdot\widetilde{a}_\ell = \overline{\widetilde{P_{\widetilde{\nu}}}(\ell)}\overline{A}\cdot\widetilde{a}_\ell = \overline{B}\cdot\widetilde{a}_\ell. \tag{2.1.16}$$

Namely,

$$\overline{\widetilde{P_{\widetilde{\nu}}}(\ell)} \int_{\widehat{H}}^{\oplus} \overline{A}\cdot\widetilde{a}_\rho d\mu(\rho) = \int_{\widehat{H}}^{\oplus} \overline{B}\cdot\widetilde{a}_\rho d\mu(\rho)$$

for generic $\ell \in \Gamma_{\widetilde{\tau}}$. For $k \in \mathcal{K}$, let $\varphi_k(\ell_k), \ell_k = \ell|_{\mathfrak{h}_k}$, be the Corwin-Greenleaf function corresponding to σ_k. Then, according to the restriction

$$\pi_\ell|_H \ \ = \ \ (\cdots((\pi_\ell|_{H_{k_m}})|_{H_{k_{m-1}}})\cdots)|_H, \tag{2.1.17}$$

where $H_k = \exp(\mathfrak{h}_k)$ for $k \in \mathcal{K}$, for μ-almost all $\rho \in \widehat{\widehat{H}}$, $A\cdot\widetilde{a}_\rho = \varphi_A(\rho)\widetilde{a}_\rho, B\cdot\widetilde{a}_\rho = \varphi_B(\rho)\widetilde{a}_\rho$ with some scalars $\varphi_A(\rho), \varphi_B(\rho)$ coming from $\varphi_k(\ell_k)(k \in \mathcal{K})$ and the decomposition formula in the restriction (2.1.17). Thus,

$$\widetilde{P_{\widetilde{\nu}}}(\ell)\varphi_A(\rho) = \varphi_B(\rho)$$

for μ-almost all $\rho \in \widehat{H}$. In the same way, $\overline{A\widetilde{\nu}}\cdot\widetilde{a}_\rho = \overline{B}\cdot\widetilde{a}_\rho$ as in (2.1.16) means

$\overline{\varphi_A(\rho)\overline{\widetilde{\nu}}} \cdot \tilde{a}_\rho = \overline{\varphi_B(\rho)}\tilde{a}_\rho$. *In other words,* $\overline{\widetilde{\nu}} \cdot \tilde{a}_\rho = \overline{P_{\widetilde{\nu}}(\ell)}\tilde{a}_\rho$ *for* μ*-almost all* $\rho \in \widehat{H}$. *Therefore,*

$$\begin{aligned} \zeta(\overline{\widetilde{\nu}}\cdot\tilde{a}_\ell) &= \int_{\widehat{H}}^{\oplus} \varphi(\rho)(\overline{\widetilde{\nu}} \cdot \tilde{a}_\rho)d\mu(\rho) = \overline{P_{\widetilde{\nu}}(\ell)} \int_{\widehat{H}}^{\oplus} (\varphi(\rho)\tilde{a}_\rho)d\mu(\rho) \\ &= \int_{\widehat{H}}^{\oplus} \overline{\widetilde{\nu}} \cdot (\varphi(\rho)\tilde{a}_\rho)d\mu(\rho) = \overline{\widetilde{\nu}}(\zeta\cdot\tilde{a}_\ell). \end{aligned}$$

Hence,

$$\begin{aligned} [\gamma X_d + \eta, \overline{\widetilde{\nu}}] \cdot \tilde{a}_\ell &= [\zeta, \overline{\widetilde{\nu}}]\cdot\tilde{a}_\ell \\ &= 0. \end{aligned}$$

for almost all $\ell \in \Gamma_{\tilde{\tau}}$ *and consequently*

$$\overline{[X_d, \widetilde{\nu}]} \cdot \tilde{a}_\ell = 0.$$

Finally

$$[X_d, \widetilde{\nu}] \in \mathcal{U}(\mathfrak{g})\mathfrak{a}_{\tilde{\tau}} \subset \mathcal{U}(\mathfrak{g})\mathfrak{a}_{\tau},$$

which leads to the fact that $\widetilde{\nu} \in \mathcal{U}(\mathfrak{g}, \tau)$ *and since* $\overline{\widetilde{\nu}}\cdot a_\ell = \overline{P_{\widetilde{\nu}}(\ell)}a_\ell$, $\ell \in \Phi_\tau$, *we can take* $\widetilde{\nu}$ *as our* ν *we are looking for.*

Suppose now that the generic H*-orbits contained in* $\Omega_\ell|_{\mathfrak{h}}$ *are saturated with respect to* $\mathfrak{h}_{d-1}$ *for almost all* $\ell \in \Gamma_{\tilde{\tau}}$. *Then, there exists an index* $1 \leq i \leq d-1$ *and an element* $\xi = \kappa X_i + \upsilon, \kappa, \upsilon \in \mathcal{U}(\mathfrak{h}_{i-1})$, *which is e-central for* $\cup_{\ell\in\Gamma_{\tilde{\tau}}}(\Omega_\ell|_{\tilde{\mathfrak{h}}})$, *but which is not e-central for* $\cup_{\ell\in\Gamma_{\tilde{\tau}}}(\Omega_\ell|_{\mathfrak{h}})$. *We take the index* i *as small as possible. Evidently,* $[\widetilde{\mathfrak{h}}, \xi] \cdot \tilde{a}_\ell = 0, \ell \in \Gamma_{\tilde{\tau}}$, *and therefore* $[\widetilde{\mathfrak{h}}, \xi] \in \mathcal{U}(\mathfrak{g})\mathfrak{a}_{\tilde{\tau}}$. *On the other hand, for generic* $\ell \in \Gamma_{\tilde{\tau}}, [X_d, \xi] \notin \ker \pi_\ell|_H$, *since otherwise* ξ *would be e-central for* $\cup_{\ell\in\Gamma_{\tilde{\tau}}}(\Omega_\ell|_{\mathfrak{h}})$. *Just as above, according to Formulae (2.1.14) and (2.1.15), we have:*

$$[X_d, \xi] \cdot \tilde{a}_\ell = \int_{\widehat{H}}^{\oplus} \varphi(\rho)\tilde{a}_\rho d\mu(\rho),$$

where φ *denotes the Corwin-Greenleaf function corresponding to* $[X_d, \xi]$. *Here,* $\varphi(\rho) \neq 0$ *almost everywhere. On the other hand, it is evident that* $[X_d, \xi] \cdot \tilde{a}_\ell = 0$, *since* $[X_d, \xi] \in \mathcal{U}(\mathfrak{h})\mathfrak{a}_{\tilde{\tau}}$. *In this way, this possibility does not occur.*

Let as before

$$W = \sum_{j=0}^{r} w_j X_n^j, \ w_j \in \mathcal{U}(\mathfrak{h}_{n-1})(0 \leq j \leq r)$$

and write again for our element ν*:*

$$\nu = aX_n + b.$$

Now $\overline{w_r}\cdot a_\ell^H = \overline{P_{w_r}(\ell)}a_\ell^H$, *where* $P_{w_r}(\ell)$ *stands for the dominant coefficient of* $P_W(\ell)$ *with respect to* $\ell(X_n)$, *which extends to a rational function on* Φ_τ, *and hence to a polynomial function on* Φ_τ *by the induction hypothesis. On the other hand, we see that* $a, \bar{w}_r$ *commute each other on* $a_\ell^H, \ell \in \Phi_\tau$. *Thus,*

$$\widetilde{W} = a^r W - w_r \nu^r$$

is of degree less than or equal to $r-1$ *with respect to* X_n. *From the induction hypothesis, the function*

$$P_{\widetilde{W}}(\ell) \quad = \quad P_a(\ell)^r P_W(\ell) - P_{w_r}(\ell) P_\nu(\ell)^r$$

extends to a polynomial function on Φ_τ. *Write again*

$$P_W \quad = \quad \frac{\beta}{\alpha}$$

for some polynomials α, β *on* Φ_τ *having no common divisors. Then*

$$\alpha(\ell) P_{\widetilde{W}}(\ell) = P_a(\ell)^r \beta(\ell) - \alpha(\ell) P_{w_r}(\ell) P_\nu(\ell)^r,$$

which leads in turn to the fact

$$\alpha(\ell)(P_{\widetilde{W}}(\ell) + P_{w_r}(\ell) P_\nu(\ell)^r) = P_a(\ell)^r \beta(\ell), \ell \in \Phi_\tau.$$

Since the polynomials α *and* P_a *have no common divisor,* P_W *must itself be a polynomial function on* Φ_τ.

The following is then straight:

Corollary 2.1.13 *Assume that* τ *is of finite multiplicities. For any* $W \in \mathcal{U}(\mathfrak{g}, \tau)$, *the function* $P_W(\ell)$ *as in (2.1.6) extends to a* H*-invariant polynomial function on* Γ_τ.

Proof 2.5 *The fact that* τ *is of finite multiplicities entails that any* $W \in \mathcal{U}(\mathfrak{g}, \tau)$ *verifies* $\overline{W} \cdot a_\ell = \overline{P_W(\ell)} a_\ell$ *with a scalar* $P_W(\ell)$ *almost everywhere on* Φ_τ, *which extends to a rational function on* Φ_τ *(cf. Theorem 2.1.11).*

We now move to tackle Question 2. Let as above Φ_τ be an H-invariant affine subspace of Γ_τ. Let $\mathcal{V} = \mathcal{V}_{\Phi_\tau}$ be the set of the elements $W \in \mathcal{U}(\mathfrak{g}, \tau)$ such that $\overline{W} \cdot a_\ell = \overline{P_W(\ell)} a_\ell, \ell \in \Phi_\tau$, with a function $P_W(\ell)$ which extends to a polynomial function on Φ_τ. Consider the image $M = M_{\Phi_\tau}$ of the map

$$\theta : \mathcal{V} \ni W \mapsto P_W(\ell) \in \mathbb{C}[\Phi_\tau]^H.$$

Proposition 2.1.14 *Retain the same circumstances of Proposition 2.1.5. Let* $Q \in \mathbb{C}[\mathfrak{g}^*]$ *be such that* $Q|_{\Phi_\tau}$ *is* H*-invariant. Suppose that there exist* $U, V \in \mathcal{V}$ *with no common* M*-divisors, such that*

$$P_U \cdot Q = P_V \in M \tag{2.1.18}$$

Then Q *is contained in* M.

Proof 2.6 *We proceed as in the proof of Proposition 2.1.5. If* $\dim \mathfrak{g} = 1$*, there is nothing to prove. We can therefore assume that* $\dim \mathfrak{g} > 1$*. If* $\mathfrak{h} = \mathfrak{g}$*, then* f *is a character of* $\mathfrak{g}$ *and* $\Gamma_\tau = \{f\}$*, there is also nothing to prove. We can therefore assume that* $\dim \mathfrak{h} < \dim \mathfrak{g} = n$ *and take an ideal* $\widetilde{\mathfrak{g}}$ *of* $\mathfrak{g}$ *of codimension 1 containing* $\mathfrak{h}$*.*

Let X_n *as before be an element of* $\mathfrak{g} \setminus \widetilde{\mathfrak{g}}$ *and write abusively* $\ell \in \mathfrak{g}^*$ *as*

$$\ell \;=\; x_n X_n^* + \widetilde{\ell},$$

where $x_n \in \mathbb{R}$ *and* $\widetilde{\ell}$ *is identified to an element of* $(\widetilde{\mathfrak{g}})^*$*. Then*

$$Q(\ell) = p_r(\widetilde{\ell})x_n^r + p_{r-1}(\widetilde{\ell})x_n^{r-1} + \cdots + p_0(\widetilde{\ell}), \tag{2.1.19}$$

where $p_r, \cdots, p_0 \in \mathbb{C}[(\widetilde{\mathfrak{g}})^*]$*.*

We still engage an induction on $d(\mathfrak{g})$*. Remark first that if* U, V *are both in* $\mathcal{U}(\widetilde{\mathfrak{g}})$*, then by the induction hypothesis* Q *is contained in* M*. This is the case if generically the* G*-orbits for* $\ell \in \Phi_\tau$ *are saturated with respect to* $\widetilde{\mathfrak{g}}$ *or if* $\dim \mathfrak{h} = 0$*. We can now suppose that* $\dim \mathfrak{h} \geq 1$ *and that* U *or* $V \notin \mathcal{U}(\widetilde{\mathfrak{g}})$ *and so the generic* G*-orbits which encounter* Φ_τ *are non-saturated with respect to* $\widetilde{\mathfrak{g}} = \mathfrak{h}_{n-1}$*. Therefore we have the e-central element* $\sigma_n = a_n X_n + b_n, a_n, b_n \in \mathcal{U}(\widetilde{\mathfrak{g}})$ *for* Φ_τ*. Similarly to the proof of Proposition 2.1.5, we arrange another induction on* $d(\mathfrak{g})$ *of the following fact: There exists in* $\mathcal{U}(\mathfrak{g}, \tau)$ *an element*

$$\nu = aX_n + b, \;\; a, b \in \mathcal{U}(\mathfrak{h}_{n-1}),$$

in $\mathcal{V}$*, such that* P_a *is not trivial on* Φ_τ *and has no common divisors with* P_U*. It is straight that equality (2.1.18) holds for* $\widetilde{\mathfrak{h}}$ *and the same procedure as in the proof of Proposition 2.1.5 terminates the process.*

Resume the writing (2.1.19) such that $Q|_{\Phi_\tau}$ *is* H*-invariant. Then* $p_r|_{\Phi_\tau}$ *is* H*-invariant too. By the induction hypothesis, there exists* $A \in \mathcal{U}(\mathfrak{g}, \tau) \subset \mathcal{V}$*, such that*

$$P_A \;=\; p_r$$

on Φ_τ*. Let now*

$$\widetilde{Q} \;:=\; P_a^r Q - P_A P_\nu^r \in \mathbb{C}[\mathfrak{g}^*].$$

Then $\widetilde{Q}$ *is* H*-invariant on* Φ_τ *and its degree with respect to* x_n *is less than* r*. So by the induction hypothesis,* $\widetilde{Q} = P_C$ *for some* $C \in \mathcal{V}$*. Hence* $P_a^r Q - P_A P_\nu^r = P_C$ *and then* $P_a^r Q = P_D$ *for some* $D \in \mathcal{V}$*. Therefore,*

$$P_a^r P_U Q = P_a^r P_V$$

by equation (2.1.18) and finally $P_U P_D = P_a^r P_V$*. This shows that the* M*-divisors of* P_U *are provided by* P_a*. Hence* U *must be trivial and so* $Q \in M$*.*

Corollary 2.1.15 *Assume that* τ *is of finite multiplicities. Then the mapping*

$$\theta : \mathcal{V}_{\Gamma_\tau} \ni W \;\mapsto\; P_W(\ell) \in \mathbb{C}[\Gamma_\tau]^H$$

is surjective

Proof 2.7 *In this case, Corwin and Greenleaf showed that M_{Γ_τ} generates the algebra $\mathbb{C}(\Gamma_\tau)^H$ of H-invariant rational functions on Γ_τ (cf. [32]). Hence the proof by Proposition 2.1.14.*

Now, Corollaries 2.1.15 and 2.1.13 are enough to establish the polynomial conjecture:

Theorem 2.1.16 *Let $G = \exp \mathfrak{g}$ be a nilpotent Lie group with Lie algebra $\mathfrak{g}$, $f \in \mathfrak{g}^*$, $\mathfrak{h} \in S(f, \mathfrak{g})$ and $\tau = ind_H^G \chi_f$ with $H = \exp \mathfrak{h}$. Suppose that τ has finite multiplicities. Then, the algebra $D_\tau(G/H)$ is isomorphic to the algebra $\mathbb{C}[\Gamma_\tau]^H$ of the H-invariant polynomial functions on Γ_τ.*

2.1.9 Fundamental tempered solutions for $D_\tau(G/H)$

At the last part of this section we add an application whose method becomes standard (cf. [5], [22], [33]) since the original work of Raïs [94]. More precisely, let us show under our hypothesis, $\tau = \mathrm{ind}_H^G \chi$ has finite multiplicities, that each non zero element $D \in D_\tau(G/H)$ admits a fundamental tempered solution, i.e. there exists a tempered distribution E on the homogeneous space G/H such that we have $DE = \delta$, where δ denotes the Dirac measure at the point H of G/H. To see this it suffices for us, for example, to follow faithfully the arguments developed in the pages 134-136 of [5], there replacing the distribution $A_s (s \in \mathbb{C}, \mathrm{Re}\, s > 0)$ by ours defined as follows.

We briefly recall what we did in [47]. We return to the disintegration

$$\tau = \mathrm{ind}_H^G \chi \simeq \int_{\hat{G}}^{\oplus} m(\pi) \pi d\mu(\pi),$$

and suppose that τ has finite multiplicities. We need the Plancherel formula for τ [92]. Let $\mathcal{D}(G)$ be the space of the C^∞-functions with compact support on G. For $\varphi \in \mathcal{D}(G)$, we fabricate an element φ^χ of $\mathcal{H}_\tau^\infty$:

$$\varphi^\chi(g) = \int_H \varphi(gh)\chi(h) dh \ (\forall\ g \in G),$$

dh denoting a fixed Haar measure on H. Then the Plancherel formula due to Penney [92] is written for τ:

$$\varphi^\chi(g) = \int_{\hat{G}} \sum_{k=1}^{m(\pi)} \langle \pi(\varphi_g) a_\pi^k, a_\pi^k \rangle d\mu(\pi) \ (\forall\ g \in G),$$

where $\varphi_g(x) = \varphi(gx)$ for all $x \in G$. The tempered distributions a_π^k coming into this formula belong to

$$\left(\mathcal{H}_\pi^{-\infty}\right)^{H,\chi} = \{a \in \mathcal{H}_\pi^{-\infty}\ ;\ \forall\ h \in H,\ \pi(h)a = \chi(h)a\},$$

and are obtained as follows. The character χ is written as $\chi = \chi_f$ with a certain $f \in \mathfrak{g}^*$. For μ-almost all π in $\widehat{G}$, the coadjoint orbit $\Omega(\pi)$ of G encounters Γ_τ in number $m(\pi)$ of H-orbits denoted by $\omega_\pi^1, \ldots, \omega_\pi^{m(\pi)}$. We take $\ell \in \omega_\pi^k$ and a polarisation $\mathfrak{b}_\ell$ at ℓ in order to realise $\pi \simeq \pi_\ell = \mathrm{ind}_{B_\ell}^G \chi_\ell$, where $B_\ell = \exp(\mathfrak{b}_\ell)$ and $\chi_\ell(\exp\ X) = e^{i\ell(X)}$ $(\forall\ X \in \mathfrak{b}_\ell)$, what allows us to describe a_π^k explicitly : there exists an invariant measure $d\dot{h}$ on homogeneous space $H/H \cap B_\ell$ such that we have

$$\langle a_\pi^k, \psi\rangle = \int_{H/H\cap B_\ell} \overline{\psi(h)\chi(h)} d\dot{h}$$

for any $\psi \in \mathcal{H}_\pi^\infty$. Applying $R(X), X \in \mathfrak{g}$ to the Plancherel formula for τ, we calculate

$$\begin{aligned}(R(X)\varphi^\chi)(g) &= (R(X)(L_{g^{-1}}\varphi^\chi))(e) = (L(-X)(L_{g^{-1}}\varphi^\chi))(e)\\ =&(L(-X)L_{g^{-1}}\varphi)^\chi(e) = \int_{\widehat{G}} \sum_{k=1}^{m(\pi)} \langle \pi(L(-X)L_{g^{-1}}\varphi)a_\pi^k, a_\pi^k\rangle d\mu(\pi)\\ =&\int_{\widehat{G}} \sum_{k=1}^{m(\pi)} \langle \pi(-X)\pi(L_{g^{-1}}\varphi)a_\pi^k, a_\pi^k\rangle d\mu(\pi)\\ =&\int_{\widehat{G}} \sum_{k=1}^{m(\pi)} \langle \pi(L_{g^{-1}}\varphi)a_\pi^k, \pi(X)a_\pi^k\rangle d\mu(\pi).\end{aligned}$$

Consequently, we get for all $\varphi \in \mathcal{D}(G)$ and $A \in \mathcal{U}(\mathfrak{g})$

$$(R(A)\varphi^\chi)(g) = \int_{\widehat{G}} \sum_{k=1}^{m(\pi)} \langle \pi(L_{g^{-1}}\varphi)a_\pi^k, \pi(\overline{A})a_\pi^k\rangle d\mu(\pi)\ (g \in G).$$

This formula again implies the following: if $P_W(\ell) = 0$ almost everywhere in Γ_τ for $W \in \mathcal{U}(\mathfrak{g}, \tau)$, then W belongs to the kernel $\mathcal{U}(\mathfrak{g})\mathfrak{a}_\tau$. As Lipsman [77] noticed, the Plancherel formula gives us an intertwining operator between τ and its canonical central decomposition:

$$\mathcal{H}_\tau^\infty \ni \varphi^\chi \mapsto \int_{\widehat{G}}^\oplus \sum_{k=1}^{m(\pi)} \pi(\varphi)a_\pi^k d\mu(\pi)$$

for $\varphi \in \mathcal{D}(G)$. Similarly, $p : \mathcal{U}(\mathfrak{g}, \tau) \to D_\tau(G/H)$ being the canonical projection, the Plancherel formula makes $p(A), A \in \mathcal{U}(\mathfrak{g}, \tau)$, diagonal:

$$p(A) \mapsto \int_{\widehat{G}}^\oplus \sum_{k=1}^{m(\pi)} \lambda_\pi^k(A) d\mu(\pi),$$

where $\lambda_\pi^k(A) = P_A(\ell_\pi^k)$ with $\ell_\pi^k \in \omega_\pi^k$.

Example 2.1.17 Let $\mathfrak{g} = \langle e_1, e_2, e_3, e_4\rangle_\mathbb{R} : [e_1, e_2] = e_3, [e_1, e_3] = e_4$, $f = e_4^*$ relative to the dual basis of $\mathfrak{g}^*$ and $\mathfrak{h} = \mathbb{R}e_2 + \mathbb{R}e_4$. We consider $f_\lambda = \lambda e_3^* + e_4^* (\lambda \in \mathbb{R})$ in $\Gamma = f + \mathfrak{h}^\perp$ and find $\mathfrak{g}(f_\lambda) = \mathbb{R}(e_2 - \lambda e_3) \oplus \mathbb{R}e_4$. Hence, $\mathfrak{h} + \mathfrak{g}(f_\lambda)$ is a Lagrangian subspace for B_{f_λ} for $\lambda \neq 0$. Let $G = \exp \mathfrak{g}$ and $H = \exp \mathfrak{h}$. For $\ell \in \Gamma$ verifying $\ell(e_3) \neq 0$, the intersection $\Gamma \cap G \cdot \ell$ consists in two straight lines each of which is a H-orbit, while it is a straight line containing infinite number of H-orbits for $\ell \in \Gamma$ vanishing at e_3. We immediately see $p(\Gamma) = \{G\cdot f_\lambda; \lambda \geq 0\}$, p being the canonical projection from $\mathfrak{g}^*$ to the orbit space $\mathfrak{g}^*/G$. Thus,

$$\mathrm{ind}_H^G \chi_f \simeq 2 \int_0^\infty \pi_\lambda d\lambda$$

with $\pi_\lambda = \theta(f_\lambda)$.

Taking a real polarisation $\mathfrak{b} = \langle e_2, e_3, e_4 \rangle_{\mathbb{R}}$ at f_λ, we realise $\pi_\lambda \in \widehat{G}$ in $L^2(\mathbb{R})$ under the identification $\mathcal{H}_{\pi_\lambda} \ni \phi \mapsto \Phi \in L^2(\mathbb{R})$ given by $\Phi(t) = \phi(\exp(te_1))(t \in \mathbb{R})$. We then confirm that the space $\mathcal{H}_{\pi_\lambda}^\infty$ is realised as the Schwartz space $\mathcal{S}(\mathbb{R})$ and consequently

$$(\mathcal{H}_{\pi_\lambda}^{-\infty})^{H,\chi_f} = \mathbb{C}a_\lambda^1 \oplus \mathbb{C}a_\lambda^2,$$

where $a_\lambda^1 : \Phi \mapsto \overline{\Phi(0)}$ while $a_\lambda^2 : \Phi \mapsto \overline{\Phi(2\lambda)}$ if $\lambda \neq 0$ and $\Phi \mapsto \overline{(d\Phi/dt)(0)}$ if $\lambda = 0$. In this manner, $\dim\left((\mathcal{H}_{\pi_\lambda}^{-\infty})^{H,\chi_f}\right) = 2$, equal to the multiplicity in $\tau = \mathrm{ind}_H^G \chi_f$.

For $\psi \in \mathcal{D}(G)$, we put

$$\psi_H^f(g) = \int_H \psi(gh)\chi_f(h)dh \ (g \in G),$$

where $dh = dx_2 dx_4$ if $h = \exp(x_2e_2)\exp(x_4e_4)$. Let $dg = \prod_{j=1}^4 dx_j$ for $g = \prod_{j=1}^4 \exp(x_je_j)$. In these situations, Penney's Plancherel formula for τ becomes:

$$\psi_H^f(e) = \frac{1}{2\pi} \int_0^\infty (\langle \pi_\lambda(\psi)a_\lambda^1, a_\lambda^1 \rangle + \langle \pi_\lambda(\psi)a_\lambda^2, a_\lambda^2 \rangle) d\lambda.$$

In fact, let $B = \exp \mathfrak{b}$ and $db = \prod_{j=2}^4 dx_j$ for $b = \prod_{j=2}^4 \exp(x_je_j)$. Then, $\Psi_\lambda^1 = \pi_\lambda(\psi)a_\lambda^1 \in \mathcal{H}_{\pi_\lambda}^\infty = \mathcal{S}(\mathbb{R})$ is obtained by

$$\Psi_\lambda^1(t) = \int_B \psi(\exp(te_1)\cdot b)\chi_{f_\lambda}(b)db \ (t \in \mathbb{R})$$

so that

$$\langle \pi_\lambda(\psi)a_\lambda^1, a_\lambda^1 \rangle = \int_B \psi(b)\chi_{f_\lambda}(b)db.$$

Similarly, $\Psi_\lambda^2 = \pi_\lambda(\psi)a_\lambda^2 \in \mathcal{S}(\mathbb{R})$ is obtained by

$$\Psi_\lambda^2(t) = \int_B \psi(\exp((t-2\lambda)e_1)\cdot b)\chi_{f_{-\lambda}}(b)db \ (t \in \mathbb{R})$$

and then

$$\langle \pi_\lambda(\psi)a_\lambda^2, a_\lambda^2 \rangle = \int_B \psi(b)\chi_{f_{-\lambda}}(b)db.$$

From these observations,

$$\frac{1}{2\pi} \int_0^\infty (\langle \pi_\lambda(\psi)a_\lambda^1, a_\lambda^1 \rangle + \langle \pi_\lambda(\psi)a_\lambda^2, a_\lambda^2 \rangle) d\lambda$$

$$=\frac{1}{2\pi}\int_0^\infty d\lambda\int_{\mathbb{R}}dx_3\int_H(e^{i\lambda x_3}+e^{-i\lambda x_3})\psi(h\exp(x_3e_3))\chi_{f_\lambda}(h)dh$$
$$=\int_H\psi(h)\chi_f(h)dh=\psi_H^f(e)$$

by the help of Plancherel formula for $\mathbb{R}$.

Denote by $\mathcal{S}(G)$ the space of the Schwartz functions on G. The Plancherel formula is generalised to $\mathcal{S}(G)$, and we reformulate it in the manner of Lipsman [76], [78]. Let $\tilde{\eta}$ be a finite measure on Γ_τ equivalent to the Lebesgue measure and η its quotient image in Γ_τ/H under the action of H. Then, these measures being appropriately normalised, we define for all $\phi\in\mathcal{S}(G)$,

$$\phi^\chi(g)=\int_H\phi(gh)\chi(h)dh\ (\forall\ g\in G),$$

and obtain

$$\phi^\chi(e)=\int_{\Gamma_\tau/H}\langle\pi_\ell(\phi)a_\ell,a_\ell\rangle d\eta(\dot{\ell}),$$

where the integral converges absolutely.

Let $D\in D_\tau(G/H)$. Recall the isomorphism of algebras $\Xi:D_\tau(G/H)\to\mathbb{C}[\Gamma_\tau]^H$ and define our distribution A_s by the formula

$$A_s(\phi^\chi)=\int_{\Gamma_\tau/H}(\Xi(D)(\ell))^s\langle\pi_\ell(\phi)a_\ell,a_\ell\rangle d\eta(\dot{\ell}).$$

We know [33] that the map

$$\mathcal{S}(G)\ni\phi\mapsto\pi_\ell(\phi)a_\ell\in\mathcal{H}_{\pi_\ell}^\infty$$

is continuous.

We now consider the coadjoint action of H in Γ_τ, where we introduce the coordinates $\{\ell_j=\ell(X_{d+j}):1\le j\le q=n-d\}$, and form the principal couch $\mathcal{V}_e$ by means of the sequence (2.1.3), which is of Jordan-Hölder between $\mathfrak{h}$ and $\mathfrak{g}$. Namely, $e=(e_1,\dots,e_q)$ is a multi-index of non-negative integers such that the set $\mathcal{V}_e$ formed by the elements $\ell\in\Gamma_\tau$ verifying $\dim H\cdot(\ell|_{\mathfrak{h}_{d+j}})=e_j$ $(1\le j\le q)$ is a non-empty Zariski open set of Γ_τ. Let $T(e)=\{1\le j\le q:e_j=e_{j-1}\}$ with $e_0=0$. Then, except a negligible set, the space of the H-orbits Γ_τ/H is identified to $\mathcal{V}_e\cap\mathcal{T}$ with $\mathcal{T}=f+\sum_{j\in T(e)}\mathbb{R}X_j{}^*$ and $\{\ell_j;j\in T(e)\}$ define there a system of coordinates.

Remark next by induction on $\dim\mathfrak{g}+\dim\mathfrak{g}/\mathfrak{h}$ that the function

$$\Gamma_\tau/H\ni\dot{\ell}\mapsto\langle\pi_\ell(\phi)a_\ell,a_\ell\rangle\in\mathbb{C}\quad(\phi\in\mathcal{S}(G))\tag{2.1.20}$$

is rapidly decreasing. In fact, keeping $\mathfrak{z}$ to denote the centre of $\mathfrak{g}$, if $\mathfrak{q}=\mathfrak{h}\cap\mathfrak{z}\cap\ker f$ is not trivial, we pass to the quotient $\tilde{\mathfrak{g}}=\mathfrak{g}/\mathfrak{q}$. When $\mathfrak{z}\not\subset\mathfrak{h}$, take $Z\in\mathfrak{z}\backslash(\mathfrak{h}\cap\mathfrak{z})$ and put $\tilde{\mathfrak{h}}=\mathfrak{h}+\mathbb{R}Z,\tilde{H}=\exp\tilde{\mathfrak{h}}$. The character χ of H extends into χ_α $(\forall\alpha\in\mathbb{R})$ by $\chi_\alpha(\exp(xZ))=e^{i\alpha x}$ $(x\in\mathbb{R})$ which induces $\tau_\alpha=\mathrm{ind}_{\tilde{H}}^G\chi_\alpha$. Then the function (2.1.20) is by induction hypothesis rapidly decreasing in $\Gamma_{\tau_\alpha}/\tilde{H}$ for generic $\alpha \in \mathbb{R}$. On the other hand, it is nothing but the Fourier transform in the direction $\tilde{H}/H$. Look at this situation more closely.

First, a simple calculation gives us

$$(\pi_\ell(\phi)a_\ell)(g) = \int_{B_\ell/B_\ell\cap H} \phi^X(gb)\chi_\ell(b)d\dot{b}$$

for all $g \in G$ with the notations already introduced, except $d\dot{b}$ denoting an invariant measure on homogeneous space $B_\ell/B_\ell \cap H$. We deduce from this, for all $m \in \mathbb{N}$,

$$\begin{aligned}&\left(1+(\ell(Z))^2\right)^m(\pi_\ell(\phi)a_\ell)\\ &=\left(1-(\pi_\ell(Z))^2\right)^m(\pi_\ell(\phi)a_\ell)=\pi_\ell\left(\left(1-(L_Z)^2\right)^m\phi\right)a_\ell,\end{aligned}$$

where L_Z denotes the left action of $Z \in \mathfrak{g}$. From [38], there exists in $\mathcal{S}(G)$ two functions ψ, ξ whose convolution is ϕ, i.e. $\phi = \psi * \xi$. Define $\psi' \in \mathcal{S}(G)$ by the formula

$$\psi'(g) = \overline{((1-(L_Z)^2)^m\,\psi(g^{-1}))}\ (\forall g \in G).$$

We have then

$$\begin{aligned}&\int_{\Gamma_\tau/H}\left(1+(\ell(Z))^2\right)^m|\langle\pi_\ell(\phi)a_\ell,a_\ell\rangle|\,d\eta(\dot{\ell})\\ &=\int_{\Gamma_\tau/H}\left|\left\langle\pi_\ell\left(\left(1-(L_Z)^2\right)^m\phi\right)a_\ell,a_\ell\right\rangle\right|d\eta(\dot{\ell})=\int_{\Gamma_\tau/H}|\langle\pi_\ell(\xi)a_\ell,\pi_\ell(\psi')a_\ell\rangle|\,d\eta(\dot{\ell})\\ &\le\left(\int_{\Gamma_\tau/H}\|\pi_\ell(\xi)a_\ell\|^2_{\mathcal{H}_{\pi_\ell}}d\eta(\dot{\ell})\right)^{1/2}\left(\int_{\Gamma_\tau/H}\|\pi_\ell(\psi')a_\ell\|^2_{\mathcal{H}_{\pi_\ell}}d\eta(\dot{\ell})\right)^{1/2}\\ &=\|\xi^X\|_{\mathcal{H}_\tau}\|(\psi')^X\|_{\mathcal{H}_\tau}<\infty,\end{aligned}$$

because the map

$$\mathcal{H}_\tau \ni \phi^X \mapsto \int_{\Gamma_\tau/H}\pi_\ell(\phi)a_\ell d\eta(\dot{\ell}) \in \int_{\Gamma_\tau/H}\mathcal{H}_{\pi_\ell}d\eta(\dot{\ell})$$

is an intertwining isometry [78].

It remains to examine the case where $\dim\mathfrak{z} = 1$ and f does not vanish on $\mathfrak{z} = \mathbb{R}Z$. Introduce a Heisenberg triplet (X, Y, Z) verifying $[X, Y] = Z$ and such that $\mathfrak{g} = \mathbb{R}X + \mathfrak{g}'$ with the centraliser $\mathfrak{g}'$ of Y in $\mathfrak{g}$. Let $G' = \exp(\mathfrak{g}')$. The irreducible representation $\pi \in \widehat{G}$ generic for the measure η is realised through a polarisation contained in $\mathfrak{g}'$, what means that there exists $\pi' \in \widehat{G'}$ such that $\pi = \mathrm{ind}_{G'}^G \pi'$. Let first $\mathfrak{h} \subset \mathfrak{g}'$. Consider $\tau' = \mathrm{ind}_H^{G'}\chi$ and $\Gamma'_{\tau'} = \{\ell' \in (\mathfrak{g}')^*; \ell'|_{\mathfrak{h}} = f|_{\mathfrak{h}}\}$. It immediately follows that Γ_τ/H is identified to $\Gamma'_{\tau'}/H'$, what gives the result. Finally, let $\mathfrak{h} \not\subset \mathfrak{g}'$. Introduce $\mathfrak{h}' = (\mathfrak{h} \cap \mathfrak{g}') + \mathbb{R}Y, H' = \exp(\mathfrak{h}'), f' \in (\mathfrak{g}')^*$ verifying $f'|_{\mathfrak{h}\cap\mathfrak{g}'} = f|_{\mathfrak{h}\cap\mathfrak{g}'}$ and $f'(Y) = 0$. Let this time $\tau' = \mathrm{ind}_{H'}^{G'}\chi_{f'}$ and $\Gamma'_{\tau'} = \{\ell' \in (\mathfrak{g}')^*; \ell'|_{\mathfrak{h}'} = f'|_{\mathfrak{h}'}\}$. Last, put $\ell' = \ell|_{\mathfrak{g}'}$ for generic $\ell \in \Gamma_\tau$. Then, Γ_τ/H is identified to $\Gamma'_{\tau'}/H'$, and taking X in

$\mathfrak{h}$, the integral relative to the coordinate along to X in the distribution a_ℓ makes us observe that the function (2.1.20) reduces to that come from the data $(\pi', \mathfrak{h}', f')$:

$$\Gamma'_{\tau'}/H' \ni \ell' \mapsto \langle \pi'_{\ell'}(\phi')a_{\ell'}, a_{\ell'}\rangle \in \mathbb{C}.$$

It follows the result from the induction hypothesis.

We see in this fashion that $A_s(s \in \mathbb{C}, \mathrm{Re}\, s > 0)$ is a tempered distribution and it is possible to follow the route arranged in [5] to the:

Theorem 2.1.18 *Suppose that $\tau = ind_H^G \chi$ has finite multiplicities. Then, any non-zero element of $D_\tau(G/H)$ admits a fundamental tempered solution.*

2.2 DUFLO'S POLYNOMIAL CONJECTURE FOR INDUCTION

What can we say on the centre of $D_\tau(G/H)$? There is a problem due to Duflo (Problem 3 in [40]). Duflo proposed the following question. Let $\mathfrak{h}$ be a subalgebra of $\mathfrak{g}$ and λ be a representation of dimension one of $\mathfrak{h}$. Let $\mathfrak{h}_\lambda$ be the Lie subalgebra of $\mathcal{U}(\mathfrak{g})$ equal to $\{X + \lambda(X), X \in \mathfrak{h}_\mathbb{C}\}$. The left ideal $\mathcal{U}(\mathfrak{g})\mathfrak{h}_\lambda$ is invariant under the adjoint action of $\mathfrak{h}$. Let $D_\lambda = (\mathcal{U}(\mathfrak{g})/\mathcal{U}(\mathfrak{g})\mathfrak{h}_\lambda)^\mathfrak{h}$ be the space of the $\mathfrak{h}$-invariant elements of $\mathcal{U}(\mathfrak{g})/\mathcal{U}(\mathfrak{g})\mathfrak{h}_\lambda$ and C_λ its centre. On the other hand, the symmetric algebra $S(\mathfrak{g})$ of $\mathfrak{g}$ has a Poisson structure $\{\cdot,\cdot\}$ well determined by the equality

$$\{X, Y\} = [X, Y],\ \ X, Y \in \mathfrak{g}.$$

Let us define

$$P_\lambda = (S(\mathfrak{g})/S(\mathfrak{g})\mathfrak{h}_{\lambda-\rho})^\mathfrak{h},$$

where $\rho(X) = -(1/2)\mathrm{tr}\, \mathrm{ad}_{\mathfrak{g}/\mathfrak{h}}(X)$ $(X \in \mathfrak{h})$. This is the algebra of $\mathfrak{h}$-invariant polynomial functions on the affine subspace $-\lambda + \rho + \mathfrak{h}^\perp$ of $\mathfrak{g}^*$ and inherits from $S(\mathfrak{g})$ a Poisson structure. We denote by Z_λ its Poisson centre. Then, Duflo asks: are the algebras C_λ and Z_λ isomorphic?

Let $G = \exp \mathfrak{g}$ be a connected and simply connected real nilpotent Lie group with Lie algebra $\mathfrak{g}$. Being given an analytic subgroup $H = \exp \mathfrak{h}$ with Lie algebra $\mathfrak{h}$ and a unitary character χ of H, we consider the monomial representation $\tau = \mathrm{ind}_H^G \chi$ of G.

Define next the algebra $D_\tau(G/H)$ of the G-invariant linear differential operators with C^∞-coefficients on the line bundle over G/H associated to the data (G, H, χ).

We focus on the so-called Duflo's polynomial problem (cf. Problem 3 in [40]). This asks about what can be said about the centre of $D_\tau(G/H)$, especially when τ decomposes into irreducibles with infinite multiplicities. Let

$$\mathfrak{a}_\tau = \{X + \sqrt{-1}f(X); X \in \mathfrak{h}\}. \tag{2.2.1}$$

Let us recall the affine subspace $\Gamma_\tau = f + \mathfrak{h}^\perp$ of the dual vector space $\mathfrak{g}^*$ of $\mathfrak{g}$, where $\mathfrak{h}^\perp = \{\ell \in \mathfrak{g}^*; \ell(\mathfrak{h}) = \{0\}\}$. Further consider the algebra

$$\mathbb{C}[\Gamma_\tau]^H = (S(\mathfrak{g})/S(\mathfrak{g})\overline{\mathfrak{a}_\tau})^H$$

of the H-invariant polynomial functions on the affine subspace Γ_τ. It inherits the Poisson structure from $S(\mathfrak{g})$.

The question consists in the following: Is the centre C_τ of $D_\tau(G/H)$ isomorphic to the Poisson centre Z_τ of $\mathbb{C}[\Gamma_\tau]^H$? We would like to positively answer this question in the general setting of nilpotent Lie groups. Such a solution was only known for some restrictive cases (cf. [100]). Following [9], we study this problem.

2.2.1 Generation families of C_τ and Z_τ

We keep all our objects and notations and assume that $H \neq G$. Take back the flag of ideals of $\mathfrak{g}$, and put

$$\mathcal{I} = \{i_1 < i_2 < \cdots < i_d\} = \{1 \leq i \leq n; \mathfrak{h} \cap \mathfrak{g}_i \neq \mathfrak{h} \cap \mathfrak{g}_{i-1}\},$$

where $d = \dim \mathfrak{h}$, and

$$\mathcal{J} = \{j_1 < j_2 < \cdots < j_q\} = \{1, 2, \ldots, n\} \setminus \mathcal{I}$$

with $q = \dim(\mathfrak{g}/\mathfrak{h})$. For $1 \leq k \leq q$, put $\mathfrak{h}_{d+k} = \mathfrak{h} + \mathfrak{g}_{j_k}$, then

$$\mathfrak{h} = \mathfrak{h}_d \subset \mathfrak{h}_{d+1} \subset \cdots \subset \mathfrak{h}_{n-1} \subset \mathfrak{h}_n = \mathfrak{g} \tag{2.2.2}$$

is a sequence of subalgebras of $\mathfrak{g}$ which is a composition sequence for the action of $\mathfrak{h}$ on $\mathfrak{g}/\mathfrak{h}$. Let $S_H, T_H \subset \{d+1, \ldots, n\}$ denote respectively the set of jump and non-jump indices for the action of H on Γ_τ.

Let us denote by $\mathcal{J}_0$ the set of $\{1 \leq k \leq q\}$ such that $j_k \in S(e)$ and that $d + k \in S_H$. We take an element

$$\kappa_k = a'_k X_{r_k} + b'_k, \quad a'_k, b'_k \in \mathcal{U}(\mathfrak{h} + \mathfrak{g}_{r_k - 1}),$$

which is e-central for $\Gamma_\tau|_{\mathfrak{h}_{d+k-1}}$ but not for $\Gamma_\tau|_{\mathfrak{h}_{d+k}}$. We choose κ_k in such a manner that r_k is as small as possible. Let

$$\mathcal{J}_0 = \{k_1 < k_2 < \cdots < k_s\}.$$

We next prove the following result which explains how some families of rational generators of C_τ and Z_τ look like.

Theorem 2.2.1 *Keep the same previous notation. We can choose $\{\overline{\kappa_{k_j}}\}_{k_j \in \mathcal{J}_0}$ in such a fashion that C_τ (resp. Z_τ) is rationally generated by $\{\overline{\sigma_j}\}_{j \in T(e)}$ (resp. $\{\overline{\varphi_j}\}_{j \in T(e)}$) and $\{\overline{\kappa_{k_j}}\}_{k_j \in \mathcal{J}_0}$ (resp. $\{P_{\overline{\kappa_{k_j}}}\}_{k_j \in \mathcal{J}_0}$).*

Proof 2.8 *Let $\mathfrak{g}'$ be an ideal of codimension 1 of $\mathfrak{g}$ containing $\mathfrak{h}$. Being possible to choose the flag S as in (2.1.1), we can suppose $\mathfrak{g}' = \mathfrak{g}_{n-1}$. Put $G' = \exp(\mathfrak{g}')$. We proceed by induction on $\dim \mathfrak{g}$ and we shall treat several cases:*

2.2.1.1 Case 1: Saturation of H-orbits

Suppose first that the generic G-orbits which encounter Γ_τ are saturated with respect to $\mathfrak{g}'$. If the generic H-orbits in Γ_τ are also saturated with respect to $\mathfrak{g}'$, then $D_\tau(G/H) = D_{\tau'}(G'/H)$ and $\mathbb{C}[\Gamma_\tau]^H = \mathbb{C}[\Gamma_{\tau'}]^H$, where $\tau' = ind_H^{G'}\chi_f$. Hence, the result follows immediately by the induction hypothesis.

2.2.1.2 Case 2: Non-saturation of H-orbits and saturation of G-orbits

Next, we investigate the essential case where the generic H-orbits in Γ_τ are non-saturated with respect to $\mathfrak{g}'$. There exists therefore an element σ' which is e-central for the H-action on $\Gamma_{\tau'}$ but not on Γ_τ, so we see that C_τ (resp. Z_τ) is contained in $D_{\tau'}(G'/H)$ (resp. $\mathbb{C}[\Gamma_{\tau'}]^H$). On the other hand, there exists (cf. [32] and [54]) in $\mathcal{U}(\mathfrak{g},\tau)$ an element

$$\nu = aX_n + b, a, b \in \mathcal{U}(\mathfrak{g}'),$$

where $a \notin \mathcal{U}(\mathfrak{g})\mathfrak{a}_\tau$, and respectively in $\mathbb{C}[\Gamma_\tau]^H$, an element

$$\zeta = \alpha X_n + \beta, \alpha, \beta \in S(\mathfrak{g}'),$$

with $\alpha \notin S(\mathfrak{g})\overline{\mathfrak{a}_\tau}$ is $\mathfrak{h}$-invariant. Taking the brackets of ν (resp. ζ) with elements of $\mathcal{U}(\mathfrak{g}',\tau')$ (resp. $S(\mathfrak{g}')^H$) if necessary, we can suppose $a \in C_{\tau'}$ (resp. $\alpha \in Z_{\tau'}$).

Now take $\kappa_{k_i}, 1 \le i \le s$ such that $\overline{\kappa_{k_i}}$ belongs to $C_{\tau'}$ by induction hypothesis. We are then done once $\overline{\kappa_{k_i}}$ belongs to C_τ, otherwise we will show that one can modify $\overline{\kappa_{k_i}}$ to be so. We first prove the following:

Lemma 2.2.2 *Assume that $[\overline{\kappa_{k_i}}, \nu] \equiv 0$ modulo $\mathcal{U}(\mathfrak{g})\mathfrak{a}_\tau$, then $\overline{\kappa_{k_i}} \in C_\tau$. Likewise, one finds $P_{\overline{\kappa_{k_i}}} \in Z_\tau$ if $\{P_{\overline{\kappa_{k_i}}}, \zeta\} \in S(\mathfrak{g})\overline{\mathfrak{a}_\tau}$.*

Proof 2.9 *Let*

$$W = \sum_{j=0}^{r} w_j X_n^j,\ w_j \in \mathcal{U}(\mathfrak{g}')\ (0 \le j \le r),$$

be any element of $\mathcal{U}(\mathfrak{g},\tau)$. It follows that $w_r \notin \mathcal{U}(\mathfrak{g})\mathfrak{a}_\tau$ is $\mathfrak{h}$-invariant and so is the case for

$$W' = a^r W - w_r \nu^r.$$

As W' is of degree smaller than or equal to $r-1$ with respect to X_n, it commutes with $\overline{\kappa_{k_i}}$ from a new induction on r. In this way, $[\overline{\kappa_{k_i}}, W] \in \mathcal{U}(\mathfrak{g})\mathfrak{a}_\tau$ and consequently $\overline{\kappa_{k_i}} \in C_\tau$.

Now, we suppose that $[\overline{\kappa_{k_i}}, \nu]$ does not belong to $\mathcal{U}(\mathfrak{g})\mathfrak{a}_\tau$ and try to modify κ_{k_i}. In order to attain this aim, we go to apply the construction used in [10]. Remark first that

$$[w, [\nu, \overline{\kappa_{k_i}}]] \equiv 0 \text{ modulo } \mathcal{U}(\mathfrak{g})\mathfrak{a}_\tau \tag{2.2.3}$$

for any $w \in \mathcal{U}(\mathfrak{g}',\tau')$, and this entails

$$[\overline{\kappa_{k_i}}, [\overline{\kappa_{k_i}}, \nu]] \equiv 0 \text{ modulo } \mathcal{U}(\mathfrak{g})\mathfrak{a}_\tau. \tag{2.2.4}$$

Taking this into account, set $\overline{\kappa_{k_i}} = \kappa$ *for simplicity and put*

$$\psi = \lambda_0 + \lambda_1\nu + \lambda_2\nu^2, \ \lambda_j \in \mathbb{C}(0 \leq j \leq 2)$$

and

$$\psi_0 = \psi, \ \psi_1 = [\kappa, \psi], \ \psi_2 = [\kappa, [\kappa, \psi]].$$

Then, we have the following:

Lemma 2.2.3 *Assume that* $[\nu, [\nu, \kappa]] \not\equiv 0$ *modulo* $\mathcal{U}(\mathfrak{g})\mathfrak{a}_\tau$*, then the element*

$$\gamma = \psi_0\psi_2 + \psi_2\psi_0 - \psi_1^2$$

is of the form $\gamma = \alpha\nu + \beta$ *for some* $\alpha \notin \mathcal{U}(\mathfrak{g})\mathfrak{a}_\tau$*, and* $\beta \in \mathcal{U}(\mathfrak{g}')$ *in such that* $[\kappa, \gamma] \equiv 0$ *modulo* $\mathcal{U}(\mathfrak{g})\mathfrak{a}_\tau$*.*

Proof 2.10 *Having in mind (2.2.4), let us calculate* γ*. We have:*

$$\begin{aligned}
\gamma &= 2\lambda_2(\lambda_0 + \lambda_1\nu + \lambda_2\nu^2)[\kappa, \nu]^2 + 2\lambda_2[\kappa, \nu]^2(\lambda_0 + \lambda_1\nu + \lambda_2\nu^2) \\
&\quad - \{\lambda_1[\kappa, \nu] + \lambda_2([\kappa, \nu]\nu + \nu[\kappa, \nu])\}^2 \\
&= 4\lambda_0\lambda_2[\kappa, \nu]^2 + 4\lambda_1\lambda_2[\kappa, \nu]^2\nu + 2\lambda_1\lambda_2[\nu, [\kappa, \nu]^2] + 4\lambda_2^2[\kappa, \nu]^2\nu^2 + 2\lambda_2^2[\nu^2, [\kappa, \nu]^2] \\
&\quad - \Big(\lambda_1^2[\kappa, \nu]^2 + \lambda_2^2[\kappa, \nu]\nu[\kappa, \nu]\nu + \lambda_2^2\nu[\kappa, \nu]\nu[\kappa, \nu] + \lambda_1\lambda_2([\kappa, \nu]^2\nu + [\kappa, \nu]\nu[\kappa, \nu]) \\
&\quad + \lambda_2^2([\kappa, \nu]\nu^2[\kappa, \nu] + \nu[\kappa, \nu]^2\nu) + \lambda_1\lambda_2([\kappa, \nu]\nu[\kappa, \nu] + \nu[\kappa, \nu]^2)\Big).
\end{aligned}$$

Pursuing the last computations, we get:

$$\begin{aligned}
\gamma &= 4\lambda_2^2[\kappa, \nu]^2\nu^2 + 4\lambda_1\lambda_2[\kappa, \nu]^2\nu + 2\lambda_2^2(\nu[\nu, [\kappa, \nu]^2] + [\nu, [\kappa, \nu]]\nu) + 4\lambda_0\lambda_2[\kappa, \nu]^2 \\
&\quad + 2\lambda_1\lambda_2[\nu, [\kappa, \nu]^2] - \Big(\lambda_1^2[\kappa, \nu]^2 + \lambda_2^2[\kappa, \nu]^2\nu^2 + \lambda_2^2[\kappa, \nu]\nu^2[\kappa, \nu] + \lambda_2^2[\nu, [\kappa, \nu]]\nu[\kappa, \nu] \\
&\quad + \lambda_1\lambda_2(2[\kappa, \nu]^2\nu + [\kappa, \nu][\nu, [\kappa, \nu]]) + \lambda_2^2[\kappa, \nu][\nu, [\kappa, \nu]]\nu \\
&\quad + \lambda_2^2([\kappa, \nu]^2\nu^2 + [\kappa, \nu][\nu^2, [\kappa, \nu]] + [\kappa, \nu]^2\nu^2 + [\nu, [\kappa, \nu]^2]\nu) \\
&\quad + \lambda_1\lambda_2([\kappa, \nu]^2\nu + [\kappa, \nu][\nu, [\kappa, \nu]] + [\kappa, \nu]^2\nu + [\nu, [\kappa, \nu]^2])\Big).
\end{aligned}$$

On the other hand, the last term leads to:

$$\begin{aligned}
\gamma &= \lambda_2^2([\nu, [\kappa, \nu]^2] - 3[\kappa, \nu][\nu, [\kappa, \nu]] + 2[\nu, [\kappa, \nu]] - [\nu, [\kappa, \nu]][\kappa, \nu])\nu \\
&\quad - 2\lambda_2^2[\kappa, \nu][\nu, [\nu, [\kappa, \nu]]] + 2\lambda_2^2[\nu, [\nu, [\kappa, \nu]^2]] + 4\lambda_0\lambda_2[\kappa, \nu]^2 \\
&\quad + 2\lambda_1\lambda_2[\nu, [\kappa, \nu]^2] - \lambda_1^2[\kappa, \nu]^2 - \lambda_2^2[\nu, [\kappa, \nu]]^2 - 2\lambda_1\lambda_2[\kappa, \nu][\kappa, [\kappa, \nu]] \\
&\quad - \lambda_1\lambda_2[\kappa, \nu][\nu, [\kappa, \nu]] - 2\lambda_2^2[\kappa, \nu][\nu, [\nu, [\kappa, \nu]]] - \lambda_1\lambda_2[\nu, [\kappa, \nu]^2].
\end{aligned}$$

Now, from the last computation, we end up with:

$$\begin{aligned}
\gamma &= \lambda_2^2([\kappa, \nu][\nu, [\kappa, \nu]] - 3[\kappa, \nu][\nu, [\kappa, \nu]] + 2[\nu, [\kappa, \nu]])\nu \\
&\quad + (4\lambda_0\lambda_2 - \lambda_1^2)[\kappa, \nu]^2 + \lambda_1\lambda_2[\nu, [\kappa, \nu]^2] - 4\lambda_2^2[\kappa, \nu][\nu, [\nu, [\kappa, \nu]]] \\
&\quad + 2\lambda_2^2[\nu, [\nu, [\kappa, \nu]^2]] - \lambda_2^2[\nu, [\kappa, \nu]]^2 - 3\lambda_1\lambda_2[\kappa, \nu][\nu, [\kappa, \nu]]
\end{aligned}$$

$$= 2\lambda_2^2(1 - [\kappa, \nu])[\nu, [\kappa, \nu]]\nu + (4\lambda_0\lambda_2 - \lambda_1^2)[\kappa, \nu]^2 + \lambda_1\lambda_2[\nu, [\kappa, \nu]^2]$$
$$- 4\lambda_2^2[\kappa, \nu][\nu, [\nu, [\kappa, \nu]]] + 2\lambda_2^2[\nu, [\nu, [\kappa, \nu]^2]] - \lambda_2^2[\nu, [\kappa, \nu]]^2 - 3\lambda_1\lambda_2[\kappa, \nu][\nu, [\kappa, \nu]].$$

We see therefore that one can take $\alpha = 2\lambda_2^2(1 - [\kappa, \nu])[\nu, [\kappa, \nu]]$ *which does not belong to* $\mathcal{U}(\mathfrak{g})\mathfrak{a}_\tau$ *for* $\lambda_2 \neq 0$*, as was to be shown. The fact that* $[\kappa, \gamma] \equiv 0$ *modulo* $\mathcal{U}(\mathfrak{g})\mathfrak{a}_\tau$ *is immediate.*

We are now in position to assume that $[\nu, [\nu, \kappa]] \equiv 0$ *modulo* $\mathcal{U}(\mathfrak{g})\mathfrak{a}_\tau$*. Take back the flag* $\mathcal{S}$*, so that* $\mathfrak{g}_{n-1} = \mathfrak{h}_{n-1}$*. There is a unique index* $2 \leq r_0 \leq n-1$ *in* $S(e)$ *which is a non-jump index for the coadjoint action of* $G' = \exp(\mathfrak{g}')$*. Then, there exists an element* $\kappa_{r_0} = a_{r_0}X_{r_0} + b_{r_0}, a_{r_0}, b_{r_0} \in \mathcal{U}(\mathfrak{g}_{r_0-1})$ *which is e-central for* $\Gamma_\tau|_{\mathfrak{g}'}$ *with respect to the action of* G'*. It must be noticed that* a_{r_0} *is e-central for* Γ_τ*.*

We now prove the following:

Lemma 2.2.4 *We have that* $r_0 < k_i$ *and the element* $\kappa' = [\nu, \overline{\kappa_{r_0}}]\kappa - [\nu, \kappa]\overline{\kappa_{r_0}}$ *verifies* $[\kappa', \nu] \equiv 0$ *modulo* $\mathcal{U}(\mathfrak{g})\mathfrak{a}_\tau$*.*

Proof 2.11 *We show that* $r_0 < k_i$ *by induction on* $\dim \mathfrak{g}$*. Let* $\mathfrak{z}$ *be the centre of* $\mathfrak{g}$ *and suppose* $\dim \mathfrak{z} = 1$*. Namely,* $\mathfrak{g}_1 = \mathfrak{z}$ *at the flag* $\mathcal{S}$*. Let* $\mathfrak{z} = \mathbb{R}Z$*. Let*

$$\Gamma'_\tau = \{\ell \in \Gamma_\tau : \ell(Z) \neq 0\}.$$

If $\Gamma'_\tau = \emptyset$*, we can pass to the quotient* $\mathfrak{g}/\mathfrak{z}$*. Otherwise, let* $\mathfrak{g}_2 = \mathbb{R}Y + \mathfrak{z}$ *and* $X \in \mathfrak{g}$ *such that* $[X, Y] = Z$*. Denote by* $\widetilde{\mathfrak{g}}$ *the centraliser of* Y *in* $\mathfrak{g}$ *so that* $\mathfrak{g} = \mathbb{R}X + \widetilde{\mathfrak{g}}$*. When* $\mathfrak{h} \not\subset \widetilde{\mathfrak{g}}$*, put* $\widetilde{\mathfrak{h}} = \mathfrak{h} \cap \widetilde{\mathfrak{g}}$ *and identify* Γ_τ *to the affine subspace* $\widetilde{\Gamma} = f|_{\widetilde{\mathfrak{h}}} + (\widetilde{\mathfrak{h}})^\perp$ *of* $(\widetilde{\mathfrak{g}})^*$*. Then, we can apply there the induction hypothesis.*

When $\mathfrak{h} \subset \widetilde{\mathfrak{g}}$*, we take* $\kappa_{r_0} = Y$*. This means that* $r_0 = 2 < k_i$*. Suppose next* $\dim \mathfrak{z} \geq 2$*. Then, for generic* $\ell \in \Gamma_\tau$*, we can pass to the quotient* $\mathfrak{g}/(\mathfrak{z} \cap \ker \ell)$ *to which apply the induction hypothesis. Now since* $[\nu, [\nu, \overline{\kappa_{r_0}}]] \equiv 0$ *modulo* $\mathcal{U}(\mathfrak{g})\mathfrak{a}_\tau$*, we find* $[\kappa', \nu] \equiv 0$ *modulo* $\mathcal{U}(\mathfrak{g})\mathfrak{a}_\tau$*, as was to be shown.*

Up to this step, Lemma 2.2.4 says that we obtain a new κ_{k_i} *such that* $\overline{\kappa_{k_i}}$ *commutes with* $\nu = aX_n + b \in \mathcal{U}(\mathfrak{g}, \tau), a \not\equiv 0$ *modulo* $\mathcal{U}(\mathfrak{g})\mathfrak{a}_\tau$*. This means that* $\overline{\kappa_{k_i}} \in C_\tau$ *as in Lemma 2.2.2.*

It remains now to show that the family $\{\overline{\kappa_{k_j}}\}_{j \in \mathcal{J}_0}$ *rationally generates* C_τ*. Let then* $W \in C_\tau$*, the induction hypothesis applied to* τ' *says that there exist two polynomials* Q_1, Q_2 *of* $\{\overline{\sigma_j}(1 \leq j \leq m), \overline{\kappa_{k_i}}(1 \leq i \leq s), \overline{\kappa_{r_0}}\}$ *verifying* $Q_1W \equiv Q_2$ *modulo* $\mathcal{U}(\mathfrak{g})\mathfrak{a}_\tau$*. Supposing that* $\overline{\kappa_{r_0}}$ *appears effectively in this relation, we rewrite it as*

$$\sum_{j=0}^{r} q_j \overline{\kappa_{r_0}}^j \equiv 0,$$

where $q_j (0 \leq j \leq r)$ *are polynomials of* $\{\overline{\sigma_j}(1 \leq j \leq m), \overline{\kappa_{k_i}}(1 \leq i \leq s), W\}$ *and suppose that* $r > 0$ *is minimal. Applying* ν *to this relation, we have*

$$(\sum_{j=1}^{r-1} j q_j \overline{\kappa_{r_0}}^{j-1})[\nu, \overline{\kappa_{r_0}}] \equiv 0,$$

which is absurd because $[\nu, \overline{\kappa_{r_0}}] \not\equiv 0$*.*

Now, we go to investigate the situation of $(S(\mathfrak{g})/S(\mathfrak{g})\overline{\mathfrak{a}_\tau})^H$ in Cases 2.2.1.1 and 2.2.1.2, and we draw the same path as above. The induction hypothesis says that $\overline{\varphi_k}, 1 \leq k \leq m$, $v_{k_i} = P_{\overline{\kappa_{k_i}}}, 1 \leq i \leq s$ and $v_{r_0} = P_{\overline{\kappa_{r_0}}}$ provide us a system of rational generators of $Z_{\tau'}$. Let $\zeta = aX_n + b(a, b \in S(\mathfrak{g}'))$ be an element of $S(\mathfrak{g})$ such that $a \notin S(\mathfrak{g})\overline{\mathfrak{a}_\tau}$ and that $[\mathfrak{h}, \zeta] \subset S(\mathfrak{g})\overline{\mathfrak{a}_\tau}$. For $1 \leq i \leq s$, we go to modify v_{k_i} as before if $\{\zeta, v_{k_i}\} \notin S(\mathfrak{g})\overline{\mathfrak{a}_\tau}$. In this case, we go to utilise once again the method used in [15]. Repeating to take the Poisson products with elements of $(S(\mathfrak{g}')/S(\mathfrak{g}')\overline{\mathfrak{a}_\tau})^H$, we can suppose that $a \in Z_{\tau'}$. Let $r \in \mathbb{N}$ be a sufficiently large integer and $F(t)$ a polynomial of one variable having the degree $2r$. Put $\psi = v_{k_i} + F(v_{r_0}), \psi_0 = \psi, \psi_1 = \{\zeta, \psi\}, \psi_2 = \{\zeta, \{\zeta, \psi\}\}$ and so on. Then, one finds $\psi_{2r} \not\equiv 0$ and $\psi_{2r+1} \equiv 0$. Here, all the $\psi_j (0 \leq j \leq 2r)$ commute with elements of $(S(\mathfrak{g}')/S(\mathfrak{g}')\overline{\mathfrak{a}_\tau})^H$ and

$$\begin{aligned} v'_{k_i} &= (\psi_0\psi_{2r} + \psi_{2r}\psi_0) - (\psi_1\psi_{2r-1} + \psi_{2r-1}\psi_1) + \cdots \\ &\quad + (-1)^{r-3}(\psi_{r-2}\psi_{r+2} + \psi_{r+2}\psi_{r-2}) \\ &\quad + (-1)^{r-2}(\psi_{r-1}\psi_{r+1} + \psi_{r+1}\psi_{r-1}) + (-1)^{r-1}\psi_r^2 \end{aligned}$$

satisfies $\{\zeta, v'_{k_i}\} \equiv 0$. This being done, a convenient choice of polynomial $F(t)$ gives us the desired new element v'_{k_i} of Z_τ.

2.2.1.3 *Case 3: Non-saturation of G-orbits*

Suppose now that the generic G-orbits which encounter Γ_τ are non-saturated with respect to $\mathfrak{g}'$. Namely, $n = j_m \in T(e)$. Let

$$W = \sum_{k=0}^{r} w_k X_n^k \; (w_k \in \mathcal{U}(\mathfrak{g}'), 0 \leq k \leq r)$$

with $w_r \notin \mathcal{U}(\mathfrak{g})\mathfrak{a}_\tau$ be an element of C_τ. This brings immediately that $w_r \in C_\tau$ and hence that $\widetilde{W} = \overline{a_m}^r W - w_r\overline{\sigma_m}^r \in C_\tau$. Because the degree of $\widetilde{W}$ relative to X_n is smaller than or equal to $r-1$, the induction hypothesis on r assures that $\widetilde{W}$ is written as a rational function of $\{\overline{\sigma_j}; 1 \leq j \leq m\}$ and $\{\overline{\kappa_{k_1}}, \ldots, \overline{\kappa_{k_s}}\}$. Hence the same for W.

Concerning Z_τ we can argue similarly. Let

$$\zeta(\ell) = \sum_{k=0}^{r} \zeta_k(\ell')\ell_n^k \; (\zeta_k \in \mathbb{C}[\Gamma_{\tau'}], 0 \leq k \leq r)$$

with $\zeta_r \neq 0$ be an element of Z_τ. This brings immediately that $\zeta_r \in Z_\tau$ and hence that $\widetilde{\zeta} = \overline{p_m}^r\zeta - \zeta_r\overline{\varphi_m}^r \in Z_\tau$. Because the degree of $\widetilde{\zeta}$ relative to ℓ_n is smaller than or equal to $r-1$, the induction hypothesis on r assures that $\widetilde{\zeta}$ is written as a rational function of $\{\overline{\varphi_j}; 1 \leq j \leq m\}$ and $\{P_{\overline{\kappa_{k_1}}}, \ldots, P_{\overline{\kappa_{k_s}}}\}$. Hence the same for ζ and this closes the proof of the theorem.

2.2.2 Case of finite multiplicities

Suppose that G is nilpotent and $\tau = \mathrm{ind}_H^G \chi_f$ has finite multiplicities. We know that the algebra $D_\tau(G/H)$ is commutative, i.e. $C_\tau = D_\tau(G/H)$, and isomorphic to the algebra $\mathbb{C}[\Gamma_\tau]^H$ of the H-invariant polynomial functions on Γ_τ. So, Duflo's problem suggests that the algebra $\mathbb{C}[\Gamma_\tau]^H$ is Poisson commutative.

Theorem 2.2.5 *Keep the setting where τ is of finite multiplicities. Then, the algebra $\mathbb{C}[\Gamma_\tau]^H$ is Poisson commutative. Thus, the algebras C_τ and Z_τ are isomorphic.*

Proof 2.12 *As usual we proceed by induction on $\dim G$. If G is commutative or if $G = H$, there is nothing to do. Suppose $G \neq H$ and put $\tilde{\mathfrak{g}}$ a codimension one ideal of $\mathfrak{g}$ containing $\mathfrak{h}$. Let*

$$\widetilde{G} = \exp\tilde{\mathfrak{g}}, \tilde{\tau} = ind_H^{\widetilde{G}}\chi_f \text{ and } \Gamma_{\tilde{\tau}} = \Gamma_\tau|_{\tilde{\mathfrak{g}}}.$$

If $\mathbb{C}[\Gamma_\tau]^H \subset \mathbb{C}[\Gamma_{\tilde{\tau}}]^H$, our claim follows from the induction hypothesis. Otherwise, $\mathcal{U}(\mathfrak{g},\tau) \not\subset \mathcal{U}(\tilde{\mathfrak{g}}) + \mathcal{U}(\mathfrak{g})\mathfrak{a}_\tau$ and the generic G-orbits which encounter Γ_τ are non-saturated with respect to $\tilde{\mathfrak{g}}$. Consequently, there is an e-central element

$$\sigma = aX_n + b, \; a, b \in \mathcal{U}(\tilde{\mathfrak{g}}) \tag{2.2.5}$$

where a is also e-central. Let

$$\varphi(\ell) = p(\ell)\ell_n + q(\ell), \; \ell_n = \ell(X_n), \ell \in \Gamma_\tau, \tag{2.2.6}$$

be the Corwin-Greenleaf function corresponding to σ. Here, $p(\ell), q(\ell)$ depend only on $\ell|_{\tilde{\mathfrak{g}}}$ and $p(\ell)$ is the Corwin-Greenleaf function corresponding to the e-central element a. Note here the following:

Lemma 2.2.6 *The functions $\varphi(\ell), p(\ell)$ belong to Poisson centre of $\mathbb{C}[\Gamma_\tau]^H$.*

Proof 2.13 *Recall the G-invariant layer U of $\mathfrak{g}^*$ constructed from the flag S of ideals of $\mathfrak{g}$. This leads to the fact that $\varphi(\ell), p(\ell)$ are G-invariant functions, and hence they satisfy as elements of $S(\mathfrak{g})$ that*

$$\{X, \varphi\} = \{X, p\} = 0, \tag{2.2.7}$$

for any $X \in \mathfrak{g}$. Now clearly any element of $(S(\mathfrak{g})/S(\mathfrak{g})\bar{\mathfrak{a}}_\tau)^H$ can written as $\sum_\alpha a_\alpha Y^\alpha$ through a well-chosen basis Y and (2.2.7) allows to conclude.

Take an arbitrary element $\psi(\ell)$ of $\mathbb{C}[\Gamma_\tau]^H$ and write

$$\psi(\ell) = \sum_{j=0}^{r} \psi_j(\ell)\ell_n^j,$$

where $\psi_j(\ell)(0 \leq j \leq r)$ depend only upon $\ell|_{\tilde{\mathfrak{g}}}$. In order to see that ψ is in the Poisson centre of $\mathbb{C}[\Gamma_\tau]^H$, we employ a new induction on r. If $r = 0$, ψ is in the Poisson centre of $\mathbb{C}[\Gamma_{\tilde{\tau}}]^H$. Let $\zeta(\ell) = \sum_{k=0}^m \zeta_k(\ell)\ell_n^k$ be an arbitrary element of $\mathbb{C}[\Gamma_\tau]^H$, where $\zeta_k(\ell)(0 \leq k \leq m)$ depend only on $\ell|_{\tilde{\mathfrak{g}}}$. We must notice that $\zeta_m(\ell) \in \mathbb{C}[\Gamma_{\tilde{\tau}}]^H$ and we do assume $m > 0$. As

$$\zeta_m(\ell)\varphi(\ell)^m - p(\ell)^m\zeta(\ell)$$

is of degree smaller than or equal to $m-1$ *relative to* ℓ_n*, it commutes, by the induction hypothesis on* m*, with* $\psi(\ell)$ *for the Poisson structure of* $\mathbb{C}[\Gamma_\tau]^H$*. Since* $\varphi(\ell), p(\ell)$ *belong to the Poisson centre of* $\mathbb{C}[\Gamma_\tau]^H$ *by Lemma 2.2.6,* $\psi(\ell)$ *commutes with* $\zeta(\ell)$ *and* $\psi(\ell)$ *finds itself in the Poisson centre of* $\mathbb{C}[\Gamma_\tau]^H$*.*

If $r > 0$*, we can repeat the same argument as precedent. Namely,* $\psi_r(\ell)\varphi(\ell)^r - p(\ell)^r\psi(\ell)$ *and hence* $\psi(\ell)$ *too belong to the Poisson centre of* $\mathbb{C}[\Gamma_\tau]^H$*. In sum,* $\mathbb{C}[\Gamma_\tau]^H$ *is Poisson commutative.*

2.2.3 Proof of Duflo's polynomial Conjecture for nilpotent inductions

We now leave the hypothesis of finiteness of the multiplicities of τ. When $H = G$, there is nothing to do. Suppose hereafter that $H \neq G$. We are ready to prove the main result of this section.

Theorem 2.2.7 *Let* $W \in C_\tau$ *and let for generic* $\ell \in \Gamma_\tau$*,* a_ℓ *be the Penney distribution as in (2.1.4). Then the identity (2.1.6) holds with some eigenfunction* $P_W \in Z_\tau$*. Further, the mapping* $C_\tau \ni W \mapsto P_W(\ell) \in Z_\tau$ *is injective and extends to an isomorphism of the abelian algebras* C_τ *and* Z_τ*.*

Proof 2.14 *We proceed through different steps, here is the first:*

Step 1*: We first show that for* $W \in C_\tau$*, (2.1.6) holds for some element* $P_W \in Z_\tau$*. We proceed by induction on* $\dim\mathfrak{g} + \dim\mathfrak{h}$*. If* $\dim\mathfrak{h} = 0$*, then* $\mathcal{U}(\mathfrak{g},\tau) = \mathcal{U}(\mathfrak{g})$*,* C_τ *is nothing but the centre* $\mathcal{Z}$ *of* $\mathcal{U}(\mathfrak{g})$ *and* a_ℓ *is the Dirac measure. Let* $\beta : S(\mathfrak{g}) \to \mathcal{U}(\mathfrak{g})$ *be the symmetrisation map. When* $W \in \mathcal{Z}$*, the result of Kirillov says that*

$$P_W(\ell) = \overline{\beta^{-1}(\bar{W})(i\ell)}$$

for generic $\ell \in \Gamma_\tau = \mathfrak{g}^*$*. Hence, our assertion is well known as Duflo's isomorphism.*

Let now $\mathfrak{g}'$ *be an ideal of codimension 1 of* $\mathfrak{g}$ *containing* $\mathfrak{h}$*. Taking the flag* $\mathcal{S}$ *passing through* $\mathfrak{g}'$*, we can suppose* $\mathfrak{g}' = \mathfrak{h}_{n-1}$*. Let* $G' = \exp(\mathfrak{g}')$ *and* $\tau' = ind_H^{G'}\chi$*. Assuming that the generic* G*-orbits which encounter* Γ_τ *are saturated with respect to* $\mathfrak{g}'$ *(which is the situation in the two coming cases), we have that* $C_\tau \subset C_{\tau'}$ *and* $Z_\tau \subset Z_{\tau'}$*. Then by the induction hypothesis on* $\dim\mathfrak{g}$*,*

$$\overline{W}\cdot a_\ell = \overline{W}\cdot a_{\ell'} = \overline{P'_W(\ell')}a_{\ell'} = \overline{P_W(\ell)}a_\ell,$$

for any $W \in C_\tau$*, with a polynomial function* $P_W \in Z_{\tau'}$*. Here, the notation prime signifies its corresponding object concerning* τ'*. We shall treat the following cases.*

Case 1*:*

Suppose first that the generic G*-orbits which encounter* Γ_τ *are saturated with respect to* $\mathfrak{g}'$*. Then there is a unique index* $j_0 \in S(e)$ *which is a non-jump index for* $\Gamma_{\tau'}$ *but not so for* Γ_τ*. Let*

$$\sigma'_{j_0} = a'X_{j_0} + b', a', b' \in \mathcal{U}(\mathfrak{g}_{j_0-1}),$$

be a corresponding e-central element for $\Gamma_{\tau'}$ and let

$$\varphi'_{j_0}(\ell) = p'(\ell')\ell_{j_0} + q'(\ell'), \ell' = \ell|_{\mathfrak{g}'},$$

be the associated Corwin-Greenleaf function. It is evident that $\overline{\sigma'_{j_0}} \in \mathcal{U}(\mathfrak{g}, \tau)$ *and* $\overline{\varphi'_{j_0}} \in \mathbb{C}[\Gamma_\tau]^H$. *Remark that* σ'_{j_0} *and* φ'_{j_0} *are both invariant by* $\mathfrak{g}'$ *but not by* $\mathfrak{g}$. *This signifies that* $C_\tau \subset \mathcal{U}(\mathfrak{g}')$ *and* $Z_\tau \subset S(\mathfrak{g}')$.

We also assume in this case that the generic H*-orbits in* Γ_τ *are saturated with respect to* $\mathfrak{g}'$, *then* $\mathcal{U}(\mathfrak{g}, \tau) \subset \mathcal{U}(\mathfrak{g}')$ *modulo the kernel* $\mathcal{U}(\mathfrak{g})\mathfrak{a}_\tau$ *and* $\mathbb{C}[\Gamma_\tau]^H \subset S(\mathfrak{g}')$. *Therefore, it is clear that* $C_\tau = C_{\tau'}$ *and* $Z_\tau = Z_{\tau'}$. *So, we are done in this situation.*

Case 2:

We assume here that the generic H*-orbits in* Γ_τ *are non-saturated with respect to* $\mathfrak{g}'$, *but still the generic* G*-orbits are. Let* $d = \dim \mathfrak{h} \geq 1$ *and put this time* $\widetilde{\tau} = ind_{\widetilde{H}}^{G}(\chi|_{\widetilde{H}})$, *where* $\widetilde{H} = \exp(\widetilde{\mathfrak{h}})$ *with* $\widetilde{\mathfrak{h}} = \mathfrak{h}_{d-1}$. *Denote by* $\widetilde{a}_\ell$ *the Penney's distribution concerning* $\widetilde{\mathfrak{h}}$ *and so on. When* $W \in C_{\widetilde{\tau}}$, *we have some* $\widetilde{P}_W \in Z_{\widetilde{\tau}}$ *by the induction hypothesis, but* $Z_{\widetilde{\tau}} \subset Z_\tau$ *on the other hand and this closes this case.*

Assume otherwise that $W \notin C_{\widetilde{\tau}}$. *In order to show that* $P_W \in Z_\tau$, *it suffices to construct a* $\mathfrak{h}$*-invariant element* $\nu = aX_n + b$ $(a, b \in S(\mathfrak{g}'))$, *such that* $a \not\equiv 0$ *modulo* $S(\mathfrak{g}')\overline{\mathfrak{a}_\tau}$ *and* $\{P_W, \nu\} \equiv 0$ *modulo* $S(\mathfrak{g}')\overline{\mathfrak{a}_\tau}$. *This reduces to argue similarly as in Lemma 2.2.2.*

Towards such a purpose, let us first construct in $\mathbb{C}[\Gamma_\tau]^{\widetilde{H}}$ *two* P_W*-invariant elements*

$$\tau_1 = \alpha_1 X_n + \beta_1, \ \tau_2 = \alpha_2 X_n + \beta_2,$$

where $\alpha_j, \beta_j (1 \leq j \leq 2)$ *are elements of* $S(\mathfrak{g}')$, *in the sense that* $\{P_W, \tau_j\} \equiv 0$ *modulo* $S(\mathfrak{g})\overline{\mathfrak{a}_\tau}$. *Let*

$$\nu_1 = a_1 X_n + b_1 (a_1, b_1 \in S(\mathfrak{g}'), a_1 \not\equiv 0 \text{ modulo } \mathrm{S}(\mathfrak{g})\overline{\mathfrak{a}_\tau})$$

be an element of $\mathbb{C}[\Gamma_\tau]^H$. *Taking the Poisson product with* P_W *several times if necessary, we can suppose that* $\{a_1, P_W\} \equiv 0$ *modulo* $S(\mathfrak{g})\overline{\mathfrak{a}_\tau}$. *Let* $\nu_2 = a_2 X_n + b_2 (a_2, b_2 \in S(\mathfrak{g}'))$ *be an element of* $\mathbb{C}[\Gamma_\tau]^{\widetilde{H}}$. *We suppose similarly that* $\{P_W, a_2\} \equiv 0$ *modulo* $S(\mathfrak{g})\overline{\mathfrak{a}_\tau}$. *Then modulo* $S(\mathfrak{g})\overline{\mathfrak{a}_\tau}$,

$$\begin{aligned} \{P_W, \nu_1\} &= a_1\{P_W, X_n\} + \{P_W, b_1\} \\ \{P_W, \nu_2\} &= a_2\{P_W, X_n\} + \{P_W, b_2\} \end{aligned}$$

Hence,

$$\{P_W, \nu_1\}a_2 - \{P_W, \nu_2\}a_1 = \{P_W, b_1\}a_2 - \{P_W, b_2\}a_1 \tag{2.2.8}$$

modulo $S(\mathfrak{g})\overline{\mathfrak{a}_\tau}$.

Let us pick now an element $\widetilde{\nu} \in S(\mathfrak{g}')^{\widetilde{H}}$, *which is not* $\mathfrak{h}$*-invariant and chosen at the lowest level in the sequence (2.2.2) of subalgebras* $\{\mathfrak{h}_j\}_{j=1}^n$ *of* $\mathfrak{g}$, *say* $\mathfrak{h}_{j_0}$. *Moreover, we suppose that* $\widetilde{\nu}$ *is written as*

$$\widetilde{\nu} = \widetilde{a} X_{j_0} + \widetilde{b} \tag{2.2.9}$$

with $\widetilde{a}, \widetilde{b} \in S(\mathfrak{h}_{j_0 - 1})$.

Now, put

$$\nu_2 = \nu_1 + \widetilde{\nu}.$$

Then we have,

$$a_2 = a_1, \; b_2 = b_1 + \widetilde{\nu}$$

and consequently equation (2.2.8) becomes

$$\{P_W, b_1\}a_1 - \{P_W, b_1 + \widetilde{\nu}\}a_1 = \{P_W, \widetilde{\nu}\}a_1 \not\equiv 0$$

modulo $S(\mathfrak{g})\overline{\mathfrak{a}_\tau}$. *Since* $\{P_W, \{P_W, \nu_j\}\} \equiv 0$ *modulo* $S(\mathfrak{g})\overline{\mathfrak{a}_\tau}$ $(1 \leq j \leq 2)$ *from the choice of* $\widetilde{\nu}$ *as in equation (2.2.9), we find a* $\widetilde{\mathfrak{h}}$*-invariant element*

$$\tau_1 = \{P_W, \nu_2\}\nu_1 - \{P_W, \nu_1\}\nu_2 = \{P_W, \widetilde{\nu}\}\nu_1 - \{P_W, \nu_1\}\widetilde{\nu} \tag{2.2.10}$$

which verifies

$$\{P_W, \tau_1\} \equiv 0 \text{ modulo } S(\mathfrak{g})\overline{\mathfrak{a}_\tau}. \tag{2.2.11}$$

In order to fabricate the second element τ_2*, we start from the element*

$$\nu_2' = \nu_1(1 + \widetilde{\nu}). \tag{2.2.12}$$

Then,

$$\{P_W, \nu_2'\} = \{P_W, \nu_1\}(1 + \widetilde{\nu}) + \nu_1\{P_W, \widetilde{\nu}\}, \; \{P_W, \{P_W, \nu_2'\}\} = 2\{P_W, \nu_1\}\{P_W, \widetilde{\nu}\}. \tag{2.2.13}$$

Further, this carries $\{P_W, \{P_W, \{P_W, \nu_2'\}\}\} \equiv 0$ *modulo* $S(\mathfrak{g})\overline{\mathfrak{a}_\tau}$. *Put finally*

$$\tau_2 = 4\{P_W, \nu_1\}\{P_W, \widetilde{\nu}\}\nu_2' - (\{P_W, \nu_1\}(1 + \widetilde{\nu}) + \nu_1\{P_W, \widetilde{\nu}\})^2. \tag{2.2.14}$$

We then see that

$$\{P_W, \tau_2\} \equiv 0 \text{ modulo } S(\mathfrak{g})\overline{\mathfrak{a}_\tau} \tag{2.2.15}$$

by a simple calculation making use of (2.2.12) and (2.2.13).

To close the proof in this case, we now show the:

Lemma 2.2.8 *There exists a* $\mathfrak{h}$*-invariant element* $\nu = aX_n + b$ $(a, b \in S(\mathfrak{g}'))$*, such that* $a \not\equiv 0$ *modulo* $S(\mathfrak{g}')\overline{\mathfrak{a}_\tau}$ *and* $\{P_W, \nu\} \equiv 0$ *modulo* $S(\mathfrak{g}')\overline{\mathfrak{a}_\tau}$.

Proof 2.15 *According to the choice of* $\tilde{\nu}$ *as in (2.2.9) and* τ_1 *as in (2.2.10), remark first*

$$\{X_d, \{X_d, \tau_1\}\} \equiv 0 \tag{2.2.16}$$

modulo $S(\mathfrak{g}')\overline{\mathfrak{a}_\tau}$ *as* $\{X_d, \tau_1\} \in S(\mathfrak{h}_{j_0-1})$. *This being so, let*

$$F(t) = \lambda_0 + \lambda_1 t + \lambda_2 t^2 \; (\lambda_j \in \mathbb{C}, 0 \leq j \leq 2).$$

and put $\psi = \tau_2 + F(\tau_1)$. *As* $\{P_W, \tau_j\} \equiv 0$ $(1 \leq j \leq 2)$ *by (2.2.11) and (2.2.15), we see that* $\{P_W, \psi\} \equiv 0$. *Moreover, it turns out that* $\{X_d, \psi\}$ *is* P_W*-invariant and further*

$$\psi_k = \{X_d, \{X_d, \{\cdots, \psi\}\cdots\}\} = (ad(X_d))^k(\psi) \; (k \in \mathbb{N})$$

are also P_W-invariant.
Then modulo $S(\mathfrak{g}')\overline{\mathfrak{a}_\tau}$, we get thanks to equation (2.2.16),

$$\begin{aligned}\psi_0 &= \tau_2 + \lambda_0 + \lambda_1\tau_1 + \lambda_2\tau_1^2,\\ \psi_1 &= \{X_d, \tau_2\} + \lambda_1\{X_d, \tau_1\} + 2\lambda_2\{X_d, \tau_1\}\tau_1\\ \psi_2 &= \{X_d, \{X_d, \tau_2\}\} + 2\lambda_2\{X_d, \tau_1\}^2.\end{aligned}$$

Since $\psi_3 \equiv 0$ modulo $S(\mathfrak{g})\overline{\mathfrak{a}_\tau}$, we find that $\tau = 2\psi_0\psi_2 - \psi_1^2$ is $\mathfrak{h}$-invariant.
On the other hand,

$$\begin{aligned}\{X_d, \tau_2\} &= 4\{P_W, \nu_1\}\{P_W, \tilde{\nu}\}\nu_1\{X_d, \tilde{\nu}\}\\ &- 2\{P_W, \nu_1\}\{X_d, \tilde{\nu}\}\Big(\{P_W, \nu_1\}(1+\tilde{\nu}) + \nu_1\{P_W, \tilde{\nu}\}\Big)\\ &= 2\{P_W, \nu_1\}\{P_W, \tilde{\nu}\}\{X_d, \tilde{\nu}\}\nu_1 - 2\{P_W, \nu_1\}^2\{X_d, \tilde{\nu}\}(1+\tilde{\nu})\\ &= 2\{P_W, \nu_1\}\{X_d, \tilde{\nu}\}(\{P_W, \tilde{\nu}\}\nu_1 - \{P_W, \nu_1\}(1+\tilde{\nu})).\end{aligned}$$

Hence

$$\{X_d, \{X_d, \tau_2\}\} = -2\{P_W, \nu_1\}^2\{X_d, \tilde{\nu}\}^2. \tag{2.2.17}$$

This being so, we get from (2.2.17) that

$$\psi_2 = -2\{P_W, \nu_1\}^2\{X_d, \tilde{\nu}\}^2 + 2\lambda_2\{X_d, \tau_1\}^2. \tag{2.2.18}$$

Let now $\eta = 2\psi_0\psi_2 - \psi_1^2$. Then by (2.2.18), one gets

$$\begin{aligned}\eta = &\left(\tau_2 + \lambda_0 + \lambda_1\tau_1 + \lambda_2\tau_1^2\right)\left(2\lambda_2\{X_d, \tau_1\}^2 - 2\{P_W, \nu_1\}^2\{X_d, \tilde{\nu}\}^2\right)\\ &- \left(\{X_d, \tau_2\} + \lambda_1\{X_d, \tau_1\} + 2\lambda_2\{X_d, \tau_1\}\tau_1\right)^2.\end{aligned}$$

So, this entails that modulo some terms of degree ≤ 1 with respect to ν_1, η reads:

$$\begin{aligned}\eta \equiv &-4\lambda_2\{P_W, \nu_1\}^2\{X_d, \tilde{\nu}\}^2{\tau_1}^2 - 4\{P_W, \nu_1\}^2\{P_W, \tilde{\nu}\}^2\{X_d, \tilde{\nu}\}^2{\nu_1}^2\\ &- 8\lambda_2\{P_W, \nu_1\}\{P_W, \tilde{\nu}\}\{X_d, \tilde{\nu}\}\{X_d, \tau_1\}\nu_1\tau_1.\end{aligned}$$

Finally, modulo some terms of degree ≤ 1 with respect to X_n, we have

$$\eta \equiv 4(\lambda_2 - 1)\{P_W, \nu_1\}^2\{P_W, \tilde{\nu}\}^2\{X_d, \tilde{\nu}\}^2{\nu_1}^2 \neq 0,$$

modulo $S(\mathfrak{g})\overline{\mathfrak{a}_\tau}$ for $\lambda_2 \neq 1$. Conclusively, we regain in this manner the element ν as in the statement of the Lemma.

Note first that if (2.1.6) holds, then the mapping $C_\tau \ni W \mapsto P_W(\ell)$ is injective by Proposition 2.1.2.

Case 3*: Suppose now that the generic G-orbits which encounter Γ_τ are non-saturated with respect to $\mathfrak{g}'$. Namely, $n = j_m \in T(e)$. We can treat this situation just as the case 2.2.1.3.*

Step 2: *This step consists in showing that the mapping $C_\tau \ni W \mapsto P_W(\ell)$ is injective. But, this derives directly from Proposition 2.1.2, where the arguments are based on the fact that (2.1.6) holds.*

Step 3: *Let us keep our situation and take inversely $\zeta \in Z_\tau$. It is clear that $\zeta \in S(\mathfrak{g}')^H$ modulo $S(\mathfrak{g})\overline{\mathfrak{a}_\tau}$. Hence $\zeta \in Z_{\tau'}$ and the induction hypothesis says that there exists $W \in C_{\tau'}$ such that $P_W = \zeta$. This being assumed, we can repeat the same argument developed above to assert $W \in C_\tau$. We content ourselves to sketch our route. Remark that everything is settled in Case 2.2.1. For the next Case 2.2.2, let $\nu_1 = a_1X_n + b_1(a_1, b_1 \in \mathcal{U}(\mathfrak{g}'), a_1 \not\equiv 0)$ be an element of $\mathcal{U}(\mathfrak{g}, \tau)$. Taking the brackets with W several times if necessary, we can assume $[a_1, W] \equiv 0$ modulo $\mathcal{U}(\mathfrak{g})\mathfrak{a}_\tau$. If $\zeta \in Z_{\tilde{\tau}}$, $W \in C_{\tilde{\tau}} \subset C_\tau$ by induction hypothesis. Suppose $W \notin C_{\tilde{\tau}}$.*

Everything as before, put

$$\nu_2 = \nu_1 + \widetilde{\nu},\ \nu_2' = \nu_1 + \widetilde{\nu}^2$$

with an element $\widetilde{\nu} \in \mathcal{U}(\mathfrak{g}, \tilde{\tau})$ which is not $\mathfrak{h}$-invariant and chosen at the lowest level in the sequence of subalgebras $\{\mathfrak{h}_j\}_{j=1}^n$ of $\mathfrak{g}$, say $\mathfrak{h}_{j_0}$. Furthermore, we suppose that $\widetilde{\nu}$ is written as $\widetilde{\nu} = \widetilde{a}X_{j_0} + \widetilde{b}$ with $\widetilde{a}, \widetilde{b} \in \mathcal{U}(\mathfrak{h}_{j_0-1})$. Next, we construct the $\widetilde{\mathfrak{h}}$-invariant elements τ_1, τ_2 and then τ which commute with W. Seeking a $\mathfrak{h}$-invariant element from τ, we regain ν_1 with its invariance by W. In other words, we prove $W \in C_\tau$.

Let us finally look at Case 3 above. Let $\zeta(\ell) = \sum_{k=0}^m \zeta_k(\ell)\ell_n^k$ be an arbitrary element of Z_τ, where $\zeta_k(\ell)(0 \le k \le m)$ depend only upon $\ell|_{\widetilde{\mathfrak{g}}}$. We must notice that $\zeta_m(\ell)$ belongs also to Z_τ and we do assume that $m > 0$. We employ an induction on the degree m of $\zeta(\ell)$. As

$$\tilde{\zeta}(\ell) = \zeta_m(\ell)\varphi(\ell)^m - p(\ell)^m\zeta(\ell)$$

is of degree smaller than or equal to $m-1$ relative to ℓ_n, $\tilde{\zeta} = P_{\widetilde{W}}$ for some $\widetilde{W} \in C_\tau$ by the induction hypothesis. Here, we take $\varphi(\ell)$, $p(\ell)$ as in (2.2.6). Let now σ be as in (2.2.5), then clearly

$$P_{a^m}(\ell)\zeta(\ell) = P_{W_0}(\ell)P_{\sigma^m}(\ell) - P_{\widetilde{W}}(\ell),$$

for some $W_0 \in C_\tau$. Hence Proposition 2.1.14 affirms that $\zeta = P_Q$ for some $Q \in \mathcal{U}(\mathfrak{g}, \tau)$ such that $a^mQ = W_0\sigma^m - \widetilde{W} \in C_\tau$ and then finally $Q \in C_\tau$. This closes the proof of the theorem.

2.3 THE POLYNOMIAL CONJECTURE FOR NILPOTENT RESTRICTIONS

Let G be a connected and simply connected nilpotent Lie group, K an analytic subgroup of G and π an irreducible unitary representation of G whose coadjoint orbit of G is denoted by $\Omega(\pi)$. Let $\mathcal{U}(\mathfrak{g})$ be the enveloping algebra of $\mathfrak{g}_{\mathbb{C}}$, $\mathfrak{g}$ designating the Lie algebra of G. We consider the algebra $D_\pi(G)^K \simeq (\mathcal{U}(\mathfrak{g})/\ker\pi)^K$ of the K-invariant elements of $\mathcal{U}(\mathfrak{g})/\ker\pi$. It turns out that this algebra is commutative if and only if the restriction $\pi|_K$ of π to K has finite multiplicities (cf. [10]). In this section, we suppose this eventuality and we provide a proof of the polynomial conjecture asserting that $D_\pi(G)^K$ is isomorphic to the algebra $\mathbb{C}[\Omega(\pi)]^K$ of K-invariant polynomial functions on $\Omega(\pi)$.

2.3.1 Differential operators and multiplicities

We know in the nilpotent case the branching laws of induced and restricted representations. Let $p : \mathfrak{g}^* \to \mathfrak{k}^*$ be the restriction mapping. For $\pi \in \widehat{G}$, we consider a finite measure μ_π on $\mathfrak{g}^*$ equivalent to the canonical measure on the orbit $\Omega_G(\pi)$ which is regarded as a measure on $\mathfrak{g}^*$. Put $\nu_\pi = (\theta_K \circ p)_*(\mu_\pi)$. The restriction $\pi|_K$ of π to K is disintegrated as:

$$\pi|_K \simeq \int_{\hat{K}}^{\oplus} m_\sigma^\pi \sigma d\nu_\pi(\sigma)$$

where the multiplicities m_σ^π is obtained as the number of the K-orbits contained in $\Omega_G(\pi) \cap p^{-1}(\Omega_K(\sigma))$ as in Theorem 1.5.10.

We also record here that the multiplicities are either uniformly bounded almost everywhere or equal to the infinity almost everywhere. According to these two eventualities, we say that the representation $\pi|_K$ has either finite or infinite multiplicities.

With the same notations and contexts, we say that $\mathfrak{k}$ is co-isotropic at $\ell \in \mathfrak{g}^*$ if $\mathfrak{k}^\ell$ is isotropic for B_ℓ. This notion allows us to characterise the finite multiplicities of $\pi|_K$.

Proposition 2.3.1 ([10]) *$\pi|_K$ has finite multiplicities if and only if $\mathfrak{k}$ is co-isotropic generically on $\Omega = \Omega(\pi)$.*

Let us define our algebra $D_\pi(G)^K$ as follows. Let $\ker \pi$ be the primitive ideal of $\mathcal{U}(\mathfrak{g})$ associated to π. Set

$$\mathcal{U}_\pi(\mathfrak{g})^{\mathfrak{k}} = \{A \in \mathcal{U}(\mathfrak{g}); [A, \mathfrak{k}] \subset \ker \pi\}$$

Then, $D_\pi(G)^K$ is the image of $\mathcal{U}_\pi(\mathfrak{g})^{\mathfrak{k}}$ by π so that

$$D_\pi(G)^K \cong \mathcal{U}_\pi(\mathfrak{g})^{\mathfrak{k}} / \ker \pi \cong (\mathcal{U}(\mathfrak{g}) / \ker \pi)^K,$$

where the last member is the algebra of the K-invariant elements of $\mathcal{U}(\mathfrak{g}) / \ker \pi$.

Let

$$\{0\} = \mathfrak{g}_0 \subset \mathfrak{g}_1 \subset \cdots \subset \mathfrak{g}_{n-1} \subset \mathfrak{g}_n = \mathfrak{g} \tag{2.3.1}$$

be a Jordan-Hölder sequence of $\mathfrak{g}$, i.e. an increasing sequence of ideals of $\mathfrak{g}$ such that $\dim(\mathfrak{g}_j) = j, j = 0, ..., n$. Let $\{Y_1, ..., Y_n\}$ be a Jordan-Hölder basis of $\mathfrak{g}$, associated to this Jordan-Hölder sequence, and $\{Y_1^*, ..., Y_n^*\}$ the basis of $\mathfrak{g}^*$ such that $Y_i^*(Y_j) = \delta_{i,j}, 1 \le i, j \le n$. The projections $p_j : \mathfrak{g}^* \to \mathfrak{g}_j^*, 1 \le j \le n$, intertwine G-actions, and for $\ell \in \mathfrak{g}^*$ we consider the dimension index

$$e(\ell) = (e_1(\ell), \ldots, e_n(\ell))$$

defined by $e_j(\ell) = \dim(G{\cdot}p_j(\ell)), 1 \leq j \leq n$. Put $\mathcal{E} = \{e(\ell); \ell \in \mathfrak{g}^*\}$. Inversely, each ordered set $e = (e_1, \ldots, e_n)$ of non-negative n-integers provides us a layer U_e of G-orbits in $\mathfrak{g}^*$ defined by

$$U_e = \{\ell \in \mathfrak{g}^*; e_j(\ell) = e_j,\ 1 \leq j \leq n\}.$$

For each $e \in \mathcal{E}$, we consider the set of jump indices

$$S(e) = \{1 \leq j \leq n; e_{j-1} \neq e_j\}$$

and that of non-jump indices

$$T(e) = \{1 \leq j \leq n; e_{j-1} = e_j\}$$

with $e_0 = 0$. Then we have $\mathfrak{g}^* = \mathfrak{g}^*_S \oplus \mathfrak{g}^*_T$, where $\mathfrak{g}^*_S = \mathbb{R} - \text{span}\{Y_j; j \in S(e)\}$ and $\mathfrak{g}^*_T = \mathbb{R} - \text{span}\{Y_j; j \in T(e)\}$. There exists an order among the elements of $\mathcal{E} = \{e^{(1)} > \cdots > e^{(k)}\}$ in such a manner that $U_{e^{(1)}}$ and $\cup_{j \leq i} U_{e^{(j)}}$ are Zariski open sets of $\mathfrak{g}^*$ for every i. In this way, all the layers U_e are semi-algebraic set, i.e. difference of two Zariski open sets of $\mathfrak{g}^*$. Let U_e be an arbitrary layer, we write $S(e) = \{j_1 < \cdots < j_r\}$, where r designates the dimension of the G-orbits in U_e. Then there exist some functions $R^e_j : U_e \times \mathbb{R}^r \to \mathbb{R},\ j = 1, ..., n$ such that:

(a) For $f \in U_e$ fixed, $x = (x_1, ..., x_r) \mapsto R^e_j(f, x) : \mathbb{R}^r \to \mathbb{R}$ is a polynomial function in x and the coefficients are G-invariant functions on U_e;

(b) $R^e_j(f, x) = x_k$ for $j = j_k \in S(e),\ f \in U_e$;

(c) If $j_k \leq j < j_{k+1}$, then $R^e_j(f, x)$ depends only on $x_1, ..., x_k$;

(d) For any $f \in U_e$, the coadjoint orbit $G{\cdot}f$ is given by:

$$G{\cdot}f = \{\sum_{j=1}^{n} R^e_j(f, x) Y^*_j; x \in \mathbb{R}^r\},$$

(see [23]).

Let $\mathcal{I} = \{i_1 < i_2 < \cdots < i_d\}$ be the set of indices $i (1 \leq i \leq n)$ such that $\mathfrak{k} \cap \mathfrak{g}_{i-1} \neq \mathfrak{k} \cap \mathfrak{g}_i$, and put

$$\mathcal{J} = \{j_1 < j_2 < \cdots < j_q\} = \{1, 2, \ldots, n\} \setminus \mathcal{I}\ (q = \dim(\mathfrak{g}/\mathfrak{k})).$$

Set $\mathfrak{k}_s = \mathfrak{k} \cap \mathfrak{g}_{i_s}, 1 \leq s \leq d$, and $\mathfrak{k}_{d+r} = \mathfrak{k} + \mathfrak{g}_{j_r}, 1 \leq r \leq q$. In this manner, we get a sequence of subalgebras:

$$\{0\} = \mathfrak{k}_0 \subset \mathfrak{k}_1 \subset \cdots \subset \mathfrak{k}_d = \mathfrak{k} \subset \mathfrak{k}_{d+1} \subset \cdots \subset \mathfrak{k}_{n-1} \subset \mathfrak{k}_n = \mathfrak{g} \tag{2.3.2}$$

with $\dim(\mathfrak{k}_j) = j (0 \leq j \leq n)$. This is a Jordan-Hölder sequence for the K-action. We extract a non-zero vector $X_j \in \mathfrak{k}_j \setminus \mathfrak{k}_{j-1}$ for $1 \leq j \leq n$. Then, the dual basis $\{X^*_1, \ldots, X^*_n\}$ is a Jordan-Hölder basis of $\mathfrak{g}^*$ for the unipotent action $\text{Ad}^*(K)$. The projections $p^{\mathfrak{k}}_j : \mathfrak{g}^* \to \mathfrak{k}^*_j, 1 \leq j \leq n$, intertwine K-actions, and for $\ell \in \mathfrak{g}^*$ we consider the dimension index

$$e_K(\ell) = (e^K_1(\ell), \ldots, e^K_n(\ell))$$

defined by $e_j^K(\ell) = \dim(K\cdot(p_j^{\mathfrak{k}}(\ell))), 1 \le j \le n$. Inversely, each ordered set $e_K = (e_1^K, \dots, e_n^K)$ of non-negative n-integers provides us a layer Ω_e of K-orbits in $\mathfrak{g}^*$ defined by

$$\Omega_e = \{\ell \in \mathfrak{g}^*; e_j^K(\ell) = e_j^K,\ 1 \le j \le n\}.$$

Let us return to our orbit Ω. There exists a unique layer $\Omega_e \subset \mathfrak{g}^*$ such that $\Omega \cap \Omega_e$ is a non-empty Zariski open set of Ω. We consider the set of jump indices

$$S_K(e_K) = \{1 \le j \le n; e_{j-1}^K \ne e_j^K\}$$

and that of non-jump indices

$$T_K(e_K) = \{1 \le j \le n; e_{j-1}^K = e_j^K\}$$

with $e_0 = 0$. Then we have $\mathfrak{g}^* = \mathfrak{g}^*_{S_K} \oplus \mathfrak{g}^*_{T_K}$.

2.3.1.1 *Coordinates on coadjoint orbits*

We take back the flag of ideals (2.3.1) of $\mathfrak{g}$ to parameterise the orbit $\Omega = \Omega(\pi)$ and denote there by S_Ω and T_Ω respectively the sets of jump and non-jump indices. Let $\{Y_1, \dots, Y_n\}$ be a Malcev basis adapted to the flag (2.3.1), $\ell_j = \ell(Y_j)$ $(1 \le j \le n)$ for $\ell \in \Omega$, $S_\Omega = \{s_1 < \cdots < s_r\}$, $r = \dim \Omega$ and $x_k = \ell_{s_k}$ for $1 \le k \le r$. We describe the orbit Ω by the polynomial relations

$$\ell_j = F_j(x_1, \dots, x_k),\ s_k < j < s_{k+1}, \tag{2.3.3}$$

where $x = (x_1, \dots, x_r)$ runs through $\mathbb{R}^r$. In these circumstances, the rational functions on Ω are nothing but the rational functions of the variables $(x_1, \dots, x_r)$.

2.3.2 The polynomial conjecture for restrictions

Let $r_j^e(f)$ be the image in $\mathcal{U}(\mathfrak{g})$ by the symmetrisation of the element

$$R_j^e(f, -iY_{j_1}, \dots, Y_{j_r})$$

in the symmetric algebra $S(\mathfrak{g})$ of $\mathfrak{g}_\mathbb{C}$, namely, we replace the variable x_k in $R_j^e(f, x)$ by $-iY_{j_k}$. Notice in particular that $r_{j_k}^e(f) = -iY_{j_k}$. Let V_e be the subspace of $S(\mathfrak{g})$ spanned by the elements of the form $Y_{j_1}^{\alpha_1} \cdots Y_{j_r}^{\alpha_r}$, $\alpha_1, \dots, \alpha_r \in \mathbb{N}$, and let F_e be the image in $\mathcal{U}(\mathfrak{g})$ of V_e by the symmetrisation. On the other hand, let E_e be the subspace of $\mathcal{U}(\mathfrak{g})$ spanned by the elements of the form $Y_{j_1}^{\alpha_1} \cdots Y_{j_r}^{\alpha_r} \alpha_1, \dots, \alpha_r \in \mathbb{N}$. If $S(e) = \emptyset$, we put $V_e = F_e = E_e = \mathbb{C}\cdot 1$. Pedersen proved that the primitive ideal $\ker \pi$, where $\pi \in \widehat{G}$ such that $f \in \Omega(\pi)$ is generated by the elements

$$u_j^e(f) = Y_j - ir_j^e(f),\ j \in T(e)$$

and that

$$\mathcal{U}(\mathfrak{g}) = \ker \pi \oplus E_e = \ker \pi \oplus F_e$$

(see Theorem 2.1.1 and Theorem 2.2.1 in [88]). In the same way, the actions of π on E_e and F_e are faithful (see Lemma 2.2.12 and Lemma 2.2.13 in [88]). In this way, identifying E_e and F_e à $\mathcal{U}(\mathfrak{g})/\ker\pi$ and abusing notations, we have

$$D_\pi(G)^K \simeq E_e^K \simeq F_e^K \simeq \mathbb{C}[Y_{j_1},\dots,Y_{j_r}]^K.$$

These isomorphisms are simply isomorphisms of vector spaces. Then we have:

Theorem 2.3.2 ([10]) *At the passage from $\mathfrak{k}_{j-1}$ to $\mathfrak{k}_j$, the algebra $\mathcal{U}_\pi(\mathfrak{g})^{\mathfrak{k}}$ extends when $j \in T_K(e^K)$, while it does not extend when $j \in S_K(e^K)$. More precisely, putting $\mathcal{U}_\pi(\mathfrak{k}_j)^{\mathfrak{k}} = \mathcal{U}_\pi(\mathfrak{g})^{\mathfrak{k}} \cap \mathcal{U}(\mathfrak{k}_j)$ for $1 \le j \le n$, we have:*
(1) If $j \in T_K(e^K)$, $\mathcal{U}_\pi(\mathfrak{k}_j)^{\mathfrak{k}} \neq \mathcal{U}_\pi(\mathfrak{k}_{j-1})^{\mathfrak{k}} + \mathcal{U}(\mathfrak{k}_j)(\mathcal{U}(\mathfrak{k}_{j-1}) \cap \ker\pi)$ and there exists $W \in \mathcal{U}_\pi(\mathfrak{k}_j)^{\mathfrak{k}}$ having the form $W = aX_j + b$ $(a, b \in \mathcal{U}(\mathfrak{k}_{j-1}))$ with $\pi(a) \neq 0$.
(2) If $j \in S_K(e^K)$, then $\mathcal{U}_\pi(\mathfrak{k}_j)^{\mathfrak{k}} = \mathcal{U}_\pi(\mathfrak{k}_{j-1})^{\mathfrak{k}} + \mathcal{U}(\mathfrak{k}_j)(\mathcal{U}(\mathfrak{k}_{j-1}) \cap \ker\pi)$.

Finally, we obtain:

Theorem 2.3.3 ([10]) *$\pi|_K$ has finite multiplicities if and only if the algebra $D_\pi(G)^K$ is commutative.*

We then substantiated in [10] the following conjecture:

Conjecture 2.3.1 *(Baklouti-Fujiwara) Let G be a connected and simply connected nilpotent Lie group, K an analytic subgroup of G. Let $\pi \in \widehat{G}$ be a unitary and irreducible representation of G such that $\pi|_K$ is of finite multiplicities. Then the algebra $D_\pi(G)^K$ is isomorphic to the algebra $\mathbb{C}[\Omega(\pi)]^K$ of the K-invariant polynomial functions on $\Omega(\pi)$.*

We positively proved Conjecture 2.3.1 in many settings, especially when K is a normal subgroup of G or where the orbit $\Omega(\pi)$ is flat in [5] and further, the case where K is abelian or where $\Omega(\pi)$ admits a normal polarising subgroup [15].

2.3.3 Penney distributions and rationality

We keep the notations introduced before. Let $G = \exp\mathfrak{g}$ be a nilpotent Lie group and $\Omega = \Omega(\pi)$ the coadjoint orbit of G associated to π. For $\ell \in \Omega$, we take the Vergne polarisation $\mathfrak{b}[\ell]$ of $\mathfrak{g}$ at ℓ constructed from the flag $\mathcal{S}$ and realise π as $\pi = \mathrm{ind}_{B[\ell]}^G \chi_\ell$, where $B[\ell] = \exp(\mathfrak{b}[\ell])$ and $\chi_\ell(\exp X) = e^{i\ell(X)}$ for $X \in \mathfrak{b}[\ell]$. Then, we use flag $\mathcal{S} \cap \mathfrak{k}$ in order to construct the Vergne polarisation $\mathfrak{b}[\ell|_{\mathfrak{k}}]$ of $\mathfrak{k}$ at the point $\ell|_{\mathfrak{k}} \in \mathfrak{k}^*$. Put $B[\ell|_{\mathfrak{k}}] = \exp(\mathfrak{b}[\ell|_{\mathfrak{k}}])$. It is easy to see that there exists a Malcev basis of $\mathfrak{b}[\ell|_{\mathfrak{k}}]$ relative to $\mathfrak{b}[\ell|_{\mathfrak{k}}] \cap \mathfrak{b}[\ell]$ which is contained in a Malcev basis of $\mathfrak{g}$ relative to $\mathfrak{b}[\ell]$. It follows immediately that the formula

$$\langle a_\ell, \varphi\rangle = \int_{B[\ell|_{\mathfrak{k}}]/(B[\ell|_{\mathfrak{k}}]\cap B[\ell])} \overline{\varphi(b)\chi_\ell(b)}d\dot{b} \ (\varphi \in \mathcal{H}_\pi^\infty), \tag{2.3.4}$$

$d\dot{b}$ being an invariant measure on the homogeneous space $B[\ell|_{\mathfrak{k}}]/(B[\ell|_{\mathfrak{k}}]\cap B[\ell])$, defines a semi-invariant generalised vector $a_\ell \in (\mathcal{H}_\pi^{-\infty})^{B[\ell|_{\mathfrak{k}}],\chi_\ell}$.

Lemma 2.3.4 *The representation $\pi|_K$ has finite multiplicities if and only if for $\tilde{\nu}$-almost every $\ell \in \Omega(\pi)$ the subspace $\mathfrak{b}[\ell|_\mathfrak{k}]+\mathfrak{g}(\ell)$ is Lagrangian for the skew-symmetric bilinear form B_ℓ, where $\mathfrak{b}[\ell|_\mathfrak{k}]$ denotes any polarisation of $\mathfrak{k}$ at $\ell|_\mathfrak{k}$.*

Proof 2.16 *We proceed by induction on $\dim G$. Let $\mathfrak{g}'$ be a subalgebra of codimension 1 containing $\mathfrak{k}$, $G' = \exp(\mathfrak{g}')$ and $p : \mathfrak{g}^* \to (\mathfrak{g}')^*$ the restriction mapping. We write $\ell' = p(\ell) \in (\mathfrak{g}')^*$. If the orbit $\Omega(\pi)$ is non- saturated with respect to $\mathfrak{g}'$, then $p(\Omega(\pi)) = G'\cdot\ell'$ and $\pi' = \pi|_{G'}$ is irreducible. Since $\dim(\mathfrak{g}(\ell)) = \dim(\mathfrak{g}'(\ell')) + 1$ and since a subspace $\mathfrak{p}'$ of $\mathfrak{g}'$ is Lagrangian for $B_{\ell'}$ if and only if $\mathfrak{p}' + \mathfrak{g}(\ell)$ is Lagrangian for B_ℓ, we see that the induction hypothesis applied to G' gives us the desired result.*

If the orbit $\Omega(\pi)$ is saturated with respect to $\mathfrak{g}'$, then $p(\Omega(\pi))$ is a disjoint union of a one parameter family $\{\omega_t\}_{t\in\mathbb{R}}$ of G'-orbits. Let $\pi'_t \in \widehat{G'}$ correspond to ω_t. It follows that $\pi|_K$ has finite multiplicities if and only if $\pi'_t|_K$ has finite multiplicities for almost all $t \in \mathbb{R}$. By the induction hypothesis we know that for every $t \in \mathbb{R}$ such that $\pi'_t|_K$ has finite multiplicities, the subspace $\mathfrak{b}[\ell'_t|_\mathfrak{k}]+\mathfrak{g}'(\ell'_t)$ is Lagrangian for $B_{\ell'_t}$ at $\tilde{\nu}'_t$-almost all $\ell'_t \in \omega_t$. We conclude from these observations that if $\pi|_K$ has finite multiplicities, then $\tilde{\nu}$- almost everywhere $\mathfrak{b}[\ell'|_\mathfrak{k}] + \mathfrak{g}'(\ell')$ is Lagrangian for $B_{\ell'}$. Furthermore, since $\Omega(\pi)$ is saturated with respect to $\mathfrak{g}'$, for every $\ell \in \Omega(\pi)$ and $\ell' = p(\ell) \in (\mathfrak{g}')^*$, we have $\mathfrak{g}'(\ell') = \mathfrak{g}(\ell) + \mathbb{R}Y_\ell$, where $Y_\ell \in \mathfrak{g}'$ depends smoothly on ℓ and $Ad^*(\exp(\mathbb{R}Y_\ell)) \cdot \ell = \ell + (\mathfrak{g}')^\perp$ for all $\ell \in \Omega(\pi)$. Hence for every ℓ such that $K \cdot \ell \supset \ell + (\mathfrak{g}')^\perp$, we can find a Z_ℓ in $\mathfrak{k}(\ell|_\mathfrak{k}) \subset \mathfrak{b}[\ell|_\mathfrak{k}]$ for which $0 \neq Z_\ell \cdot \ell = ad^*(Z_\ell)(\ell) \in (\mathfrak{g}')^\perp$. It follows from these observations that if $\pi|_K$ has finite multiplicities,

$$\mathfrak{b}[\ell|_\mathfrak{k}] + \mathfrak{g}(\ell) = \mathfrak{b}[\ell'|_\mathfrak{k}] + \mathfrak{g}'(\ell')$$

is Lagrangian for B_ℓ almost everywhere for $\tilde{\nu}$ on $\Omega(\pi)$.

If on the other hand $\mathfrak{b}[\ell|_\mathfrak{k}] + \mathfrak{g}(\ell)$ is Lagrangian almost everywhere for $\tilde{\nu}$ on $\Omega(\pi)$, then we can choose a $Z_\ell \in \mathfrak{k}$ such that $\langle \ell, [Z_\ell, \mathfrak{g}']\rangle = \{0\}$ and $\langle \ell, [X, Z_\ell]\rangle = 1$, where $\mathfrak{g} = \mathbb{R}X + \mathfrak{g}'$, so that $K \cdot \ell$ is saturated with respect to $\mathfrak{g}'$ for all these ℓ''s. Hence we also have $\mathfrak{b}[\ell'|_\mathfrak{k}] + \mathfrak{g}'(\ell') = \mathfrak{b}[\ell|_\mathfrak{k}] + \mathfrak{g}(\ell)$ is Lagrangian for $B_{\ell'}$ at those ℓ''s. The induction hypothesis and the structure of $\Omega(\pi)$ tell us that for almost all $t \in \mathbb{R}$, $\pi'_t|_K$ has finite multiplicities. Whence $\pi|_K$ has finite multiplicities too.

Suppose for a while that $\pi|_K$ has finite multiplicities. According to the last lemma, we know that $\mathfrak{b}[\ell|_\mathfrak{k}]+\mathfrak{g}(\ell)$ is a Lagrangian subspace for B_ℓ almost everywhere for $\tilde{\nu}$ on $\Omega(\pi)$. For such a ℓ, we know that the choice of the polarisation $\mathfrak{b}[\ell]$ does not matter for the definition of the distribution a_ℓ (cf. [65]). In fact, if $\mathfrak{b}'[\ell]$ is another polarisation of $\mathfrak{g}$ at ℓ, we can form the distribution a'_ℓ as before and if T denotes an intertwining operator between these two realisations of π_ℓ, then $a'_\ell \circ T = c a_\ell$ with some $0 \neq c \in \mathbb{C}$.

Theorem 2.3.5 *Suppose that $\pi|_K$ has finite multiplicities. Then $\tilde{\nu}$-almost everywhere on $\Omega(\pi)$, the distribution a_ℓ is an eigenvector for every element in $D_\pi(G)^K$. In other words, for $W \in \mathcal{U}_\pi(\mathfrak{g})^\mathfrak{k}$, we have $\pi_\ell(W)a_\ell = P_W(\ell)a_\ell$ for $\tilde{\nu}$-almost all $\ell \in \Omega(\pi)$ with certain complex scalars $P_W(\ell)$. Furthermore the function $\ell \mapsto P_W(\ell)$ is K-invariant.*

Proof 2.17 *If $G = K$, then the algebra $D_\pi(G)^K$ is reduced to $\mathbb{C}Id$ according to Schur's lemma and there is nothing to prove. Let now $q = \dim(G/K) > 0$ and we proceed by induction on $\dim G$. Let $G' = \exp(\mathfrak{k}_{n-1})$ in the sequence (2.3.2) and $\ell' = \ell|_{\mathfrak{k}_{n-1}}$. Suppose first that $\Omega(\pi)$ is saturated with respect to $\mathfrak{k}_{n-1}$. As we remarked above, we can choose the polarisation $\mathfrak{b}[\ell]$ as we want, so we can assume $\mathfrak{b}[\ell] \subset \mathfrak{k}_{n-1}$. Let $\pi' = ind_{B[\ell]}^{G'}\chi_\ell$ with $B[\ell] = \exp(\mathfrak{b}[\ell])$. Then a_ℓ can be identified with $a_{\ell'}$. We know from [10] that $\mathcal{U}_{\pi_\ell}(\mathfrak{g})^{\mathfrak{k}} \subset \mathcal{U}(\mathfrak{k}_{n-1}) + \ker(\pi_\ell)$. Hence we can apply the induction hypothesis to $(\pi', a_{\ell'})$.*

Suppose now that $\Omega(\pi)$ is non-saturated with respect to $\mathfrak{k}_{n-1}$. Then

$$\pi' = \pi|_{G'} = ind_{B[\ell']}^{G'}\chi_{\ell'} \in \widehat{G'},$$

where $B[\ell'] = \exp(\mathfrak{b}[\ell] \cap \mathfrak{k}_{n-1})$, and a result of Pedersen [88] says that there exists an element $A \in \ker(\pi_\ell)$ of the form $A = X_n + V$ with $V \in \mathcal{U}(\mathfrak{k}_{n-1})$. Let $W = \sum_{j=0}^{r} V_j X_n^j$, where $V_j \in \mathcal{U}(\mathfrak{k}_{n-1}) (0 \leq j \leq r)$, be an element of $\mathcal{U}_\pi(\mathfrak{g})^{\mathfrak{k}}$. Replacing in this expression of W the vector X_n by the element A, we see that $W \in \mathcal{U}_{\pi'}(\mathfrak{k}_{n-1})^{\mathfrak{k}} + \ker(\pi_\ell)$. On the other hand, since $\mathfrak{k} \subset \mathfrak{k}_{n-1}$, we can identify a_ℓ with $a_{\ell'} \in \mathcal{H}_{\pi'}^{-\infty}$. These considerations allow us to descend to G', where the induction hypothesis applies. The function $P_W(\ell)$ is easily checked to be K-invariant.

Let $W \in \mathcal{U}(\mathfrak{g})$, where $\mathcal{U}(\mathfrak{g})$ designates the enveloping algebra of $\mathfrak{g}_{\mathbb{C}}$, such that $[\mathfrak{k}, W] \subset \ker \pi$ and that $W \cdot a_\ell$ is a multiple of a_ℓ almost everywhere on Ω for the invariant measure. Namely, there exists a function $P_W(\ell)$ defined almost everywhere on Ω such that $W \cdot a_\ell = P_W(\ell) a_\ell$. Suppose also that the function $P_W(\ell)$ extends into a rational function on Ω.

Lemma 2.3.6 *$P_W(\ell) = 0$ almost everywhere on Ω if and only if $W \in \ker \pi$.*

Proof 2.18 *If $W \in \ker \pi$, $P_W(\ell) \equiv 0$ because $a_\ell \in \mathcal{H}_\pi^{-\infty}$. Suppose that $P_W(\ell) = 0$ almost everywhere on Ω and we go to see $W \in \ker \pi$ by induction on $\dim G$.*

Let $p : \mathfrak{g}^* \to (\mathfrak{k}_{n-1})^*$ be the restriction mapping and $K_{n-1} = \exp(\mathfrak{k}_{n-1})$. If Ω is non-saturated with respect to $\mathfrak{k}_{n-1}$, there exists in $\ker \pi$ an element A having the form $A = X_n + V$ with a certain $V \in \mathcal{U}(\mathfrak{k}_{n-1})$. Making use of A to chase from W the part which is found outside of $\mathcal{U}(\mathfrak{k}_{n-1})$, we can suppose that $W \in \mathcal{U}(\mathfrak{k}_{n-1})$. Since $p(\Omega)$ is a K_{n-1}-orbit, the induction hypothesis gives us immediately the desired result.

Suppose now that Ω is saturated with respect to $\mathfrak{k}_{n-1}$. This needs that W belongs to $\mathcal{U}(\mathfrak{k}_{n-1})$. The restriction $\pi|_{K_{n-1}}$ is disintegrated as $\pi|_{K_{n-1}} \simeq \int_{\mathbb{R}}^{\oplus} \pi_t dt$ into a one parameter family $\{\pi_t\}_{t \in \mathbb{R}}$ of irreducible unitary representations of K_{n-1} and accordingly the restriction $p(\Omega) = \Omega|_{\mathfrak{k}_{n-1}}$ is decomposed as $p(\Omega) = \sqcup_{t \in \mathbb{R}} \omega_t$, where ω_t is the coadjoint orbit of K_{n-1} associated to π_t. Then, the induction hypothesis says that W belongs to $\ker(\pi_t)$ for almost all $t \in \mathbb{R}$ and hence $W \in \ker \pi$.

Suppose for a while that $\pi|_K$ has finite multiplicities. Concerning to the sequence (2.3.2), we put $K_j = \exp(\mathfrak{k}_j)$ for $1 \leq j \leq n$ so that $K_d = K$ and $K_n = G$. Let $j \in T_K(e^K)$. Since $\pi|_K$ has finite multiplicities, the subalgebra $\mathfrak{k}_j(\ell|_{\mathfrak{k}_j})$ is not contained in $\mathfrak{k}_{j-1}$ for $\tilde{\nu}$-almost all $\ell \in \Omega(\pi)$. Therefore, there exists an e-central element $W_j \in \mathcal{U}_\pi(\mathfrak{k}_j)^{\mathfrak{k}_j}$ having the form

$$W_j = a_j X_j + b_j \ (a_j, b_j \in \mathcal{U}(\mathfrak{k}_{j-1}), a_j \notin \ker \pi),$$

which leads $a_j \in \mathcal{U}_\pi(\mathfrak{k}_{j-1})^{\mathfrak{k}_j}$. Furthermore, $P_{W_j}(\ell) = \varphi_j(\ell)\ell_j + \psi_j(\ell)$, where $\ell_k = \ell(X_k)(1 \leq k \leq n)$ for $\ell \in \Omega(\pi)$ and φ_j, ψ_j are rational functions in $(\ell_1, \ldots, \ell_{j-1})$.

Theorem 2.3.7 *Suppose that $\pi|_K$ has finite multiplicities. Let $W \in \mathcal{U}_\pi(\mathfrak{g})^{\mathfrak{k}}$. Then the function $P_W(\ell)$, which is defined $\tilde{\nu}$-almost everywhere on $\Omega(\pi)$ by the last theorem, extends to a K-invariant rational function on $\Omega(\pi)$. Moreover, the homomorphism $\mathcal{U}_\pi(\mathfrak{g})^{\mathfrak{k}} \ni W \mapsto P_W(\ell)$ defines an imbedding of $D_\pi(G)^K$ into the field $\mathbb{C}(\Omega(\pi))^K$ of the K-invariant rational functions on $\Omega(\pi)$.*

Proof 2.19 *The definition of the function $P_W(\ell)$ implies $P_{WW'}(\ell) = P_W(\ell)P_{W'}(\ell)$ for any $W, W' \in \mathcal{U}_\pi(\mathfrak{g})^{\mathfrak{k}}$.*

Let $1 \leq m \leq n$ be the smallest index in the sequence (2.3.2) such that $W \in \mathcal{U}(\mathfrak{k}_m) + \ker \pi$. We proceed by induction on m to show that $P_W(\ell)$ is rational on $\Omega(\pi)$. We first remark that in order to compute the eigenvalue $P_W(\ell)$ we can consider $W \in \mathcal{U}(\mathfrak{k}_m)$ and replace $a_\ell \in \mathcal{H}_\pi^{-\infty}$ by $a_{\ell|_{\mathfrak{k}_m}} \in \mathcal{H}_\sigma^{-\infty}$, where σ is the irreducible unitary representation of $K_m = \exp(\mathfrak{k}_m)$ corresponding to $K_m \cdot \ell|_{\mathfrak{k}_m}$, repeating step by step the observation made in the proof of the last theorem to go down from $\mathfrak{g}$ to $\mathfrak{k}_{n-1}$.

If $m = 1$, then W is in the centre of $\mathcal{U}(\mathfrak{k})$ modulo $\ker \pi$ and $P_W(f) = \widehat{W}(f)$, a K-invariant polynomial function on $\mathfrak{k}^*$ (cf. [31]).

Let us write $W = \sum_{k=0}^r X_m^k w_k$ with $r > 0$ and $w_k \in \mathcal{U}(\mathfrak{k}_{m-1})$ for $0 \leq k \leq r$ satisfying $w_r \notin \ker \pi$. We shall use another induction on the degree r of W relative to X_m. If the element w_r is not contained in $\mathcal{U}_\pi(\mathfrak{g})^{\mathfrak{k}}$, then we can find elements $X_1, \ldots, X_p$ in $\mathfrak{k}$ such that $[X_1, [\cdots, [X_p, w_r] \cdots]] \notin \ker \pi$, but $[X, [X_1, [\cdots, [X_p, w_r] \cdots]]] \in \ker \pi$ for all $X \in \mathfrak{k}$, i.e. such that $u_r = [X_1, [\cdots, [X_p, w_r] \cdots]]$ is contained in $\mathcal{U}_\pi(\mathfrak{g})^{\mathfrak{k}} \cap \mathcal{U}(\mathfrak{k}_{m-1})$ outside of $\ker \pi$. The element $V = [X_1, [\cdots, [X_p, W] \cdots]]$ of $\mathcal{U}(\mathfrak{g})$ is then contained in $\ker \pi$ and has the form $V = X_m^r u_r + V'$ with a certain $V' \in \sum_{k=0}^{r-1} X_m^k \mathcal{U}(\mathfrak{k}_{m-1})$. We apply the induction hypothesis to $\widetilde{W} = u_r W - V w_r \in \mathcal{U}_\pi(\mathfrak{g})^{\mathfrak{k}}$ which is of degree $< r$ modulo $\ker \pi$ in X_m. Hence, by the induction hypothesis on m and r and the multiplicative property of $P_W(\ell)$, the function

$$P_{\widetilde{W}}(\ell) = P_{u_r}(\ell)P_W(\ell) - P_V(\ell)P_{w_r}(\ell) = P_{u_r}(\ell)P_W(\ell)$$

and then also $P_W(\ell) = P_{\widetilde{W}}(\ell)/P_{u_r}(\ell)$ admit an extension to a rational function on $\Omega(\pi)$. Hence we can assume now that $w_r \in \mathcal{U}_\pi(\mathfrak{g})^{\mathfrak{k}}$. Since $W \in \mathcal{U}(\mathfrak{k}_m)$, we know $m \in T_K(e^K)$ and so generically $\dim(K_m \cdot \ell|_{\mathfrak{k}_m}) = \dim(K_m \cdot \ell|_{\mathfrak{k}_{m-1}})$ on $\Omega(\pi)$. Therefore, we have an e-central element $W_m = a_m X_m + b_m$ in $\mathcal{U}_\pi(\mathfrak{k}_m)^{\mathfrak{k}}$ with $a_m, b_m \in \mathcal{U}(\mathfrak{k}_{m-1}), a_m \notin \ker \pi$ and $a_m \in \mathcal{U}_\pi(\mathfrak{k}_{m-1})^{\mathfrak{k}}$. Hence the element

$$\widetilde{W} = a_m^r W - W_m^r w_r$$

of $\mathcal{U}_\pi(\mathfrak{k}_m)^{\mathfrak{k}}$ is of degree $< r$ in X_m modulo $\ker\pi$ *and we can apply the induction hypothesis to it. Hence*

$$P_{\widetilde{W}}(\ell) = (P_{a_m}(\ell))^r P_W(\ell) - (P_{W_m}(\ell))^r P_{w_r}(\ell),$$

$P_{a_m}(\ell), P_{w_r}(\ell)$ *and therefore*

$$P_W(\ell) = \frac{P_{\widetilde{W}}(\ell) + (P_{W_m}(\ell))^r P_{w_r}(\ell)}{(P_{a_m}(\ell))^r}$$

are also rational functions on $\Omega(\pi)$.

2.3.4 K-diagonal elements

We leave the assumption of finite multiplicities in the proof for many technical reasons. We shall just handle the elements $W \in \mathcal{U}_\pi(\mathfrak{g})^{\mathfrak{k}}$ such that $W \cdot a_\ell = P_W(\ell) a_\ell$ almost everywhere for $\tilde{\nu}$ on $\Omega(\pi)$, assuming further that the function $P_W(\ell)$ extends to a rational function on $\Omega(\pi)$. Let us first define the following:

Definition 2.1 1. We say that $W \in \mathcal{U}_\pi(\mathfrak{g})^{\mathfrak{k}}$ is K-diagonal, if

$$\pi(W) a_\ell \quad = \quad P_W(\ell) a_\ell$$

for a certain scalar $P_W(\ell) \in \mathbb{C}$ independent of the polarisations chosen to describe the distribution a_ℓ and $\ell \mapsto P_W(\ell)$ extends to a rational function on Ω.

2. Let $\mathcal{U}$ be the set of K-diagonal elements of $\mathcal{U}_\pi(\mathfrak{g})^{\mathfrak{k}}$. Let

$$\Theta : \mathcal{U} \ni W \mapsto P_W. \tag{2.3.5}$$

Remark 2.3.2 1. From ([62], Theorem 4.1), any K-diagonal element of $\mathcal{U}(\mathfrak{g})$ belongs to $\mathcal{U}_\pi(\mathfrak{g})^{\mathfrak{k}}$.

2. Definition 2.1 is posed independently from the fact that $\pi|_K$ has finite multiplicities or not. In the case of finiteness, any element of $\mathcal{U}_\pi(\mathfrak{g})^{\mathfrak{k}}$ is K-diagonal (cf. Theorem 2.3.7).

Next, we can easily adapt the arguments of ([5], Lemma 3.2) to prove the following:

Lemma 2.3.3 *Let $W \in \mathcal{U}(\mathfrak{g})$ be K-diagonal. Then P_W is identically zero if and only if $W \in \ker\pi$.*

Proof 2.20 *If $W \in \ker\pi$, $P_W(\ell) \equiv 0$ because $a_\ell \in \mathcal{H}_\pi^{-\infty}$. Suppose that $P_W(\ell) = 0$ almost everywhere on Ω and let us prove that $W \in \ker\pi$ by induction on* $\dim G$.

Let $p : \mathfrak{g}^* \to (\mathfrak{k}_{n-1})^*$ be the restriction mapping and $K_{n-1} = \exp(\mathfrak{k}_{n-1})$. If Ω is non-saturated with respect to $\mathfrak{k}_{n-1}$, there exists in $\ker\pi$ an element A having the form $A = X_n + V$ with a certain $V \in \mathcal{U}(\mathfrak{k}_{n-1})$. Making use of A to cancel from W the part which is found outside of $\mathcal{U}(\mathfrak{k}_{n-1})$, we can suppose that $W \in \mathcal{U}(\mathfrak{k}_{n-1})$. Since $p(\Omega)$ is a K_{n-1}-orbit, the induction hypothesis gives us immediately the desired result.

Suppose now that Ω is saturated with respect to $\mathfrak{k}_{n-1}$. This implies that W belongs to $\mathcal{U}(\mathfrak{k}_{n-1})$. The restriction $\pi|_{K_{n-1}}$ is disintegrated as $\pi|_{K_{n-1}} \simeq \int_{\mathbb{R}}^{\oplus} \pi_t dt$ into a one parameter family $\{\pi_t\}_{t\in\mathbb{R}}$ of irreducible unitary representations of K_{n-1} and accordingly the restriction $p(\Omega) = \Omega|_{\mathfrak{k}_{n-1}}$ is decomposed as $p(\Omega) = \sqcup_{t\in\mathbb{R}}\omega_t$, where ω_t is the coadjoint orbit of K_{n-1} associated to π_t. Then, the induction hypothesis says that W belongs to $\ker \pi_t$ for almost all $t \in \mathbb{R}$ and hence $W \in \ker \pi$.

The first step to prove Conjecture 2.3.1 consists in proving the following Theorem:

Theorem 2.3.4 *Let $\pi \in \hat{G}$ and let $W \in \mathcal{U}(\mathfrak{g})$ be K-diagonal. The function P_W extends to a K-invariant polynomial function on Ω.*

The proof of Theorem 2.3.4 will be achieved through different steps. Let us start with the following:

2.3.4.1 *A preliminary inductive proof*

In order to prove Theorem 2.3.4, we proceed by induction on $\delta(G,K) = \dim G + \dim G/K$. For small $\delta(G,K)$, G turns out to be abelian and the answer is immediate. Consider the flag of algebras (2.3.2) of $\mathfrak{g}$ and for the sake of simplicity of notation, denote $\mathfrak{g}' = \mathfrak{k}_{n-1}$ which contains $\mathfrak{k}$. Put $G' = \exp(\mathfrak{g}')$ and suppose that Theorem 2.3.4 holds for G'.

2.3.4.2 *Case where the ideal $\mathfrak{g}'$ is of non-saturation*

Suppose that the orbit Ω is non-saturated with respect to $\mathfrak{g}'$, namely that $n \in T_\Omega$. Then the projection $pr : \mathfrak{g}^* \to (\mathfrak{g}')^*$ turns out to be a K-equivariant homeomorphism between Ω and $\omega = pr(\Omega)$ which is a G'-orbit. Hence, $\mathbb{C}[\Omega]^K \cong \mathbb{C}[\omega]^K$. On the other hand, $\pi' = \pi|_{G'}$ is irreducible and there exists in $\ker \pi$ an element W' having the form $W' = X_n + A$ with $A \in \mathcal{U}(\mathfrak{g}')$ which allows us to identify $D_\pi(G)^K$ with $D_{\pi'}(G')^K$. Since ω is the coadjoint orbit of G' associated to π' and since $a_\ell = a_{\ell|_{\mathfrak{g}'}}$, the induction hypothesis proves Theorem 2.3.4 in this case.

2.3.4.3 *Case where the ideal $\mathfrak{g}'$ is of saturation*

Suppose now that Ω is saturated with respect to $\mathfrak{g}'$, namely that $n \in S_\Omega$. We have the following lemma:

Lemma 2.3.5 *([15], Lemma 4.1) There exists one and only one index $2 \le j \le n-1$ belonging to S_Ω and $b \in \mathcal{U}(\mathfrak{g}_{j-1})$ such that $Y_j + b \in \mathcal{U}_\pi(\mathfrak{g}_j)^{\mathfrak{g}'}$.*

Likewise, if $j = s_i$ $(1 \le i \le r-1)$, there exists a G'-invariant polynomial function

$$\alpha = x_i + \varphi(x_1, \ldots, x_{i-1}) \tag{2.3.6}$$

on Ω, which separates the G'-orbits $w_\alpha = \{\ell \in \Omega : \alpha(\ell) = \alpha\}$ contained in $pr(\Omega)$. This means that $pr(\Omega) = \coprod_{\alpha\in\mathbb{R}} \omega_\alpha$, the disjoint union of G'-orbits ω_α. Accordingly,

$$\pi|_{G'} \simeq \int_{\mathbb{R}}^{\oplus} \pi_\alpha d\alpha \tag{2.3.7}$$

with $\pi_\alpha = \theta_{G'}(\omega_\alpha)$ for all $\alpha \in \mathbb{R}$.

Since the orbit Ω is saturated with respect to $\mathfrak{g}'$, for any $\ell \in \Omega$ there exists then a polarisation $\mathfrak{b}[\ell]$ at ℓ contained in $\mathfrak{g}'$, which is also a polarisation at $\ell|_{\mathfrak{g}'}$. Furthermore we can suppose that $W \in \mathcal{U}(\mathfrak{g}')$, since $\pi(W)a_\ell = P_W(\ell)a_\ell, \ell \in \Omega$, and $\pi_\ell = \mathrm{ind}_{G'}^G \pi_{\ell|_{\mathfrak{g}'}}$. It follows then from the definition of $a_\ell, \ell \in \Omega$, that

$$P_W^G(\ell) \quad = \quad P_W^{G'}(\ell|_{\mathfrak{g}'}), \ell \in \Omega,$$

where the index ^{G} (resp. $^{G'}$) indicates the action of W on a_ℓ (resp. on $a_{\ell|_{\mathfrak{g}'}}$). We apply the induction hypothesis to W and G'. Then it follows that the function P_W, which is rational on Ω restricts to the G'-orbits $\omega_\alpha, \alpha \in \mathbb{R}$, as a polynomial function. Let ℓ be a point of Ω, for each real number t, let α be such that $Ad^*(\exp tX_n)\ell \in \omega_\alpha$, then:

$$P_W(Ad^*(\exp(tX_n)g')\ell) = P_W(\alpha, g') \quad = \quad \frac{A(\alpha, g')}{B(\alpha, g')}, g' \in G',$$

for two polynomial functions A, B. Since $P_W|_{\omega_\alpha}$ is polynomial, we have that B is independent of the variable g' and so P_W is given by a polynomial function A divided by a polynomial function in α.

The following consequence is then immediate.

Corollary 2.3.6 *Suppose that $\mathfrak{k}$ is contained in an ideal of codimension 2. Then for every K-diagonal $W \in \mathcal{U}_\pi(\mathfrak{g})^{\mathfrak{k}}$, the function $\ell \mapsto P_W(\ell)$ is polynomial.*

Proof 2.21 *Let $\mathfrak{h}_j, j = 1, 2$ be two distinct ideals of codimension 1 containing $\mathfrak{k}$. Accordingly to Subsection 4.1.1, we can assume that the orbit Ω is saturated with respect to $\mathfrak{h}_j, j = 1, 2$. We fix the flag (2.3.1) such that $\mathfrak{g}_{n-2} = \mathfrak{h}_1 \cap \mathfrak{h}_2$, and $\mathfrak{g}_{n-1} = \mathfrak{h}_1$, thus, if w is the G_{n-2}-orbit of $l|\mathfrak{g}_{n-2}$, the set of jump indices are $S_\Omega = S_w \cup \{n-1, n\}$, or $S_\Omega = S_w \cup \{i, k, n-1, n\}$.*
In the first case, we have $n - 1 = s_i$, $n = s_{i+1}$, by (8), there is a H_1-invariant polynomial function $\alpha_1 = x_i + \varphi_1(x_1, ...x_{i-1})$ separating the H_1-orbits, and replacing $\mathfrak{h}_1$ by $\mathfrak{h}_2$ in the flag (2.3.1), there is a H_2-invariant polynomial function $\alpha_2 = x_{i+1} + \varphi_2(x_1, ...x_{i-1})$ separating the H_2-orbits. Moreover, for any complex numbers c_j , there is no common divisor for $\alpha_1 + c_1$ and $\alpha_2 + c_2$.
In the second case, suppose the jump indices for the H_1-orbit w_1 of $l|\mathfrak{h}_1$ are $S_w \cup \{i, n-1\}$, with $i = s_{i_1}$, and by (2.3.8), there is a G_{n-2}-invariant polynomial function $\beta_1 = x_{i_1} + \varphi_1(x_1, ...x_{i_1-1})$ separating the G_{n-2}-orbits in the H_1-orbit w_1. Suppose X_n be in $\mathfrak{h}_2 \setminus \mathfrak{h}_1$, and $k = s_{i_2}$, by (2.3.8) there is $\alpha_2 = x_{i_2} + \varphi_2(x_1, ...x_{i_2-1})$ separating the H_2-orbits in Ω. Fix X_n such that $\alpha_2(\exp(tX_n)l) = t$ for each l in Ω such that $\alpha_2(l) = 0$. Finally put: $\alpha_1(\exp(tX_n)l) = \beta_1(l|\mathfrak{h}_1)$ or

$$\alpha_1(x_i) = (e^{-\alpha_2(x_i)ad^*(X_n)}\beta_1)(x_i) = \sum_m \frac{(-\alpha_2(x_i))^m}{m!}\beta_1((ad^*(X_n))^m(x_i)).$$

The function α_2, polynomial on Ω is H_2-invariant and separates H_2-orbits in Ω. Moreover, since for any complex numbers c_1 and c_2,

$$\alpha_1 + c_1 = e^{c_2 ad^*(X_n)}\beta_1 + c_1 + \sum_{m>0} \frac{(-\alpha_2 - c_2)^m}{m!} e^{c_2 ad^*(X_n)}\beta_1((ad^*(X_n))^m \cdot),$$

and $e^{c_2 ad^*(X_n)}\beta_1 = x_{i_1} + \psi(x_1, ..., x_{i_1 - 1})$ *there is no common divisor for* $\alpha_1 + c_1$ *and* $\alpha_2 + c_2$.
In both cases, applying the induction hypothesis to W *and* H_j, *we can write* P_W *as a quotient of a polynomial function* A_j *by a function* $B_j(\alpha_j)$, *polynomial in* α_j. *Thus:* $P_W = \frac{A_1}{B_1(\alpha_1)} = \frac{A_2}{B_2(\alpha_2)}$, *and*

$$B_2(\alpha_2)A_1 = B_1(\alpha_1)A_2.$$

Since $\alpha_1 + c_1$ *and* $\alpha_2 + c_2$ *have no common divisor,* P_W *itself is a polynomial function.*

On the other hand, let $W \in \mathcal{U}_\pi(\mathfrak{k}_v)^{\mathfrak{k}}$. If $v \leq d$, W belongs to $\mathcal{U}_\pi(\mathfrak{k})^{\mathfrak{k}}$ and the operator $\sigma(W)$ is a scalar for almost all $\sigma \in \widehat{K}$ with respect to the measure ν_π used in the irreducible decomposition of $\pi|_K$. Then, we can apply Theorem 2.1.1 in [11] to get:

Proposition 2.3.7 *For any* $W \in \mathcal{U}_\pi(\mathfrak{k})^{\mathfrak{k}}$, *the function* $\ell \mapsto P_W(\ell)$ *is polynomial on* Ω.

2.3.4.4 Proof of Theorem 2.3.4

As usual, we can assume that the centre $\mathfrak{z}$ of $\mathfrak{g}$ has dimension 1, that $\pi (= \pi_\ell)$ is not 0 on $\mathfrak{z}$ and that $\mathfrak{z} \subset \mathfrak{k}$. Also according to 2.3.4.2, we can assume that for every subalgebra $\mathfrak{g}'$ of codimension one containing $\mathfrak{k}$, Ω is saturated with respect to $\mathfrak{g}'$. In particular, a polarisation $\mathfrak{b}[\ell]$ with $B[\ell] = \exp(\mathfrak{b}[\ell])$ of ℓ can always be found in $\mathfrak{g}'$ and $W \in \mathcal{U}(\mathfrak{g}')$.

We make now a further induction on j_0, the smallest index $j \in \{1, ..., n\}$, such that $W \in \mathcal{U}(\mathfrak{k}_{j_0})$. If $j_0 \leq d$, then W is an e-central element of Corwin-Greenleaf for the projection of Ω on $\mathfrak{k}^*$ and hence the function $P_W(\ell)$ is polynomial as in Proposition 2.3.7. We can therefore assume that $j_0 \geq d + 1$.

Let now $\mathfrak{l} = \mathfrak{k}_{d+1}$ and $L = \exp \mathfrak{l}$. If the generic L-orbits in $\Omega|_{\mathfrak{l}}$ are non-saturated with respect to $\mathfrak{k}$, there exists a $\nu = aX_{d+1} + b$, $a, b \in \mathcal{U}(\mathfrak{k})$ which is e-central for $\Omega|_{\mathfrak{l}}$. Applying W, ν to the Penney distribution $a_\ell (\ell \in \Omega)$, we see that they commute modulo $\ker \pi$ and so W is also L-invariant. If we use the Penney distributions a_ℓ^L (as in formula (2.3.4)) and if $\mathfrak{b}[\ell|_{\mathfrak{l}}] \cap \mathfrak{b}[\ell] = \mathfrak{b}[\ell|_{\mathfrak{k}}] \cap \mathfrak{b}[\ell]$, we see that for some $X_{d+1} \in (\mathfrak{l}(\ell|_{\mathfrak{l}}) \cap \ker \ell) \setminus \mathfrak{k}$, we have for any $\varphi \in \mathcal{H}_\pi^\infty$:

$$\begin{aligned}
\langle W \cdot a_\ell^L, \varphi \rangle &= \int_{B[\ell|_{\mathfrak{l}}]/(B[\ell|_{\mathfrak{l}}]\cap B[\ell])} \overline{\pi(W^*)\varphi(b)\chi_\ell(b)}\dot{d}b \\
&= \int_{\mathbb{R}} \int_{B[\ell|_{\mathfrak{k}}]/(B[\ell|_{\mathfrak{k}}]\cap B[\ell])} \overline{\pi(W^*)\varphi(\exp(sX_{d+1})b)\chi_\ell(\exp(sX_{d+1})b)}\dot{d}bds \\
&= \int_{\mathbb{R}} \int_{B[\ell|_{\mathfrak{k}}]/(B[\ell|_{\mathfrak{k}}]\cap B[\ell])} \overline{\pi(W^*)(\pi(\exp(-sX_{d+1}))\varphi)(b)\chi_\ell(\exp b)}\dot{d}bds \\
&= \int_{\mathbb{R}} P_W^K(\ell|_{\mathfrak{k}}) \int_{B[\ell|_{\mathfrak{l}}]/(B[\ell|_{\mathfrak{l}}]\cap B[\ell])} \overline{\varphi(\exp(sX_{d+1})b)\chi_\ell(\exp b)}\dot{d}bds \\
&= P_W^K(\ell)\langle a_\ell^L, \varphi \rangle.
\end{aligned}$$

Therefore W is L-diagonal. Since $\delta(G,L) < \delta(G,K)$, the induction hypothesis implies that $P_W^K = P_W^L$ is polynomial.

Recall now that we are in the situation where the orbit Ω is saturated with respect to $\mathfrak{k}_{n-1}$. There exists then by Lemma 2.3.5, a unique index $2 \leq r_0 \leq n-1$ belonging to S_Ω and $b \in \mathcal{U}(\mathfrak{g}_{r_0-1})$ such that

$$\kappa = Y_{r_0} + b \in \mathcal{U}_\pi(\mathfrak{g}_{r_0})^{\mathfrak{g}'} \tag{2.3.8}$$

and $[X_n, \kappa] \neq 0 \mod \ker \pi$. The polynomial function P_κ on $\Omega|_{\mathfrak{k}_{n-1}}$ then separates the K_{n-1}-orbits $\omega_y, y \in \mathbb{R}$ and, as we have seen in Subsection 2.3.4.3, W belongs to $\mathcal{U}(\mathfrak{k}_{n-1})$ and P_W can be written as $\frac{A}{B}$ for a polynomial function A on Ω divided by a polynomial B in the variable P_κ.

Let now $\widetilde{\mathfrak{g}}$ be another ideal of $\mathfrak{g}$ of codimension 1. If $\mathfrak{k} \subset \widetilde{\mathfrak{g}}$, then Theorem 2.3.4 holds by Corollary 2.3.6. Hence we assume that $\mathfrak{k} \not\subset \widetilde{\mathfrak{g}}$. Let us treat first the case where Ω is not saturated with respect to $\widetilde{\mathfrak{g}}$. Write $\mathfrak{g} = \mathbb{R}\widetilde{X} + \widetilde{\mathfrak{g}}$ and $\widetilde{G} = \exp \widetilde{\mathfrak{g}}$. We can again assume as in 2.3.4.2 that $W \in \mathcal{U}(\widetilde{\mathfrak{g}})$. Let $\widetilde{\mathfrak{k}} = \mathfrak{k} \cap \widetilde{\mathfrak{g}}$ and $\widetilde{K} = \exp \widetilde{\mathfrak{k}}$. If $\mathfrak{b}[\ell|_\mathfrak{k}] \subset \widetilde{\mathfrak{g}}$ almost everywhere on Ω, then $a_\ell = a_{\ell|_{\widetilde{\mathfrak{g}}}}$ and the induction hypothesis tells us that $P_W(\ell)$ is a polynomial function on the $\widetilde{G}$-orbit $\widetilde{\Omega} = \widetilde{p}(\Omega)$, where $\widetilde{p} : \mathfrak{g}^* \to (\widetilde{\mathfrak{g}})^*$ is the restriction map. Hence P_W is also a polynomial function on Ω.

If $\mathfrak{b}[\ell|_\mathfrak{k}] \not\subset \widetilde{\mathfrak{g}}$ for almost all $\ell \in \Omega$, let us write $\mathfrak{b}[\ell|_\mathfrak{k}] = \mathbb{R}\widetilde{X(\ell)} + \mathfrak{b}[\widetilde{\ell}|_{\widetilde{\mathfrak{k}}}]$, where $\widetilde{\ell} = \widetilde{p}(\ell)$. We remark that we can take $\mathfrak{b}[\widetilde{\ell}|_{\widetilde{\mathfrak{k}}}]$ to be the Vergne polarisation at $\widetilde{\ell}|_{\widetilde{\mathfrak{k}}} \in (\widetilde{\mathfrak{k}})^*$ built from a Jordan-Hölder sequence $\mathcal{S} \cap \widetilde{\mathfrak{g}}$ of $\widetilde{\mathfrak{g}}$, $\mathcal{S}$ denoting the flag (2.3.1) of $\mathfrak{g}$. As W is K-invariant, we see that

$$\langle W \cdot a_\ell, \varphi \rangle = \int_\mathbb{R} \langle W \cdot a_{\widetilde{\ell}}, \varphi(\exp(t\widetilde{X(\ell)}) \cdot) \rangle dt \ (\ell \in \Omega)$$

for $\varphi \in \mathcal{H}_\pi^\infty$. We identify $\mathcal{H}_\pi^{-\infty}$ with $\mathcal{H}_{\widetilde{\pi}}^{-\infty}$. Fixing a generic $\ell \in \Omega$ and taking a Malcev basis in $\mathfrak{g}$ relative to $\mathfrak{b}[\ell]$, which contains a Malcev basis in $\mathfrak{b}[\ell|_\mathfrak{k}]$ relative to $\mathfrak{b}[\ell|_\mathfrak{k}] \cap \mathfrak{b}[\ell]$, we identify the space $\mathcal{H}_\pi$ of π with $\mathbb{R}^m, m = \dim(\mathfrak{g}/\mathfrak{b}[\ell])$. Since

$$B[\ell|_\mathfrak{k}]/(B[\ell|_\mathfrak{k}] \cap B[\ell]) \simeq B[\ell|_\mathfrak{k}]B[\ell]/B[\ell] = \exp(\mathbb{R}\widetilde{X(\ell)})B[\widetilde{\ell}|_{\widetilde{\mathfrak{k}}}]B[\ell]/B[\ell],$$

we finally get the following two eventualities: either

$$B[\ell|_\mathfrak{k}]/(B[\ell|_\mathfrak{k}] \cap B[\ell]) \simeq B[\widetilde{\ell}|_{\widetilde{\mathfrak{k}}}]/(B[\widetilde{\ell}|_{\widetilde{\mathfrak{k}}}] \cap B[\widetilde{\ell}])$$

or

$$\begin{aligned} B[\ell|_\mathfrak{k}]/(B[\ell|_\mathfrak{k}] \cap B[\ell]) &\simeq \exp(\mathbb{R}\widetilde{X(\ell)})B[\widetilde{\ell}|_{\widetilde{\mathfrak{k}}}]B[\ell]/B[\ell] \\ &\simeq \exp(\mathbb{R}\widetilde{X(\ell)}) \times B[\widetilde{\ell}|_{\widetilde{\mathfrak{k}}}]/(B[\widetilde{\ell}|_{\widetilde{\mathfrak{k}}}] \cap B[\widetilde{\ell}]). \end{aligned}$$

In the first case, the distribution a_ℓ associated to π can be identified with the generalised vector $a_{\widetilde{\ell}}$ of $\widetilde{\pi} = \pi|_{\widetilde{G}}$. In the second case, for $\varphi \in \mathcal{H}_\pi^\infty$ satisfying

$$\varphi(\exp(t\widetilde{X(\ell)})\widetilde{g}) = \phi(t)\psi(\widetilde{g}), \ t \in \mathbb{R}, \widetilde{g} \in \widetilde{G},$$

with $\phi \in C_c^\infty(\mathbb{R}), \psi \in \mathcal{H}_{\widetilde{\pi}}^\infty$, the $\mathfrak{k}$-invariance of W implies that

$$\begin{aligned}\langle W{\cdot}a_\ell, \varphi\rangle &= (\int_{\mathbb{R}} \overline{\phi(t)e^{it\ell(\widetilde{X(\ell)})}dt})\langle W{\cdot}a_{\widetilde{\ell}}, \psi\rangle \\ &= P_W(\ell)(\int_{\mathbb{R}} \overline{\phi(t)e^{it\ell(\widetilde{X(\ell)})}dt})\langle a_{\widetilde{\ell}}, \psi\rangle.\end{aligned}$$

In both cases, we see that $W{\cdot}a_{\widetilde{\ell}} = P_W(\ell)a_{\widetilde{\ell}}$. According to the induction hypothesis $P_W(\ell) = P_W(\widetilde{\ell})$ is a polynomial function on $\widetilde{\Omega}$ and hence also on Ω.

We can now assume, as we have seen before, that Ω is saturated with respect to $\widetilde{\mathfrak{g}}$, that for generic $\ell \in \Omega$, the L-orbits of $\ell|_{\mathfrak{l}}$ are saturated with respect to $\mathfrak{k}$ and that $\mathfrak{k} \not\subset \widetilde{\mathfrak{g}}$.

Recall again $\widetilde{\mathfrak{k}} := \mathfrak{k} \cap \widetilde{\mathfrak{g}}$. If $W \in \mathcal{U}(\widetilde{\mathfrak{g}})$, then the last computation tells us that

$$W \cdot a^{\widetilde{K}}_{\ell|_{\widetilde{\mathfrak{g}}}} = P_W(\ell|_{\widetilde{\mathfrak{g}}})a^{\widetilde{K}}_{\ell|_{\widetilde{\mathfrak{g}}}},$$

Since $\delta(\tilde{G}, \tilde{K}) < \delta(G, K)$, by the induction hypothesis, $P_W(\ell|_{\widetilde{\mathfrak{g}}})$ is a polynomial on the $\tilde{G}$-orbit of $\tilde{\ell}$.

Suppose that $P_\kappa(\ell) \neq 0$ and $ad^*(X_n)P_\kappa = 1$. Let $\widetilde{\kappa}_1 = Y_{\widetilde{r_0}} + \widetilde{U}, \widetilde{U} \in \mathcal{U}(\mathfrak{g}_{\widetilde{r_0}-1})$ be the e-central element of Corwin-Greenleaf in $\mathcal{U}(\widetilde{\mathfrak{g}})$ associated to $\mathfrak{k}_{n-1} \cap \widetilde{\mathfrak{g}}$ and $\tilde{G}$-orbit of $\tilde{\ell}$ as in (2.3.8). Then as in the proof of Corollary 2.3.6, we conclude that the denominator of the rational function P_W is a polynomial in $P_{\widetilde{\kappa}_1}(Ad^*(\exp(-P_\kappa(\ell)X_n))\ell)$. Since the denominator is also a polynomial in $P_\kappa(\ell)$, it follows that P_W is in fact a polynomial function.

Therefore we can finally assume that W is not contained in $\mathcal{U}(\widetilde{\mathfrak{g}})$. This means that $\mathfrak{b}[\ell|_{\mathfrak{k}}] \not\subset \widetilde{\mathfrak{k}}$ for generic $\ell \in \Omega$. This being assumed, we suppose that the denominator of the rational function $P_W(\ell)$ is not trivial. We are brought to the case where this denominator is equal to $P_{\kappa-c}(\ell)$ for some $c \in \mathbb{C}$. Take $\widetilde{X}$ in $\mathfrak{k}$. In these circumstances, there exists in $\mathcal{U}(\mathfrak{k})$ an element

$$\sigma = \bar{a}\widetilde{X} + \bar{b}, \ \bar{a}, \bar{b} \in \mathcal{U}(\widetilde{\mathfrak{k}}) \tag{2.3.9}$$

which is e-central for $\Omega|_{\mathfrak{k}}$. If W is of degree m relatively to $\widetilde{X}$ with the dominant term $w_m\widetilde{X}^m, w_m \in \mathcal{U}(\widetilde{\mathfrak{g}})$, we saw in Subsection 2.3.1.1 that w_m and $\bar{a}$ are $\mathfrak{k}$-invariant. Then, applying $\bar{a}$ and w_m to $a_\ell(\ell \in \Omega)$, we see that they commute each other modulo $\ker\pi$. Thus,

$$W_1 = \bar{a}^m W - w_m \sigma^m \tag{2.3.10}$$

is of degree inferior to m relatively to $\widetilde{X}$. Repeating this process, we build an element $\widetilde{W} \in \mathcal{U}(\widetilde{\mathfrak{g}})$ such that $P_{\widetilde{W}}(\ell)$ is a polynomial function on Ω. This means that α is a factor of $\bar{a}$.

Recall that j_0 is the smallest index such that $W \in \mathcal{U}(\mathfrak{k}_{j_0})$ modulo $\ker\pi$. We now prove the following:

Lemma 2.3.8 *There exists a K-diagonal element*

$$\nu = \beta X_{j_0} + \gamma, \ \beta, \gamma \in \mathcal{U}(\mathfrak{k}_{j_0-1}),$$

in $\mathcal{U}_\pi(\mathfrak{k}_{j_0})^{\mathfrak{k}}$ such that $P_\nu(\ell)$ extends to a polynomial function on Ω and such that β is not divisible by α modulo $\ker\pi$.

Proof 2.22 *We proceed by induction on* $\dim\mathfrak{k}$. *Let first* $\dim\mathfrak{k} = 1$, *namely $\mathfrak{k}$ is abelian. At each point $\ell \in \Omega$, the Penney's distribution a_ℓ is nothing but the Dirac measure at the unit element of G. Put $\mathfrak{b} = \cap_{\ell\in\Omega}\mathfrak{b}[\ell]$, which is an ideal of $\mathfrak{g}$. Then, the existence of W allows us to take X_{j_0} in $\mathfrak{b}$. This being done, $\nu = X_{j_0}$ suits us. Suppose now that* $\dim\mathfrak{k} > 1$. *Let us repeat the above construction of the element $\widetilde{W} \in \mathcal{U}_\pi(\widetilde{\mathfrak{g}})^{\mathfrak{k}}$ such that $P_{\widetilde{W}}(\ell) = \widetilde{P}_{\widetilde{W}}(\widetilde{\ell})$ extends to a polynomial function on Ω. Here, $\widetilde{\ell} = \ell|_{\widetilde{\mathfrak{g}}}$ and $\widetilde{P}$ designates the object obtained from the pair $(a_{\widetilde{\ell}}, \widetilde{\mathfrak{k}})$.*

In the first step of construction, if $w_m \notin \mathcal{U}(\widetilde{\mathfrak{k}}_{j_0-1})$, then we put $W' = w_m$ which belongs to $\mathcal{U}_\pi(\widetilde{\mathfrak{g}})^{\mathfrak{k}}$ but not in $\mathcal{U}(\widetilde{\mathfrak{k}}_{j_0-1})$, where $\widetilde{\mathfrak{k}}_{j_0-1} = \mathfrak{k}_{j_0-1} \cap \widetilde{\mathfrak{g}}$. Otherwise, the element W_1 defined in equation (2.3.10) does not belong to $\mathcal{U}(\mathfrak{k}_{j_0-1})$, and we replace W by W_1, and continue the construction of $\widetilde{W}$. At the end of this process, we get an element W' in $\mathcal{U}_\pi(\widetilde{\mathfrak{g}})^{\mathfrak{k}}$ but not in $\mathcal{U}(\widetilde{\mathfrak{k}}_{j_0-1})$.

Hence, by the induction hypothesis, there exists a $\widetilde{K}$-diagonal element

$$\widetilde{\nu} = \widetilde{a}X_{j_0} + \widetilde{b},\ \widetilde{a}, \widetilde{b} \in \mathcal{U}(\widetilde{\mathfrak{k}}_{j_0-1}),$$

in $\mathcal{U}_\pi(\widetilde{\mathfrak{k}}_{j_0})^{\widetilde{\mathfrak{k}}}$ such that $\widetilde{P}_{\widetilde{\nu}}(\widetilde{\ell})$ extends to a polynomial function on Ω and that $\widetilde{a}$ is not divisible by α modulo $\ker\pi$. *Since σ is e-central for $\Omega|_{\mathfrak{k}}$, it gives us the polynomial function $P_\sigma(\ell)$ when it is applied to Penney's distributions for $\widetilde{\mathfrak{k}}$. It follows that $[\sigma, \widetilde{\nu}] \in$* $\ker\pi$. *Thus, $\widetilde{\nu}$ turns out to be $\mathfrak{k}$-invariant and $P_{\widetilde{\nu}}(\ell) = \widetilde{P}_{\widetilde{\nu}}(\widetilde{\ell})$.*

We continue the proof of Theorem 2.3.4. Let us write

$$W = \sum_{j=0}^{r} w_j X_{j_0}^j,\ w_j \in \mathcal{U}(\mathfrak{k}_{j_0-1})(0 \le j \le r). \tag{2.3.11}$$

We go now to engage a double induction on the index $j_0 > d$ and on the degree r of X_{j_0} in the expression of W. As w_r is $\mathfrak{k}$-invariant, it follows from the induction hypothesis that $w_r \cdot a_\ell = P_{w_r}(\ell)a_\ell$ for $\ell \in \Omega$ with a function $P_{w_r}(\ell)$ which extends into a polynomial function on Ω. Next, in the expression (2.3.11), let us suppose our assertion established for the elements whose degree relative to X_{j_0} is inferior or equal to $r-1$. We see that

$$\widetilde{W} = \beta^r W - w_r \nu^r$$

is of degree inferior to r relative to X_{j_0} and hence $P_{\widetilde{W}}(\ell)$ is a polynomial function on Ω. One deduces from this that $P_W(\ell)$ is polynomial because β is not divisible by α.
□

Corollary 2.3.9 *Suppose that $\pi|_K$ has finite multiplicities. Then the rational function $\ell \mapsto P_W(\ell) = \Theta(W)(\ell)$ extends to a polynomial function on Ω, where Θ is defined as in equation (4.2.4) .*

2.3.5 Proof of Conjecture 2.3.1

Recall first the flag of subalgebras (2.3.2), where $\mathfrak{k} = \mathfrak{k}_d$, $j_0 \geq d+1$ the smallest index such that $W \in \mathcal{U}(\mathfrak{k}_{j_0})$ and α as given in equation (2.3.6). Let us first prove the following result, which could be regarded as a substitute to Lemma 2.3.8. Repeating this process, we get the element $\widetilde{\nu}$ in Lemma 2.3.8.

Proposition 2.3.10 *Let $m \leq d$ such that the generic K_m-orbits in $\Omega|_{\mathfrak{k}_m}$ are saturated with respect to $\mathfrak{k}_{m-1}$. Write $\mathfrak{k}_m = \mathbb{R}X_m + \mathfrak{k}_{m-1}$ for some $X_m \in \mathfrak{k}_m \setminus \mathfrak{k}_{m-1}$ and let*

$$\tau_m = a'_m X_{k_m} + b'_m, a'_m, b'_m \in \mathcal{U}(\mathfrak{k}_{k_m-1})$$

be an e-central element for $\Omega|_{\mathfrak{k}_{m-1}}$ which is not e-central for $\Omega|_{\mathfrak{k}_m}$ with the index k_m as small as possible. Then:

1. τ_m and $[X_m, \tau_m]$ can be chosen in a way that they are not divisible by α modulo $\ker\pi$.

2. Suppose that $\mathfrak{h}'_m = \mathfrak{k}_m + \mathfrak{g}_{j_0-1}$ is strictly included in $\mathfrak{h}_m = \mathfrak{k}_m + \mathfrak{g}_{j_0}$ and there exists $W_m \in \mathcal{U}_\pi(\mathfrak{h}_m)^{\mathfrak{k}_m}$ such that $W_m \notin \mathcal{U}_\pi(\mathfrak{h}'_m)^{\mathfrak{k}_m}$, which gives us a rational function on Ω when it is applied to Penney's distributions for $\mathfrak{k}_m$, then there exists an element

$$\nu_m = a_m X_{j_0} + b_m, \ a_m, b_m \in \mathcal{U}(\mathfrak{h}'_m),$$

where $\mathfrak{g}_{j_0} = \mathbb{R}X_{j_0} + \mathfrak{g}_{j_0-1}$, which is $\mathfrak{k}_m$-invariant and gives us a polynomial function on Ω when it is applied to Penney's distributions for $\mathfrak{k}_m$ and such that a_m is not divisible by α modulo $\ker\pi$.

Proof 2.23 *Let us proceed by induction on* $\dim\mathfrak{k}$. *The claim is trivial when* $\dim\mathfrak{k} \leq 3$. *We prove both the assertions at the same time in case of saturation. Let $4 \leq m \leq d$ and suppose that the generic orbits by $K_m = \exp(\mathfrak{k}_m)$ in $\Omega|_{\mathfrak{k}_m}$ are saturated with respect to $\mathfrak{k}_{m-1}$. Let*

$$\tau_m = a'_m X_{k_m} + b'_m, a'_m, b'_m \in \mathcal{U}(\mathfrak{k}_{k_m-1})$$

be a e-central element for $\Omega|_{\mathfrak{k}_{m-1}}$ which is not e-central for $\Omega|_{\mathfrak{k}_m}$ and which is not divisible by α. Choose the index k_m as small as possible.

Replacing $\mathfrak{k}$ by $\mathfrak{k}_m$, Lemma 2.3.8 gives us the element $\tau_m = a'_m X_{k_m} + b'_m$, with $a'_m, b'_m \in \mathcal{U}(\mathfrak{k}_{k_m-1})$ and a'_m is not divisible by α modulo $\ker\pi$. *Now $[X_m, \tau_m]$ is by construction in $\mathcal{U}(\mathfrak{k}_{k_m-1})$, thus it is not divisible by α modulo* $\ker\pi$.

This being done, suppose that there exists $W_m \in \mathcal{U}_\pi(\mathfrak{h}_m)^{\mathfrak{k}_m} \backslash \mathcal{U}_\pi(\mathfrak{h}'_m)$, where $\mathfrak{h}_m = \mathfrak{k}_m + \mathfrak{g}_{j_0}$, which gives us a rational function on Ω when it is applied to Penney's distributions for $\mathfrak{k}_m$ and let us build the element ν_m with the properties cited in the proposition.

By the saturation argument, we see that $W_m \in \mathcal{U}(\mathfrak{h}_{m-1})$ and that the Penney's distributions for $\mathfrak{k}_m$ are the same as those for $\mathfrak{k}_{m-1}$. Therefore, by the induction hypothesis, there exists a K_{m-1}-diagonal element

$$\nu_{m-1} = a_{m-1} X_{j_0} + b_{m-1}, \ a_{m-1}, b_{m-1} \in \mathcal{U}(\mathfrak{h}'_{m-1})$$

in $\mathcal{U}(\mathfrak{h}_{m-1})$ which gives us a polynomial function on Ω when it is applied to Penney's distributions for $\mathfrak{k}_{m-1}$ and such that a_{m-1} is not divisible by α. If ν_{m-1} is $\mathfrak{k}_m$-invariant, it is qualified as our desired ν_m. Suppose that ν_{m-1} is not $\mathfrak{k}_m$-invariant and retake the construction of our ν introduced in [15]. For a sufficiently large integer $v \in \mathbb{N}$, we consider

$$\psi = \nu_{m-1} + F(\tau_m),$$

where $F(t)$ is a polynomial in one variable t of degree $2v$. For $k \in \mathbb{N}$, put

$$\psi_0 = \psi, \ \psi_k = (ad(X_m))^k(\psi).$$

Remark that $[X_m, [X_m, \tau_m]] \in \ker \pi$. Therefore, if v is sufficiently large, then

$$\psi_{2v} \notin \ker \pi, \ \psi_{2v+1} \in \ker \pi.$$

We now build an element of $\mathcal{U}(\mathfrak{h}_{m-1})^{\mathfrak{k}_m}$ by the formula

$$\begin{aligned}\nu_m =& (\psi_0\psi_{2v} + \psi_{2v}\psi_0) - (\psi_1\psi_{2v-1} + \psi_{2v-1}\psi_1) + \cdots \\ &+ (-1)^{v-2}(\psi_{v-2}\psi_{v+2} + \psi_{v+2}\psi_{v-2}) \\ &+ (-1)^{v-1}(\psi_{v-1}\psi_{v+1} + \psi_{v+1}\psi_{v-1}) + (-1)^v\psi_v^2.\end{aligned}$$

Remark once again the fact that v is sufficiently large. This assures that ν_m is of degree 1 with respect to X_{j_0}. Moreover, $[X_m, \nu_{m-1}]$ applied to a_ℓ gives us a polynomial function on Ω. Indeed, we see by definition that

$$P_{[X_m,\nu_{m-1}]}(\ell) = \frac{d}{dt}P_{\nu_{m-1}}(\exp(tX_m)\cdot\ell)\big|_{t=0}, \ \ell \in \Omega.$$

It follows that $\nu_m \cdot a_\ell = P_{\nu_m}(\ell)a_\ell$ for generic $\ell \in \Omega$ with a polynomial function $P_{\nu_m}(\ell)$ on Ω.

Finally, since, for any k, $(ad(X_n))^k F(\tau_m)$ belongs to $\mathcal{U}(\mathfrak{h}'_m)$, we can choose the polynomial F such that $\nu_m = a_m X_{j_0} + b_m$, $a_m, b_m \in \mathcal{U}(\mathfrak{h}'_m)$ and a_m not divisible by α modulo $\ker \pi$. Indeed, let

$$F(t) = \lambda_0 + \lambda_1 t + \cdots + \lambda_{2v-1}t^{2v-1} + \lambda_{2v}t^{2v}, \ \lambda_j \in \mathbb{C} \ (0 \leq j \leq 2v).$$

Suppose that $(ad(X_m))^k(a_{m-1})$ $(0 \leq k \leq n_0)$ are not divisible by α modulo $\ker \pi$, but that $(ad(X_m))^{n_0+1}(a_{m-1})$ and hence all the elements $(ad(X_m))^k(a_{m-1})$, $k \geq n_0 + 1$ are divisible by α modulo $\ker \pi$.

Considering the λ_j as variables, and supposing that for any choice of these variables, the coefficient a_m of X_{j_0} in ν_m is divisible by α modulo $\ker \pi$, thus for any j, the coefficient of $\lambda_j X_{j_0}$ of ν_m is divisible by α modulo $\ker \pi$. Remark now that the terms $\lambda_{2v-n_0}X_{j_0}$ in ν_m appear only in the sum:

$$\sum_{k \geq n_0}(-1)^k(\psi_k\psi_{2v-k} + \psi_{2v-k}\psi_k) \equiv 2\sum_{k \geq n_0}(-1)^k\psi_{2v-k}\psi_k \ (mod \ \ker \pi)$$

and they are modulo $\ker \pi$:

$$\Big(\sum_{k \geq n_0} c_k(adX_m)^{2v-k}({\tau_m}^{2v-n_0})(adX_m)^k a_{m-1}\Big)\lambda_{2v-n_0}X_{j_0},$$

where c_k is a numerical constant. Each term in this sum is divisible by α except the first one, by definition of n_0. This proves that there is a polynomial F such that the conditions of the proposition hold for ν_m.

Now, suppose that the generic K_m-orbits in $\Omega|_{\mathfrak{k}_m}$ are non-saturated with respect to $\mathfrak{k}_{m-1}$. Then, there exists an element

$$\sigma_m = c_m X_m + d_m,\ \ c_m, d_m \in \mathcal{U}(\mathfrak{k}_{m-1})$$

which is e-central for $\Omega|_{\mathfrak{k}_m}$. If

$$W_m = v_r X_m^r + v_{r-1} X_m^{r-1} + \cdots + v_1 X_m + v_0,\ \ v_j \in \mathcal{U}(\mathfrak{h}_{m-1})(0 \leq j \leq r)$$

with $v_r \notin \ker\pi (r > 0)$, $W_m' = c_m^r W_m - \sigma_m^r v_r$ is $\mathfrak{k}_m$-invariant and of degree smaller or equal to $r-1$ relative to X_m because v_r, c_m are also $\mathfrak{k}_m$-invariant and commute each other modulo $\ker\pi$. Repeating these manipulations if necessary, we arrive to a $\mathfrak{k}_m$-invariant element $W_{m-1} \in \mathcal{U}(\mathfrak{h}_{m-1})$ which gives us a rational function on Ω when it is applied to Penney's distributions. From the induction hypothesis there exists a $\mathfrak{k}_{m-1}$-invariant element ν_{m-1} which satisfies the required conditions as above. Applying ν_{m-1}, σ_m to Penney's distributions for $\mathfrak{k}_{m-1}$, we confirm that they commute each other modulo $\ker\pi$. In this way, ν_{m-1} turns out to be $\mathfrak{k}_m$-invariant and is qualified as our desired ν_m.

Corollary 2.3.11 *Let $\mathfrak{h}$ be a subalgebra of $\mathfrak{k}$ and $\mathfrak{h}'$ an ideal of codimension 1 in $\mathfrak{h}$ such that the generic orbits by $H = \exp\mathfrak{h}$ in $\Omega|_{\mathfrak{h}}$ are saturated with respect to $\mathfrak{h}'$. Let*

$$\tau = a' X_{k'} + b',\ \ a', b' \in \mathcal{U}(\mathfrak{k}_{k'-1}),\ \ a' \notin \ker\pi,$$

be a e-central element for $\Omega|_{\mathfrak{h}'}$, which is not e-central for $\Omega|_{\mathfrak{h}}$ for which k' is minimal. Then τ and $[X, \tau]$ can be chosen on a way that they are not divisible by α, where $\mathfrak{h} = \mathbb{R}X + \mathfrak{h}'$.

We now look at the surjectivity of the homomorphism Θ defined by equation (4.2.4). We first record the following, which will be of use later

Proposition 2.3.12 *([15], Proposition 4.4) Keep the same notations and hypotheses and let us denote by y' the variable corresponding to the polynomial function defined as in equation (2.3.6). Then for every polynomial $\zeta(x) \in \mathbb{C}[\Omega]^K$, there exists a polynomial $s(y')$ of y' such that the product $s(y')\zeta(x)$ is in the image of Θ.*

Let $\mathcal{V}$ be the set of K-diagonal elements $W \in \mathcal{U}_\pi(\mathfrak{g})^{\mathfrak{k}}$ such that $W \cdot a_\ell = P_W(\ell) a_\ell$ with a function $P_W(\ell)$ which extends to a polynomial function on Ω. We consider the image M of the mapping

$$\Theta_{\mathcal{V}} : \mathcal{V} \ni W \mapsto P_W \in \mathbb{C}[\Omega]^K.$$

We now prove the following:

Proposition 2.3.13 *Let $q(\ell) \in \mathbb{C}[\Omega]^K$. If there exists $0 \neq u(\ell) \in M$ such that the product $u(\ell)q(\ell)$ belongs to M, then the function $q(\ell)$ itself belongs to M.*

Proof 2.24 *We proceed by induction on $\dim G + \dim(G/K)$. Let $u(\ell) = P_{W_1}(\ell)$ and $u(\ell)q(\ell) = P_{W_2}(\ell)$ with $W_1, W_2 \in \mathcal{V}$. Examine first the case where $\mathfrak{k} = \{0\}$. Put $\mathfrak{b} = \cap_{\ell\in\Omega}\mathfrak{b}[\ell]$, which is an ideal of $\mathfrak{g}$. It is seen that $\mathcal{U}(\mathfrak{b})$ is identified modulo $\ker\pi$ to the symmetric algebra $S(\mathfrak{b})$ of $\mathfrak{b}$ because $[\mathfrak{b}, \mathfrak{b}] \subset \ker\pi$. Then, W_1, W_2 belong to $\mathcal{U}(\mathfrak{b}) \simeq S(\mathfrak{b})$ and W_2 is divisible by W_1, namely that there exists $W \in S(\mathfrak{b}) \simeq \mathcal{U}(\mathfrak{b})$ such that $W_2 = W_1W$. It is clear that $W \in \mathcal{U}_\pi(\mathfrak{g})^{\mathfrak{k}}$ and $P_W(\ell) = q(\ell)$. In sum, $q(\ell) \in M$.*

Suppose that $\dim\mathfrak{k} \geq 1$. Keep the notations introduced before. When Ω is non-saturated with respect to $\mathfrak{k}_{n-1}$, W_1, W_2 are taken in $\mathcal{U}_\pi(\mathfrak{k}_{n-1})^{\mathfrak{k}}$ and the result derives immediately from the induction hypothesis.

Suppose that Ω is saturated with respect to $\mathfrak{k}_{n-1}$. It follows that $W_1, W_2 \in \mathcal{U}(\mathfrak{k}_{n-1})$ and that $q(\ell)$ depends only on $\ell' = \ell|_{\mathfrak{k}_{n-1}}$. For almost all $t \in \mathbb{R}$, there exists by the induction hypothesis an element $W_t \in \mathcal{U}_{\pi_t}(\mathfrak{k}_{n-1})^{\mathfrak{k}}$ verifying $P_{W_t}(\ell') = q(\ell')$ for almost all $\ell' \in \omega_t$. Here, W_t depends rationally on $t \in \mathbb{R}$. By Proposition 2.3.12, there exists a polynomial $s(y')$ of $y' = P_\kappa(\ell)$ such that $s(y')q(\ell) \in M$.

Now take an ideal $\widetilde{\mathfrak{g}} \neq \mathfrak{k}_{n-1}$ of codimension 1 in $\mathfrak{g}$. Suppose first that Ω is non-saturated with respect to $\widetilde{\mathfrak{g}}$. Then W_1, W_2 are in $\mathcal{U}(\widetilde{\mathfrak{g}})$ modulo $\ker\pi$. If $\mathfrak{k} \subset \widetilde{\mathfrak{g}}$, the induction hypothesis provides us the result. If $\mathfrak{k} \not\subset \widetilde{\mathfrak{g}}$, put $\widetilde{\mathfrak{k}} = \mathfrak{k} \cap \widetilde{\mathfrak{g}}$ and $\widetilde{K} = \exp\widetilde{\mathfrak{k}}$. The induction hypothesis assures that there exists a $\widetilde{K}$-diagonal $\widetilde{W} \in \mathcal{U}_\pi(\widetilde{\mathfrak{g}})^{\widetilde{\mathfrak{k}}}$ so that we have $\widetilde{P}_{\widetilde{W}}(\ell) = q(\ell)$. Since $q(\ell)$ is $\mathfrak{k}$-invariant, $\widetilde{W}$ turns out to be $\mathfrak{k}$-invariant and hence $P_{\widetilde{W}}(\ell) = q(\ell)$. In this way, $q(\ell) \in M$.

Recall now our previous notations: $\mathfrak{g}' = \mathfrak{k}_{n-1}$, κ its corresponding e-central element and y' as in equation (2.3.6). Suppose that Ω is saturated with respect to $\widetilde{\mathfrak{g}}$. If $\mathfrak{k} \subset \widetilde{\mathfrak{g}}$, W_1, W_2 belong to $\mathcal{U}(\widetilde{\mathfrak{g}})$. As above, there exists a polynomial $\widetilde{s}(\widetilde{y})$ of $\widetilde{y} = P_{\widetilde{\kappa}}(\ell)$ such that $\widetilde{s}(\widetilde{y})q(\ell) \in M$. Let $s(y')q(\ell) = P_{W'}(\ell)$ and $\widetilde{s}(\widetilde{y})q(\ell) = P_{\widetilde{W}}(\ell)$ for some $W', \widetilde{W} \in \mathcal{V}$. Then, $\widetilde{s}(\widetilde{\kappa})W' \equiv s(\kappa)\widetilde{W}$ modulo $\ker\pi$. Therefore, W' must be divisible modulo $\ker\pi$ by $s(\kappa)$ and $W' \equiv s(\kappa)W$ modulo $\ker\pi$ with a certain K-diagonal $W \in \mathcal{U}_\pi(\mathfrak{g})^{\mathfrak{k}}$. Thus, $q(\ell) = P_W(\ell)$.

Finally, suppose that $\mathfrak{k}$ is not found in $\widetilde{\mathfrak{g}}$. We shall argue similarly as in the proof of Theorem 2.3.4. If $\mathfrak{b}[\ell|_{\mathfrak{k}}] \subset \widetilde{\mathfrak{k}}$ almost everywhere on Ω, W_1, W_2 belong to $\mathcal{U}(\widetilde{\mathfrak{g}})$ and hence $q(\ell)$ depends only on $\ell|_{\widetilde{\mathfrak{g}}}$. From the induction hypothesis applied to $\widetilde{\mathfrak{k}}$, there exists a $\widetilde{K}$-diagonal $\widetilde{W} \in \mathcal{U}_\pi(\widetilde{\mathfrak{g}})^{\widetilde{\mathfrak{k}}}$ such that $q(\ell) = \widetilde{P}_{\widetilde{W}}(\ell)$. Since $q(\ell)$ is $\mathfrak{k}$-invariant, $\widetilde{W}$ is $\mathfrak{k}$-invariant too and $\widetilde{P}_{\widetilde{W}}(\ell) = P_{\widetilde{W}}(\ell)$. Therefore, $q(\ell) \in M$.

We place in the last possibility where $\mathfrak{b}[\ell|_{\mathfrak{k}}] \not\subset \widetilde{\mathfrak{k}}$ almost everywhere on Ω. It is sufficient for us to treat the case where $s(y') = \alpha$ which is a polynomial in y' of degree 1.

Let j_0 be the smallest index such that $q(\ell)$ belongs to the symmetric algebra $S(\mathfrak{k}_{j_0}) = \mathbb{C}[\mathfrak{k}_{j_0}^*]$ of $\mathfrak{k}_{j_0}$ with respect to the sequence (2.3.2) of subalgebras. Aligning back to Subsection 2.3.1.1, let $\{Y_k\}_{k=1}^n$ be a Jordan-Hölder basis of $\mathfrak{g}$ adapted to the flag (2.3.1) and let

$$S = \{s_1 < \cdots < s_r\}$$

be the set of jump indices for Ω *with respect to the flag (2.3.1) which appear in* $\mathfrak{k}_{j_0}$. *Set* $x_i = \ell(Y_{s_i})$ *for* $1 \leq i \leq r$, *where* $Y_{s_r} = X_{j_0}$ *changing the ordering. So,* $q(\ell)$ *depends on* $\{x_1, \ldots, x_r\}$. *Write*

$$q(\ell) = \sum_{j=0}^{v} q_j(\ell) x_r^j,$$

where $q_j(\ell)(0 \leq j \leq v)$ *are polynomial functions of* $x_1, \ldots, x_{r-1}$.

Everything as in the proof of Lemma 2.3.8, we now prove by induction on the dimension of $\mathfrak{k}$ *that there exists in* M *an element*

$$\nu(\ell) = \beta(\ell) x_r + \gamma(\ell),$$

where $\beta(\ell), \gamma(\ell)$ *are polynomials of* $\{x_1, \ldots, x_{r-1}\}$ *and where* $\beta(\ell) \in M$ *is not divisible by* α. *Indeed, assume first that* $j_0 > d$. *Making use of the e-central element* σ *for* $\Omega|_{\mathfrak{k}}$ *as in equation (2.3.9), one finds in* $S(\mathfrak{k}_{j_0} \cap \tilde{\mathfrak{g}}) \cap \mathbb{C}[\Omega]^{\tilde{K}}$, *an element* $\tilde{q}(\ell)$ *outside* $S(\mathfrak{k}_{j_0-1})$ *such that* $\alpha\tilde{q}(\ell) \in \tilde{M}$, *the corresponding set for* $\tilde{\mathfrak{k}}$. *By the induction hypothesis, there exists in* $\tilde{M}$ *an element*

$$\tilde{\nu}(\ell) = \tilde{\beta}(\ell) x_r + \tilde{\gamma}(\ell),$$

where $\tilde{\beta}(\ell), \tilde{\gamma}(\ell)$ *are polynomials of* $\{x_1, \ldots, x_{r-1}\}$ *and where* $\tilde{\beta}(\ell) \in \tilde{M}$ *is not divisible by* α. *Now, using the element* σ *as in (2.3.9),* $\tilde{\nu}(\ell)$ *turns out to be* K*-invariant and hence belongs to* M *as is to be shown.*

When $j_0 \leq d$, *we first prove the following lemma.*

Lemma 2.3.14 *We regard the symmetric algebra* $S(\mathfrak{k})$ *of* $\mathfrak{k}$ *as the algebra of polynomial functions on* $\Omega|_{\mathfrak{k}}$ *through the evaluation* $\Omega|_{\mathfrak{k}} \ni \ell \mapsto \sqrt{-1}\ell(X)$ *for* $X \in \mathfrak{k}$. *Let* $\zeta : S(\mathfrak{k}) \to \mathcal{U}(\mathfrak{k})$ *be the symmetrisation map. Then,* $\zeta(q)$ *is* K*-diagonal and*

$$\zeta(q) \cdot a_\ell = q(\ell) a_\ell, \ \ \ell \in \Omega|_{\mathfrak{k}}.$$

Proof 2.25 *We proceed by induction on* $\dim \mathfrak{k}$. *When* $\dim \mathfrak{k} = 1$, *the claim is trivial. Let* $\mathfrak{z}(\mathfrak{k})$ *be the centre of* $\mathfrak{k}$. *If* $\dim \mathfrak{z}(\mathfrak{k}) = 1$, $\mathfrak{z}(\mathfrak{k})$ *is nothing but the centre* $\mathfrak{z}$ *of* $\mathfrak{g}$. *As* $\pi|_{\mathfrak{z}} \neq 0$, $q \in S(\mathfrak{k}')$ *where* $\mathfrak{k}'$ *denotes the centraliser of* $\mathfrak{k}_2$ *in* $\mathfrak{k}$, *where* $\mathfrak{k}_2$ *is as in the flag (2.3.1). Since* $\mathfrak{b}[\ell|_{\mathfrak{k}}] \subset \mathfrak{k}'$ *for* $\ell \in \Omega$, *we can apply the induction hypothesis to* $\mathfrak{k}'$. *Suppose* $\dim \mathfrak{z}(\mathfrak{k}) \geq 2$. *For* $\ell \in \Omega|_{\mathfrak{k}}$, *we put* $\mathfrak{a} = \mathfrak{z}(\mathfrak{k}) \cap \ker \ell$, $\bar{\mathfrak{k}} = \mathfrak{k}/\mathfrak{a}$ *and* $\bar{\ell} \in (\bar{\mathfrak{k}})^*$ *such that* $\bar{\ell} \circ p = \ell$ *with the canonical projection* $p : \mathfrak{k} \to \bar{\mathfrak{k}}$. *Let* $a_{\bar{\ell}}$ *be the Penney distribution of* $\bar{\mathfrak{k}}$ *at* $\bar{\ell}$. *Then, we have* $\bar{\zeta}(\bar{q}) \cdot a_{\bar{\ell}} = \bar{q}(\bar{\ell}) a_{\bar{\ell}}$ *from the induction hypothesis applied to* $\bar{\mathfrak{k}}$. *Here,* $\bar{\zeta} : S(\bar{\mathfrak{k}}) \to \mathcal{U}(\bar{\mathfrak{k}})$ *denotes the symmetrisation map and* $\bar{q} \in S(\bar{\mathfrak{k}})$ *is such that* $\bar{q} \circ p = q$. *Thus, we get the claim.*

Now if $j_0 > d$, *we use assertion 2 of Proposition 2.3.10 to argue similarly as in the previous case.*

We now utilise a new induction on the degree v *of* q *relatively to* x_r. *If so,*

$$\beta(\ell)^v q(\ell) - q_v(\ell) \nu(\ell)^v$$

is of degree smaller than v relatively to x_r and hence belongs to M. Thus, $\beta(\ell)^v q(\ell) \in M$. Let

$$\beta(\ell)^v = P_{W_3}(\ell), \ \beta(\ell)^v q(\ell) = P_{W_4}(\ell)$$

with $W_3, W_4 \in \mathcal{V}$. Then,

$$\alpha\beta(\ell)^v q(\ell) = P_{W'}(\ell)P_{W_3}(\ell) = P_{\kappa'}(\ell)P_{W_4}(\ell),$$

where κ' is the polynomial in κ of degree 1 such that $P_{\kappa'}(\ell) = \alpha$. In other words,

$$W'W_3 \equiv \kappa' W_4$$

modulo $\ker \pi$. *Because W_3 is not divisible by κ', W' must be divisible by κ'. Consequently, $q(\ell)$ belongs to M.*

Remark 2.3.15 It is worth-noting here that by a result of M. Duflo (cf. [39]), for any $\sigma \in \widehat{K}$ and any $q \in S(\mathfrak{k})^K$, $\sigma(\zeta(q)) = q(\ell)id$ for any ℓ in the orbit associated to σ in $\mathfrak{k}^*$. It remains unclear to us whether this results provides directly a proof of Lemma 2.3.14.

Corollary 2.3.16 *Keep the same notation and assume that $\pi|_K$ has finite multiplicities, then the mapping Θ defined by equation (4.2.4) is surjective.*

Corollaries 2.3.9 and 2.3.16 allow to complete the proof of Conjecture 2.3.1. We have the following:

Theorem 2.3.17 *Let $G = \exp \mathfrak{g}$ be a connected and simply connected nilpotent Lie group. Then Conjecture 2.3.1 holds. That is, when $\pi|_K$ has finite multiplicities, the mapping Θ gives by passing to the quotient an isomorphism of algebras from $D_\pi(G)^K$ to the algebra $\mathbb{C}[\Omega(\pi)]^K$ of the K-invariant polynomial functions on the orbit $\Omega(\pi)$.*

Remark 2.3.18 If c is purely imaginary, then we can take $x_{i_0} = P_{\sqrt{-1}\kappa}(\ell)$ as a coordinate on Ω (cf. [15]) and the polynomial function $P_{\alpha W}(\ell)$ on Ω vanishes on the section

$$\{\ell \in \Omega; x_{i_0} = -\sqrt{-1}c\}$$

Hence, the polynomial function $P_{\alpha W}(\ell)$ must be divisible by $P_\alpha(\ell)$ and the function $P_W(\ell)$ itself is polynomial. More in detail, let $\{X, Y, Z\}$ be the triple of Kirillov such that $\mathfrak{z} = \mathbb{R}Z$. Suppose for example that $\mathfrak{k}_{n-1}$ is equal to the centraliser $\mathfrak{g}'$ of Y in $\mathfrak{g}$. Hence, $\kappa = Y$ and $\alpha = Y - c$. Put $G' = \exp(\mathfrak{g}')$. For $\ell' \in (\mathfrak{g}')^*$, we denote by $\rho_{\ell'}$ the irreducible representation of $\mathcal{U}(\mathfrak{g}')$ associated to ℓ'. Then, the theory of induced representations of Lie algebras says that $\rho_{\ell'}$ vanishes at αW if $\ell'(Y) = c$. Now, denote by $\pi'_{\ell'}$ the irreducible unitary representation of G' associated to ℓ'. Suppose that c is purely imaginary. As $\ker(\pi'_{\ell'}) = \ker(\rho_{i\ell'})$, $\pi'_{\ell'}(\alpha W)$ vanishes on the section

$$\{\ell \in \Omega; i\ell(Y) = c\} = \{\ell \in \Omega; \ell(Y) = -ic\}.$$

This means that $P_{\alpha W}(\ell)$ must be divisible by $P_\alpha(\ell) = i(\ell(Y) + ic)$. In consequence, $P_W(\ell)$ turns out to be polynomial on Ω.

2.3.6 Examples

Here, we give some examples.

Example 2.3.8 Let $\mathfrak{g} = \langle X, P_1, P_2, Q, Y, Z\rangle_{\mathbb{R}}$ with the non-zero brackets

$$[X, P_1] = P_2, [X, Y] = Z, [P_1, Q] = Y, [P_2, Q] = Z.$$

Let $\pi \in \widehat{G}$ such that $Z^* \in \Omega(\pi)$ and $\mathfrak{k} = \langle P_1, Y, Z\rangle_{\mathbb{R}}$. Then, a direct computation shows that an element $\ell \in \Omega(\pi)$ is written as

$$\ell = Z^* + x_1 Y^* + x_2 Q^* + x_3 P_2^* + x_1 x_3 P_1^* + x_4 X^*,\ x_j \in \mathbb{R}\ (1 \leq j \leq 4).$$

From this, $P_1 - x_1 P_2 - x_3 Y \in \mathfrak{g}(\ell)$. On the other hand, we have

$$\langle a_\ell, \varphi\rangle = \int_{\mathbb{R}} \overline{\varphi(\exp(tP_1))} e^{itx_1x_3} dt$$

for $\varphi \in \mathcal{H}_\pi^\infty$. So,

$$\begin{aligned}
\langle P_2 a_\ell, \varphi\rangle &= \frac{d}{ds} \int_{\mathbb{R}} \overline{\varphi(\exp(sP_2)\exp(tP_1))} e^{itx_1x_3} dt\Big|_{s=0} \\
&= \frac{d}{ds} \int_{\mathbb{R}} \overline{\varphi\left(\exp\left(\left(t + \frac{s}{x_1}\right) P_1\right) \exp\left(sP_2 - \frac{s}{x_1} P_1\right)\right)} e^{itx_1x_3} dt\Big|_{s=0} \\
&= \frac{d}{ds} \int_{\mathbb{R}} \overline{\varphi\left(\exp\left(\left(t + \frac{s}{x_1}\right) P_1\right)\right)} e^{itx_1x_3} dt\Big|_{s=0} \\
&= \frac{d}{ds} \int_{\mathbb{R}} \overline{\varphi(\exp(tP_1))} e^{i\left(t - \frac{s}{x_1}\right)x_1x_3} dt\Big|_{s=0} \\
&= \left(\frac{d}{ds} e^{-isx_3}\right)\Big|_{s=0} \int_{\mathbb{R}} \overline{\varphi(\exp(tP_1))} e^{itx_1} dt = -ix_3 \langle a_\ell, \varphi\rangle.
\end{aligned}$$

for $\ell \in \Omega(\pi)$ satisfying $x_1 = \ell(Y) \neq 0$. Put $\Phi(x,t) = \varphi(\exp(xX)\exp(tP_2))$ for $(x,t) \in \mathbb{R}^2$ and compute:

$$\begin{aligned}
(P_1\Phi)(x,t) &= \frac{d}{ds}\varphi(\exp(-sP_1)\exp(xX)\exp(tP_2))|_{s=0} \\
&= \frac{d}{ds}\varphi(\exp(xX)\exp(-sP_1 + sxP_2)\exp(tP_2))|_{s=0} \\
&= \frac{d}{ds}\varphi(\exp(xX)\exp((t + sx)P_2)\exp(-sP_1 + sx_1P_2 \\
&\quad + sx_3Y)\exp(-sx_1P_2 - sx_3Y))|_{s=0} \\
&= \frac{d}{ds}\varphi(\exp(xX)\exp((t + sx - sx_1)P_2))|_{s=0} = (x - x_1)\frac{\partial\Phi}{\partial t}(x,t), \\
(P_2\Phi)(x,t) &= \frac{d}{ds}\varphi(\exp(-sP_2)\exp(xX)\exp(tP_2))|_{s=0} = -\frac{\partial\Phi}{\partial t}(x,t), \\
(Y\Phi)(x,t) &= \frac{d}{ds}\varphi(\exp(-sY)\exp(xX)\exp(tP_2))|_{s=0} \\
&= \frac{d}{ds}\varphi(\exp(xX)\exp(-sY + xsZ)\exp(tP_2))|_{s=0} = i(x_1 - x)\Phi(x,t).
\end{aligned}$$

Taking these observations into account, we check that $P_1 + iYP_2$ belongs to $\ker\pi$.

Example 2.3.9 Let $\mathfrak{g} = \langle X, P_1, P_2, P_3, Q_1, Q_2, Y, Z\rangle_{\mathbb{R}}$ with the non-zero brackets

$$[X, P_1] = P_2, [X, P_2] = P_3, [P_1, Q_1] = Q_2,$$
$$[X, Q_2] = [P_2, Q_1] = Y, [X, Y] = [P_3, Q_1] = Z.$$

Let $\pi \in \widehat{G}$ such that $Z^* \in \Omega(\pi)$ and $\mathfrak{k} = \langle P_1, Y, Z\rangle_{\mathbb{R}}$. Then, a direct computation shows that an element $\ell \in \Omega(\pi)$ is written as

$$\ell = Z^* + x_1 Y^* + \frac{x_1^2}{2} Q_2^* + x_2 Q_1^* + x_3 P_3^* + x_1 x_3 P_2^* + \frac{x_1^2 x_3}{2} P_1^* + x_4 X^*, \ x_j \in \mathbb{R}\ (1 \leq j \leq 4).$$

From this, $P_1 - \frac{x_1^2}{2} P_3 - x_1 x_3 Y, P_2 - x_1 P_3 - x_3 Y$ belong to $\mathfrak{g}(\ell)$. On the other hand, we have

$$\langle a_\ell, \varphi\rangle = \int_{\mathbb{R}} \overline{\varphi(\exp(tP_1))e^{\frac{ix_1^2 x_3}{2}t}}\, dt$$

for $\varphi \in \mathcal{H}_\pi^\infty$. So,

$$\begin{aligned}
\langle P_3 a_\ell, \varphi\rangle &= \frac{d}{ds}\int_{\mathbb{R}} \overline{\varphi(\exp(sP_3)\exp(tP_1))e^{\frac{ix_1^2 x_3}{2}t}}\, dt\Big|_{s=0} \\
&= \frac{d}{ds}\int_{\mathbb{R}} \overline{\varphi\left(\exp(tP_1)\exp\left(\frac{2s}{x_1^2}P_1\right)\exp\left(sP_3 - \frac{2s}{x_1^2}P_1\right)\right)e^{\frac{ix_1^2 x_3}{2}t}}\, dt\Big|_{s=0} \\
&= \frac{d}{ds}\int_{\mathbb{R}} \overline{\varphi\left(\exp\left(\left(t + \frac{2s}{x_1^2}\right)P_1\right)\right)e^{\frac{ix_1^2 x_3}{2}t}}\, dt\Big|_{s=0} \\
&= \frac{d}{ds}\int_{\mathbb{R}} \overline{\varphi(\exp(tP_1))e^{\frac{ix_1^2 x_3}{2}\left(t - \frac{2s}{x_1^2}\right)}}\, dt\Big|_{s=0} \\
&= \left(\frac{d}{ds}e^{-isx_3}\right)\Big|_{s=0}\int_{\mathbb{R}} \overline{\varphi(\exp(tP_1))e^{\frac{ix_1^2 x_3}{2}t}}\, dt = -ix_3\langle a_\ell, \varphi\rangle,
\end{aligned}$$

for $\ell \in \Omega(\pi)$ satisfying $x_1 = \ell(Y) \neq 0$. Put $\Phi(x, t) = \varphi(\exp(xX)\exp(tP_3))$ for $(x, t) \in \mathbb{R}^2$ and compute:

$$\begin{aligned}
(P_1\Phi)(x, t) &= \frac{d}{ds}\varphi(\exp(-sP_1)\exp(xX)\exp(tP_3))\Big|_{s=0} \\
&= \frac{d}{ds}\varphi\left(\exp(xX)\exp\left(-sP_1 + sxP_2 - \frac{sx^2}{2}P_3\right)\exp(tP_3)\right)\Big|_{s=0} \\
&= \frac{d}{ds}\varphi\left(\exp(xX)\exp\left(\left(t - \frac{sx^2}{2}\right)P_3\right)\exp(sxP_2 - sP_1)\right)\Big|_{s=0} \\
&= \frac{d}{ds}\varphi\left(\exp(xX)\exp\left(\left(t - \frac{sx^2}{2} + s\left(xx_1 - \frac{x_1^2}{2}\right)\right)P_3\right)\times\right. \\
&\quad \left.\times \exp\left(sxP_2 - sP_1 - s\left(xx_1 - \frac{x_1^2}{2}\right)P_3\right)\right)\Big|_{s=0}
\end{aligned}$$

$$= \frac{d}{ds}\varphi\left(\exp(xX)\exp\left(\left(t - \frac{sx^2}{2} + s\left(xx_1 - \frac{x_1^2}{2}\right)\right)P_3\right)\right)\Big|_{s=0}$$
$$= \left(xx_1 - \frac{x_1^2}{2} - \frac{x^2}{2}\right)\frac{\partial\Phi}{\partial t},$$
$$(P_3\Phi)(x,t) = \frac{d}{ds}\varphi(\exp(-sP_3)\exp(xX)\exp(tP_3))\Big|_{s=0} = -\frac{\partial\Phi}{\partial t},$$
$$(Y\Phi)(x,t) = \frac{d}{ds}\varphi(\exp(-sY)\exp(xX)\exp(tP_3))\Big|_{s=0}$$
$$= \frac{d}{ds}\varphi(\exp(xX)\exp(tP_3)\exp(-sY + sxZ))\Big|_{s=0}$$
$$= \left(\frac{d}{ds}e^{-i(-sx_1+sx)}\right)\Big|_{s=0}\Phi(x,t)$$
$$= i(x_1 - x)\Phi(x,t).$$

Taking these observations into account, we check that $P_1 - \frac{Y^2}{2}P_3$ belongs to $\ker\pi$. Finally,

$$(P_2\Phi)(x,t) = \frac{d}{ds}\varphi(\exp(-sP_2)\exp(xX)\exp(tP_3))\Big|_{s=0}$$
$$= \frac{d}{ds}\varphi(\exp(xX)\exp(-sP_2 + sxP_3)\exp(tP_3))\Big|_{s=0}$$
$$= \frac{d}{ds}\varphi(\exp(xX)\exp((t + sx + -sx_1)P_3)\exp(-sP_2 + sx_1P_3))\Big|_{s=0}$$
$$= \frac{d}{ds}\Phi(x, t + s(x - x_1))\Big|_{s=0} = (x - x_1)\frac{\partial\Phi}{\partial t}.$$

Accordingly, $P_2 + iYP_3$ belongs to $\ker\pi$.

Now we give some illustrative examples on the polynomial conjecture.

Example 2.3.10 Let $\mathfrak{g} = \langle X,T,U,V,Y,Z\rangle_{\mathbb{R}}$ with the non-zero brackets $[X,Y] = Z, [X,T] = U, [T,V] = Y, [U,V] = Z$. Let $G = \exp\mathfrak{g}, f = Z^* \in \mathfrak{g}^*, \Omega = G{\cdot}f, \mathfrak{k} = \langle T,Y,Z\rangle_{\mathbb{R}}$ and $K = \exp\mathfrak{k}$. Then, $\mathfrak{k}$ is abelian and $\mathfrak{g}(f) = \mathbb{R}T \oplus \mathbb{R}Z$. It follows that $\dim\Omega = 4$ and $\mathfrak{b} = \mathbb{R}V + \mathfrak{k}$ is a polarisation of $\mathfrak{g}$ at f. We construct the irreducible unitary representation π of G associated to Ω as $\pi = \mathrm{ind}_B^G\chi_f$ where $B = \exp\mathfrak{b}$. Simple calculations show

$$\Omega = \{\ell = \alpha X^* + \beta\lambda T^* + \beta U^* + \gamma V^* + \lambda Y^* + Z^*; \alpha,\beta,\gamma,\lambda \in \mathbb{R}\}.$$

For $\ell \in \Omega$, we have

$$\mathfrak{g}(\ell) = \mathbb{R}(T - \ell(Y)U - \ell(U)Y) + \mathbb{R}Z.$$

While,

$$\mathfrak{b}[\ell] = \langle T - \ell(Y)U, V, Y, Z\rangle_{\mathbb{R}}$$

is a polarisation of $\mathfrak{g}$ at ℓ and $\mathfrak{b}[\ell|_{\mathfrak{k}}] = \mathfrak{k}$ is that of $\mathfrak{k}$ at the restriction $\ell|_{\mathfrak{k}} \in \mathfrak{k}^*$. Let $p : \mathfrak{g}^* \to \mathfrak{k}^*$ be the restriction mapping. We immediately see that

$$p(\Omega) = \mathbb{R}T^* + \mathbb{R}Y^* + Z^*$$

except the subset $(\mathbb{R} \setminus \{0\})T^* + Z^*$. Furthermore, for $\delta = aT^* + bY^* + Z^* \in p(\Omega)$ $(a \in \mathbb{R}, b \in \mathbb{R} \setminus \{0\})$, $p^{-1}(\delta)$ is a K-orbit. In this way, the restriction $\pi|_K$ is without multiplicities. In more details, for $\delta = aT^* + bY^* + Z^* \in \mathfrak{k}^*$ $(a, b \in \mathbb{R})$, let χ_δ be the unitary character of K defined by

$$\chi_\delta(\exp(tT + yY + zZ)) = e^{i(at+by+z)}.$$

Then,

$$\pi|_K \simeq \int_{\mathbb{R}^2}^{\oplus} \chi_\delta dadb.$$

Besides, we define the Penney distribution a_ℓ at $\ell \in \Omega$ verifying $\ell(Y) \neq 0$ by the formula

$$a_\ell(\varphi) = \int_{\mathbb{R}} \overline{\varphi(\exp(tT))e^{i\ell(T)t}}dt$$

for $\varphi \in \mathcal{H}_\pi^\infty$.

Now we consider a Jordan-Hölder sequence

$$\begin{aligned} &\mathfrak{g}_1 = \mathbb{R}Z \subset \mathfrak{g}_2 = \mathbb{R}Z + \mathbb{R}Y \subset \mathfrak{g}_3 = \langle V, Y, Z\rangle_{\mathbb{R}} \\ &\subset \mathfrak{g}_4 = \langle U, V, Y, Z\rangle_{\mathbb{R}} \subset \mathfrak{g}_5 = \langle Z, Y, V, U, T\rangle_{\mathbb{R}} \subset \mathfrak{g}_6 = \mathfrak{g}. \end{aligned}$$

of $\mathfrak{g}$ and the associated sequence of subalgebras

$$\mathfrak{k}_1 = \mathfrak{g}_1 \subset \mathfrak{k}_2 = \mathfrak{g}_2 \subset \mathfrak{k}_3 = \mathfrak{k} \subset \mathfrak{k}_4 = \mathfrak{k} + \mathbb{R}V \subset \mathfrak{k}_5 = \mathfrak{k}_4 + \mathbb{R}U \subset \mathfrak{k}_6 = \mathfrak{g}.$$

When we pass from $\mathfrak{k}_4$ to $\mathfrak{k}_5$, the element

$$\sigma = YU - ZT$$

appears in the centre of the enveloping algebra $\mathcal{U}(\mathfrak{g})$ and satisfies the relation $\pi(\sigma) = 0$. Moreover, U is $\mathfrak{k}$-invariant and

$$\pi(Z)\cdot a_\ell = ia_\ell, \pi(Y)\cdot a_\ell = i\ell(Y)a_\ell, \pi(T)\cdot a_\ell = i\ell(T)a_\ell$$

for all $\ell \in \Omega$ verifying $\ell(Y) \neq 0$. It results from this that $\pi(U)\cdot a_\ell = i\ell(U)a_\ell$ and the algebra $D_\pi(G)^K$ is isomorphic to the algebra $\mathbb{C}[\Omega]^K$.

Example 2.3.11 Let $\mathfrak{g} = \langle X, T, U, V, Y, Z\rangle_{\mathbb{R}}$ with non-zero brackets

$$[X, Y] = Z, [X, T] = U, [U, T] = Z, [T, V] = Y, [U, V] = Z.$$

Let $G = \exp \mathfrak{g}, f = Z^* \in \mathfrak{g}^*, \Omega = G\cdot f, \mathfrak{k} = \langle T, Y, Z\rangle_{\mathbb{R}}$ and $K = \exp \mathfrak{k}$. Then, $\mathfrak{k}$ is abelian and $\mathfrak{g}(f) = \mathbb{R}(T - V) \oplus \mathbb{R}Z$. It follows that $\dim \Omega = 4$ and $\mathfrak{b} = \mathbb{R}V + \mathfrak{k}$ is a polarisation of $\mathfrak{g}$ at f. We construct the irreducible unitary representation π of G associated to Ω as $\pi = \mathrm{ind}_B^G \chi_f$ where $B = \exp \mathfrak{b}$. Simple calculations show

$$\Omega = \{\ell = \alpha X^* + (\beta\lambda + \gamma)T^* + \beta U^* + \gamma V^* + \lambda Y^* + Z^*; \alpha, \beta, \gamma, \lambda \in \mathbb{R}\}.$$

Let $p : \mathfrak{g}^* \to \mathfrak{k}^*$ be the restriction mapping. We immediately see that

$$p(\Omega) = \mathbb{R}T^* + \mathbb{R}Y^* + Z^*.$$

Furthermore, for $\delta = aT^* + bY^* + Z^* \in p(\Omega)$ $(a, b \in \mathbb{R})$, $p^{-1}(\delta)$ is a K-orbit. In this way, the restriction $\pi|_K$ is without multiplicities.

Now we consider a Jordan-Hölder sequence

$$\begin{aligned} \mathfrak{g}_1 &= \mathbb{R}Z \subset \mathfrak{g}_2 = \mathbb{R}Z + \mathbb{R}Y \subset \mathfrak{g}_3 = \langle V, Y, Z\rangle_{\mathbb{R}} \\ &\subset \mathfrak{g}_4 = \langle U, V, Y, Z\rangle_{\mathbb{R}} \subset \mathfrak{g}_5 = \langle Z, Y, V, U, T\rangle_{\mathbb{R}} \subset \mathfrak{g}_6 = \mathfrak{g} \end{aligned}$$

of $\mathfrak{g}$ and the associated sequence of subalgebras

$$\mathfrak{k}_1 = \mathfrak{g}_1 \subset \mathfrak{k}_2 = \mathfrak{g}_2 \subset \mathfrak{k}_3 = \mathfrak{k} \subset \mathfrak{k}_4 = \mathfrak{k} + \mathbb{R}V \subset \mathfrak{k}_5 = \mathfrak{k}_4 + \mathbb{R}U \subset \mathfrak{k}_6 = \mathfrak{g}.$$

From $\mathfrak{k}_4$ to $\mathfrak{k}_5$, the element

$$\sigma = YU - ZT + VZ$$

appears in the centre of the enveloping algebra $\mathcal{U}(\mathfrak{g})$. The Hilbert space $\mathcal{H}_\pi$ of π is identified to $L^2(\mathbb{R}^2)$ by the mapping $\mathcal{H}_\pi \ni \varphi \mapsto \Phi \in L^2(\mathbb{R}^2)$ given by

$$\Phi(x, s) = \varphi(\exp(xX)\exp(sU)) \ (x, s \in \mathbb{R}).$$

In this situation, simple calculations show

$$\begin{aligned} &\pi(Z) = i, (\pi(Y)\Phi)(x, s) = -ix\Phi(x, s), \\ &(\pi(V)\Phi)(x, s) = -is\Phi(x, s), (\pi(T)\Phi)(x, s) = (-is + x\frac{\partial}{\partial s})\Phi(x, s) \\ &(\pi(U)\Phi)(x, s) = -\frac{\partial\Phi}{\partial s}(x, s). \end{aligned}$$

Hence, we confirm $\pi(\sigma) = 0$ and the $\mathfrak{k}$-invariant element $w = YU + VZ$ is nothing but the element ZT modulo $\ker \pi$.

Example 2.3.12 Let finally $\mathfrak{g} = \langle X, S, T, R, U, V, Y, Z\rangle_{\mathbb{R}}$ with non-zero brackets

$$\begin{aligned} &[X, Y] = Z, [X, U] = V, [X, V] = Y, [T, U] = V \\ &[X, T] = S, [X, S] = R, [S, U] = Y, [R, U] = Z. \end{aligned}$$

Then,

$$[\mathfrak{g}, \mathfrak{g}] = \langle S, R, V, Y, Z\rangle_{\mathbb{R}}$$

is abelian. Let

$$G = \exp \mathfrak{g}, f = Z^* \in \mathfrak{g}^*, \Omega = G{\cdot}f,$$
$$\mathfrak{k} = \langle T, U, V, Y, Z\rangle_{\mathbb{R}}$$

and $K = \exp \mathfrak{k}$. We have

$$\mathfrak{g}(f) = \langle T, S, V, Z\rangle_{\mathbb{R}},$$

$\dim \Omega = 4$ and $\mathfrak{b} = \mathbb{R}T + [\mathfrak{g}, \mathfrak{g}]$ is a polarisation of $\mathfrak{g}$ at f. We realise the irreducible unitary representation π of G associated to Ω as $\pi = \mathrm{ind}_B^G \chi_f$ where $B = \exp \mathfrak{b}$. Next, take a Jordan-Hölder sequence

$$\begin{aligned} \mathfrak{g}_1 = \mathbb{R}Z \subset \mathfrak{g}_2 = \mathbb{R}Z + \mathbb{R}Y \subset \mathfrak{g}_3 = \langle Z, Y, V\rangle_{\mathbb{R}} \\ \subset \mathfrak{g}_4 = \langle Z, Y, V, R\rangle_{\mathbb{R}} \subset \mathfrak{g}_5 = [\mathfrak{g}, \mathfrak{g}] \\ \subset \mathfrak{g}_6 = [\mathfrak{g}, \mathfrak{g}] + \mathbb{R}U \subset \mathfrak{g}_7 = [\mathfrak{g}, \mathfrak{g}] + \mathbb{R}U + \mathbb{R}T \subset \mathfrak{g}_8 = \mathfrak{g} \end{aligned}$$

of $\mathfrak{g}$ and the corresponding sequence of subalgebras

$$\begin{aligned} \mathfrak{k}_1 = \mathfrak{g}_1 \subset \mathfrak{k}_2 = \mathfrak{g}_2 \subset \mathfrak{k}_3 = \mathfrak{g}_3 \subset \mathfrak{k}_4 = \mathfrak{k}_3 + \mathbb{R}U \subset \mathfrak{k}_5 = \mathfrak{k} \\ \subset \mathfrak{k}_6 = \mathfrak{k} + \mathbb{R}R \subset \mathfrak{k}_7 = \mathfrak{k} + \mathbb{R}R + \mathbb{R}S \subset \mathfrak{k}_8 = \mathfrak{g}. \end{aligned}$$

For

$$\ell = aT^* + bU^* + cV^* + dY^* + Z^* \in \mathfrak{k}^* \ (a, b, c, d \in \mathbb{R})$$

satisfying $c \neq 0$, $\mathfrak{b}[\ell] = \langle U, V, Y, Z\rangle_{\mathbb{R}}$ is the Vergne polarisation relative to the Jordan-Hölder sequence $\{\mathfrak{k}_j\}_{j=1}^5$ of $\mathfrak{k}$. Now the Hilbert space $\mathcal{H}_\pi$ of π is identified to $L^2(\mathbb{R}^2)$ by the mapping $\mathcal{H}_\pi \ni \varphi \mapsto \Phi \in L^2(\mathbb{R}^2)$ given by

$$\Phi(a, b) = \varphi(\exp(aX)\exp(bU)) \ (a, b \in \mathbb{R}).$$

So, simple calculations show

$$\begin{aligned} &\pi(Z) = i, (\pi(Y)\Phi)(a, b) = -ia\Phi(a, b), \\ &(\pi(V)\Phi)(a, b) = \frac{ia^2}{2}\Phi(a, b), (\pi(R)\Phi)(a, b) = ib\Phi(a, b) \\ &(\pi(S)\Phi)(a, b) = -iab\Phi(a, b), (\pi(U)\Phi)(a, b) = -(\frac{\partial}{\partial b} + \frac{ia^3}{6})\Phi(a, b), \\ &(\pi(T)\Phi)(a, b) = \frac{ia^2 b}{2}\Phi(a, b), (\pi(X)\Phi)(a, b) = -\frac{\partial \Phi}{\partial a}(a, b). \end{aligned}$$

From this, $\ker \pi$ is spanned by

$$Z - i, \ 2ZV - Y^2, \ S + iYR, \ T + iVR.$$

On the other hand, our algebra $\mathcal{U}_\pi(\mathfrak{g})^{\mathfrak{k}}$ is generated modulo $\ker \pi$ by $\{YT - VS, V, Y, Z\}$. But,

$$\begin{aligned} YT - VS \equiv YT - \frac{Y^2}{2i}S \equiv \frac{Y}{2i}(2iT - YS) \\ \equiv \frac{Y}{2i}(2iT + iY^2R) \equiv \frac{Y}{2}(2T + Y^2R) \equiv Y(T + iVR) \equiv 0 \end{aligned}$$

modulo $\ker\pi$. Other computations show

$$\Omega = \{\ell = aX^* + \frac{cd^2}{2}T^* + bU^* + cdS^* + cR^* + \frac{d^2}{2}V^* + dY^* + Z^* \ (a,b,c,d \in \mathbb{R})\}.$$

Here also the restriction $\pi|_K$ turns out to be without multiplicities. For $\ell \in \Omega$ verifying $\ell(V) \neq 0$, the Penney distribution a_ℓ is defined by the formula

$$a_\ell(\varphi) = \int_{\mathbb{R}} \overline{\varphi(\exp(\lambda U))\chi_\ell(\exp(\lambda U))}d\lambda \ (\varphi \in \mathcal{H}_\pi^\infty).$$

Lastly, we immediately see

$$\pi(Z)a_\ell = ia_\ell, \ \pi(Y)a_\ell = i\ell(Y)a_\ell, \ \pi(V)a_\ell = i\ell(V)a_\ell.$$

In this way, we confirm once again that our algebra $D_\pi(G)^K$ is isomorphic to the algebra $\mathbb{C}[\Omega]^K$.

2.4 DUFLO'S POLYNOMIAL CONJECTURE FOR NILPOTENT RESTRICTIONS

We take back all the notations of the previous section. Note first that the algebra $\mathbb{C}[\Omega]^K$ naturally inherits the Poisson structure. Let $Z(\mathbb{C}[\Omega]^K)$ be the Poisson centre of $\mathbb{C}[\Omega]^K$ and $C(\pi|_K)$ the centre of $D_\pi(G)^K$. The aim here is to show that $C(\pi|_K)$ and $Z(\mathbb{C}[\Omega]^K)$ are isomorphic. This is what we call Duflo's polynomial Conjecture for restrictions.

Let

$$\mathcal{S} : \{0\} = \mathfrak{g}_0 \subset \mathfrak{g}_1 \subset \cdots \subset \mathfrak{g}_{n-1} \subset \mathfrak{g}_n = \mathfrak{g}$$

be an ideal sequence of $\mathfrak{g}$ such that $\dim(\mathfrak{g}_j/\mathfrak{g}_{j-1}) = 1$ for $1 \leq j \leq n$ and $\{X_1, \ldots, X_n\}$ a basis of $\mathfrak{g}$ adapted to $\mathcal{S}$. Let $S = \{i_1 < \cdots < i_m\}$ be the set of jump-indices of Ω. Put $V = \sum_{k=1}^m \mathbb{R}X_{i_k}$ and denote by $S(V)$ the symmetric algebra of V which is identified with the algebra $\mathbb{C}[\Omega]$ of the polynomial functions on Ω. Next, for $\ell \in \Omega$, we denote by $\mathfrak{k}^\ell$ the orthogonal of $\mathfrak{k}$ with respect to the bilinear form B_ℓ. Then, the algebra $\mathbb{C}[\Omega]^K$ of the K-invariant polynomials on Ω is constituted of the elements of $S(V)$ which is written as polynomial of $\mathfrak{k}^\ell$ at almost all the points ℓ of Ω. If this is true, this algebra is Poisson commutative if $\pi|_K$ has finite multiplicities, because the Poisson product is given by $\{X,Y\}(\ell) = B_\ell([X,Y])$ for $X, Y \in V$ and $\ell \in \Omega$ and that we know $\pi|_K$ has finite multiplicities if and only if $\mathfrak{k}^\ell$ is isotropic for B_ℓ almost everywhere on Ω. Let, in the flag $\mathcal{S}$,

$$\mathcal{I} = \{1 \leq i \leq n; \mathfrak{g}_i \cap \mathfrak{k} \neq \mathfrak{g}_{i-1} \cap \mathfrak{k}\} = \{i_1 < \cdots < i_d\}$$

with $d = \dim\mathfrak{k}$. For $1 \leq m \leq d$, put $\mathfrak{k}_m = \mathfrak{k} \cap \mathfrak{g}_{i_m}$. This being done,

$$\{0\} = \mathfrak{k}_0 \subset \mathfrak{k}_1 \subset \cdots \subset \mathfrak{k}_{d-1} \subset \mathfrak{k}_d = \mathfrak{k} \tag{2.4.1}$$

is a composition sequence of $\mathfrak{k}$, which gives us the Vergne polarisation $\mathfrak{b}[\ell|_{\mathfrak{k}}]$ at point $\ell|_{\mathfrak{k}} \in \mathfrak{k}^*(\ell \in \Omega)$. Put $B[\ell|_{\mathfrak{k}}] = \exp(\mathfrak{b}[\ell|_{\mathfrak{k}}])$. For $\ell \in \Omega$, we realise π at ℓ by means of the Vergne polarisation $\mathfrak{b}[\ell]$ constructed starting from $\mathcal{S}$. Let $B[\ell] = \exp(\mathfrak{b}[\ell])$. Now the distribution a_ℓ (2.3.4) at point $\ell \in \Omega$ is defined by the formula

$$\langle a_\ell, \varphi \rangle = \int_{B[\ell|_{\mathfrak{k}}]/(B[\ell|_{\mathfrak{k}}] \cap B[\ell])} \overline{\varphi(k)\chi_\ell(k)} d\dot{k}$$

for $\varphi \in \mathcal{H}_\pi^\infty$. If the line mentioned above is correct, we would have, for $W \in C(\pi|_K)$,

$$W \cdot a_\ell = P_W(\ell) a_\ell$$

almost everywhere on Ω with a certain $P_W(\ell) \in Z(\mathbb{C}[\Omega]^K)$ and the mapping $C(\pi|_K) \ni W \mapsto P_W \in Z(\mathbb{C}[\Omega]^K)$ would provide us an isomorphism of algebras between the two centres. Let

$$\mathcal{J} = \{1, 2, \ldots, n\} \setminus \mathcal{I} = \{j_1 < \cdots < j_q\},$$

where $q = \dim(\mathfrak{g}/\mathfrak{k}) = n - d$, in the flag $\mathcal{S}$. Set $\mathfrak{k}_{d+m} = \mathfrak{k} + \mathfrak{g}_{j_m}$, then

$$\{0\} = \mathfrak{k}_0 \subset \mathfrak{k}_1 \subset \cdots \subset \mathfrak{k}_d = \mathfrak{k} \subset \mathfrak{k}_{d+1} \subset \cdots \subset \mathfrak{k}_n = \mathfrak{g} \tag{2.4.2}$$

is a sequence of subalgebras of $\mathfrak{g}$.

2.4.1 Case of finite multiplicities restrictions

Theorem 2.4.1 *If $\pi|_K$ has finite multiplicities, then $\mathbb{C}[\Omega]^K$ is Poisson commutative. Consequently, our mapping $W \mapsto P_W$ provides us an isomorphism of algebras between $C(\pi|_K)$ and $Z(\mathbb{C}[\Omega]^K)$.*

Proof 2.26 *Suppose that $\pi|_K$ has finite multiplicities. Let us proceed by induction on $\dim G$. When $\mathfrak{g}$ is commutative or $\mathfrak{g} = \mathfrak{k}$, there is nothing to prove. Let $\mathfrak{g}' = \mathfrak{k}_{n-1}$ and $G' = \exp(\mathfrak{k}_{n-1})$. Suppose first Ω is non-saturated with respect to $\mathfrak{k}_{n-1}$. Then, $\Omega' = \Omega|_{\mathfrak{g}'}$ is a G'-orbit associated to $\pi' = \pi|_{G'}$ which is an irreducible representation of G'. We identify Ω to Ω' by the restriction mapping. Let $Q \in \mathbb{C}[\Omega']^K$. By induction hypothesis, $Q(\ell')$ is written in a polynomial of $\mathfrak{k}^{\ell'}$ almost everywhere on Ω'. Let ω' be the symplectic form on Ω' defined by*

$$\omega'_{\ell'}(X, Y) = \ell'([X, Y]), \ X, Y \in \mathfrak{g}', \ \ell' \in \Omega'.$$

Now, the Poisson product of $\mathbb{C}[\Omega']^K$ is given by $\{X, Y\}(\ell') = \omega'_{\ell'}(X, Y)$ for $X, Y \in \mathfrak{g}'$. Since we know [10] that $\mathfrak{k}^{\ell'}$ is isotropic for $\omega'_{\ell'}$ almost everywhere on Ω', the algebra $\mathbb{C}[\Omega']^K$ is Poisson commutative.

Suppose next Ω is saturated with respect to $\mathfrak{g}'$. Then,

$$\pi|_{G'} \simeq \int_{\mathbb{R}}^{\oplus} \pi'_t dt, \Omega|_{\mathfrak{g}'} = \sqcup_{t \in \mathbb{R}} \Omega'_t,$$

where π_t is an irreducible representation of G' whose coadjoint orbit of G' is Ω'_t for *all* $t \in \mathbb{R}$. *Let* $W \in \mathcal{U}_\pi(\mathfrak{g})^{\mathfrak{k}}$. *Since* $\pi|_K$ *has finite multiplicities, the generic* K*-orbits in* Ω *are saturated with respect to* $\mathfrak{g}'$, *what means* $W \in \mathcal{U}(\mathfrak{g}')$ *modulo* $\ker \pi$. *As* $\pi'_t|_K$ *have finite multiplicities and* $[\mathfrak{k}, W] \subset \ker(\pi'_t)$ *for almost all* $t \in \mathbb{R}$, *the induction hypothesis says that* W *is written in a polynomial of* $\mathfrak{k}^{\ell'}$ *for almost all* $\ell' \in \Omega'_t$ *at generic* $t \in \mathbb{R}$. *If so, we can repeat the argument developed above in order to see that the algebra* $\mathbb{C}[\Omega]^K$ *is Poisson commutative.*

2.4.2 Generation families on $Z(\mathbb{C}[\Omega]^K)$ and $C(\pi|_K)$

Now we leave the hypothesis of the finite multiplicities. When $H = G$, there is nothing to do. Suppose hereafter that $H \neq G$ and go the same way as in the induction case. Let us retake the sequence (2.4.1) and let $p_j : \mathfrak{k}^* \to (\mathfrak{k}_j)^*(1 \leq j \leq d)$ the canonical projections which intertwine the actions of K. For $\ell \in \mathfrak{k}^*$, we set $e_j(\ell) = \dim(K{\cdot}p_j(\ell)), e(\ell) = (e_1(\ell), \ldots, e_d(\ell))$ and $\mathcal{E} = \{e(\ell); \ell \in \mathfrak{k}^*\}$. For $e \in \mathcal{E}$, we define the K-invariant layer $U_e = \{\ell \in \mathfrak{k}^*; e(\ell) = e\}$. Let $p : \mathfrak{g}^* \to \mathfrak{k}^*$ be the restriction map. Then, there exists a unique element $e \in \mathcal{E}$ such that $p(\Omega) \cap U_e$ is a non-empty Zariski open set of $p(\Omega)$. Let $T(e) = \{r_1 < \cdots < r_m\}$ be the set of the non-jump indices of e. For $1 \leq k \leq m$, there is an element $\sigma_k = a_k Y_{r_k} + b_k, a_k, b_k \in \mathcal{U}(\mathfrak{k}_{r_k-1})$ which is e-central for U_e. Next, we look at the action of K on the sequence (2.4.2). We there consider a jump index $k, d+1 < k < n$, such that there is no element in $\mathcal{U}(\mathfrak{k})$ which is e-central for $\Omega|_{\mathfrak{k}_{k-1}}$. Let $\mathcal{K} = \{k_1 < \cdots < k_s\}$ be the set of such indices k. For $1 \leq u \leq s$, we take an element $\kappa_u = c_u Y_{i_u} + d_u, c_u, d_u \in \mathcal{U}(\mathfrak{k}_{i_u-1})$ which is e-central for $\Omega|_{\mathfrak{k}_{k_u-1}}$. We take such a κ_u in the manner that i_u would be as small as possible. Denote by $\varphi_k(1 \leq k \leq m)$ and $\phi_u(1 \leq u \leq s)$ Corwin-Greenleaf functions corresponding to σ_k and κ_u.

Theorem 2.4.2 *We can choose* $\{\kappa_1, \ldots, \kappa_s\}$ *(resp.* $\{\phi_1, \ldots, \phi_s\}$*) so that*

$$\{\kappa_1, \ldots, \kappa_s, \sigma_1, \ldots, \sigma_m\} \text{ (resp. } \{\phi_1, \ldots, \phi_s, \varphi_1, \ldots, \varphi_m\})$$

is a system of rational generators of $C(\pi|_K)$ *(resp.* $Z(\mathbb{C}[\Omega]^K)$*).*

Proof 2.27 *Let us keep the notations introduced above. We proceed by induction on* $\dim G$. *If* G *is commutative or* $K = G$, *there is nothing to do. Suppose* $\dim G \geq 3$ *and* $K \neq G$. *If* Ω *is non-saturated with respect to* $\mathfrak{g}'$, *we identify* Ω *to* Ω' *and the induction hypothesis immediately gives us the result.*

Suppose that Ω *is saturated with respect to* $\mathfrak{g}'$. *We may suppose* $\mathfrak{g}' = \mathfrak{g}_{n-1}$. *There is a unique jump index* $2 \leq i_0 \leq n-1$ *where we find an element* $\sigma_{i_0} = a_0 X_{i_0} + b_0, a_0, b_0 \in \mathcal{U}(\mathfrak{g}_{i_0-1})$ *which is e-central for* Ω' *but not for* Ω. *This means that the centre* $C(\pi|_K)$ *is contained in* $\mathcal{U}(\mathfrak{g}')$. *Similarly, the commutativity with the function of Corwin-Greenleaf* φ_0 *corresponding to* σ_{i_0} *implies that the centre* $Z(\mathbb{C}[\Omega]^K)$ *is contained in the symmetric algebra* $S(\mathfrak{g}')$ *of* $\mathfrak{g}'$.

If generic K*-orbits in* Ω *are saturated with respect to* $\mathfrak{g}'$, $\kappa_1, \ldots, \kappa_s$ *(resp.* $\phi_1, \ldots, \phi_s$*) are central elements of* $D_\pi(G)^K$ *(resp.* $\mathbb{C}[\Omega]^K$*). Indeed,* $\mathcal{U}_\pi(\mathfrak{g})^{\mathfrak{k}} \subset \mathcal{U}(\mathfrak{g}')$

and the induction hypothesis finds $[\mathcal{U}_\pi(\mathfrak{g})^{\mathfrak{k}}, \kappa_j] \subset \ker \pi'_t$ *for* $1 \leq j \leq s$ *and almost all* $t \in \mathbb{R}$*. This says that* $\kappa_j (1 \leq j \leq s)$ *belong to* $C(\pi|_K)$*. Same for* $\phi_j (1 \leq j \leq s)$*. Moreover, the induction hypothesis says that* $\{\kappa_1, \ldots, \kappa_s, \sigma_1, \ldots, \sigma_m\}$ *(resp.* $\{\phi_1, \ldots, \phi_s, \varphi_1, \ldots, \varphi_m\}$*) is a system of rational generators of* $C(\pi'_t|_K)$ *(resp.* $Z(\mathbb{C}[\Omega'_t]^K)$*) for almost all* $t \in \mathbb{R}$*. This gives us the desired result.*

Suppose that generic K*-orbits in* Ω *are non-saturated with respect to* $\mathfrak{g}'$*. Let* $\nu = aY_n + b, a, b \in \mathcal{U}(\mathfrak{g}'), \pi(a) \neq 0$*, be an element of* $\mathcal{U}_\pi(\mathfrak{g})^{\mathfrak{k}}$*. Taking the crochet with elements of* $\mathcal{U}_\pi(\mathfrak{g})^{\mathfrak{k}} \cap \mathcal{U}(\mathfrak{g}')$*, we can suppose that* a *belongs to the centre of* $\mathcal{U}_\pi(\mathfrak{g})^{\mathfrak{k}} \cap \mathcal{U}(\mathfrak{g}')$*. For* $1 \leq u \leq s$*, suppose* $[\nu, \kappa_u] \notin \ker \pi$*. If* $[\nu, [\nu, \kappa_u]] \in \ker \pi$*, we replace* κ_u *by* $[\nu, \sigma_{i_0}]\kappa_u - [\nu, \kappa_u]\sigma_{i_0}$*. Here, we can verify* $i_0 < i_u$ *by induction on* $\dim \mathfrak{g}$ *as in the case of induction. Indeed, let* $\mathfrak{z}$ *be the centre of* $\mathfrak{g}$*. If* $\dim \mathfrak{z} \geq 2$*, we can descend to the quotient* $\mathfrak{g}/(\mathfrak{z} \cap \ker \pi)$*. Suppose* $\dim \mathfrak{z} = 1, \pi|_{\mathfrak{z}} \neq 0$ *and let* $\mathfrak{h}$ *be the centraliser of* $\mathfrak{g}_2$ *in* $\mathfrak{g}$*. If* $\mathfrak{k} \subset \mathfrak{h}$*, we see* $i_0 = 2 < i_u$*. If* $\mathfrak{k} \not\subset \mathfrak{h}$*, let* $H = \exp \mathfrak{h}$ *and*

$$\pi|_H \simeq \int_{\mathbb{R}}^{\oplus} \rho_t dt, \;\; \rho_t \in \widehat{H} \;(t \in \mathbb{R}),$$

and we can apply the induction hypothesis to (H, ρ_t)*. Suppose* $[\nu, [\nu, \kappa_u]] \notin \ker \pi$*. Let* v *be a sufficiently large integer. We utilise a method already employed several times. Let* $F(t)$ *be a polynomial of one variable with degree* $2v$*. Put* $\psi = \kappa_u + F(\sigma_{i_0})$ *and*

$$\psi_0 = \psi, \psi_1 = [\nu, \psi], \psi_2 = [\nu, [\nu, \psi]]$$

and so on. Then, $\psi_{2v} \neq 0, \psi_{2v+1} \equiv 0$ *and* $\psi_j (0 \leq j \leq 2v)$ *commute with the elements of* $\mathcal{U}_\pi(\mathfrak{g})^{\mathfrak{k}} \cap \mathcal{U}(\mathfrak{g}')$*. The new* κ_u *is obtained by*

$$\begin{aligned}\kappa_u =& (\psi_0\psi_{2v} + \psi_{2v}\psi_0) - (\psi_1\psi_{2v-1} + \psi_{2v-1}\psi_1) + \cdots \\ &+ (-1)^{v-1}(\psi_{v-1}\psi_{v+1} + \psi_{v+1}\psi_{v-1}) + (-1)^v(\psi_v)^2.\end{aligned}$$

This being done, the modified elements $\kappa_j (1 \leq j \leq s)$ *are found in the centre* $C(\pi|_K)$ *and we obtain the result as above. Same for the side of* $Z(\mathbb{C}[\Omega]^K)$*.*

2.4.3 Proof of Dulfo's Conjecture for restrictions

Now, the main result of this subsection says:

Theorem 2.4.3 *The mapping* $W \mapsto P_W$ *gives us an isomorphism of algebras between* $C(\pi|_K)$ *and* $Z(\mathbb{C}[\Omega]^K)$*.*

Proof 2.28 *Let us begin by repeating the beginning of the proof of Theorem 2.4.2. Keep the notations introduced above. We proceed by induction on* $\dim G$*. If* G *is commutative or* $K = G$*, there is nothing to do. Suppose* $\dim G \geq 3$ *and* $K \neq G$*. If* Ω *is non-saturated with respect to* $\mathfrak{g}'$*, we identify* Ω *to* Ω' *and the induction hypothesis immediately gives us the result.*

Suppose that Ω *is saturated with respect to* $\mathfrak{g}'$*. We can suppose* $\mathfrak{g}' = \mathfrak{g}_{n-1}$*. There is a unique jump index* $2 \leq i_0 \leq n-1$ *where we find an element* $\sigma_{i_0} = a_0 X_{i_0} + b_0, a_0, b_0 \in \mathcal{U}(\mathfrak{g}_{i_0 - 1})$ *which is e-central for* Ω' *but not for* Ω*. This means that the centre*

$C(\pi|_K)$ is contained in $\mathcal{U}(\mathfrak{g}')$. Similarly, the commutativity with the Corwin-Greenleaf function φ_0 corresponding to σ_{i_0} implies that the centre $Z(\mathbb{C}[\Omega]^K)$ is contained in the symmetric algebra $S(\mathfrak{g}')$ of $\mathfrak{g}'$. Let $W \in C(\pi|_K)$. We go to see that $W{\cdot}a_\ell = P_W(\ell)a_\ell$ almost everywhere on Ω with a function P_W which belongs to the Poisson centre $Z(\mathbb{C}[\Omega]^K)$. From Theorem 3 in [10], the algebra $D_{\pi_t'}(G')^K$ is rationally generated by the image under π_t' of the algebra $\mathcal{U}_\pi(\mathfrak{g})^{\mathfrak{k}}$ for almost all $t \in \mathbb{R}$. This being assumed, $\pi_t'(W)$ belongs to the centre of $D_{\pi_t'}(G')^K$ for almost all $t \in \mathbb{R}$. Hence, the induction hypothesis assures that $P_W|_{\Omega_t'}$ belongs to the Poisson centre of the algebra $\mathbb{C}[\Omega_t']^K$ for almost all $t \in \mathbb{R}$. On the other hand, the proof of Lemma 5.3.24 in [19] shows that the function P_W extends into a K-invariant rational function and hence [18] polynomial function on Ω. Since the Poisson structure on Ω is given by the symplectic form ω whose restriction on Ω_t' gives there the Poisson product, one verifies that P_W belongs to the Poisson centre $Z(\mathbb{C}[\Omega]^K)$ of $\mathbb{C}[\Omega]^K$.

Inversely, let $Q \in Z(\mathbb{C}[\Omega]^K)$. We want to verify that there is $W \in C(\pi|_K)$ such that $P_W = Q$. From the induction hypothesis, there exists $W_t' \in \mathcal{U}_{\pi_t'}(\mathfrak{g}')^{\mathfrak{k}}$ such that $Q|_{\Omega_t'} = P_{W_t'}$ for generic $t \in \mathbb{R}$.

Let $\tilde{\mathfrak{g}}$ be an ideal of codimension 1 of $\mathfrak{g}$ different from $\mathfrak{g}'$. Suppose first $\mathfrak{k} \subset \tilde{\mathfrak{g}}$. If Ω is non-saturated with respect to $\tilde{\mathfrak{g}}$, the result is directly obtained by induction hypothesis. When Ω is saturated with respect to $\tilde{\mathfrak{g}}$, let $\tilde{G} = \exp\tilde{\mathfrak{g}}$ and

$$\pi|_{\tilde{G}} \simeq \int_{\mathbb{R}}^{\oplus} \tilde{\pi}_t dt,\ \ \Omega|_{\tilde{\mathfrak{g}}} = \sqcup_{t\in\mathbb{R}}\tilde{\Omega}_t$$

as before. From the induction hypothesis, there exists $\tilde{W}_t \in \mathcal{U}_{\tilde{\pi}_t}(\tilde{\mathfrak{g}})^{\mathfrak{k}}$ such that $Q|_{\tilde{\Omega}_t} = P_{\tilde{W}_t}$ for generic $t \in \mathbb{R}$. For generic s, t,

$$W_s'{\cdot}a_\ell = \tilde{W}_t{\cdot}a_\ell = Q(\ell)a_\ell$$

almost everywhere on $\Omega_s' \cap \tilde{\Omega}_t$. This means that

$$W_s' - \tilde{W}_t \in \ker\pi_s' + \ker\tilde{\pi}_t,$$

and hence that there exists $W \in \mathcal{U}_\pi(\mathfrak{g})^{\mathfrak{k}} \cap \mathcal{U}(\mathfrak{g}')$ such that $P_W(\ell) = Q(\ell)$ almost everywhere on Ω.

We go to see this in a different way. If π vanishes on an ideal $\mathfrak{a} \neq \{0\}$ of $\mathfrak{g}$, we can pass as usual to the quotient $\mathfrak{g}/\mathfrak{a}$ to which applies the induction hypothesis. Suppose that there exists no such ideal $\mathfrak{a}$. This means that $\mathfrak{g}_1$ is nothing but the centre of $\mathfrak{g}$ and that π does not vanish there. Let $\mathfrak{g}_2 = \mathbb{R}Y \oplus \mathfrak{g}_1$ and denote by $\mathfrak{h}$ the centraliser of $\mathfrak{g}_2$ in $\mathfrak{g}$. Here, $\mathfrak{h}$ is an ideal of codimension 1 of $\mathfrak{g}$.

Suppose first $\mathfrak{k} \subset \mathfrak{h}$ and put $s = \ell(Y)$ for $\ell \in \Omega$. Then, our element W_s' is a polynomial of s and we obtain the element W in search by substituting $-\sqrt{-1}Y$ to s in W_s'. When $\mathfrak{k} \not\subset \mathfrak{h}$, $\mathfrak{h}$ will take the place of $\tilde{\mathfrak{g}}$ in what follows.

Therefore, if generic K-orbits in Ω are saturated with respect to $\mathfrak{g}'$, $\mathcal{U}_\pi(\mathfrak{g})^{\mathfrak{k}} \subset \mathcal{U}(\mathfrak{g}')$ and our mapping $W \mapsto P_W$ provides us an isomorphism of algebras between $C(\pi|_K)$ and $Z(\mathbb{C}[\Omega]^K)$.

Suppose that generic K-orbits in Ω are non-saturated with respect to $\mathfrak{g}'$. Let $\nu = aY_n + b, a, b \in \mathcal{U}(\mathfrak{g}'), \pi(a) \neq 0$, be an element of $\mathcal{U}_\pi(\mathfrak{g})^{\mathfrak{k}}$. Taking the crochet with *elements of* $\mathcal{U}_\pi(\mathfrak{g})^{\mathfrak{k}} \cap \mathcal{U}(\mathfrak{g}')$*, we can suppose that* a *belongs to the centre of* $\mathcal{U}_\pi(\mathfrak{g})^{\mathfrak{k}} \cap \mathcal{U}(\mathfrak{g}')$*. From the induction hypothesis and what we have seen, there exists in* $\mathcal{U}_\pi(\mathfrak{g})^{\mathfrak{k}} \cap \mathcal{U}(\mathfrak{g}')$ *an element* W' *which verifies* $P_{W'} = Q$ *and which commutes modulo* $\ker \pi$ *with all the elements of* $\mathcal{U}_\pi(\mathfrak{g})^{\mathfrak{k}} \cap \mathcal{U}(\mathfrak{g}')$*. By putting this time* $\psi = W' + F(\sigma_{i_0})$*, we apply to the pair* (ν, ψ) *the same manipulation as one used in the proof of Theorem 2.4.2 in order to get an element* $\tilde{W}$ *of* $C(\pi|_K)$*.*

Now, let j_0 *be the smallest index such that* Q *belongs to the symmetric algebra* $S(\mathfrak{k}_{j_0})$ *of* $\mathfrak{k}_{j_0}$*. Here, we utilise another new induction on the degree* r *of* Q *with respect to* Y_{j_0}*. Let* $\alpha Y_{j_0}^r$ *be the dominant term of* Q *and* $\beta Y_{j_0}^r$ *that of* $\tilde{W}$*. This leads* $\alpha \in Z(\mathbb{C}[\Omega]^K)$ *and* $\beta \in C(\pi|_K)$*. Because* $Q' = P_\beta Q - \alpha P_{\tilde{W}}$ *is of degree smaller than or equal to* $r-1$ *with respect to* Y_{j_0}*, the induction hypothesis says that* Q' *and then* $P_\beta Q$ *belong to the image of our mapping* $P : C(\pi|_K) \ni W \mapsto P_W \in Z(\mathbb{C}[\Omega]^K)$*. Namely, there exists* $W'' \in C(\pi|_K)$ *such that* $P_{W''} = P_\beta Q = P_{\beta W'}$*. Consequently,* $W'' - \beta W' \in \ker \pi$ *and hence* $W' \in C(\pi|_K)$*.*

Finally, suppose $\mathfrak{k} \not\subset \tilde{\mathfrak{g}}$ *and put* $\tilde{K} = \exp \tilde{\mathfrak{k}}$ *with* $\tilde{\mathfrak{k}} = \mathfrak{k} \cap \tilde{\mathfrak{g}}$*. If* Ω *is non-saturated with respect to* $\tilde{\mathfrak{g}}$*, the induction hypothesis says that there exists* $\tilde{W} \in C(\tilde{\pi}|_{\tilde{K}})$ *such that* $P_{\tilde{W}} = Q$*. Further,* $\tilde{W}$ *is* $\mathfrak{k}$*-invariant because* Q *is so. At last, suppose that* Ω *is saturated with respect to* $\tilde{\mathfrak{g}}$*. If* $Q \in S(\mathfrak{g}')$*, the induction hypothesis says that there exists* $W_s \in C(\tilde{\pi}_s|_{\tilde{K}})$ *verifying* $P_{W_s} = Q|_{\tilde{\Omega}_s}$ *for almost all* $s \in \mathbb{R}$*. Since* Q *is* $\mathfrak{k}$*-invariant,* $W = W_s$ *does not depend on* $s \in \mathbb{R}$*. This shows* $W \in C(\pi|_K)$ *verifying* $P_W = Q$*. In particular, this is the case if* $\mathfrak{b}[\ell] \subset \mathfrak{g}'$ *for almost all* $\ell \in \Omega$*. Indeed, there is in* $\mathcal{U}(\mathfrak{k}_{d-1})$ *a non-trivial element which is e-central for* $\Omega|_{\mathfrak{k}_{d-1}}$*. Then, if* $\mathcal{U}_\pi(\mathfrak{g})^{\mathfrak{k}} \not\subset \mathcal{U}(\tilde{\mathfrak{g}})$*, there is an index* $d+1 \leq k_0$ *such that there exists a non-trivial element in* $\mathcal{U}(\mathfrak{k}_{k_0}) \cap \mathcal{U}(\tilde{\mathfrak{g}})$ *which is e-central for* $\tilde{\Omega}$*. It follows that there comes in* $\mathcal{U}_\pi(\mathfrak{g})^{\mathfrak{k}}$ *a new element at the level of the index* k_0 *and consequently* $C(\pi|_K)$ *is contained in* $\mathcal{U}(\tilde{\mathfrak{g}})$*.*

It remains for us to examine the case where $\mathfrak{b}[\ell] \not\subset \tilde{\mathfrak{g}}$ *and where* $Q \notin S(\tilde{\mathfrak{g}})$*. But it suffices for us to imitate the proof of Proposition 5 in [18]. In fact, if* $j_0 \leq d$*, the result comes from Lemma 4 in [18]. Suppose* $d+1 \leq j_0$*. Let* m *be the degree of* Q *with respect to* Y_d *and* $\sigma = \bar{a}Y_d + \bar{b}, \bar{a}, \bar{b} \in \mathcal{U}(\mathfrak{k}_{d-1})$ *an e-central element for* $\Omega|_{\mathfrak{k}}$*. Everything as before,* $P_{\bar{a}^m} Q$ *belongs to the image of our mapping* $P : C(\pi|_K) \to Z(\mathbb{C}[\Omega]^K)$*. On the other hand, we see by induction on* $\dim \mathfrak{k}$ *that there is in* $C(\pi|_K)$ *an element* $\nu = a'Y_{j_0} + b', a', b' \in \mathcal{U}(\mathfrak{k}_{j_0-1})$ *such that* a' *has no common factor with* $\bar{a}$*. This being assumed,* $P_{a'}^r Q$ *is found in the image of the mapping* P*. Let*

$$P_{\bar{a}}^m Q = P_{W_1}, \; P_{a'}^r Q = P_{W_2}$$

with $W_1, W_2 \in C(\pi|_K)$*. Hence,*

$$P_{\bar{a}}^m P_{a'}^r Q = P_{a'}^r P_{W_1} = P_{\bar{a}}^m P_{W_2}.$$

It follows that $(a')^r W_1 \equiv (\bar{a})^m W_2$ *modulo* $\ker \pi$*. Since* $a', \bar{a}$ *have no common factor,* W_1 *must be divided by* $\bar{a}^m$ *and hence* Q *is found in the image of the mapping* P*.*

CHAPTER 3

Primitive zero ideals and Casimir elements

Let $G = \exp \mathfrak{g}$ be an exponential solvable Lie group with Lie algebra $\mathfrak{g}$, $\mathfrak{g}^*$ the dual vector space of $\mathfrak{g}$, $\widehat{G}$ the unitary dual of G and $\mathcal{U}(\mathfrak{g})$ the enveloping algebra of the complexified Lie algebra $\mathfrak{g}_{\mathbb{C}}$. The orbit method for G investigates the correspondence between a coadjoint orbit $\Omega \subset \mathfrak{g}^*$ and an irreducible unitary representation $(\pi, \mathcal{H})$ of G. Since the Hilbert space $\mathcal{H}$ is regarded as a quantisation of the symplectic manifold Ω, it is natural to expect a close relation between the ideal $I(\Omega) := \{F \in S(\mathfrak{g}); F|_\Omega = 0\}$ and the annihilator $\ker d\pi := \{W \in \mathcal{U}(\mathfrak{g}) \ : \ d\pi(W) = 0\}$ (cf. [7] and [88] in the nilpotent context).

One technique to tackle this problem is to study the existence of Casimir generators at the level $\mathfrak{g}_j$ related to a composition series $(\mathfrak{g}_j)_j$ of ideals of $\mathfrak{g}$ and a layering $(U_e)_{e\in\mathcal{E}}$ of $\mathfrak{g}^*$. These are elements of the enveloping algebra $\mathcal{U}(\mathfrak{g})$ which diagonalise the actions of the derivative of unitary and irreducible representations relatively to a Zariski open set $\mathcal{Z}$ on each layer U_e. We shall accurately determine when there is a new comer A_j of Casimir elements on $\Sigma_e = U_e \cap \mathcal{Z}$ when passing through $\mathfrak{g}_j$. We show that this happens if and only if, j runs a set of indices $\mathcal{J}_0$ adequately defined. Moreover, we show that the corresponding Casimir functions $\{\varphi_{A_j},\ j \in \mathcal{J}_0\}$ turn out to be polynomials on $\mathcal{Z} \cap U_e$.

We shall make use of such a construction, to produce a family of generators of the kernel of an irreducible and unitary representation of an exponential solvable Lie group, which is quite scarce in literature except in the nilpotent situation. The construction also helps to show that in the case of the maximal element $e \in \mathcal{E}$, the Poisson centre of the Poisson symmetric algebra $S(\mathfrak{g})$ coincides with the set of all Casimir functions on $\mathcal{Z} \cap U_e$.

Beyond that, Casimir elements are very used to label irreducible representations of Lie groups. Since the Casimir operator acts as a scalar in any irreducible representation of the Lie algebra, it helps to separate non-equivalent irreducible representations. As such, the eigenvalues of Casimir operators provide an invariant that can classify them as will be developed later in the chapter.

DOI: 10.1201/9781003685739-3

3.1 CASIMIR OPERATORS OF THE ENVELOPING ALGEBRAS

Let

$$\mathcal{S} : \mathfrak{g}_1 \subset \cdots \subset \mathfrak{g}_{m-1} \subset \mathfrak{g}_m = \mathfrak{g} \tag{3.1.1}$$

be a composition series of ideals of $\mathfrak{g}$ which passes the nilpotent radical $\mathfrak{n}$ of $\mathfrak{g}$, say $\mathfrak{g}_{j_0} = \mathfrak{n}$. Hence, $1 \leq \dim(\mathfrak{g}_j/\mathfrak{g}_{j-1}) \leq 2$. Recall back the layering of the dual vector space $\mathfrak{g}^*$ as in Subsection 2.1.3. Let further $e \in \mathcal{E}$ be the maximal element. Then, U_e is a G-invariant Zariski open set of $\mathfrak{g}^*$. More generally, take any $e \in \mathcal{E}$. We then see that $\{\ell \in \mathfrak{g}^*; e \leq e(\ell)\}$ is a Zariski open set of $\mathfrak{g}^*$.

We now introduce the following definition of e-Casimir elements of $\mathcal{U}(\mathfrak{g})$.

Definition 3.1.1 Let $e \in \mathcal{E}$. We say that $A \in \mathcal{U}(\mathfrak{g})$ is an e-Casimir element, if there exists a G-invariant Zariski open set $\mathcal{Z}$ of $\mathfrak{g}^*$ verifying $\Sigma_e = \mathcal{Z} \cap U_e \neq \emptyset$ and such that $\pi_\ell(A)$ is a scalar operator for $\ell \in \Sigma_e$. We shall also say in the sequel that A is a Casimir element for Σ_e. We write $\pi_\ell(A) = \varphi_A(\ell)I$ and we denote by $\varphi_A(\ell)$ the e-Casimir function on Σ_e corresponding to the e-Casimir element A.

Let also $Cas(\Sigma_e)$ and $FCas(\Sigma_e)$ denote the sets of e-Casimir elements and functions respectively.

3.1.1 Minimal non-central ideals and Casimir elements

We here refer to Subsection 1.1.2 which develops some structure details on minimal non-central ideals of exponential solvable Lie algebras.

Let $\ell \in \mathfrak{g}^*$ be a linear form on $\mathfrak{g}$ which does not vanish on any non-trivial ideal of $\mathfrak{g}$. This means that $\dim \mathfrak{z} \leq 1$, where $\mathfrak{z}$ denotes the centre of $\mathfrak{g}$, and that $\ell|_\mathfrak{z} \neq 0$ if $\mathfrak{z} \neq \{0\}$.

Take a minimal non-central ideal $\mathfrak{a}$ of $\mathfrak{g}$, which is abelian and of dimension ≤ 3. Let $\mathfrak{m}$ be the centraliser of $\mathfrak{a}$. Let as in Subsection 1.1.2 $\mathfrak{a}^\ell = \{X \in \mathfrak{g}; \ell([X, \mathfrak{a}]) = \{0\}\}$ and $\mathfrak{g}_0 = \{X \in \mathfrak{g}; [X, \mathfrak{a}] \subset \mathfrak{z}\}$. We shall treat the following situations:

Case 1: $\dim \mathfrak{a} = 1$ or 2 and $\mathfrak{z} = \{0\}$.

(i) When $\dim \mathfrak{a} = 1$, we write $\mathfrak{a} = \mathbb{R}Y$ with $\ell(Y) = 1$. Then there exists a non-zero vector $X \in \mathfrak{g}$ such that $\mathfrak{g} = \mathfrak{m} \oplus \mathbb{R}X$ and $[X, Y] = Y$. Let for instance $\mathfrak{m} = \langle Y \rangle_\mathbb{R}$ and $G = \exp \mathfrak{g}$. Let $\ell = aY^* \in \mathfrak{g}^*(0 \neq a \in \mathbb{R}), \mathfrak{b} = \mathbb{R}Y, B = \exp \mathfrak{b}, \chi_\ell(\exp T) = e^{i\ell(T)}(T \in \mathfrak{b})$ and $\pi_\ell = \mathrm{ind}_B^G \chi_\ell$. As $\mathfrak{b}$ is a (real) polarisation of $\mathfrak{g}$ at ℓ satisfying the Pukanszky condition, π_ℓ is an irreducible unitary representation of G.

This being assumed,

$$\mathfrak{g}_1 = \mathfrak{b} \subset \mathfrak{g}_2 = \mathfrak{g}$$

is a composition series of ideals of $\mathfrak{g}$ and the dimension index $e(\ell) = (1, 2)$ gives us a Zariski open set

$$\mathcal{Z} = \{\ell \in \mathfrak{g}^*; \ell|_\mathfrak{b} \neq 0\}.$$

Let $W = yY \in \mathfrak{b}(y \in \mathbb{R})$ and $\varphi \in C_0^\infty(\mathbb{R})$. Then,

$$(\pi_\ell(W)\varphi)(t) = iaye^{-t}\varphi(t) \ (t \in \mathbb{R}).$$

This shows that there is no Casimir elements on $\mathcal{Z}$ except the constants.

(ii) Assume now that $\dim \mathfrak{a} = 2$ and write $\mathfrak{a} = \mathbb{R}Y_1 \oplus \mathbb{R}Y_2$. Then as in the previous case, one can write $\mathfrak{g} = \mathfrak{m} \oplus \mathbb{R}X$ with $[X, Y_1] = Y_1 - \alpha Y_2$ and $[X, Y_2] = \alpha Y_1 + Y_2$, for some non-zero real number α. In this case, $(\ell(Y_1), \ell(Y_2)) \neq (0, 0)$.
A simple example in this situation is the following: Let $\mathfrak{m} = \mathfrak{a}$ and $G = \exp \mathfrak{g}$. Let $0 \neq \ell = aY_1^* + bY_2^* \in \mathfrak{g}^* (a, b \in \mathbb{R}), \mathfrak{b} = \mathbb{R}Y_1 + \mathbb{R}Y_2, B = \exp \mathfrak{b}, \chi_\ell(\exp X) = e^{i\ell(X)} (X \in \mathfrak{b})$ and $\pi_\ell = \mathrm{ind}_B^G \chi_\ell$. As $\mathfrak{b}$ is a (real) polarisation of $\mathfrak{g}$ at ℓ verifying the Pukanszky condition, π_ℓ is an irreducible unitary representation of G.

This being assumed,

$$\{0\} = \mathfrak{g}_0 \subset \mathfrak{g}_1 = \mathfrak{b} \subset \mathfrak{g}_2 = \mathfrak{g}$$

is a composition series of ideals of $\mathfrak{g}$ and the dimension index $e(\ell) = (1, 2)$ provides us with the Zariski open set

$$\mathcal{Z} = \{\ell \in \mathfrak{g}^*; \ell|_{\mathfrak{b}} \neq 0\}.$$

Let $W = xY_1 + yY_2 \in \mathfrak{b} (x, y \in \mathbb{R})$ and $\varphi \in C_0^\infty(\mathbb{R})$. Then,

$$(\pi_\ell(W)\varphi)(t) = ie^{-t}((ax + by)\cos(\alpha t) + (ay - bx)\sin(\alpha t))\varphi(t) \ (t \in \mathbb{R}).$$

This proves that there is no Casimir element on $\mathcal{Z}$ except the constants.

Case 2: We now assume that $\dim \mathfrak{z} = 1$. Let first $\dim \mathfrak{a} = 2$. We write $\mathfrak{z} = \mathbb{R}Z$ with $\ell(Z) = 1$, and $\mathfrak{a} = \mathbb{R}Z \oplus \mathbb{R}Y$ with $\ell(Y) = 0$. Assume further $[\mathfrak{g}, \mathfrak{a}] = \mathfrak{a}$. As in Subsection 1.1.2, there exist non-zero vectors $X, T \in \mathfrak{g}$ such that $\mathfrak{g} = \mathfrak{m} \oplus \mathbb{R}X \oplus \mathbb{R}T$ with the commutators $[X, Y] = Z$, $[T, X] = -X$, and $[T, Y] = Y$.

Let us consider again the situation where $\mathfrak{m} = \mathfrak{a}$. This is the case of the Boidol group $G = \exp \mathfrak{g}$. Let $\ell = \ell(\alpha, \beta) = \alpha T^* + \beta Z^* \in \mathfrak{g}^*$, $O(\alpha, \beta) = G \cdot f(\alpha, \beta)$ and $\pi_\ell \in \widehat{G}$ corresponding to $O(\alpha, \beta)$ for $\beta \neq 0$. Put

$$\mathfrak{g}_1 = \mathfrak{z} = \mathbb{R}Z, \mathfrak{g}_2 = \mathfrak{z} \oplus \mathbb{R}Y, \mathfrak{g}_3 = \mathfrak{g}_2 + \mathbb{R}X,$$

where $\mathfrak{z}$ denotes the centre of $\mathfrak{g}$. Then,

$$\mathfrak{g}_1 \subset \mathfrak{g}_2 \subset \mathfrak{g}_3 \subset \mathfrak{g}$$

is a composition series of ideals of $\mathfrak{g}$ and the dimension index $e(\ell) = (0, 1, 2, 2)$ gives us a Zariski open set $\mathcal{Z}$. Then the element

$$W = ZT - YX$$

gives us a Casimir element on $\mathcal{Z}$.

(iii) We finally look at the case where $\dim \mathfrak{a} = 3$. Then there exist non-zero vectors X_1, X_2, T, Y_1, Y_2 such that $\mathfrak{a} = \mathbb{R}Z \oplus \mathbb{R}Y_1 \oplus \mathbb{R}Y_2$ with $\ell(Y_1) = \ell(Y_2) = 0$, and

$$\mathfrak{g} = \mathfrak{m} \oplus \mathbb{R}X_1 \oplus \mathbb{R}X_2 \oplus \mathbb{R}T \tag{3.1.2}$$

with the commutators

$$[X_i, Y_j] = \delta_{i,j} Z (1 \leq i, j \leq 2), \quad [T, Y_2] = Y_2 + \alpha Y_1 \text{ and } [T, Y_1] = Y_1 - \alpha Y_2,$$

for some non-zero $\alpha \in \mathbb{R}$. Likewise, let $\mathfrak{a} = \mathfrak{m}$ and $\ell = \beta T^* + \gamma Z^* \in \mathfrak{g}^* (\gamma \neq 0)$, $\Omega = G{\cdot}\ell \subset \mathfrak{g}^*$. Denote by $\pi_\ell \in \widehat{G}$ the irreducible unitary representation of G corresponding to Ω. Let

$$\mathfrak{g}_1 = \mathfrak{z} = \mathbb{R}Z, \mathfrak{g}_2 = \mathfrak{z} \oplus \mathbb{R}Y_1 \oplus \mathbb{R}Y_2, \mathfrak{g}_3 = \mathfrak{g}_2 + \mathbb{R}X_1 + \mathbb{R}X_2.$$

Then,

$$\mathfrak{g}_1 \subset \mathfrak{g}_2 \subset \mathfrak{g}_3 \subset \mathfrak{g}_4 = \mathfrak{g}$$

is a composition series of ideals of $\mathfrak{g}$. The dimension index $e(\ell) = (0, 2, 4, 4)$ defines a Zariski open set $\mathcal{Z}$ of $\mathfrak{g}^*$. It is then easy to check that

$$W = ZT - (Y_1 - \alpha Y_2) X_1 - (Y_2 + \alpha Y_1) X_2$$

provides us with a Casimir element on $\mathcal{Z}$.

3.1.2 Existence of Casimir elements

Let

$$S' : \mathfrak{k}_1 \subset \mathfrak{k}_2 \subset \cdots \subset \mathfrak{k}_n = \mathfrak{g}, \ \dim(\mathfrak{k}_i) = i (1 \leq i \leq n)$$

be a good sequence of subalgebras of $\mathfrak{g}$ associated to the composition series (3.1.1). This means that, if $\mathfrak{k}_i$ is not an ideal of $\mathfrak{g}$, there is an index $1 \leq j \leq m$ verifying $\mathfrak{k}_{i+1} = \mathfrak{g}_j$ and $\mathfrak{k}_{i-1} = \mathfrak{g}_{j-1}$ with the convention $\mathfrak{k}_0 = \mathfrak{g}_0 = \{0\}$ and $\mathfrak{g}_{j_0}$ stands for the nil radical of $\mathfrak{g}$. At $\ell \in \mathfrak{g}^*$, we consider the Vergne polarisation $\mathfrak{b}[\ell]$ constructed from the good sequence S' to induce the irreducible monomial representation $\pi_\ell = \mathrm{ind}_{B[\ell]}^G \chi_\ell$, where $B[\ell] = \exp(\mathfrak{b}[\ell])$ and $\chi_\ell(\exp X) = e^{i\ell(X)} (X \in \mathfrak{b}[\ell])$.

We shall discuss in this section the existence of Casimir elements related to a layer U_e of a fixed $e = (e_1, \ldots, e_m) \in \mathcal{E}$ as in Definition 3.1.1. More precisely, how to decide that when passing from $\mathfrak{g}_{j-1}$ to $\mathfrak{g}_j$, new Casimir elements on Σ_e show up? Write $\mathfrak{g}_j = \mathbb{R}X_j + \mathfrak{g}_{j-1}$, then in the positive case, a new comer should have the form

$$W = \sum_{k=0}^{p} a_k X_j^k$$

for some elements $a_k, 0 \leq k \leq p$ in $\mathcal{U}(\mathfrak{g}_{j-1})$. On the other hand, in some situations, one may encounter a setting where $p > 1$ and no element of the form $a_j X_j + b_j$ $(a_j, b_j \in \mathcal{U}(\mathfrak{g}_{j-1}))$ appears, as reveals the following example.

Example 3.1.2 Let $G = \exp \mathfrak{g}, \mathfrak{g} = \langle T, X, Y \rangle_{\mathbb{R}} : [T, Y] = 2Y, [T, X] = -X$. Set

$$\mathfrak{g}_1 = \mathbb{R}Y, \ \mathfrak{g}_2 = \mathfrak{g}_1 \oplus \mathbb{R}X.$$

Then,

$$\mathfrak{g}_1 \subset \mathfrak{g}_2 \subset \mathfrak{g}$$

is a flag of ideals of $\mathfrak{g}$ and the dimension index $e(\ell) = (1, 1, 2)$ defines a Zariski open set $\mathcal{Z}$ of $\mathfrak{g}^*$. We see that the element YX^2 belongs to the centre of $\mathcal{U}(\mathfrak{g})$. Furthermore, there is no Casimir element having the form

$$aX + b, \; a, b \in \mathbb{C}[Y].$$

On the other hand, when the root of the adjoint endomorphism on the level of $\mathfrak{g}_j$ is complex, then a new comer may intervene both the generators of $\mathfrak{g}_j/\mathfrak{g}_{j-1}$. In other words, let in this case $\mathfrak{g}_j = \mathbb{R}X_{j,1} + \mathbb{R}X_{j,2} + \mathfrak{g}_{j-1}$, then a Casimir element may have the form

$$W = \sum_{k=0}^{p} a_k X_{j,1}^k + b_k X_{j,2}^k$$

for some elements $a_k, b_k, 0 \leq k \leq p$ in $\mathcal{U}(\mathfrak{g}_{j-1})$ as produced in the following example:

Example 3.1.3 Let $G_5 = \exp(\mathfrak{g}_5), \mathfrak{g}_5 = \langle T, X_1, X_2, Y_1, Y_2 \rangle_{\mathbb{R}} : [T, X_1] = -X_1 - \alpha X_2, [T, X_2] = -X_2 + \alpha X_1, [T, Y_1] = Y_1 - \alpha Y_2, [T, Y_2] = Y_2 + \alpha Y_1 (0 \neq \alpha \in \mathbb{R})$. Let

$$\mathfrak{g}_1 = \mathbb{R}Y_1 \oplus \mathbb{R}Y_2, \mathfrak{g}_2 = \mathfrak{g}_1 \oplus \mathbb{R}X_1 \oplus \mathbb{R}X_2.$$

Then,

$$\mathfrak{g}_1 \subset \mathfrak{g}_2 \subset \mathfrak{g}_3 = \mathfrak{g}$$

is a composition series of ideals of $\mathfrak{g}$. The dimension index $e(\ell) = (1, 1, 2)$ defines a Zariski open set $\mathcal{Z}$ of $\mathfrak{g}^*$. The element

$$W = (Y_1 - \alpha Y_2)X_1 + (Y_2 + \alpha Y_1)X_2$$

belongs to the centre of $\mathcal{U}(\mathfrak{g})$ and hence provides us with a Casimir element on $\mathcal{Z}$.

Having in mind all the above, let us mention that unlike the nilpotent setting one can have a situation where $\dim(\mathfrak{g}_j/\mathfrak{g}_{j-1}) = 2$, $e_j = e_{j-1} + 1$, but still a new comer appears at the level of $\mathfrak{g}_j$. We have the following example:

Example 3.1.4 *Let $G = \exp \mathfrak{g}$ with $\mathfrak{g} = \langle T, P, X_1, X_2, Y_1, Y_2 \rangle_{\mathbb{R}} : [T, P] = 2P, [T, X_1] = -X_1 - \alpha X_2, [T, X_2] = -X_2 + \alpha X_1, [T, Y_1] = Y_1 - \alpha Y_2, [T, Y_2] = Y_2 + \alpha Y_1, [P, X_1] = Y_2 + \alpha Y_1, [P, X_2] = \alpha Y_2 - Y_1 (0 \neq \alpha \in \mathbb{R})$. Let*

$$\mathfrak{g}_1 = \mathbb{R}Y_1 + \mathbb{R}Y_2, \mathfrak{g}_2 = \mathfrak{g}_1 \oplus \mathbb{R}X_1 \oplus \mathbb{R}X_2, \mathfrak{g}_3 = \mathfrak{g}_2 \oplus \mathbb{R}P.$$

Then,

$$\mathfrak{g}_1 \subset \mathfrak{g}_2 \subset \mathfrak{g}_3 \subset \mathfrak{g}_4 = \mathfrak{g}$$

is a composition series of ideals of $\mathfrak{g}$. The dimension index $e(\ell) = (1, 2, 3, 4)$ defines a Zariski open set $\mathcal{Z}$ of $\mathfrak{g}^*$. The element

$$W = (Y_1 - \alpha Y_2)X_1 + (Y_2 + \alpha Y_1)X_2$$

belongs to the centre of $\mathcal{U}(\mathfrak{g})$ and hence provides us with a Casimir element on $\mathcal{Z}$.

3.1.3 A result of existence

Let $\mathcal{K}$ be the set of indices $j_0 < k \leq m, e_k = e_{k-1} + 1$, and

$$\mathfrak{b}[\ell] \cap \mathfrak{g}_k = \mathfrak{b}[\ell] \cap \mathfrak{g}_{k-1} \tag{3.1.3}$$

for generic $\ell \in \Sigma_e$. Put $\mathfrak{g}_k = \mathbb{R}X_k + \mathfrak{g}_{k-1}$ for $k \in \mathcal{K}$. These being supposed, we propose the following definition:

Definition 3.1.5 *Let $e = (e_1, \ldots, e_m) \in \mathcal{E}$, and let $\mathcal{J}_0 := \mathcal{J}_0(e)$ be the set of indices $1 \leq j \leq m-1$ verifying the following conditions:*

1. $e_j \neq e_{j-1} + \dim(\mathfrak{g}_j/\mathfrak{g}_{j-1})$.
2. *There exists a Casimir element $W \in \mathcal{U}(\mathfrak{g}_j) \setminus \mathcal{U}(\mathfrak{g}_{j-1})$ on $\Sigma_e|_{\mathfrak{g}_{m-1}}$, such that for any $k \in \mathcal{K}$ there exists $n_k \in \mathbb{N}$ satisfying $\pi_\ell((adX_k)^{n_k}(W)) = 0$ for $\ell \in \Sigma_e$.*

We now prove our first main upshot:

Theorem 3.1.6 *Let $G = \exp \mathfrak{g}$ be an exponential solvable Lie group. Retain all the notations and definitions as above, and take $e \in \mathcal{E}$. We have the following:*

1. Fix $1 \leq j \leq m-1$, then at the passage from $\mathfrak{g}_{j-1}$ to $\mathfrak{g}_j$, it appears a new e-Casimir element W_j on Σ_e if and only if $j \in \mathcal{J}_0$.

2. For $j \in \mathcal{J}_0$, we can write $W_j = \sum_{k=0}^{p} a_k X_j^k$, for some elements $a_k, 0 \leq k \leq p$ of $\mathcal{U}(\mathfrak{g}_{j-1})$, (alternatively, $W = \sum_{k=0}^{p} a_k X_{j,1}^k + b_k X_{j,2}^k$ for some elements $a_k, b_k, 0 \leq k \leq p$ of $\mathcal{U}(\mathfrak{g}_{j-1})$).

3. The e-Casimir functions $\{\varphi_{W_j},\ j \in \mathcal{J}_0\}$ are polynomial functions on Σ_e.

Proof 3.1 *The proof will be done through several steps using induction on the dimension of G, and for suitability in notations we just avoid the indexation of the layer e. Here is the first:*

Step 1: Let $m \in \mathcal{K}$ and $j \in \mathcal{J}_0$. This is basically the situation where the generic orbits of Σ are saturated with respect to $\mathfrak{g}_{m-1}$. Then there exists a Casimir element $W \in \mathcal{U}(\mathfrak{g}_j)$ for $\Sigma|_{\mathfrak{g}_{m-1}}$ such that for some $r \in \mathbb{N}$, we have:

$$\pi_\ell((ad(X_m))^r(W)) = 0, \tag{3.1.4}$$

for $\ell \in \Sigma$. For simplicity we assume $\mathfrak{g}_j = \mathbb{R}X + \mathfrak{g}_{j-1}$. Then the element W takes the form

$$W = \sum_{k=0}^{p'} a_k X^k, \tag{3.1.5}$$

for some elements $a_k, 0 \leq k \leq p$ of $\mathcal{U}(\mathfrak{g}_{j-1})$. Let r be minimal verifying the condition in equation (3.1.4). Then if $r = 1$, there is nothing to prove. Indeed, one can write *using Vergne-like polarisation as in (3.1.3) that* $\pi_\ell = ind_{G'}^{G}\pi'_{\ell'}$, *where* $G' = \exp(\mathfrak{g}_{m-1})$ *and* $\pi'_{\ell'}$ *is the irreducible unitary representation of* G' *corresponding to* $\ell' = \ell|_{\mathfrak{g}_{m-1}}$. *We know that* $\pi'_{\ell'}(W)$ *is a scalar operator for any* $\ell \in \Sigma$, *which leads to*

$$\pi_\ell([V, W]) = \int_{\mathbb{R}}^{\oplus} \pi'_{\exp(tX_m)\ell'}([V, W])dt = 0 \tag{3.1.6}$$

for any $V \in \mathfrak{g}_{m-1}$. *Then (3.1.4) and (3.1.6) allow to conclude.*

Suppose that $r = 2$. *For* $\ell \in \Sigma$, *one can write using Vergne-like polarisation as in (3.1.3) that* $\pi_\ell = ind_{G'}^{G}\pi'_{\ell'}$, *where* $G' = \exp(\mathfrak{g}_{m-1})$ *and* $\pi'_{\ell'}$ *is the irreducible unitary representation of* G' *corresponding to* $\ell' = \ell|_{\mathfrak{g}_{m-1}}$. *Given the covariance relation (1.2.2), we regard the space* $\mathcal{H}_{\pi_\ell}$ *of* π_ℓ *as the space of functions* $\varphi : G \to \mathcal{H}_{\pi'_{\ell'}}$ *satisfying the covariance relation*

$$\varphi(gg') = \pi'_{\ell'}(g')^{-1}(\varphi(g)) \ (g \in G, g' \in G').$$

Then using the fact that $r = 2$, *we get for* $g = \exp T \in G$ *and* $\varphi \in \mathcal{H}^{\infty}_{\pi_\ell}$,

$$\begin{aligned}(\pi_\ell(W)\cdot\varphi)(g) = \varphi(W^*\cdot g) &= \varphi(g(W^* - [T, W^*])) &(3.1.7)\\ &= \pi'_{\ell'}(W - [T, W]) \cdot (\varphi(g)), &(3.1.8)\end{aligned}$$

since $\pi_\ell([T, [T, W]]) = 0$, *which means that* $\pi_\ell([T, W])$ *is constant in the direction transversal to the* G'*-orbit* $G'\cdot\ell$. *As* $\pi_\ell([\mathfrak{g}, [T, W]]) = 0$, $\pi_\ell([T, W])$ *is a scalar operator. Thus,* $\pi'_{\ell'}(W)$ *depends only on* $G\cdot\ell$. *Hence,* $\pi_\ell(W)$ *is a scalar operator.*

In this way, we find $r \geq 3$ *and we shall prove that we can generate new Casimir elements. Let now* W *as in (3.1.5), and*

$$W_p = (ad(X_m))^{p-1}(W), p \geq 1,$$

then remark first all the W_p*'s are Casimir elements on* $\Sigma|_{\mathfrak{g}_{m-1}}$. *Indeed,*

$$\pi'_{\exp(sX_m)\ell'}(W) = \pi'_{\ell'}(Ad(\exp(-sX_m))W),$$

are scalar operators for any $s \in \mathbb{R}$ *and* $\ell \in \Sigma$, *which is enough to conclude. We have the following:*

Lemma 3.1.7 *Assume* $r > 1$, *then the generator (3.1.5) can be taken as* $W = aX + b$ $(a, b \in \mathcal{U}(\mathfrak{g}_{j-1}))$.

Proof 3.2 *We have* $(adX_m)^d(W) \in \mathcal{U}(\mathfrak{g}_{j-1})$ *for some* $d \in \mathbb{N}$. *If not,* $(adX_m)^{r-1}(W)$ *is a Casimir element at the level of* $\mathfrak{g}_j$. *Remark first that we can suppose* $[X_m, W] \in \mathcal{U}(\mathfrak{g}_{j-1})$ *replacing* W *by* $(adX_m)^q(W)$, *if necessary, where* $q \in \mathbb{N}$ *is maximal such that* $(adX_m)^q(W) \notin \mathcal{U}(\mathfrak{g}_{j-1})$. *If* $[X_m, a_{p'}X] \in \ker \pi_\ell$ *for almost all* $\ell \in \Sigma$, *we are done. Otherwise, it is immediate that* $p' = 1$.

Let now W as in Lemma 3.1.7, and assume first that $r = 2q+1$ is odd. We fabricate the new element

$$\begin{aligned}\widetilde{W} =&(W_1W_r + W_rW_1) - (W_2W_{r-1} + W_{r-1}W_2) + \cdots \\ &+ (-1)^{q-2}(W_{q-1}W_{q+3} + W_{q+3}W_{q-1}) \\ &+ (-1)^{q-1}(W_qW_{q+2} + W_{q+2}W_q) + (-1)^q W_{q+1}^2.\end{aligned}$$

Then, $\widetilde{W}$ is Casimir on Σ and we remark that

$$\widetilde{W} = (aW_r + W_ra)X + b'$$

with a certain $b' \in \mathcal{U}(\mathfrak{g}_{j-1})$.

Second, suppose that r is even. We replace W by

$$W' = W(W_{r-1} + cW_r)\ (0 \neq c \in \mathbb{C}).$$

Putting $W'_p = (ad(X_m))^{p-1}(W')$ for $1 \leq p \leq r+1$, we have

$$\begin{aligned}&W'_2 \equiv W_2(W_{r-1} + cW_r) + W_1W_r, \ldots, \\ &W'_r \equiv W_r(W_{r-1} + cW_r) + (r-1)W_{r-1}W_r,\ W'_{r+1} \equiv rW_r^2\end{aligned}$$

modulo $\ker(\pi_\ell), \ell \in \Sigma$. *Hence our new element is obtained by*

$$\begin{aligned}\widetilde{W} =&r\{W_1(W_{r-1} + cW_r)W_r^2 + W_r^2W_1(W_{r-1} + cW_r)\} \\ &- W_1W_r\{W_r(W_{r-1} + cW_r) + (r-1)W_{r-1}W_r\} \\ &- \{W_r(W_{r-1} + cW_r) + (r-1)W_{r-1}W_r\}W_1W_r + \widetilde{V}\end{aligned}$$

with a certain $\widetilde{V} \in \mathcal{U}(\mathfrak{g}_{j-1})$. As we have

$$\begin{aligned}\widetilde{W} \equiv&r(W_1W_{r-1}W_r^2 + W_r^2W_1W_{r-1}) \\ &- (r-1)(W_1W_rW_{r-1} + W_{r-1}W_rW_1)W_r \\ &- (W_1W_r^2W_{r-1} + W_rW_{r-1}W_1W_r) + c(r-1)(W_1W_r^2 + W_r^2W_1)W_r\end{aligned}$$

modulo $\mathcal{U}(\mathfrak{g}_{j-1})$, we can visibly choose $0 \neq c \in \mathbb{C}$ so that $\widetilde{W} \notin \mathcal{U}(\mathfrak{g}_{j-1}) + \ker(\pi_\ell)$ because of $\pi_\ell((W_1W_r^2 + W_r^2W_1)W_r) \neq 0$. Indeed, if $\pi_\ell((W_1W_r^2 + W_r^2W_1)W_r) = 0$, then $\pi_\ell(W_1W_r^2 + W_r^2W_1) = 0$ because $\pi_\ell(\mathcal{U}(\mathfrak{g}))$ has no zero-divisor. This means there is a Casimir element at the level of $\mathfrak{g}_j$ as $W_1 = W$.

Step 2: Now we go to establish the assertion mentioned above. If all elements of Σ vanish on a non-trivial ideal of $\mathfrak{g}$, we can pass to the quotient of $\mathfrak{g}$ by this ideal to which applies the induction hypothesis. If $\mathfrak{g}_1$ is central in $\mathfrak{g}$, then $\mathfrak{g}_1 = \mathbb{R}Z$ is of dimension 1, the element Z is a Casimir element on Σ and $\ell(Z) \neq 0$ for $\ell \in \Sigma$.
If $\mathfrak{g}_1$ is not central in $\mathfrak{g}$, it is a minimal ideal of $\mathfrak{g}$ and the computations in [Subsection 1.1.2: Case 1, (i) and (ii)] show that there is no Casimir element on Σ which appears at this stage. Furthermore, all Casimir elements belong to $\mathcal{U}(\widetilde{\mathfrak{g}})$, where $\widetilde{\mathfrak{g}}$ denotes the centraliser of $\mathfrak{g}_1$ in $\mathfrak{g}$ which is a saturation ideal of codimension 1 for generic linear

forms. Hence, it suffices for us to apply the induction hypothesis to the action of G *on* $\widetilde{\mathfrak{g}}$. *Indeed, let* $\widetilde{\mathfrak{g}}$ *be an ideal of codimension 1 in* $\mathfrak{g}$ *such that the coadjoint orbits of* G *contained in* Σ *are saturated with respect to* $\widetilde{\mathfrak{g}}$. *In this case, the Casimir elements on* Σ *are contained in* $\mathcal{U}(\widetilde{\mathfrak{g}})$. *Cutting* $\mathcal{S}$ *by* $\widetilde{\mathfrak{g}}$, *we have the composition series* $\widetilde{\mathcal{S}}$ *and the index set* $(\widetilde{\mathcal{J}})_0$ *for* $\widetilde{\mathfrak{g}}$. *As* $\mathcal{J}_0 \subset (\widetilde{\mathcal{J}})_0$ *and we can consider that* $\mathfrak{g} = \mathbb{R}X_m + \widetilde{\mathfrak{g}}$, *the result comes from the induction hypothesis applied to* $\widetilde{\mathfrak{g}}$ *followed by what we have seen above.*

Step 3: Let j_1 *be the minimal index such that* $\mathfrak{g}_{j_1}$ *is not central in* $\mathfrak{g}$. *Then,* $\mathfrak{a} = \mathfrak{g}_{j_1}$ *is a minimal non-central ideal of* $\mathfrak{g}$. *Taking into account what we have seen above, we can place ourselves in the following situation. If*

$$\mathfrak{a} = \mathbb{R}Y + \mathfrak{g}_{j_1-1},$$

there is an element $T \in \mathfrak{g}$ *verifying* $[T, Y] = Y$. *If* $\dim(\mathfrak{a}/\mathfrak{g}_{j_1-1}) = 2$, *there are* $Y_1, Y_2 \in \mathfrak{a}, T \in \mathfrak{g}, 0 \neq \alpha \in \mathbb{R}$ *such that*

$$\mathfrak{a} = \mathbb{R}Y_1 + \mathbb{R}Y_2 + \mathfrak{g}_{j_1-1}, [T, Y_1] = Y_1 - \alpha Y_2, [T, Y_2] = Y_2 + \alpha Y_1.$$

In these two eventualities $[\mathfrak{g}, \mathfrak{a}] \cap \mathfrak{g}_{j_1-1} \neq \{0\}$, *and in hand the setting of Step 2, we can suppose that the coadjoint* G*-orbits in* U_e *are non-saturated with respect to the ideal* $\widetilde{\mathfrak{g}} = \{X \in \mathfrak{g}; [X, \mathfrak{a}] \subset \mathfrak{g}_{j_1-1}\}$ *of codimension 1. Denote by* $\mathfrak{g}_0$ *the centraliser of* $\mathfrak{a}$ *in* $\mathfrak{g}$ *and put* $\mathfrak{k} = \mathbb{R}T + \mathfrak{g}_0$. *Provided that we set* $S = \{\ell \in U_e; \ell(Y) = 0\}$ *or* $S = \{\ell \in U_e; \ell(Y_1) = \ell(Y_2) = 0\}$ *according to the case,* S *gives us a section of the coadjoint* G*-orbits in* U_e. *Remark on the other hand that* G*-generic orbits are saturated with respect to* $\mathfrak{k}$. *Therefore* $\mathcal{J}_0 \subset (\widetilde{\mathcal{J}})_0$ *and any Casimir element for* Σ *is contained in* $\mathcal{U}(\mathfrak{k})$. *For* $\ell \in S$, $\mathfrak{k} \subset \mathfrak{a}^\ell = \{X \in \mathfrak{g}; \ell([X, \mathfrak{a}]) = \{0\}\}$ *as in (1.1.5) in Subsection 1.1.2. Besides, putting* $\mathfrak{d} = \mathbb{R}Y$ *or* $\mathfrak{d} = \mathbb{R}Y_1 + \mathbb{R}Y_2$, $\mathfrak{d}$ *is an idcal of* $\mathfrak{k}$ *on which* $\pi'_{\ell|_\mathfrak{k}} \in \widehat{K}$ *vanishes for* $\ell \in S$. *Here,* $\pi'_{\ell|_\mathfrak{k}}$ *stands for the irreducible unitary representation of* K *associated to the coadjoint orbit* $K\cdot(\ell|_\mathfrak{k})$ *of* K.

This being assumed, the induction hypothesis applied to K/D, *where* $K = \exp \mathfrak{k}$ *and* $D = \exp \mathfrak{d}$, *says that there exists a Casimir element* W *on the corresponding set* $\widetilde{\Sigma}$ *of* $\mathfrak{k}^*$. *Further, the Casimir function* φ_W *is a polynomial function on* $\widetilde{\Sigma}$. *Now, let* $\{V_1, V_2, \ldots, V_q\}$ *be a co-exponential basis of* $\mathfrak{g}$ *to* $\mathfrak{k}$, *namely the mapping*

$$\mathbb{R}^q \times K \ni (v_q, v_{q-1}, \ldots, v_1, k) \mapsto \exp(v_q V_q)\exp(v_{q-1}V_{q-1})\cdots\exp(v_1 V_1)k \in G$$

is a diffeomorphism. Remark that these elements $\{V_1, \cdots, V_q\}$ *are taken in* $\mathfrak{g}_{j_0}$. *Moreover, when* $j = m$, *the minimal integer* $r \in \mathbb{N}$ *satisfying* $(\pi_\ell(V_1))^r(W) = 0$, *for example, almost everywhere on* Σ *is not equal to 2 because of* $[V_1, W] \notin \mathcal{U}(\mathfrak{a}^\ell)$. *Starting from* W, *we are ready to repeat the manipulation described in Step 1 applied to* $V_1, \ldots, V_q$ *in turn to arrive to a Casimir element on* Σ.

Step 4: It remains to tackle the case where all coadjoint G*-orbits contained in* Σ *are non-saturated with respect to* $\mathfrak{g}' = \mathfrak{g}_{m-1}$ *and* $j < m$ *belongs to* $\mathcal{J}_0$. *In this case, the induction hypothesis implies that there exists an element* $W' \in \mathcal{U}(\mathfrak{g}')$ *which is Casimir on* $\Sigma|_{\mathfrak{g}'}$. *Furthermore we find ourselves in the last situation of*

Step 3 just examined above. Namely, $\mathfrak{g} = \mathbb{R}T + \mathfrak{g}'$ *and there is the Casimir element* $W_m = PT + Q, P, Q \in \mathcal{U}(\mathfrak{g}')$ *on* Σ. *Thus,* $[W_m, W'] \in \ker \pi_\ell$ *for* $\ell \in \Sigma$. *As* $\pi_\ell|_{G'} \simeq \pi'_{\ell|\mathfrak{g}'}$, *the unitary and irreducible representation of* G' *associated to* $\ell|\mathfrak{g}'$, $[P, W'], [Q, W']$ *belong to* $\ker \pi_\ell$, *and so does* $[T, W']$ *too. This means that* W' *is Casimir on* Σ. *In this way, we establish the assertion.*

Conversely, if there exists s new Casimir element at the level of $\mathfrak{g}_j$, *the second condition for* $j \in \mathcal{J}_0$ *is trivial. On the other hand, if* $e_j = e_{j-1} + \dim(\mathfrak{g}_j/\mathfrak{g}_{j-1})$, *the* G*-orbit of* $\ell|_{\mathfrak{g}_j}$ *is saturated with respect to* $\mathfrak{g}_{j-1}$ *and there is no Casimir element at this stage. In this way, we establish the assertion.*

We therefore arrive to the following:

Corollary 3.1.8 *For* $j \in \mathcal{J}_0$, *let* $\{W_{j,1}, \ldots, W_{j,r_j}\}$ *be a system of generators of* $Cas(\Sigma_e) \setminus (Cas(\Sigma_e) \cap \mathcal{U}(\mathfrak{g}_{j-1}))$. *A generator family of* $Cas(\Sigma_e)$ *is obtained as* $\{W_m, W_{j,k}(1 \le k \le r_j), j \in \mathcal{J}_0\}$, *where* W_m *is the generator which may appear when* $m \in T(e)$. *Further if* $m \in S(e)$, $Cas(\Sigma_e)$ *is generated by* $\{W_{j,k}(1 \le k \le r_j), j \in \mathcal{J}_0\}$. *Accordingly, the set* $\{\ell \mapsto \varphi_{W_m}(\ell), \varphi_{W_{j,k}}(\ell)(1 \le k \le r_j), j \in \mathcal{J}_0\}$ *generates* $FCas(\Sigma_e)$.

Remark 3.1.9 1. These reasoning are applied even to the case where $\mathfrak{d}$ is a minimal ideal of $\mathfrak{g}$ (cf. Subsection 1.1.2). In this case, $\mathfrak{k}$ is an ideal of codimension 1 in $\mathfrak{g}$ and the generic coadjoint orbits of G are saturated with respect to $\mathfrak{k}$, the composition series $\mathcal{S}$ cut by $\mathfrak{k}$ provides us with the composition series $\mathcal{S}_\mathfrak{k}$ and the index set $(\mathcal{J}_0)_\mathfrak{k}$ for $\mathfrak{k}$. The Casimir elements of $\mathcal{U}(\mathfrak{g})$ are contained in $\mathcal{U}(\mathfrak{k})$ and $\mathcal{J}_0 \subset (\mathcal{J}_0)_\mathfrak{k}$. All taken into account, we can apply the induction hypothesis to $\mathfrak{k}$ and reason as before.

2. Suppose the setting where $\dim(\mathfrak{g}_j/\mathfrak{g}_{j-1}) = 2$ and $e_j = e_{j-1} + 2$ in Definition 3.1.5. Put $\mathfrak{g}' = \{X \in \mathfrak{g}; [X, \mathfrak{g}_j] \subset \mathfrak{g}_{j-1}\}$ and $G' = \exp(\mathfrak{g}')$, which is a normal subgroup of codimension 1 in G. If there appears a new Casimir element at the passage from $\mathfrak{g}_{j-1}$ to $\mathfrak{g}_j$ for $\pi \in \widehat{G}$, the orbit $\Omega(\pi)$ turns out to be non-saturated with respect to $\mathfrak{g}'$. Then, we find a contradiction by induction hypothesis applied to $\pi|_{G'}$.

3.1.4 Examples

This subsection aims to provide explicit computations of some examples illustrating Theorem 3.1.6. The following examples show that hypothesis (2) of $\mathcal{J}_0$ in Definition 3.1.5 is necessary.

Example 3.1.10 Let $G = \exp \mathfrak{g}, \mathfrak{g} = \langle T, X, Y, Z \rangle_\mathbb{R}$:

$$[T, X] = X, [T, Y] = Z.$$

Let $\ell = \ell(\alpha, \beta) = \alpha X^* + \beta Z^* \in \mathfrak{g}^*$, $O(\alpha, \beta) = G{\cdot}\ell(\alpha, \beta)$ and $\pi_\ell \in \widehat{G}$ corresponding to $O(\alpha, \beta)$ for $\alpha\beta \neq 0$. Put

$$\mathfrak{g}_1 = \mathfrak{z} = \mathbb{R}Z, \mathfrak{g}_2 = \mathfrak{z} \oplus \mathbb{R}Y, \mathfrak{g}_3 = \mathfrak{g}_2 \oplus \mathbb{R}X.$$

Then,

$$\mathfrak{g}_1 \subset \mathfrak{g}_2 \subset \mathfrak{g}_3 \subset \mathfrak{g}$$

is a composition series of ideals of $\mathfrak{g}$ and the dimension index $e(\ell) = (0,1,1,2)$ gives us a Zariski open set $\mathcal{Z}$. For $\varphi \in \mathcal{H}^\infty_{\pi_\ell}$, we put $\Phi(t) = \varphi(\exp(tT))(t \in \mathbb{R})$. Then, we have

$$\begin{aligned}(X\cdot\Phi)(t) &= \frac{d}{ds}\varphi(\exp(-sX)\exp(tT))\big|_{s=0}\\ &= \frac{d}{ds}\varphi(\exp(tT)\exp(-se^{-t}X))\big|_{s=0}\\ &= \frac{d}{ds}e^{i\alpha se^{-t}}\Phi(t)\big|_{s=0} = i\alpha e^{-t}\Phi(t).\end{aligned}$$

This means that there is no Casimir element on $\mathcal{Z}$ at the level $\mathfrak{g}_3$. In this example, the Casimir elements are nothing but the polynomials of Z.

Example 3.1.11 Let $G = \exp\mathfrak{g}, \mathfrak{g} = \langle T, X_1, X_2, Y, Z\rangle_{\mathbb{R}} : [T, X_1] = X_1 - \alpha X_2, [T, X_2] = X_2 + \alpha X_1, [T, Y] = Z$ with a non-zero real number α. Put $\ell = \beta X_1^* + \gamma X_2^* + \delta Z^* \in \mathfrak{g}^*$ with real numbers β, γ, δ verifying $\beta^2 + \gamma^2 \neq 0, \delta \neq 0$ and denote $\pi_\ell \in \widehat{G}$ the irreducible unitary representation of G corresponding to ℓ. Let

$$\mathfrak{g}_1 = \mathfrak{z} = \mathbb{R}Z, \mathfrak{g}_2 = \mathfrak{z} \oplus \mathbb{R}Y, \mathfrak{g}_3 = \mathfrak{g}_2 \oplus \mathbb{R}X_1 \oplus \mathbb{R}X_2.$$

Then,

$$\mathfrak{g}_1 \subset \mathfrak{g}_2 \subset \mathfrak{g}_3 \subset \mathfrak{g}_4 = \mathfrak{g}$$

is a composition series of ideals of $\mathfrak{g}$. The dimension index $e(\ell) = (0,1,1,2)$ defines a Zariski open set $\mathcal{Z}$ of $\mathfrak{g}^*$. For $\varphi \in \mathcal{H}^\infty_{\pi_\ell}$ we put $\Phi(t) = \varphi(\exp(tT))(t \in \mathbb{R})$. Then,

$$\begin{aligned}(X_1\cdot\Phi)(t) &= \frac{d}{ds}\varphi(\exp(-sX_1)\exp(tT))\big|_{s=0}\\ &= \frac{d}{ds}\varphi(\exp(tT)\exp(-se^{-t}(\cos\alpha t)X_1 - \sin(\alpha t)X_2))\big|_{s=0}\\ &= \frac{d}{ds}e^{ise^{-t}(\beta\cos(\alpha t) - \gamma\sin(\alpha t))}\Phi(t)\big|_{s=0}\\ &= ie^{-t}(\beta\cos(\alpha t) - \gamma\sin(\alpha t))\Phi(t)\end{aligned}$$

for example. This means that there is no Casimir element on $\mathcal{Z}$ at the level $\mathfrak{g}_3$. In this example too, the Casimir elements are nothing but the polynomials of Z.

Remark that, if we put

$$\mathfrak{g}_1 = \mathbb{R}Z, \mathfrak{g}_2 = \mathfrak{z} \oplus \mathbb{R}X_1 \oplus \mathbb{R}X_2, \mathfrak{g}_3 = \mathfrak{g}_2 \oplus \mathbb{R}Y,$$

we have another composition series

$$\mathfrak{g}_1 \subset \mathfrak{g}_2 \subset \mathfrak{g}_3 \subset \mathfrak{g}$$

of $\mathfrak{g}$. In this case, the dimension index is $(0,1,1,2)$ and there is no Casimir element at the level $\mathfrak{g}_3$.

Example 3.1.12 Let $G = \exp \mathfrak{g}, \mathfrak{g} = \langle T, X, Y_1, Y_2, Y_3, Z \rangle_{\mathbb{R}}$:

$$[T, X] = X, [T, Y_1] = Y_2, [T, Y_2] = Y_3, [T, Y_3] = Z.$$

Put

$$\mathfrak{g}_1 = \mathbb{R}Z, \mathfrak{g}_2 = \mathfrak{g}_1 \oplus \mathbb{R}Y_3, \mathfrak{g}_3 = \mathfrak{g}_2 \oplus \mathbb{R}Y_2,$$
$$\mathfrak{g}_4 = \mathfrak{g}_3 \oplus \mathbb{R}Y_1, \mathfrak{g}_5 = \mathfrak{g}_4 \oplus \mathbb{R}X.$$

Then,

$$\mathfrak{g}_1 \subset \mathfrak{g}_2 \subset \mathfrak{g}_3 \subset \mathfrak{g}_4 \subset \mathfrak{g}_5 \subset \mathfrak{g}_6 = \mathfrak{g}$$

is a composition series of ideals of $\mathfrak{g}$. The dimension index $e(\ell) = (0, 1, 1, 1, 1, 2)$ defines a Zariski open set $\mathcal{Z}$ of $\mathfrak{g}^*$. At the level $\mathfrak{g}_4$, we have a Casimir element

$$A = Z^2 Y_1 - Z Y_3 Y_2 + \frac{1}{3} Y_3^3$$

on $\mathcal{Z}$. Indeed, A belongs to the centre of $\mathcal{U}(\mathfrak{g})$ and the situation is essentially nilpotent.

3.1.5 The Poisson centre of the symmetric algebra

We here regard the symmetric algebra $S(\mathfrak{g})$ of $\mathfrak{g}_{\mathbb{C}}$ as the algebra of all polynomial functions on $\mathfrak{g}^*$ by $X(\ell) = \sqrt{-1}\ell(X)$ for $X \in \mathfrak{g}$ and $\ell \in \mathfrak{g}^*$. Note that $S(\mathfrak{g})$ is equipped with the Poisson structure $\{.,.\}$ defined by $\{X, Y\} = [X, Y]$ for X, Y in $\mathfrak{g}$. Let $\sigma : S(\mathfrak{g}) \to \mathcal{U}(\mathfrak{g})$ be the symmetrisation map. We start by proving the following:

Proposition 3.1.1 *Retain all the previous notation. Let $Q \in S(\mathfrak{g})$ be an element of $\mathbb{C}[\Sigma_e]^G$, then $W = \sigma(Q) \in Cas(\Sigma_e)$ and $\varphi_W \equiv Q$ on Σ_e. Conversely, let $W \in Cas(\Sigma_e)$, then $Q = \sigma^{-1}(W) \in \mathbb{C}[\Sigma_e]^G$ and $\varphi_W \equiv Q$ on Σ_e (i.e. $Q \in FCas(\Sigma_e)$).*

Proof 3.3 *Suppose that $Q \in S(\mathfrak{g})$ is an element of $\mathbb{C}[\Sigma]^G$, namely a G-invariant polynomial function on Σ. Then $W = \sigma(Q)$ is a Casimir element on Σ and $\varphi_W(\ell) = Q(\ell)$ for $\ell \in \Sigma$. Indeed, if $e_m = e_{m-1} + 1$, $Q \in S(\mathfrak{g}_{m-1})$ and we can apply the induction hypothesis to $\mathfrak{g}_{m-1}$ and use the $\mathfrak{g}$-invariance to obtain the result, modulo the computations (cf. equation (3.1.7)) in Step 1 of the proof of Theorem 3.1.6.*

Suppose now that $e_m = e_{m-1}$. If $Q \in S(\mathfrak{g}_{m-1})$, the result comes directly from the induction hypothesis, as $\pi_\ell|_{G_{m-1}} \simeq {\pi'}_{\ell|\mathfrak{g}_{m-1}}$, the unitary and irreducible representation of $G_{m-1} = \exp(\mathfrak{g}_{m-1})$ associated to $\ell|\mathfrak{g}_{m-1}$ thanks to Subsection 2.3.4.3. If $Q \notin S(\mathfrak{g}_{m-1})$, let $\mathfrak{a}$ be a minimal non-central ideal of $\mathfrak{g}$ and $\mathfrak{z}$ the centre of $\mathfrak{g}$. We can therefore assume that $\mathfrak{g}_{m-1} = \{X \in \mathfrak{g}; [X, \mathfrak{a}] \subset \mathfrak{z}\}$, and that the generic coadjoint orbits of G in Σ are non-saturated with respect to $\mathfrak{g}_{m-1}$ according to the above. Hence $\mathfrak{a}$ is not a minimal ideal of $\mathfrak{g}$ and hence $\mathfrak{a} \cap \mathfrak{z}$ is not trivial. If every $\ell \in \Sigma$ vanishes on some non-trivial ideal of $\mathfrak{g}$, we can pass to the quotient of $\mathfrak{g}$ by this ideal to which applies the induction hypothesis. Suppose finally that $\ell \in \Sigma$ does not vanish on the non-trivial ideal $[\mathfrak{g}, \mathfrak{a}]$ of $\mathfrak{g}$, we find that $\mathfrak{a}^\ell = \{X \in \mathfrak{g}; \ell([X, \mathfrak{a}]) = \{0\}\}$ is a proper subalgebra of $\mathfrak{g}$. This is nothing but the stabiliser of $\ell|_\mathfrak{a} \in \mathfrak{a}^*$ in $\mathfrak{g}$ with *respect to the coadjoint action. For $\ell \in \Sigma$ we fix $\alpha = \ell|_{\mathfrak{a}}$ and consider the section $\Sigma_\alpha = \{\ell \in \Sigma; \ell|_{\mathfrak{a}} = \alpha\}$ and put $\mathfrak{a}^\alpha = \mathfrak{a}^\ell$ for $\ell \in \Sigma_\alpha$. Then, as $A{\cdot}\ell = \ell + (\mathfrak{a}^\alpha)^\perp$, $A = \exp \mathfrak{a}$, for $\ell \in \Sigma_\alpha$, $Q \in S(\mathfrak{a}^\alpha)$ on Σ_α. The induction hypothesis applied to $\mathfrak{a}^\alpha$ says that $W = \sigma(Q)$ is a Casimir element and $\varphi_W(\ell) = Q(\ell)$ on Σ_α. Extending this result on Σ by using G-invariance, we get the result all as in Step 3 of Theorem 3.1.6. Conversely, let $W \in \mathcal{U}(\mathfrak{g})$ be a Casimir element on Σ. If $e_m = e_{m-1}+1$, then $\pi(X_m)$ is a derivation operator, where $X_m \in \mathfrak{g} \setminus \mathfrak{g}_{m-1}$. Hence $Q = \sigma^{-1}(W) \in S(\mathfrak{g}_{m-1})$ is $\mathfrak{g}$-invariant and the induction hypothesis gives $\varphi_W(\ell) = Q(\ell)$ on Σ. Suppose $e_m = e_{m-1}$. If $W \in \mathcal{U}(\mathfrak{g}_{m-1})$, the result comes directly from the induction hypothesis. If $W \notin \mathcal{U}(\mathfrak{g}_{m-1})$, we can follow inversely the same argument as above. Since we can regard $\mathfrak{g}_{m-1} = \{X \in \mathfrak{g}; [X, \mathfrak{a}] \subset \mathfrak{z}\}$, $\mathfrak{a}$ is not a minimal ideal of $\mathfrak{g}$ and hence $\mathfrak{a} \cap \mathfrak{z}$ is not trivial. If every $\ell \in \Sigma$ vanishes on some non-trivial ideal of $\mathfrak{g}$, we can pass to the quotient of $\mathfrak{g}$ by this ideal to which applies the induction hypothesis. Suppose finally that $\ell \in \Sigma$ does not vanish on the non-trivial ideal $[\mathfrak{g}, \mathfrak{a}]$ of $\mathfrak{g}$, we find that $\mathfrak{a}^\ell$ is a proper subalgebra of $\mathfrak{g}$. Using all the notations introduced above, we notice $W \in \mathcal{U}(\mathfrak{a}^\alpha)$ because of $A{\cdot}\ell = \ell + (\mathfrak{a}^\alpha)^\perp$ for $\ell \in \Sigma_\alpha$. The induction hypothesis applied to $\mathfrak{a}^\alpha$ says that $Q = \sigma^{-1}(W) \in S(\mathfrak{a}^\alpha)$ is $\mathfrak{a}^\alpha$-invariant and $\varphi_W(\ell) = Q(\ell)$ on Σ_α. Extending this result on Σ by using G-invariance, we get the result.*

The following is then immediate:

Theorem 3.1.13 *Let $G = \exp \mathfrak{g}$ be an exponential solvable Lie group with Lie algebra $\mathfrak{g}$. Let*

$$\mathfrak{g}_1 \subset \cdots \subset \mathfrak{g}_{m-1} \subset \mathfrak{g}_m = \mathfrak{g}$$

be a composition series of ideals of $\mathfrak{g}$,

$$e(\ell) = (e_1(\ell), e_2(\ell), \ldots, e_m(\ell))$$

the dimension indices for $\ell \in \mathfrak{g}^*$, and $\mathcal{E} = \{e(\ell); \ell \in \mathfrak{g}^*\}$. Let e be the maximal element of $\mathcal{E}$ and $U_e = \{\ell \in \mathfrak{g}^*; e(\ell) = e\}$. There exists a Zakiski open set $\mathcal{Z}$ of $\mathfrak{g}^*$ verifying the following:

1. $\Sigma_e = \mathcal{Z} \cap U_e$ is a non-empty G-invariant set.

2. Under the symmetrisation map $\sigma : S(\mathfrak{g}) \to \mathcal{U}(\mathfrak{g})$, $\mathbb{C}[\Sigma_e]^G$ corresponds to the set $Cas(\Sigma_e)$ of Casimir elements on Σ_e.

3. Let φ_W be the Casimir function on Σ_e associated to the Casimir element W. Then, $\varphi_{\sigma(Q)}(\ell) = Q(\ell)$ for $Q \in \mathbb{C}[\Sigma_e]^G, \ell \in \Sigma_e$.

As e is maximal, the layer U_e itself is a G-invariant Zariski open set in this case. Concerning the existence of such a G-invariant Zariski open set, everything is reduced to the G-action on $(\mathfrak{g}_{m-1})^*$ if $e_m = e_{m-1} + 1$. In the case where $e_m = e_{m-1}$, making use of the Casimir element introduced in Proposition 3.1.1, we are reduced also in this case to the G-action on $(\mathfrak{g}_{m-1})^*$.

The following is an immediate consequence from Theorem 3.1.13:

Corollary 3.1.14 *Keep the notations and hypotheses of Theorem 3.1.13. Let* $\pi \in \widehat{G}$ *whose coadjoint orbit* Ω *is contained in* Σ. *If* $W \in Cas(\Sigma_e)$ *belongs to* $\ker \pi$, *then* $\sigma^{-1}(W)$ *vanishes on* Ω.

Indeed, it is enough to write $\pi_\ell(W) = \varphi_W(\ell)I = \sigma^{-1}(W)(\ell)I = 0$, for any $\ell \in \Omega$.

We now prove the main result in this section:

Theorem 3.1.15 *Let* $e \in \mathcal{E}$ *be the maximal element. The Poisson centre* $C(S(\mathfrak{g}))$ *of the Poisson algebra* $S(\mathfrak{g})$ *coincides with the set* $FCas(\Sigma_e)$ *of all Casimir functions on* $\Sigma_e = \mathcal{Z} \cap U_e$.

Proof 3.4 *Evidently, each Casimir function on* Σ *belongs to* $C(S(\mathfrak{g}))$. *Let* $Q \in C(S(\mathfrak{g}))$ *and show that* Q *is a Casimir function on* Σ. *Let* $\widetilde{\mathfrak{g}}$ *be a codimension 1 ideal of* $\mathfrak{g}$, *which could just be taken as the ideal* $\mathfrak{g}_{m-1}$ *in the sequence (3.1.1). If the generic coadjoint orbits of* G *are saturated with respect to* $\widetilde{\mathfrak{g}}$, *we have* $Q \in S(\widetilde{\mathfrak{g}})$ *and any Casimir element on* Σ *belongs to* $\mathcal{U}(\widetilde{\mathfrak{g}})$. *Let* $\widetilde{\mathcal{S}}$ *be the composition series of* $\widetilde{\mathfrak{g}}$ *given by* $\mathcal{S} \cap \widetilde{\mathfrak{g}}$, $\widetilde{e}, U_{\widetilde{e}}$ *and* $\widetilde{\mathcal{Z}}$ *our objects corresponding to* $\widetilde{\mathfrak{g}}$. *Then, the induction hypothesis applied to* $\widetilde{\mathfrak{g}}$ *implies that* Q *is a Casimir function on* $\widetilde{\mathcal{Z}} \cap \widetilde{U_{\widetilde{e}}}$. *Since* Q *is* $\mathfrak{g}$*-invariant, we can extend* $\widetilde{\mathcal{Z}}$ *to a* G*-invariant non-empty Zariski open set* $\mathcal{Z}$ *of* $\mathfrak{g}^*$ *so that* Q *is a Casimir function on* Σ.

Suppose now that the generic coadjoint orbits of G *are non-saturated with respect to* $\widetilde{\mathfrak{g}}$. *Then, we follow the same reasoning as in Proposition 3.1.1 by introducing a minimal non-central ideal* $\mathfrak{a}$.

Remark 3.1.16 The symmetrisation map gives for any Lie algebra $\mathfrak{g}$ a vector space isomorphism from the polynomial algebra $S(\mathfrak{g})$ to the universal enveloping algebra $\mathcal{U}(\mathfrak{g})$. This map is not an algebra homomorphism. It is equivariant with respect to the natural representation of $\mathfrak{g}$ on these spaces, so it restricts to a vector space isomorphism $\sigma\colon S(\mathfrak{g})^{\mathfrak{g}} \to U(\mathfrak{g})^{\mathfrak{g}}$, where the superscript indicates the subspace annihilated by the action of $\mathfrak{g}$. Both $S(\mathfrak{g})^{\mathfrak{g}}$ and $U(\mathfrak{g})^{\mathfrak{g}}$ are commutative subalgebras, indeed $U(\mathfrak{g})^{\mathfrak{g}}$ is the centre of $\mathcal{U}(\mathfrak{g})$.

Let $\Theta : S(\mathfrak{g})^{\mathfrak{g}} \to \mathcal{U}(\mathfrak{g})^{\mathfrak{g}}$ be the Duflo isomorphism (cf. [39]):

$$\Theta = \sigma \circ J(D)^{1/2},$$

where $J(D)^{1/2}$ is the differential operator of infinite order with constant coefficients corresponding to the formal series :

$$J(x)^{1/2} = \Big(\det \frac{sh\mathrm{ad}\frac{1}{2}x}{\mathrm{ad}\frac{1}{2}x}\Big)^{1/2}.$$

Actually, Θ is an algebra isomorphism. In accordance with the above, we get the following diagram:

$$\begin{array}{ccc} Cas(\Sigma_e) & \xrightarrow{\ \varphi\ } & C(S(\mathfrak{g})) \\ \sigma \big\uparrow & & \big\downarrow \Theta \\ FCas(\Sigma_e) & \xleftarrow{\ \varphi\ } & \mathfrak{z}(\mathcal{U}(\mathfrak{g})) \end{array}$$

which is not commutative, unless σ and Θ coincide as in the nilpotent setting for instance. As such, Proposition 3.1.1 shows that σ turns out to be an algebra isomorphism when restricted to $FCas(\Sigma_e)$.

3.2 GENERATING FAMILIES OF PRIMITIVE IDEALS

We solve here a longstanding problem about the construction of a generators family of the kernel of an irreducible and unitary representation of an exponential solvable Lie group. This is quite scarce in literature except in the nilpotent situation, for which an algorithm provided by the work of Pedersen turns out to be efficient and simple to describe (cf. [88]).

3.2.1 Case of flat orbits

Let us come back to the general setting where G is a connected real Lie group, and let $\mathfrak{g}$ be its Lie algebra. We shall show here that, if G is exponential solvable and Ω is open in an affine subspace Γ of $\mathfrak{g}^*$, there is a very natural relation of this type. Take $\ell \in \Omega \subset \Gamma$ and let m be the codimension of Γ in $\mathfrak{g}^*$. If $m \geq 1$, one can take $T_1, \dots, T_m \in \mathfrak{g}$ for which $\Gamma = \{\ell' \in \mathfrak{g}^* \ : \ \ell'(T_k) = \ell(T_k) \ \ (k = 1, \dots, m)\}$. Under the canonical identification of the polynomial algebra $\mathrm{Pol}(\mathfrak{g}^*)$ with the symmetric algebra $S(\mathfrak{g})$, the ideal $I(\Omega)$ is generated by $T_k - \ell(T_k) \ \ (k = 1, \dots, m)$. Let π_ℓ be the irreducible unitary representation of G corresponding to Ω via the Kirillov-Bernat-Vergne mapping. Then we show in Theorem 3.2.2 below that $\ker d\pi_\ell \subset \mathcal{U}(\mathfrak{g})$ is generated by $T_k - i\ell(T_k) \ \ (k = 1, \dots, m)$. If $m = 0$, then Ω is open in $\mathfrak{g}^*$ and $I(\Omega) = \{0\}$. In this case, Theorem 3.2.2 tells us that $\ker d\pi_\ell = \{0\}$.

We first prove the following:

Proposition 3.2.1 *Let $\ell \in \mathfrak{g}^*$. The radical $\mathfrak{g}(\ell)$ is an ideal of $\mathfrak{g}$ if and only if the coadjoint orbit Ω_ℓ through ℓ is relatively open in an affine subspace Γ of $\mathfrak{g}^*$. In this case, one has*

$$\Gamma = \ell + \mathfrak{g}(\ell)^\perp = \{\ell' \in \mathfrak{g}^* : \ell'(T) = \ell(T) \ \textit{for all}\ T \in \mathfrak{g}(\ell)\}.$$

Proof 3.5 *We shall show the 'if' part. We note that the tangent space $T_\ell\Omega_\ell = \{\ell \circ \mathrm{ad}(X) \ : \ X \in \mathfrak{g}\}$ of the coadjoint orbit at ℓ equals $\mathfrak{g}(\ell)^\perp = \{\ell' \in \mathfrak{g}^* \ : \ \ell'(T) = 0 \ \textit{for all}\ T \in \mathfrak{g}(\ell)\}$. Indeed, we see easily that $T_\ell\Omega_\ell \subset \mathfrak{g}(\ell)^\perp$, whereas the codimension of the both subspaces is $\dim \mathfrak{g}(\ell)$. Since Ω_ℓ is open in Γ, we have*

$$\Gamma = \ell + T_\ell\Omega_\ell = \ell + \mathfrak{g}(\ell)^\perp = \{\ell' \in \mathfrak{g}^* : \ell'(T) = \ell(T) \ \textit{for all}\ T \in \mathfrak{g}(\ell)\}.$$

In order to show that $\mathfrak{g}(\ell)$ is an ideal of $\mathfrak{g}$, it is sufficient to show that $[X, T] \in \mathfrak{g}(\ell)$ for any $X \in \mathfrak{g}$ and $T \in \mathfrak{g}(\ell)$. Let Y be another element of $\mathfrak{g}$. Since $\ell' := \mathrm{Ad}^*(\exp tX \exp sY)\ell \in \Omega_\ell$ belongs to Γ for $s, t \in \mathbb{R}$, we have

$$\ell(T) = \ell'(T) = \ell(\mathrm{Ad}(\exp(-sY)\exp(-tX))T).$$

Differentiating at $s = t = 0$, we obtain $0 = \ell([Y, [X, T]])$, so that $[X, T] \in \mathfrak{g}(\ell)$.

Therefore the 'if' part is verified. Next we show the 'only if' part. Assume that $\mathfrak{g}(\ell)$ is an ideal. For $T \in \mathfrak{g}(\ell)$ and $X \in \mathfrak{g}$, we have $\mathrm{Ad}(\exp X)T = \sum_{j=0}^{\infty} \frac{1}{j!}\mathrm{ad}(X)^j T \in \mathfrak{g}(\ell)$. Thus, thanks to the connectedness of G, we have $\mathrm{Ad}(g)T \in \mathfrak{g}(\ell)$ for any $g \in G$. It follows that, for $\ell_1 = \mathrm{Ad}^*(g)\ell \in \Omega_\ell$ we have $\mathfrak{g}(\ell_1) = \mathrm{Ad}(g)\mathfrak{g}(\ell) = \mathfrak{g}(\ell)$. On the other hand, the tangent space of the coadjoint orbit at ℓ_1 equals $\mathfrak{g}(\ell_1)^\perp = \mathfrak{g}(\ell)^\perp$. Namely, the tangent space of the connected manifold Ω_ℓ at any point equals $\mathfrak{g}(\ell)^\perp$, so that Ω_ℓ is open in the affine space $\Gamma = \ell + \mathfrak{g}(\ell)^\perp$.

We next prove the following lemma which will be utilised repeatedly in many situations.

Lemma 3.2.1 *Let $H = \exp\mathfrak{h}$ be a normal subgroup of G of codimension one. Take $X \in \mathfrak{g} \setminus \mathfrak{h}$ so that $\mathfrak{g} = \mathbb{R}X \oplus \mathfrak{h}$. For $t \in \mathbb{R}$, let $\alpha_t : H \to H$ be the automorphism given by $\alpha_t(h) := (\exp tX)^{-1}h(\exp tX)$ $(h \in H)$. Let τ be a unitary representation of H, and π the induced unitary representation $\mathrm{ind}_H^G\tau$. For $t \in \mathbb{R}$, denote by τ_t the unitary representation $\tau \circ \alpha_t$ of H. Assume that $V = \sum_{p=0}^{m} X^p V_p \in \mathcal{U}(\mathfrak{g})$ with $V_p \in \mathcal{U}(\mathfrak{h})$ $(p = 0, 1, \ldots, m)$ belongs to $\ker d\pi$.*
(i) *For all $t \in \mathbb{R}$ and $p = 0, \ldots, m$, one has $V_p \in \ker d\tau_t$.*
(ii) *For $p = 0, \ldots, m$, one has $V_p \in \ker d\pi$.*

Proof 3.6 *Let $\mathcal{H}_\tau$ be the representation space of τ. Noting that $G = (\exp\mathbb{R}X) \ltimes H$, we realise $\pi = \mathrm{ind}_H^G\tau$ on the Hilbert space $L^2(\mathbb{R}, \mathcal{H}_\tau)$ of $\mathcal{H}_\tau$-valued L^2 functions on $\mathbb{R}$ by*

$$\pi(\exp t'X)\phi(t) := \phi(t - t'), \tag{3.2.1}$$

$$\pi(h)\phi(t) := \tau_t(h)\phi(t) \quad (t, t' \in \mathbb{R},\ h \in H,\ \phi \in L^2(\mathbb{R}, \mathcal{H}_\tau)). \tag{3.2.2}$$

Since (3.2.1) implies that $d\pi(X) = -\frac{d}{dt}$, we see from (3.2.2) that

$$0 = d\pi(V)\phi(t) = \sum_{p=0}^{m}\Big(-\frac{d}{dt}\Big)^p d\tau_t(V_p)\phi(t) \quad (t \in \mathbb{R}) \tag{3.2.3}$$

if $\phi \in L^2(\mathbb{R}, \mathcal{H}_\tau)$ is a smooth vector for the representation π. For any $t_0 \in \mathbb{R}$, $0 \le p_0 \le m$, and a smooth vector $v_0 \in \mathcal{H}_\tau$ for the representation τ, we take $f_0 \in C_c^\infty(\mathbb{R})$ which takes the value 1 on a neighbourhood of t_0, and consider (3.2.3) at $t = t_0$ with $\phi(t) = (p_0!)^{-1}(t_0 - t)^{p_0} f_0(t)v_0$. Noting that $v_0 \in \mathcal{H}_\tau$ is smooth for the representation $\tau_t = \tau \circ \alpha_t$ for any t, we obtain

$$0 = d\pi(V)\phi(t_0) = \sum_{p=0}^{m}\Big(-\frac{d}{dt}\Big)^p_{t=t_0}\Big[\frac{(t_0 - t)^{p_0}}{p_0!} f_0(t)\Big] d\tau_{t_0}(V_p)v_0 = d\tau_{t_0}(V_{p_0})v_0,$$

which means that $V_{p_0} \in \ker d\tau_{t_0}$. Thus the assertion (i) is verified. The assertion (ii) follows immediately from (i) and (3.2.2).

Now we shall discuss the case where G is a simply connected real exponential solvable Lie group. We prove the following:

Theorem 3.2.2 *Let G be a simply connected real exponential solvable Lie group, and π_ℓ the irreducible unitary representation of G corresponding to the coadjoint orbit Ω_ℓ through a linear form $\ell \in \mathfrak{g}^*$. Assume that $\mathfrak{g}(\ell)$ is an ideal of $\mathfrak{g}$. Then one has*

$$\ker d\pi_\ell = \mathcal{U}(\mathfrak{g})\text{-span}\{T - i\ell(T), T \in \mathfrak{g}(\ell)\}.$$

In particular, $\ker d\pi_\ell = \{0\}$ *if Ω_ℓ is open.*

Proof 3.7 *We shall use an induction on* $\dim G$. *If* $\dim G = 0$ *or* 1, *then the statement is trivial. We suppose therefore that* $\dim G \geq 2$. *Furthermore, remark that we can suppose that ℓ does not vanish on any non-trivial ideal of $\mathfrak{g}$, unless we can apply the induction hypothesis. Thus, let us assume that the centre $\mathfrak{z}(\mathfrak{g})$ of $\mathfrak{g}$ satisfies* $\dim \mathfrak{z}(\mathfrak{g}) \leq 1$ *and that $\ell|_{\mathfrak{z}(\mathfrak{g})} \neq 0$ if $\mathfrak{z}(\mathfrak{g}) \neq \{0\}$.*

Take a non-central minimal ideal $\mathfrak{a}$ of $\mathfrak{g}$, which is abelian and of dimension ≤ 3 by [70, Page 34]. Let $\mathfrak{g}_0$ be the centraliser of $\mathfrak{a}$. We see easily that $\mathfrak{g}_0$ is an ideal of $\mathfrak{g}$.

Assume first that $\mathfrak{z}(\mathfrak{g}) = \{0\}$, then $\dim \mathfrak{a} = 1$ *or* 2 *in this case.*

(i-1) When $\dim \mathfrak{a} = 1$, *we write $\mathfrak{a} = \mathbb{R}Y$ with $\ell(Y) = 1$. Then there exists a non-zero vector $X \in \mathfrak{g}$ such that $\mathfrak{g} = \mathfrak{g}_0 \oplus \mathbb{R}X$ and $[X, Y] = Y$. If $C \in \mathfrak{g}(\ell)$, we can write $C = C_0 + cX$ with $c \in \mathbb{R}$ and $C_0 \in \mathfrak{g}_0$. Then $0 = \ell([C, Y]) = \ell(c[X, Y]) = c\ell(Y) = c$, so that $C = C_0 \in \mathfrak{g}_0$. It means that $\mathfrak{g}(\ell) \subset \mathfrak{g}_0$, and this is a saturated case. Thus $\pi = \pi_\ell \in \hat{G}$ is induced from the irreducible unitary representation $(\pi_0, \mathcal{H}_0)$ of G_0 corresponding to the coadjoint orbit through the linear form $\ell_0 := \ell|_{\mathfrak{g}_0} \in \mathfrak{g}_0^*$.*

For $t \in \mathbb{R}$, let $\alpha_t : G \to G$ be the inner automorphism as defined in Lemma 3.2.1. As is expressed in Lemma 1.5.8, we have

$$\pi|_{G_0} \simeq \int_{\mathbb{R}}^{\oplus} \pi_t^0 dt, \tag{3.2.4}$$

where $\pi_t^0 := \pi_0 \circ \alpha_t$ is the irreducible unitary representation of G_0 corresponding to the coadjoint orbit through the linear form $\ell_t^0 := (\mathrm{Ad}^*(\exp tX)\ell)|_{\mathfrak{g}_0} \in \mathfrak{g}_0^*$.

(i-2) Let

$$V = \sum_{p=0}^{m} X^p V_p$$

be an element of $\ker d\pi$ *with $V_p \in \mathcal{U}(\mathfrak{g}_0)$. By Lemma 3.2.1 (i), we obtain* $V_p \in \bigcap_{t\in\mathbb{R}} \ker d\pi_t^0$ *for $p = 0, \ldots, m$, so that*

$$\ker d\pi = \mathcal{U}(\mathfrak{g}) \cdot \left(\bigcap_{t\in\mathbb{R}} \ker d\pi_t^0 \right). \tag{3.2.5}$$

(i-3) As is already seen, we have a saturated situation $\mathfrak{g}(\ell) \subset \mathfrak{g}_0$, so that $\mathfrak{g}(\ell)$ is a codimension-one subalgebra of $\mathfrak{g}_0(\ell_0)$. On the other hand, we see easily that $\mathbb{R}Y \oplus \mathfrak{g}(\ell) \subset \mathfrak{g}_0(\ell_0)$. Thus we obtain $\mathbb{R}Y \oplus \mathfrak{g}(\ell) = \mathfrak{g}_0(\ell_0)$. Applying $\mathrm{Ad}(\exp tX)$ *and keeping in mind that $\mathfrak{g}(\ell)$ is an ideal of $\mathfrak{g}$, we get also $\mathfrak{g}_0(\ell_t^0) = \mathbb{R}Y \oplus \mathfrak{g}(\ell) = \mathfrak{g}_0(\ell_0)$ for all*

$t \in \mathbb{R}$. *In particular,* $\mathfrak{g}_0(\ell_t^0)$ *are ideals in* $\mathfrak{g}_0$, *and we can apply the induction hypothesis to the representations* π_t^0. *Then we have*

$$\begin{aligned}\ker d\pi_t^0 &= \mathcal{U}(\mathfrak{g}_0)\text{-span}\{T - i\ell_t^0(T), T \in \mathfrak{g}_0(\ell_t^0)\} &(3.2.6)\\ &= \mathcal{U}(\mathfrak{g}_0)\text{-span}\{Y - i\ell_t^0(Y), T - i\ell_t^0(T), T \in \mathfrak{g}(\ell)\} &(3.2.7)\\ &= \mathcal{U}(\mathfrak{g}_0)\text{-span}\{Y - ie^{-t}, T - i\ell(T), T \in \mathfrak{g}(\ell)\}. &(3.2.8)\end{aligned}$$

Indeed, since $\mathrm{Ad}^*(\exp tX)\ell \in \Omega_\ell \subset \ell + \mathfrak{g}(\ell)^\perp$ *by Proposition 3.2.1, we have* $\ell_t^0(T) = \mathrm{Ad}^*(\exp tX)\ell(T) = \ell(T)$ *for* $T \in \mathfrak{g}(\ell)$, *whence (3.2.8) follows.*
(i-4) We see from (3.2.5) and (3.2.8) that $\ker d\pi \supset \mathcal{U}(\mathfrak{g})\text{-span}\{T - i\ell(T), T \in \mathfrak{g}(\ell)\}$. *We shall prove the converse inclusion. Put* $a := \dim \mathfrak{g}(\ell)$ *and let* $(\tilde{T}_1, \dots, \tilde{T}_a)$ *be a basis of* $\mathfrak{g}(\ell)$. *Put* $b := \dim \mathfrak{g}_0 - \dim \mathfrak{g}_0(\ell_0)$ *and take elements* $\tilde{X}_1, \dots, \tilde{X}_b$ *of* $\mathfrak{g}_0$ *for which* $(\tilde{X}_1, \dots, \tilde{X}_b, Y, \tilde{T}_1, \dots, \tilde{T}_a)$ *forms a basis of* $\mathfrak{g}_0$. *By the Poincaré-Birkhoff-Witt Theorem, any element* A *of the enveloping algebra* $\mathcal{U}(\mathfrak{g}_0)$ *is expressed uniquely as a* $\mathbb{C}$*-linear combinations of monomials of the form*

$$\tilde{X}^\beta Y^k \tilde{T}^\alpha = \tilde{X}_1^{\beta_1} \dots \tilde{X}_b^{\beta_b} Y^k \tilde{T}_1^{\alpha_1} \dots \tilde{T}_a^{\alpha_a}$$

with $\alpha = (\alpha_1, \dots, \alpha_a) \in \mathbb{N}^a$, $\beta = (\beta_1, \dots, \beta_b) \in \mathbb{N}^b$ *and* $k \in \mathbb{N}$, *where* $\mathbb{N}$ *denotes the set of non-negative integers. Since*

$$\tilde{T}_j^{\alpha_j} = \{(\tilde{T}_j - i\ell(\tilde{T}_j)) + i\ell(\tilde{T}_j)\}^{\alpha_j} = \sum_{p=0}^{\alpha_j} \binom{\alpha_j}{p} (i\ell(\tilde{T}_j))^{\alpha_j - p} (\tilde{T}_j - i\ell(\tilde{T}_j))^p$$

for $j = 1, \dots, a$, *the element* $A \in \mathcal{U}(\mathfrak{g}_0)$ *is also expressed uniquely as a* $\mathbb{C}$*-linear combination of*

$$\tilde{X}^\beta Y^k (\tilde{T} - i\ell(\tilde{T}))^\alpha = \tilde{X}_1^{\beta_1} \dots \tilde{X}_b^{\beta_b} Y^k (\tilde{T}_1 - i\ell(\tilde{T}_1))^{\alpha_1} \dots (\tilde{T}_a - i\ell(\tilde{T}_a))^{\alpha_a}.$$

Moreover, we have

$$\begin{aligned}&\mathcal{U}(\mathfrak{g}_0)\text{-span}\{T - i\ell(T),\, T \in \mathfrak{g}(\ell)\}\\ &= \mathbb{C}\text{-span}\{\tilde{X}^\beta Y^k (\tilde{T} - i\ell(\tilde{T}))^\alpha,\, \alpha \in \mathbb{N}^a \setminus \{(0, \dots, 0)\},\, \beta \in \mathbb{N}^b,\, k \in \mathbb{N}\},\end{aligned} \tag{3.2.9}$$

which can be deduced from the following observation: Let $\{\tilde{c}_{ij}^k\}_{i,j,k=1}^a$ *be the structure constants of the Lie algebra* $\mathfrak{g}(\ell)$ *with respect to the basis* $(\tilde{T}_1, \dots, \tilde{T}_a)$, *which means that*

$$[\tilde{T}_i, \tilde{T}_j] = \sum_{k=1}^a \tilde{c}_{ij}^k \tilde{T}_k.$$

Since $\tilde{T}_i \in \mathfrak{g}(\ell)$, *we have*

$$0 = \ell([\tilde{T}_i, \tilde{T}_j]) = \sum_{k=1}^a \tilde{c}_{ij}^k \ell(\tilde{T}_k).$$

Then we observe

$$\begin{aligned}&(\tilde{T}_i - i\ell(\tilde{T}_i))(\tilde{T}_j - i\ell(\tilde{T}_j)) - (\tilde{T}_j - i\ell(\tilde{T}_j))(\tilde{T}_i - i\ell(\tilde{T}_i)) = [\tilde{T}_i, \tilde{T}_j] = [\tilde{T}_i, \tilde{T}_j] - \ell([\tilde{T}_i, \tilde{T}_j])\\ &= \sum_{k=1}^a \tilde{c}_{ij}^k (\tilde{T}_k - i\ell(\tilde{T}_k)),\end{aligned}$$

so that

$$(\tilde{T}_i - i\ell(\tilde{T}_i))(\tilde{T}_j - i\ell(\tilde{T}_j)) = (\tilde{T}_j - i\ell(\tilde{T}_j))(\tilde{T}_i - i\ell(\tilde{T}_i)) + \sum_{k=1}^{a} \tilde{c}_{ij}^k(\tilde{T}_k - i\ell(\tilde{T}_k)).$$

Therefore the product of a monomial $(\tilde{T} - i\ell(\tilde{T}))^\alpha = (\tilde{T}_1 - i\ell(\tilde{T}_1))^{\alpha_1} \cdots (\tilde{T}_a - i\ell(\tilde{T}_a))^{\alpha_a}$ *with any* $\tilde{T}_j - i\ell(\tilde{T}_j)$ *is expressed as a* $\mathbb{C}$*-linear combination of monomials* $(\tilde{T} - i\ell(\tilde{T}))^{\alpha'}$ *with* $\alpha' \neq (0, \dots, 0)$*, whence (3.2.9) follows.*

Let $L \subset \mathcal{U}(\mathfrak{g}_0)$ *be the* $\mathbb{C}$*-linear subspace spanned by monomials* $\tilde{X}^\beta$ *with* $\beta \in \mathbb{N}^b$*, we see from (3.2.9) that*

$$L \cap (\mathcal{U}(\mathfrak{g}_0)\text{-span}\{T - i\ell(T), T \in \mathfrak{g}(\ell)\}) = \{0\}. \tag{3.2.10}$$

Suppose that there exists $W \in (\ker d\pi \cap \mathcal{U}(\mathfrak{g}_0)) \backslash (\mathcal{U}(\mathfrak{g}_0)\text{-span}\{T - i\ell(T), T \in \mathfrak{g}(\ell)\})$*. Without loss of generality, we can suppose that* $W = \sum_{k=0}^m W_k Y^k$ *with* $W_k \in L$*. For any* $t \in \mathbb{R}$*, we have*

$$0 = d\pi_t^0(W) = \sum_{k=0}^{m} (ie^{-t})^k d\pi_t^0(W_k) = d\pi_t^0(\sum_{k=0}^{m} (ie^{-t})^k W_k).$$

Thus the element $\sum_{k=0}^m (ie^{-t})^k W_k$ *of* L *equals 0 by (3.2.8) and (3.2.10) for all* $t \in \mathbb{R}$*. In particular, taking mutually different* $t_0, t_1, \dots, t_m$*, we have* $\sum_{k=0}^m (ie^{-t_j})^k W_k = 0$ *for* $j = 0, 1, \dots, m$*, which yields* $W_k = 0$ *for* $k = 0, \dots, m$*. Therefore we get* $W = 0$*, which contradicts* $W \notin \mathcal{U}(\mathfrak{g}_0)\text{-span}\{T - i\ell(T), T \in \mathfrak{g}(\ell)\}$*. Hence* $\ker d\pi \cap \mathcal{U}(\mathfrak{g}_0) \subset \mathcal{U}(\mathfrak{g}_0)\text{-span}\{T - i\ell(T), T \in \mathfrak{g}(\ell)\}$*, and we conclude that* $\ker d\pi = \mathcal{U}(\mathfrak{g})\text{-span}\{T - i\ell(T), T \in \mathfrak{g}(\ell)\}$ *by (3.2.5).*

(ii) Assume now that $\dim \mathfrak{a} = 2$ *and write* $\mathfrak{a} = \mathbb{R}Y_1 \oplus \mathbb{R}Y_2$*. Then as in the previous case, one can write* $\mathfrak{g} = \mathfrak{g}_0 \oplus \mathbb{R}X$ *with* $[X, Y_1] = Y_1 - \alpha Y_2$ *and* $[X, Y_2] = \alpha Y_1 + Y_2$ *for some non-zero real number* α *([70, Page 34]). We may and do assume that* $(\ell(Y_1), \ell(Y_2)) \neq (0, 0)$*. We observe*

$$\begin{aligned}\ell([X, Y_1])^2 + \ell([X, Y_2])^2 &= (\ell(Y_1) - \alpha\ell(Y_2))^2 + (\alpha\ell(Y_1) + \ell(Y_2))^2 \\ &= (1 + \alpha^2)(\ell(Y_1)^2 + \ell(Y_2)^2) \neq 0,\end{aligned}$$

so that we have $\ell([X, Y_h]) \neq 0$ *for either* $h = 1$ *or* 2*. We fix this* h *in what follows. Let* $C = C_0 + cX$ *be an element of* $\mathfrak{g}(\ell)$ *with* $C_0 \in \mathfrak{g}_0$ *and* $c \in \mathbb{R}$*. Then we have*

$$\begin{aligned}0 &= \ell([C, Y_1])^2 + \ell([C, Y_2])^2 = c^2(\ell([X, Y_1])^2 + \ell([X, Y_2])^2) \\ &= c^2(1 + \alpha^2)(\ell(Y_1)^2 + \ell(Y_2)^2),\end{aligned}$$

so that $c = 0$*, which means that* $\mathfrak{g}(\ell) \subset \mathfrak{g}_0$ *and we have again the saturated situation. Indeed, we have* $\mathfrak{g}_0(\ell_0) = \mathbb{R}Y_h \oplus \mathfrak{g}(\ell)$*, which is an ideal of* $\mathfrak{g}_0$*. Since* $G = (\exp \mathbb{R}X) \ltimes G_0$*, we obtain (3.2.4) and (3.2.5) by the same way as in (i). Furthermore, we have*

$$\ker d\pi_t^0 = \mathcal{U}(\mathfrak{g}_0)\text{-span}\{Y_h - i\ell_t^0(Y_h),\ T - i\ell(T), T \in \mathfrak{g}(\ell)\}.$$

Since $(\frac{d}{dt})_{t=0}\ell_t^0(Y_h) = -\ell([X, Y_h]) \neq 0$*, for any positive integer* m*, we can take*

$t_0, t_1, \ldots, t_m$ for which $\ell^0_{t_i}(Y_h) \neq \ell^0_{t_j}(Y_h)$ if $i \neq j$. Then we apply the argument in (i-4) with replacing e^{-t_i} by $\ell^0_{t_i}(Y_h)$ to obtain

$$\ker d\pi \subset \mathcal{U}(\mathfrak{g}) \cdot \left(\bigcap_{t \in \mathbb{R}} \ker d\pi^0_t \right) = \mathcal{U}(\mathfrak{g})\text{-span}\{T - i\ell(T), T \in \mathfrak{g}(\ell)\},$$

and the statement is verified in the case (ii).

We assume henceforth that $\dim \mathfrak{z}(\mathfrak{g}) = 1$. We write $\mathfrak{z}(\mathfrak{g}) = \mathbb{R}Z$ with $\ell(Z) = 1$ and take a minimal non-central ideal $\mathfrak{a}$ in $\mathfrak{g}$. As before, let $\mathfrak{g}_0$ be the centraliser of $\mathfrak{a}$. If $\dim \mathfrak{a} = 2$, then we write $\mathfrak{a} = \mathbb{R}Z \oplus \mathbb{R}Y$ with $\ell(Y) = 0$.
(iii) If $[\mathfrak{g}, \mathfrak{a}] = \mathfrak{z}(\mathfrak{g})$, then one can write $\mathfrak{g} = \mathfrak{g}_0 \oplus \mathbb{R}X$ with $[X, Y] = Z$. We can show that $\mathfrak{g}_0(\ell_0) = \mathfrak{g}(\ell) \oplus \mathbb{R}Y$. And we can apply almost the same argument as in the first case, where the only difference is that $\ell^0_t(Y) = -t$.
(iv) Let us assume that $[\mathfrak{g}, \mathfrak{a}] = \mathfrak{a}$. There exist non-zero vectors $X_1, X_2 \in \mathfrak{g}$ such that $\mathfrak{g} = \mathfrak{g}_0 \oplus \mathbb{R}X_1 \oplus \mathbb{R}X_2$ with the commutators $[X_1, Y] = Z$ and $[X_2, Y] = Y$. By Jacobi's identity, one gets

$$[X_2, X_1] = -X_1 \ mod(\mathfrak{g}_0).$$

Indeed, we have

$$[[X_2, X_1], Y] = [[X_2, Y], X_1] + [X_2, [X_1, Y]] = [Y, X_1] + [X_2, Z] = -[X_1, Y],$$

so that $[[X_2, X_1] + X_1, Y] = 0$, which means that $[X_2, X_1] + X_1 \in \mathfrak{g}_0$. It follows that $\mathfrak{g} = \mathbb{R}X_2 \ltimes (\mathbb{R}X_1 \ltimes \mathfrak{g}_0)$. In particular, $\mathfrak{g}_1 := \mathbb{R}X_1 \ltimes \mathfrak{g}_0$ forms an ideal of codimension one in $\mathfrak{g}$.

Let $C = C_0 + c_1X_1 + c_2X_2$ $(C_0 \in \mathfrak{g}_0, c_1, c_2 \in \mathbb{R})$ be an element of $\mathfrak{g}(\ell)$. Then we have

$$0 = \ell([C, Y]) = \ell(c_1Z + c_2Y) = c_1,$$

so that $[C, Y] = c_2Y$. Since $\mathfrak{g}(\ell)$ is an ideal, the element $c_2Y = [C, Y]$ belongs to $\mathfrak{g}(\ell)$, and we have

$$0 = \ell([X_1, c_2Y]) = c_2\ell(Z) = c_2.$$

Then we get $c_1 = c_2 = 0$ and we have $\mathfrak{g}(\ell) \subset \mathfrak{g}_0 \subset \mathfrak{g}_1$. Because of the saturated situation, we get $\pi = \mathrm{ind}_{G_1}^G \pi_1$, where π_1 is the unitary irreducible representation of $G_1 := \exp \mathfrak{g}_1$ corresponding to the coadjoint orbit through linear form $\ell|_{\mathfrak{g}_1} \in \mathfrak{g}_1^*$. Furthermore, let $C' = C'_0 + c'_1X_1$ $(C'_0 \in \mathfrak{g}_0,\ c'_1 \in \mathbb{R})$ be an element of $\mathfrak{g}_1(\ell_1)$. As in the same way as above, we see that $0 = \ell_1([C', Y]) = \ell(c'_1Z) = c'_1$, which leads us to $\mathfrak{g}_1(\ell_1) \subset \mathfrak{g}_0$. Thus we have again the saturation, so that $\pi_1 = \mathrm{ind}_{G_0}^{G_1} \pi_0$, where π_0 is the unitary irreducible representation of $G_0 = \exp \mathfrak{g}_0$ corresponding to the coadjoint orbit through the linear form $\ell|_{\mathfrak{g}_0} \in \mathfrak{g}_0^*$. Therefore we obtain $\pi = \mathrm{ind}_{G_1}^G \mathrm{ind}_{G_0}^{G_1} \pi_0 = \mathrm{ind}_{G_0}^G \pi_0$.

We realise π_1 on $\mathcal{H}_1 := L^2(\mathbb{R}, \mathcal{H}_0)$ as

$$\pi_1(\exp(t'X_1))\psi(t) := \psi(t - t'), \tag{3.2.11}$$

$$\pi_1(g_0)\psi(t) := \pi^0_t(g_0)\psi(t) \quad (t, t' \in \mathbb{R},\ g_0 \in G_0,\ \psi \in \mathcal{H}_1), \tag{3.2.12}$$

where $\pi_t^0(g_0) := \pi_0((\exp(tX_1))^{-1}g_0(\exp(tX_1)))$. *On the other hand, we realise the representation* π *on the space* $\mathcal{H} = L^2(\mathbb{R}, \mathcal{H}_1)$ *as*

$$\pi(\exp(s'X_2))\phi(s) := \phi(s - s'), \tag{3.2.13}$$
$$\pi(g_1)\phi(s) := \pi_s^1(g_1)\phi(s) \quad (s, s' \in \mathbb{R},\ g_1 \in G_1,\ \phi \in \mathcal{H}), \tag{3.2.14}$$

where $\pi_s^1 := \pi_1 \circ \alpha_s$ *with* $\alpha_s(g_1) := (\exp(sX_2))^{-1}g_1(\exp(sX_2))$ $(g_1 \in G_1)$. *Let* $V \in \mathcal{U}(\mathfrak{g})$ *be an element of* $\ker d\pi$ *with an expression*

$$V = \sum_{p=0}^{m}\sum_{q=0}^{n} X_2^p X_1^q V_{p,q} = \sum_{p=0}^{m} X_2^p\Big(\sum_{q=0}^{n} X_1^q V_{p,q}\Big) \qquad (V_{p,q} \in \mathcal{U}(\mathfrak{g}_0)).$$

Applying Lemma 3.2.1 (i) to the representation $\pi = \mathrm{ind}_{G_1}^{G}\pi_1$ *of the group* $G = \exp(\mathbb{R}X_2) \ltimes G_1$, *we have*

$$\sum_{q=0}^{n} X_1^q V_{p,q} \in \ker d\pi_s^1 \tag{3.2.15}$$

for all $s \in \mathbb{R}$ *and* $p = 0, \ldots, m$. *On the other hand, since* $\pi_1 = \mathrm{ind}_{G_0}^{G_1}\pi_0$ *and the normal subgroup* G_0 *of* G_1 *is stable under the automorphism* $\alpha_s : G_1 \to G_1$, *we see that* $\pi_s^1 = \pi_1 \circ \alpha_s$ *is equivalent to* $\mathrm{ind}_{G_0}^{G_1}(\pi_0 \circ \alpha_s)$. *Then applying Lemma 3.2.1 (ii) to* π_s^1, *we obtain* $V_{p,q} \in \ker d\pi_s^1$ *for all* $q = 0, \ldots, n$. *Therefore, thanks to (3.2.14), we get*

$$V_{p,q} \in \ker d\pi \quad (0 \le p \le m,\ 0 \le q \le n). \tag{3.2.16}$$

It follows that

$$\ker d\pi = \mathcal{U}(\mathfrak{g})(\ker d\pi \cap \mathcal{U}(\mathfrak{g}_0)). \tag{3.2.17}$$

Let $\mathfrak{g}_2$ *be the subalgebra* $\mathbb{R}X_2 \ltimes \mathfrak{g}_0$ *of* $\mathfrak{g}$, *and* G_2 *the subgroup* $\exp(\mathfrak{g}_2)$ *of* G. *Let us consider the restriction* $\pi|_{G_2}$ *of the representation* $\pi = \pi_\ell$.

Recalling that $\mathcal{H}_1 = L^2(\mathbb{R}, \mathcal{H}_0)$, *we identify the representation space* $\mathcal{H} = L^2(\mathbb{R}, \mathcal{H}_1)$ *of* π *with* $L^2(\mathbb{R}^2, \mathcal{H}_0)$ *via the correspondence* $L^2(\mathbb{R}, \mathcal{H}_1) \ni \phi \mapsto \Phi \in L^2(\mathbb{R}^2, \mathcal{H}_0)$ *given by* $\Phi(t, s) := \phi(s)(t) \in \mathcal{H}_0$ $(s, t \in \mathbb{R})$. *Then we have by (3.2.13)*

$$\pi(\exp(s'X_2))\Phi(t, s) = \pi(\exp(s'X_2))\phi(s)(t) = \phi(s - s')(t) = \Phi(t, s - s'),$$

while we see from (3.2.14) and (3.2.12) that

$$\begin{aligned}\pi(g_0)\Phi(t, s) &= \pi_s^1(g_0)\phi(s)(t) = \pi_1((\exp(sX_2))^{-1}g_0(\exp(sX_2)))\phi(s)(t)\\ &= \pi_t^0((\exp(sX_2))^{-1}g_0(\exp(sX_2)))\Phi(t, s).\end{aligned}$$

These formulas tell us that, for each $t \in \mathbb{R}$, *the group* G_2 *acts on* $\Phi(t, \cdot) \in L^2(\mathbb{R}, \mathcal{H}_0)$ *in the same way as the induced representation* $\pi_t^2 := \mathrm{ind}_{G_0}^{G_2}\pi_t^0$. *In other words, we get a disintegration*

$$\pi|_{G_2} \simeq \int_{\mathbb{R}}^{\oplus} \pi_t^2\, dt.$$

On the other hand, since the representation π_t^0 *corresponds to the coadjoint orbit through* $\ell_t^0 = (\mathrm{Ad}^*(\exp(tX_1))\ell)|_{\mathfrak{g}_0} \in \mathfrak{g}_0^*$, *we see that* π_t^2 *is the irreducible* *unitary representation of* G_2 *corresponding to the coadjoint orbit through* $\ell_t^2 := (\mathrm{Ad}^*(\exp(tX_1))\ell)|_{\mathfrak{g}_2} \in \mathfrak{g}_2^*$. *Since* $\mathfrak{g}(\ell)$ *is an ideal, we have*

$$\mathfrak{g}(\mathrm{Ad}^*(\exp(tX_1))\ell) = \mathrm{Ad}(\exp(tX_1))\mathfrak{g}(\ell) = \mathfrak{g}(\ell) \subset \mathfrak{g}_0 \subset \mathfrak{g}_2.$$

Thus we have a saturated situation and $\mathfrak{g}_2(\ell_t^2) = \mathbb{R}Y \oplus \mathfrak{g}(\ell)$. *Then* $\mathfrak{g}_2(\ell_t^2)$ *is an ideal of* $\mathfrak{g}_2$, *so that we can apply the induction hypothesis for the representation* π_t^2 *of* G_2. *It means that*

$$\ker d\pi_t^2 = \mathcal{U}(\mathfrak{g}_2)\text{-span}\{Y - i\ell_t^2(Y),\, T - i\ell_t^2(T),\, T \in \mathfrak{g}(\ell)\}.$$

We observe that $\ell_t^2(Y) = \ell(\mathrm{Ad}(\exp tX_1)^{-1}Y) = -t$ *and that* $\ell_t^2(T) = \ell(T)$ *for* $T \in \mathfrak{g}(\ell)$. *Then, by the same argument as in (i-4), we conclude that*

$$\ker d\pi \cap \mathcal{U}(\mathfrak{g}_0) \subset \bigcap_{t\in\mathbb{R}} \ker d\pi_t^2 = \mathcal{U}(\mathfrak{g}_2)\text{-span}\{T - i\ell(T),\, T \in \mathfrak{g}(\ell)\},$$

which together with (3.2.17) completes the proof for the case (iv).
(v) We tackle finally the case where $\dim \mathfrak{a} = 3$. *Then there exist non-zero vectors* X_1, X_2, X_3, Y_1, Y_2 *such that* $\mathfrak{a} = \mathbb{R}Z \oplus \mathbb{R}Y_1 \oplus \mathbb{R}Y_2$ *with* $\ell(Y_1) = \ell(Y_2) = 0$, *and* $\mathfrak{g} = \mathfrak{g}_0 \oplus \mathbb{R}X_1 \oplus \mathbb{R}X_2 \oplus \mathbb{R}X_3$ *with the commutators* $[X_i, Y_j] = \delta_{i,j}Z$ $(i,j = 1,2)$ *and*

$$[X_3, Y_1] = Y_1 - \alpha Y_2, \quad [X_3, Y_2] = Y_2 + \alpha Y_1$$

for some non-zero $\alpha \in \mathbb{R}$. *Let us observe that*

$$[[X_2, X_1], Y_1] = [[X_2, Y_1], X_1] + [X_2, [X_1, Y_1]] = [X_2, Z] = 0,$$

and similarly we get $[[X_2, X_1], Y_2] = 0$. *Thus* $[X_2, X_1] \in \mathfrak{g}_0$. *On the other hand, we have*

$$\begin{aligned} [[X_3, X_1], Y_1] &= [[X_3, Y_1], X_1] + [X_3, [X_1, Y_1]] = [Y_1 - \alpha Y_2, X_1] + [X_3, Z] = -Z \\ &= [-X_1 - \alpha X_2, Y_1], \\ [[X_3, X_1], Y_2] &= [[X_3, Y_2], X_1] + [X_3, [X_1, Y_2]] = [Y_2 + \alpha Y_1, X_1] + [X_3, Z] = -\alpha Z \\ &= [-X_1 - \alpha X_2, Y_2], \end{aligned}$$

so that

$$[X_3, X_1] = -X_1 - \alpha X_2 \ mod(\mathfrak{g}_0).$$

Similarly, we get

$$[X_3, X_2] = \alpha X_1 - X_2 \ mod(\mathfrak{g}_0).$$

These formulas tell us that $\mathfrak{g} = \mathbb{R}X_3 \ltimes (\mathbb{R}X_2 \ltimes (\mathbb{R}X_1 \ltimes \mathfrak{g}_0))$. *Let* $\mathfrak{g}_1$ *and* $\mathfrak{g}_2$ *be the subalgebras* $\mathbb{R}X_1 \ltimes \mathfrak{g}_0$ *and* $\mathbb{R}X_2 \ltimes (\mathbb{R}X_1 \ltimes \mathfrak{g}_0)$ *of* $\mathfrak{g}$, *respectively. We write* G_i $(i = 0,1,2)$ *for the subgroups* $\exp(\mathfrak{g}_i)$ $(i = 0,1,2)$ *of* G. *Clearly we have a sequence* $G_0 \subset G_1 \subset G_2 \subset G$.

Let $C = C_0 + c_1X_1 + c_2X_2 + c_3X_3$ $(C_0 \in \mathfrak{g}_0,\, c_1, c_2, c_3 \in \mathbb{R})$ *be an element of* $\mathfrak{g}(\ell)$. *Then we have for* $i = 1,2$

$$0 = \ell([C, Y_i]) = \ell(c_iZ + c_3[X_3, Y_i]) = c_i.$$

Thus $c_1 = c_2 = 0$. *Moreover, since* $\mathfrak{g}(\ell)$ *is an ideal, the element* $[C, Y_1] = [C_0 + c_3X_3, Y_1] = c_3(Y_1 - \alpha Y_2)$ *belongs to* $\mathfrak{g}(\ell)$. *Then we have*

$$0 = \ell([X_1, c_3(Y_1 - \alpha Y_2)]) = c_3,$$

so that $C = C_0 \in \mathfrak{g}_0$. *Therefore we have a saturated situation* $\mathfrak{g}(\ell) \subset \mathfrak{g}_0 \subset \mathfrak{g}_2$. *Then we get* $\pi_\ell = \mathrm{ind}_{G_2}^G \pi_2$, *where* π_2 *is the irreducible unitary representation of* $G_2 = \exp(\mathfrak{g}_2)$ *corresponding to the coadjoint orbit through* $\ell_2 = \ell|_{\mathfrak{g}_2} \in \mathfrak{g}_2^*$.

Let $C' = C_0' + c_1'X_1 + c_2'X_2$ $(C_0' \in g_0,\ c_1', c_2' \in \mathbb{R})$ *be an element of* $\mathfrak{g}_2(\ell_2)$. *By the same way as above, we see that* $c_1' = c_2' = 0$ *and that* $\mathfrak{g}_2(\ell_2) \subset \mathfrak{g}_0 \subset \mathfrak{g}_1$. *Furthermore, by the almost same argument, we can verify that* $\mathfrak{g}_1(\ell_1) \subset \mathfrak{g}_0$, *where* $\ell_1 := \ell|_{\mathfrak{g}_1} \in \mathfrak{g}_1^*$. *Thus we have the saturated situations, so that* $\pi_{i+1} = \mathrm{ind}_{G_i}^{G_{i+1}} \pi_i$ *for* $i = 0, 1$, *where* π_i *is the irreducible representation of* G_i *corresponding to the coadjoint orbit through* $\ell_i = \ell_{|\mathfrak{g}_i} \in \mathfrak{g}_i^*$. *Then we conclude that* $\pi = \mathrm{ind}_{G_2}^G \mathrm{ind}_{G_1}^{G_2} \mathrm{ind}_{G_0}^{G_1} \pi_0 = \mathrm{ind}_{G_0}^G \pi_0$.

Let $\mathcal{H}_2$ *be the representation space of* π_2. *We realise* $\pi = \mathrm{ind}_{G_2}^G \pi_2$ *on* $L^2(\mathbb{R}, \mathcal{H}_2)$ *as*

$$\begin{aligned}&\pi(\exp(s'X_3))\phi(s) := \phi(s - s'),\\ &\pi(g_2)\phi(s) := \pi_s^2(g_2)\phi(s) \quad (s, s' \in \mathbb{R},\ g_2 \in G_2,\ \phi \in \mathcal{H}),\end{aligned} \tag{3.2.18}$$

where $\pi_s^2 := \pi_2 \circ \alpha_s$ *with* $\alpha_s(g_2) := (\exp(sX_3))^{-1} g_2 (\exp(sX_3))$ *for* $g_2 \in G_2$. *If*

$$V = \sum_{p=0}^{k}\sum_{q=0}^{m}\sum_{r=0}^{n} X_3^p X_2^q X_1^r V_{p,q,r} = \sum_{p=0}^{k} X_3^p \Big(\sum_{q=0}^{m}\sum_{r=0}^{n} X_2^q X_1^r V_{p,q,r}\Big) \quad (V_{p,q,r} \in \mathcal{U}(\mathfrak{g}_0))$$

is an element of $\ker d\pi$, *applying Lemma 3.2.1 (i), we have* $\sum_{q=0}^{m}\sum_{r=0}^{n} X_2^q X_1^r V_{p,q,r} \in \ker d\pi_s^2$ *for all* $s \in \mathbb{R}$ *and* $p = 0, \ldots, k$. *Then by the same way as we get (3.2.16) in (iv), we obtain*

$$V_{p,q,r} \in \ker d\pi_s^2 \quad (0 \le q \le m,\ 0 \le r \le n).$$

By (3.2.18), we see that each of $V_{p,q,r}$ *belongs to* $\ker d\pi$. *Therefore (3.2.17) holds again.*

Let G_3 *be the subgroup of* G *corresponding to a subalgebra* $\mathfrak{g}_3 := \mathbb{R}X_3 \ltimes \mathfrak{g}_0$. *Similarly to the previous case (iv), we see that the restriction* $\pi_{|G_3}$ *of the unitary representation* π *is described on the space* $L^2(\mathbb{R}^3, \mathcal{H}_0)$ *as*

$$\pi(\exp(u'X_3))\Phi(s,t,u) := \Phi(s,t,u-u'), \tag{3.2.19}$$

$$\pi(g_0)\Phi(s,t,u) := \pi_{s,t}^0((\exp(uX_3))^{-1} g_0 (\exp(uX_3)))\Phi(s,t,u) \tag{3.2.20}$$

$$(s,t,u,u' \in \mathbb{R},\ \Phi \in L^2(\mathbb{R}^3, \mathcal{H}_0),\ g_0 \in G_0),$$

where $\pi_{s,t}^0(h) := \pi_0((\exp(sX_1))^{-1}(\exp(tX_2))^{-1} h (\exp(tX_2))(\exp(sX_1)))$ *for* $h \in G_0$. *The formulas (3.2.19) and (3.2.20) mean that, for each* $s, t \in \mathbb{R}$, *the group* G_3 *acts on* $\Phi(s,t,\cdot) \in L^2(\mathbb{R}, \mathcal{H}_0)$ *in the same way as the induced representation* $\pi_{s,t}^3 := \mathrm{ind}_{G_0}^{G_3} \pi_{s,t}^0$. *Thus we have a disintegration*

$$\pi|_{G_3} \simeq \int_{\mathbb{R}^2}^{\oplus} \pi_{s,t}^3 \, dsdt.$$

We note that $\pi^3_{s,t}$ is the irreducible unitary representation of G_3 corresponding to the coadjoint orbit through $\ell^3_{s,t} := (\mathrm{Ad}^*((\exp(tX_2))(\exp(sX_1)))\ell)|_{\mathfrak{g}_3} \in \mathfrak{g}_3^*$.

Similarly to (iv) again, we get $\mathfrak{g}_3(\ell^3_{s,t}) = \mathbb{R}Y_1 \oplus \mathbb{R}Y_2 \oplus \mathfrak{g}(\ell)$, which forms an ideal of $\mathfrak{g}_3$. Thus we can apply the induction hypothesis to the representation $\pi^3_{s,t}$ to obtain

$$\ker d\pi^3_{s,t} = \mathcal{U}(\mathfrak{g}_3)\text{-span}\{Y_1 - i\ell^3_{s,t}(Y_1),\ Y_2 - i\ell^3_{s,t}(Y_2),\ T - i\ell^3_{s,t}(T),\ T \in \mathfrak{g}(\ell)\}.$$

We observe that $\ell^3_{s,t}(Y_1) = -s$, $\ell^3_{s,t}(Y_2) = -t$ and that $\ell^3_{s,t}(T) = \ell(T)$ for all $T \in \mathfrak{g}(\ell)$. Then, by the same argument as in (i-4), we get

$$\ker d\pi \cap \mathcal{U}(\mathfrak{g}_0) \subset \bigcap_{(s,t)\in\mathbb{R}^2} \ker d\pi^3_{s,t} = \mathcal{U}(\mathfrak{g}_3)\text{-span}\{T - i\ell(T),\ T \in \mathfrak{g}(\ell)\}.$$

Hence, by (3.2.17), the proof of Theorem 3.2.2 is completed.

Remark 3.2.3 For a given $\ell \in \mathfrak{g}^*$, the subalgebra

$$\mathfrak{a}(\ell) = \bigcap_{\phi\in\mathrm{Ad}^*(G)\ell} \mathfrak{g}(\phi)$$

is the greatest ideal of $\mathfrak{g}$ contained in $\mathfrak{g}(\ell)$. Then Theorem 3.2.2 shows that $\mathfrak{a}(\ell) = \{0\}$ whenever $\ker d\pi_\ell$ is trivial. The converse turns out to be false. To see this, we can record the following example from [6, Example 8.3.17] and [7, 4.4. Example].

Example 3.2.4 Let $\mathfrak{g}$ be the eight-dimensional exponential solvable Lie algebra with basis $(X_1, \ldots, X_8)$ and brackets:

$$[X_1, X_2] = X_2 - X_3, [X_1, X_4] = X_4 - X_5, [X_1, X_6] = X_6 - X_7,$$

$$[X_1, X_3] = X_2 + X_3, [X_1, X_5] = X_4 + X_5, [X_1, X_7] = X_6 + X_7,$$

$$[X_2, X_4] = X_6 + \frac{1}{2}X_8, [X_3, X_5] = -X_6 + \frac{1}{2}X_8, [X_2, X_5] = [X_3, X_4] = X_7,$$

and the other brackets being zero. Then the group $G = \exp\mathfrak{g}$ admits a unitary irreducible representation $\pi = \pi_\ell$ such that:

$$\pi(X_1) = -\frac{\partial}{\partial t_1}, \pi(X_2) = -e^{t_1}\cos(t_1)\frac{\partial}{\partial t_2},$$

$$\pi(X_3) = -e^{t_1}\sin(t_1)\frac{\partial}{\partial t_2}, \pi(X_4) = -ie^{t_1}\cos(t_1)t_2,$$

$$\pi(X_5) = -ie^{t_1}\sin(t_1)t_2, \pi(X_6) = -i\frac{1}{2}e^{2t_1}\cos(2t_1),$$

$$\pi(X_7) = -i\frac{1}{2}e^{2t_1}\sin(2t_1) \text{ and } \pi(X_8) = -ie^{2t_1}.$$

The annihilator of π is generated by the three elements $Q_1, Q_2, Q_3 \in \mathcal{U}(\mathfrak{g})$ defined by:

$$Q_1 = X_2X_5 - X_3X_4, Q_2 = 4(X_6^2 + X_7^2) - X_8^2$$

and

$$Q_3 = X_6(X_3^2 - X_2^2) - 2X_2X_3X_7 + \frac{1}{2}X_8(X_2^2 + X_3^2).$$

On the other hand, the Kirillov coadjoint orbit associated to π is given by the parametric representation $(p_1, p_2, q_1, q_2) \mapsto \xi(p_1, p_2, q_1, q_2) \in \mathfrak{g}^*$, with the following coordinates in the dual basis $(X_1^*, \ldots, X_8^*)$:

$$\xi_1 = p_1, \xi_2 = (e^{q_1} \cos q_1)p_2,$$

$$\xi_3 = (e^{q_1} \sin q_1)p_2, \xi_4 = (e^{q_1} \cos q_1)q_2,$$

$$\xi_5 = (e^{q_1} \sin q_1)q_2, \xi_6 = \frac{1}{2}(e^{2q_1} \cos 2q_1),$$

$$\xi_7 = \frac{1}{2}(e^{2q_1} \sin 2q_1) \text{ and } \xi_8 = e^{2q_1}.$$

This finally entails that the associated orbit $\mathrm{Ad}^*(G)\ell$ is not open, $\mathfrak{g}(\ell)$ is not an ideal of $\mathfrak{g}$ and $\mathfrak{a}(\ell) = \{0\}$.

3.2.2 Generators of primitive ideals via non-saturation ideals

We give here a significant result about the construction of generators of the kernel of irreducible unitary representations of exponential solvable Lie groups, which will be used later. Keeping the notations as introduced above, we first prove the following:

Proposition 3.2.5 *Let $G = \exp \mathfrak{g}$ be an exponential solvable Lie group with Lie algebra $\mathfrak{g}$, $\pi \in \widehat{G}$ and Ω its coadjoint orbit which is not saturated with respect to a codimension one normal subgroup $G_0 = \exp(\mathfrak{g}_0)$ of G. Write $\mathfrak{g} = \mathfrak{g}_0 \oplus \mathbb{R}X$ for some $X \in \mathfrak{g} \setminus \mathfrak{g}_0$. Then there exists an element $W \in \mathcal{U}(\mathfrak{g})$ in $\ker \pi$ taking the form $X + Q$ for some $Q \in \mathcal{U}(\mathfrak{g}_0)$.*

Proof 3.8 *We use an induction on the dimension of G. Nothing to show if G is of a small dimension. If there exists in $\ker \pi$ an ideal $\mathfrak{a} \neq \{0\}$ of $\mathfrak{g}$, we can assume $\mathfrak{a} \subset \mathfrak{g}_0$ and pass to the quotient $\mathfrak{g}/\mathfrak{a}$ to which applies the induction hypothesis. Hence, if $\mathfrak{z}$ is the centre of $\mathfrak{g}$, then $\dim \mathfrak{z} \leq 1$ and $\pi|_{\exp \mathfrak{z}}$ is not trivial if $\dim \mathfrak{z} = 1$. Let $\mathfrak{a}$ be a minimal non-central ideal of $\mathfrak{g}$ containing $\mathfrak{z}$. Let $f \in \Omega$, $\mathfrak{g}' = \mathfrak{a}^f$ and $G' = \exp(\mathfrak{g}')$. We take a polarisation $\mathfrak{b}[f]$ of $\mathfrak{g}$ at f verifying the Pukanszky condition and contained in $\mathfrak{g}'$. Hence $\pi = ind_{B[f]}^G \chi_f$ with $B[f] = \exp(\mathfrak{b}[f])$. Put $\pi' = ind_{B[f]}^{G'} \chi_f$ so that $\pi = ind_{G'}^G \pi'$. Let Ω' be the coadjoint orbit of G' associated with π'. We remark that Ω' is non-saturated with respect to $\mathfrak{g}' \cap \mathfrak{g}_0$. From the induction hypothesis applied to π', there exists W in $\ker \pi'$ taking the form $W = \tilde{X} + Q$ for some $Q \in \mathcal{U}(\mathfrak{g}' \cap \mathfrak{g}_0)$.*

Note that G' is a maximal non-normal Lie subgroup of G. As in the proof of Theorem 3.2.2, we have either

$$\pi|_{G'} \simeq \int_{\mathbb{R}}^{\oplus} \pi'_{\exp(tX)\ell|_{\mathfrak{g}'}} dt,$$

or

$$\pi|_{G'} \simeq \pi'_+ \oplus \pi'_-,$$

where $\pi'_\pm$ designate the representations associated to the forms $\pi'_{\exp(\pm X)\ell|_{\mathfrak{g}'}}$. We then arrive to the fact that $W \in \ker \pi$.

We next prove the following:

Theorem 3.2.6 *Let $G = \exp \mathfrak{g}$ be an exponential solvable Lie group with Lie algebra $\mathfrak{g}$, $\mathfrak{z}$ the centre of $\mathfrak{g}$ and $\mathfrak{a}$ a minimal non-central ideal of $\mathfrak{g}$. Suppose $\dim \mathfrak{z} = 1$ and $\mathfrak{z} \subset \mathfrak{a}$. Put $\mathfrak{g}_0$ as in (1.1.6), and let $\pi \in \widehat{G}$ verifying $\pi|_{\mathfrak{z}} \neq 0$ and Ω the coadjoint orbit of G associated with π. Suppose finally $\mathfrak{g}_0 \neq \mathfrak{g}$ and that Ω is not saturated with respect to $\mathfrak{g}_0$. Let $\mathfrak{m}$ be the centraliser of $\mathfrak{a}$ in $\mathfrak{g}$. Then:*

1. There exists in $\ker \pi$ an element

$$W = ZT - YX + A. \tag{3.2.21}$$

if $\dim \mathfrak{a} = 2$, or

$$W = ZT - (Y_1 - \alpha Y_2)X_1 - (Y_2 + \alpha Y_1)X_2 + A \tag{3.2.22}$$

if $\dim \mathfrak{a} = 3$ with a certain element $A \in \mathcal{U}(\mathfrak{m})$. Here the elements T, X, Y, Z and X_1, X_2, Y_1, Y_2 as well as $\mathfrak{m}$ are the same ones as in Subsection 1.1.2.

2. The polynomial $\sigma^{-1}(W)$ vanishes on the orbit Ω in both the cases.

Proof 3.9 *Let $\pi \in \widehat{G}$ verifying $\pi|_{\exp \mathfrak{z}} \neq Id$ and Ω the coadjoint orbit of G associated with π. Under the hypotheses that $\mathfrak{g}_0 \neq \mathfrak{g}$ and Ω is non-saturated with respect to $\mathfrak{g}_0$, we take X or X_1, X_2 in $[\mathfrak{g}, \mathfrak{g}]$ to make a co-exponential basis of $\mathfrak{g}$ to $\widetilde{\mathfrak{g}}$.*

When $\dim \mathfrak{a} = 2$, we take the elements T, X, Y, Z as Subsection 1.1.2 where $\mathfrak{z} = \mathbb{R}Z$ and $\mathfrak{a} = \mathbb{R}Y \oplus \mathfrak{z}$ and T, X_1, X_2, Y_1, Y_2, Z such that $\mathfrak{z} = \mathbb{R}Z$ and $\mathfrak{a} = \mathbb{R}Y_1 \oplus \mathbb{R}Y_2 \oplus \mathfrak{z}$ if $\dim \mathfrak{a} = 3$. More precisely,

$$[T, X] \equiv -X \text{ or } [T, X_1] \equiv -X_1 - \alpha X_2,\ [T, X_2] \equiv -X_2 + \alpha X_1$$

modulo $\mathfrak{m}$.

Recall the Mackey machine and consider the action of G on $\mathfrak{a}^*$. Let $g \in G$, $f \in \Omega$, and $\ell = g{\cdot}f \in \mathfrak{g}^*$. Hence, $g{\cdot}(f|_{\mathfrak{a}}) = \ell|_{\mathfrak{a}}$, $\widetilde{\mathfrak{g}} = \mathfrak{a}^f$ and $\widetilde{G} = \exp \widetilde{\mathfrak{g}}$. We take a polarisation $\mathfrak{b}[f]$ of $\mathfrak{g}$ at f verifying the Pukanszky condition and contained in $\widetilde{\mathfrak{g}}$.

Hence $\pi = ind_{B[f]}^{G}\chi_f$ *with* $B[f] = \exp(\mathfrak{b}[f])$. *Put* $\widetilde{\pi} = ind_{B[f]}^{\widetilde{G}}\chi_f$ *so that* $\pi = ind_{\widetilde{G}}^{G}\widetilde{\pi}$. *Here* $\widetilde{\pi} \in \widehat{\widetilde{G}}$ *corresponding to* $\widetilde{\ell} = \ell|_{\widetilde{\mathfrak{g}}} \in (\widetilde{\mathfrak{g}})^*$.

We take X *or* X_1, X_2 *in* $[\mathfrak{g},\mathfrak{g}]$ *to make a co-exponential basis of* $\mathfrak{g}$ *to* $\widetilde{\mathfrak{g}}$. *Let* $\widetilde{\Omega}$ *be the coadjoint orbit of* $\widetilde{G}$ *associated with* $\widetilde{\pi}$. *By Proposition 3.2.5, there exists in* $\ker\widetilde{\pi}$ *an element*

$$\widetilde{W} = f(Z)T - f(Y)X + A$$

if $\dim\mathfrak{a} = 2$, *or*

$$\widetilde{W} = f(Z)T - (f(Y_1) - \alpha f(Y_2))X_1 - (f(Y_2) + \alpha f(Y_1))X_2 + A$$

if $\dim\mathfrak{a} = 3$ *with a certain element* $A \in \mathcal{U}(\mathfrak{m})$. *Then,*

$$W = ZT - YX + \sqrt{-1}A$$

if $\dim\mathfrak{a} = 2$ *or*

$$W = ZT - (Y_1 - \alpha Y_2)X_1 - (Y_2 + \alpha Y_1)X_2 + \sqrt{-1}A$$

if $\dim\mathfrak{a} = 3$ *belongs to* $\ker\pi$.

We now look at the second point. Since $\pi = ind_{\widetilde{G}}^{G}\widetilde{\pi}$, *we remove this passage from* W, *namely the contribution of the modular function* $\frac{1}{2}Tr\ ad_{\mathfrak{g}/\widetilde{\mathfrak{g}}}T$ *from* W *and get* $W + (1/2)Z$ *or* $W + Z$. *Then, we replace* Y, Z *or* Y_1, Y_2, Z *by* $if(Y), if(Z)$ *or* $if(Y_1), if(Y_2), if(Z)$ *to get the element* $\widetilde{W}_1$ *which belongs to* $\ker\widetilde{\pi}$. *Indeed, suppose* $\dim\mathfrak{a} = 2$ *for example. We replace* Y, Z *by* $if(Y), if(Z)$ *in* W *to get the element* $\widetilde{W}_2 \in \mathcal{U}(\widetilde{\mathfrak{g}})$. *Let* $\varphi \in \mathcal{H}_\pi^\infty$. *We have*

$$(\pi(W))\varphi(e) = \varphi(W^*) = (\widetilde{\pi}(\widetilde{W}_2) + (i/2))(\varphi(e)) = (\widetilde{W}_1)(\varphi(e)).$$

Thus $\widetilde{W}_1 \in \ker\widetilde{\pi}$. *The adjusting term* $\frac{1}{2}Z$ *or* Z *is regained by the difference between* $\sigma^{-1}(YX)$ *and* $if(Y)X$ *or* $\sigma^{-1}(Y_1X_1 + Y_2X_2)$ *and* $if(Y_1)X_1 + if(Y_2)X_2$. *Here,* $\sigma = \sigma_{\mathfrak{g}} : S(\mathfrak{g}) \to \mathcal{U}(\mathfrak{g})$ *denotes the symmetrisation map. Thus, by the induction hypothesis applied to* $\widetilde{W}$, *we have*

$$\sigma_{\widetilde{\mathfrak{g}}}^{-1}(\widetilde{W}_1)|_{\widetilde{\Omega}} \equiv 0.$$

We regard $(\widetilde{\mathfrak{g}})^* \subset \mathfrak{g}^*$ *as* $\mathfrak{g}^* = \mathbb{R}X^* + (\widetilde{\mathfrak{g}})^*$ *or* $\mathfrak{g}^* = \mathbb{R}X_1^* + \mathbb{R}X_2^* + (\widetilde{\mathfrak{g}})^*$. *Since* $(\exp\mathfrak{a})\cdot f = f + (\widetilde{\mathfrak{g}})^\perp$, *namely* $\Omega = \sqcup_{\ell\in\Omega|_{\mathfrak{a}}}(\exp\mathfrak{a})\cdot\widetilde{\Omega}_\ell$. *This is an analogy of the decomposition of* Ω *which is saturated with respect to a codimension one ideal* $\mathfrak{g}'$. *In fact, let* $\ell \in \Omega$ *be any element. Then* $\widetilde{\Omega}_\ell$ *is nothing but the slice*

$$\{f \in \Omega; f|_{\mathfrak{a}} = \ell|_{\mathfrak{a}}\}$$

and there exists a unique element $a \in \exp\mathfrak{a}$ *such that* $a\cdot\ell \in (\widetilde{\mathfrak{g}})^*$. *Namely,* $a\cdot\ell \in \widetilde{\Omega}_\ell$. *In this fashion, we have the theorem which generalises the work of Pedersen [88] on* $\ker\pi$ *in the nilpotent case.*

Remark 3.2.7 1. The expression formula $\Omega = \sqcup_{\ell\in\Omega|_{\mathfrak{a}}}(\exp\mathfrak{a})\cdot\widetilde{\Omega}_\ell$ is nothing but an analogy of the decomposition of Ω which is saturated with respect to a codimension 1 ideal $\mathfrak{g}'$.

2. Theorem 3.2.6 generalises the work of Pedersen (cf. [88]) on the determination of a generator set of $\ker\pi$ in the nilpotent case, based on the polynomials family describing the associated coadjoint orbit.

Remark 3.2.8 Looking closely at the proof of Proposition 3.2.5, $\sigma^{-1}(W)$ vanishes on the orbit Ω. Indeed, one remarks that at the level of equations (3.2.21) and (3.2.22) given by Theorem 3.2.6, the value of $f(Z)$ is constant on Ω.

We next prove the following:

Proposition 3.2.2 *Let $G = \exp\mathfrak{g}$ be an exponential solvable Lie group with Lie algebra $\mathfrak{g}$. Suppose that the coadjoint orbits of Σ_e are generically non-saturated with respect to a codimension one normal subgroup $\widetilde{G} = \exp(\widetilde{\mathfrak{g}})$ of G. Write $\mathfrak{g} = \widetilde{\mathfrak{g}} \oplus \mathbb{R}\widetilde{X}$ for some $\widetilde{X} \in \mathfrak{g}\setminus\widetilde{\mathfrak{g}}$. Then there exists a Casimir element $\eta \in \mathcal{U}(\mathfrak{g})$ for Σ_e, taking the form $\eta = \widetilde{X} + Q$ for some $Q \in \mathcal{U}(\widetilde{\mathfrak{g}})$. Moreover, the Casimir function φ_η turns out to be polynomial on Σ_e.*

Proof 3.10 *Let $\mathfrak{a}$ be a minimal non-central ideal of $\mathfrak{g}$ as in Subsection 1.1.2. We are then led to the case where $\mathfrak{a}$ contains $\mathfrak{z}$ which is of dimension 1, due to the non-saturation hypothesis. Taking into account everything observed above, we can place ourselves in the following situation. If*

$$\mathfrak{a} = \mathbb{R}Y + \mathfrak{z},$$

there is an element $T \in \mathfrak{g}$ verifying the commutation relation (1.1.7). If $\dim(\mathfrak{a}/\mathfrak{z}) = 2$, there are $Y_1, Y_2 \in \mathfrak{a}, T \in \mathfrak{g}, 0 \neq \alpha \in \mathbb{R}$ such that

$$\mathfrak{a} = \mathbb{R}Y_1 + \mathbb{R}Y_2 + \mathfrak{z}, [T, Y_1] = Y_1 - \alpha Y_2, [T, Y_2] = Y_2 + \alpha Y_1,$$

as in (1.1.8). In these two eventualities $[\mathfrak{g}, \mathfrak{a}] \cap \mathfrak{z} \neq \{0\}$ and we suppose that the coadjoint G-orbits of Σ_e are generically non-saturated with respect to the ideal $\widetilde{\mathfrak{g}}$ of codimension 1. Recall that $\mathfrak{k}$ denotes the centraliser of $\mathfrak{a}$ in $\mathfrak{g}$ and put $\mathfrak{m} = \mathbb{R}T + \mathfrak{k}$ and $M = \exp\mathfrak{m}$. If we put $S = \{\ell \in \Sigma_e; \ell(Y) = 0\}$ or $S = \{\ell \in \Sigma_e; \ell(Y_1) = \ell(Y_2) = 0\}$ following the cases, S gives us a section of the coadjoint G-orbits in Σ_e. Remark that, for $\ell \in S$, $\mathfrak{m}$ is nothing but $\mathfrak{a}^\ell = \{X \in \mathfrak{g}; \ell([X, \mathfrak{a}]) = \{0\}\}$. Besides, $\mathfrak{d} = \mathbb{R}Y$ or $\mathfrak{d} = \mathbb{R}Y_1 + \mathbb{R}Y_2$ is an ideal of $\mathfrak{m}$ on which $\pi'_{\ell|_\mathfrak{m}} \in \widehat{M}$ vanishes for $\ell \in S$, where $\pi'_{\ell|_\mathfrak{m}}$ designates the irreducible unitary representation of M associated to the coadjoint orbit $M\cdot(\ell|_\mathfrak{m})$ of M. This being assumed, the induction hypothesis applied to M/D, where $D = \exp\mathfrak{d}$, says that there exists a Casimir element

$$W = T + w, w \in \mathcal{U}(\mathfrak{m} \cap \widetilde{\mathfrak{g}})$$

on the corresponding set $\widetilde{\Sigma_e}$ of $\mathfrak{m}^*$. Finally, the Casimir function φ_W is polynomial on $\widetilde{\Sigma_e}$, which ends the proof of the proposition.

3.2.3 The general case

We denote by $\mathrm{Cas}(\Sigma_e)$ (resp. $\mathrm{FCas}(\Sigma_e)$) the set of Casimir elements (resp. Casimir functions) on Σ_e. As a direct aftermath, we can generate from Theorem 3.1.6 the following:

Theorem 3.2.9 *Let $G = \exp \mathfrak{g}$ be an exponential solvable Lie group. Retain all the notations and definitions as above, and take $e \in \mathcal{E}$. Let $\pi \in \widehat{G}$ and $\Omega(\pi)$ the coadjoint orbit of G corresponding to π. Suppose $\Omega(\pi) \subset \Sigma_e = \mathcal{Z} \cap U_e$. Let for $j \in \mathcal{J}_0$, $W_j = \sum_{k=0}^{p} a_k X_j^k$, $a_k (0 \leq k \leq p)$ of $\mathcal{U}(\mathfrak{g}_{j-1})$, (alternatively, $W_j = \sum_{k=0}^{p} a_k X_{j,1}^k + b_k X_{j,2}^k$, $a_k, b_k (0 \leq k \leq p)$ of $\mathcal{U}(\mathfrak{g}_{j-1})$) be the elements of e-Casimir generators of the set $Cas(\Sigma_e)$. Let also $\varphi_{W_j}(\ell)$ be the corresponding elements of the set $FCas(\Sigma_e)$ of e-Casimir functions. Then for $j \in \mathcal{J}_0$, we have in $\ker \pi$ the elements*

$$A_j = W_j - \varphi_{W_j}(\ell), \; \ell \in \Omega(\pi)$$

In addition, the element $\sigma^{-1}(A_j)$ of $S(\mathfrak{g})$ vanishes on the orbit $\Omega(\pi)$ for any $j \in \mathcal{J}_0$.

We now generate a solution to the problem in the general setting of exponential solvable Lie groups, thanks to the results of the two last sections.

Definition 3.2.10 *Take back the composition series $\mathcal{S}$ of $\mathfrak{g}$ and let $e = (e_1, \ldots, e_m) \in \mathcal{E}$. Let $\pi \in \widehat{G}$ be such that the coadjoint orbit $\Omega(\pi) \subset U_e$. Let $\mathcal{J}_0' := \mathcal{J}_0'(\pi)$ be the set of indices $1 \leq j \leq m-1$ verifying the following conditions:*

1. *$e_j \neq e_{j-1} + \dim(\mathfrak{g}_j/\mathfrak{g}_{j-1})$.*
2. *There exists $W \in \mathcal{U}(\mathfrak{g}_j) \setminus \mathcal{U}(\mathfrak{g}_{j-1})$ in $\ker \pi|_{G_{m-1}}$ such that there is $n_k \in \mathbb{N}$ verifying $(adX_k)^{n_k}(W) \in \ker \pi$, for any $k \in \mathcal{K}$. Here $\mathcal{K}$ and the X_k's are in Subsection 3.1.3.*

Theorem 3.2.11 *Let $G = \exp \mathfrak{g}$ be an exponential solvable Lie group. Let $\pi \in \widehat{G}$ and $\Omega(\pi)$ the coadjoint orbit of G corresponding to π. Then with all the notations above, the following hold:*

1. At the passage from $\mathfrak{g}_{j-1}$ to $\mathfrak{g}_j (1 \leq j \leq m-1)$, there is a new comer $W_j \in \mathcal{U}(\mathfrak{g}_j)$ in $\ker \pi$ if and only if $j \in \mathcal{J}_0'$.

2. The polynomial $\sigma^{-1}(W_j)$ of $S(\mathfrak{g})$ vanishes on the orbit $\Omega(\pi)$ for any $j \in \mathcal{J}_0'$.

Proof 3.11 *Let us first prove the assertion 1. If $\ker \pi$ contains a non-trivial ideal of $\mathfrak{g}$, we can pass to the quotient by this ideal to which applies the induction hypothesis. Thus, $\mathfrak{z}$ being the centre of $\mathfrak{g}$, we can suppose $\dim \mathfrak{z} \leq 1$.*

If $\Omega(\pi)$ is saturated with respect to the codimension 1 ideal $\mathfrak{g}_{m-1}$, we can repeat the arguments of Step 1 and Step 2 in the proof of Theorem 3.1.6. Otherwise, we *get* $\dim \mathfrak{z} = 1$ *and we take a minimal non-central ideal* $\mathfrak{a}$ *which contains* $\mathfrak{z}$. *We can therefore assume that*

$$\mathfrak{g}_{m-1} = \{X \in \mathfrak{g}; [X, \mathfrak{a}] \subset \mathfrak{z}\}.$$

Let $W \in \ker \pi$ *be as in equations (3.2.21) and (3.2.22) given by Theorem 3.2.6. Hence it suffices for us to remark* $[W, W_j] \in \ker \pi$, *which leads to* $[T, W_j] \in \ker \pi$ *as needed.*

For the second assertion, let $\{W_1^j, \dots, W_{r_j}^j\}$ *be a system of generators of* $\ker \pi \setminus (\ker \pi \cap \mathcal{U}(\mathfrak{g}_{j-1}))$ *for* $j \in \mathcal{J}_0'$. *Let* $\mathscr{A}_0 :=< W_1^j, \dots, W_{r_j}^j, j \in \mathcal{J}_0' >$, $W \in \mathcal{U}(\mathfrak{g}_{m-1})\mathscr{A}_0$ *the ideal of* $\mathcal{U}(\mathfrak{g}_{m-1})$, *and* $Q \in I(\Omega(\pi)) \cap S(\mathfrak{g}_{m-1})$. *If* $\Omega(\pi)$ *est saturated with respect to* $\mathfrak{g}_{m-1}$, *let*

$$\pi|_{G_{m-1}} \simeq \int_{t\in\mathbb{R}}^{\oplus} \rho_t dt \quad \text{and} \quad \Omega(\pi)|_{\mathfrak{g}_{m-1}} = \sqcup_{t\in\mathbb{R}}\omega_t$$

be the decompositions as in Subsection 1.2.4. Since $W \in \ker(\rho_t)$ *for almost all* $t \in \mathbb{R}$, *the induction hypothesis says* $\sigma^{-1}(W) \in I(\omega_t)$ *for almost all* $t \in \mathbb{R}$, *namely* $\sigma^{-1}(W) \in I(\Omega(\pi))$. *Conversely, since* $Q \in I(\omega_t)$ *for arbitrary* $t \in \mathbb{R}$, *the induction hypothesis says* $\sigma(Q) \in \ker(\rho_t)$ *for any* $t \in \mathbb{R}$. *Namely,* $\sigma(Q) \in \ker \pi$.

When otherwise $\Omega(\pi)$ *is non-saturated with respect to* $\mathfrak{g}_{m-1}$, *fixing* $\ell \in \Omega(\pi)$ *arbitrarily we can go to the situation where* $\dim \mathfrak{z} = 1$ *and apply the same reasoning to the subalgebra* $\mathfrak{a}^\ell \neq \mathfrak{g}$. *Indeed,* $\mathscr{A}_0 \subset \mathcal{U}(\mathfrak{a}^\ell)$ *and* $\sigma^{-1}(\mathscr{A}_0) \subseteq I(\Omega(\pi)) \cap S(\mathfrak{a}^\ell)$.

As a direct consequence, we now prove the following result, which answers the relationship between $\ker \pi$ and $I(\Omega(\pi))$:

Proposition 3.2.3 *We keep the same hypotheses and notations. Then:*

1. The primitive ideal $\ker \pi$ *is generated by the family* $\mathscr{A} = \{W_m, \mathscr{A}_0\}$, *where* W_m *is the generator which appears when* $m \in T(e)$ *thanks to Proposition 3.2.5. Further, if* $m \in S(e)$, *then* $\ker \pi = \langle \mathscr{A}_0 \rangle$.

2. The set $I(\Omega(\pi))$ *of the polynomial functions on* $\mathfrak{g}^*$ *which vanish on* $\Omega(\pi)$ *is generated by the family* $\sigma^{-1}(\mathscr{A})$.

Proof 3.12 *When* $\Omega(\pi)$ *est saturated with respect to* $\mathfrak{g}_{m-1}$, *let*

$$\pi|_{G_{m-1}} \simeq \int_{t\in\mathbb{R}}^{\oplus} \rho_t dt \quad \text{and} \quad \Omega(\pi)|_{\mathfrak{g}_{m-1}} = \bigsqcup_{t\in\mathbb{R}} \omega_t \tag{3.2.23}$$

be the usual decompositions. We fix $t_0 \in \mathbb{R}$ *and let us introduce the index set* $\mathcal{J}_0''$ *for* $(\mathfrak{g}_{m-1}, \rho_{t_0})$ *equivalently to the definition* $\mathcal{J}_0'$. *Then, we see* $\mathcal{J}_0'' = \mathcal{J}_0' \cup \{i_0\}$ *with some* $1 \leq i_0 \leq m-1$ *or* $\mathcal{J}_0'' = \mathcal{J}_0'$. *Moreover,* $\ker \pi \cap \mathcal{U}(\mathfrak{g}_{m-1}) \subset \ker(\rho_{t_0})$ *and* $I(\Omega(\pi)) \cap S(\mathfrak{g}_{m-1}) \subset I(\omega_{t_0})$. *Let* $W \in \ker \pi$ *(resp.* $Q \in I(\Omega(\pi))$*). Since* $[X_m, W] \in \ker \pi$ *(resp.* $[X_m, Q] \in I(\Omega(\pi))$*),* W *and* Q *do not contain the generators of* $\ker(\rho_{t_0})$ *(resp.* $I(\omega_{t_0})$*) which appear at the level* $\mathfrak{g}_{i_0}$.

Let $W \in \ker \pi$ and write

$$W = \sum_k a_k X_m^k$$

for some $a_k \in \mathcal{U}(\mathfrak{g}_{m-1})$. As $\pi(X_m)$ is a derivation operator with respect to the last variable of functions, elements of $\mathcal{H}_\pi^\infty$, all the a_k's belong to $\ker \pi$ and therefore in $\ker(\rho_{t_0})$. By the induction hypothesis, a_k belongs to $\mathcal{U}(\mathfrak{g}_{m-1})\mathscr{A}_0$ for any k as needed. Hence also the result for point 2.

When $\Omega(\pi)$ is non-saturated with respect to $\mathfrak{g}_{m-1}$, then by Proposition 3.2.5, for some $Q \in \mathcal{U}(\mathfrak{g}_0)$, $W_m = X_m + Q$ belongs to $\ker \pi$, for which $\sigma^{-1}(W_m)$ vanishes on $\Omega(\pi)$. Here it is also sufficient to apply the induction hypothesis to $\pi|_{G_{m-1}}$.

3.2.4 Generators of primitive ideals and Casimir elements

Let G be an exponential solvable Lie group with Lie algebra $\mathfrak{g}$. Take the composition series $\mathcal{S}$ of $\mathfrak{g}$ and let $e = (e_1, \ldots, e_m) \in \mathcal{E}$. Let $\pi \in \widehat{G}$ be such that its coadjoint orbit $\Omega(\pi) \subset U_e$. This section aims to show that a generating family of the primitive ideal $\ker \pi$ is shown to come from $Cas(\Sigma_e)$ as in Corollary 3.1.8.

Let us start with the following direct consequence from Theorem 3.1.6:

Corollary 3.2.4 *Let $G = \exp \mathfrak{g}$ be an exponential solvable Lie group. Retain all the notations and definitions as above, and take $e \in \mathcal{E}$. Let $\pi \in \widehat{G}$ and $\Omega(\pi)$ the coadjoint orbit of G corresponding to π. Suppose $\Omega(\pi) \subset \Sigma_e = \mathcal{Z} \cap U_e$. Let for $j \in \mathcal{J}_0$, $W_j = \sum_{k=0}^p a_k X_j^k$, $a_k (0 \leq k \leq p)$ in $\mathcal{U}(\mathfrak{g}_{j-1})$, (alternatively, $W_j = \sum_{k=0}^p a_k X_{j,1}^k + b_k X_{j,2}^k$, $a_k, b_k (0 \leq k \leq p)$ in $\mathcal{U}(\mathfrak{g}_{j-1})$) be an element of e-Casimir generators of the set $Cas(\Sigma_e)$. Let also $\varphi_{W_j}(\ell)$ be the corresponding element of the set $FCas(\Sigma_e)$ of e-Casimir functions. Then we have in $\ker \pi$ the element $A_j = W_j - \varphi_{W_j}(\ell)$, $\ell \in \Omega(\pi)$,*

In addition, the element $\sigma^{-1}(A_j)$ of $S(\mathfrak{g})$ vanishes on the orbit $\Omega(\pi)$.

We next prove the following:

Theorem 3.2.5 *Retain the same notations of Corollary 3.2.4. For any $W \in \mathscr{A}$, there exists $u \in Cas(\Sigma_e)$ such that $W = u - \varphi_u(\ell) \in \ker \pi_\ell$ for any $\ell \in \Sigma_e$. In other words, there exists a generation family of $\ker \pi$ consisting of Casimir elements for any $\pi \in \widehat{G}$.*

Proof 3.13 *We argue using an induction on the dimension of G. For small dimensions, the result is obvious. Let $K = G_{m-1} = \exp(\mathfrak{k})$ be the normal subgroup of G of codimension 1 of the series of composition of $\mathfrak{g}$. Assume first that the orbit $\Omega(\pi)$ is saturated with respect to K. Then $\mathscr{A} = \mathscr{A}_0$ as in Proposition 3.2.3. Then $W \in \ker \rho_t$ for almost all $t \in \mathbb{R}$ according to the decomposition (1.2.4). Now we can realise the representation π_ℓ as $\mathrm{ind}_K^G \pi'$ acting on $L^2(\mathbb{R}, \mathcal{H}_{\pi'})$, where $\pi' = \pi'_{\ell'} \in \widehat{K}$.*

By the induction hypothesis on K, there exists $u \in Cas(\Sigma_e|_\mathfrak{k})$ such that $W = u - \varphi_u(\ell') \in \ker \pi'_{\ell'}$ for any $\ell' \in \Sigma_e|_\mathfrak{k}$. Besides, we have the following: Let ξ be a

C^∞-vector of $L^2(\mathbb{R}, \mathcal{H}_{\pi'})$, then:

$$\begin{aligned}(\pi_l(u)\xi)[\exp(tX)] &= \xi[u^* \exp(tX)]\\ &= \pi'_{\ell'}(\exp(-tX)u\exp(tX)).\xi[\exp(tX)]\\ &= \pi'_{\ell'}(\exp(-tX)(W+\varphi_u(\ell'))\exp(tX)).\xi[\exp(tX)]\\ &= \varphi_u(\ell')\xi[\exp(tX)],\end{aligned}$$

as W belongs to the kernel of $\pi'_{\exp(-tX)\cdot\ell'}$ for any $t \in \mathbb{R}$. The non-saturated case is settled thanks to Proposition 3.2.2. This achieves the proof

3.3 FROBENIUS EXPONENTIAL LIE ALGEBRAS

In general, a Lie algebra V over a field $\mathbb{K}$ is called a Frobenius Lie algebra if there exists a linear form $\ell \in V^*$ for which the alternating form $B_\ell(X,Y) := \ell([X,Y])$ $(X, Y \in V)$ on V is non-degenerate. The notion of Frobenius Lie algebra is introduced by Ooms [86] in his study of primitive ideals (see also [87]). He showed that the zero ideal of a Frobenius Lie algebra is primitive, while the permittivity of the zero ideal does not imply the Frobeniusness of the Lie algebra. Then the characterisation of a Frobenius Lie algebra from its irreducible representations or its primitive ideals is a significant problem.

We note that a real Lie algebra $\mathfrak{g}$ is Frobenius if and only if it admits an open coadjoint orbit. We show in Proposition 3.2.1 that the coadjoint orbit $\Omega_\ell = \mathrm{Ad}^*(G)\ell \subset \mathfrak{g}^*$ is relatively open in an affine subspace Γ if and only if the radical $\mathfrak{g}(\ell)$ of $B_\ell : \mathfrak{g}\times\mathfrak{g} \to \mathbb{R}$ is an ideal of $\mathfrak{g}$.

3.3.1 Characterisation of Frobenius exponential algebras via primitive ideals

Towards a characterisation criterion of exponential solvable Frobenius Lie algebras in terms of primitive ideals, the following upshot is a nothing but a direct consequence from Proposition 3.2.1 and Theorem 3.2.2.

Theorem 3.3.1 *Let $\mathfrak{g}$ be an exponential solvable Lie algebra. Then the following are equivalent:*

1. *$\mathfrak{g}$ is a Frobenius Lie algebra.*
2. *There exists $\ell \in \mathfrak{g}^*$ such that $\mathfrak{g}(\ell)$ is an ideal of $\mathfrak{g}$ and $\ker d\pi_\ell$ is trivial.*
3. *There exists $\ell \in \mathfrak{g}^*$ such that the coadjoint orbit Ω_ℓ through ℓ is relatively open in an affine subspace of $\mathfrak{g}^*$ and $\ker d\pi_\ell$ is trivial.*

We remark that, owing to Moscovici and Verona [85], a relatively open coadjoint orbit in an affine subspace corresponds to a square-integrable representation (see also [66]), while the square-integrability is characterised in terms of the kernel of its infinitesimal representation for the nilpotent case in [84].

Remark 3.3.2 In what follows, we produce a counter-example of an exponential solvable Lie group with a Lie algebra $\mathfrak{g}$ admitting a linear form $\ell \in \mathfrak{g}^*$ such that $\ker d\pi_\ell$ is trivial but the corresponding orbit fails to be open in $\mathfrak{g}^*$. In this case, $\mathfrak{g}(\ell)$ fails to be an ideal of $\mathfrak{g}$. Actually, we have an example of 3-dimensional $\mathfrak{g}$ with the primitive zero ideal [6, Page 438], which is automatically non-Frobenius because of its odd dimension. We shall give below an example of the lowest even dimension.

Example 3.3.3 Let $\mathfrak{g} = \mathbb{R}\text{-span}\{X_1, X_2, X_3, X_4\}$ be a real Lie algebra of dimension 4 defined by the following bracket relations:

$$[X_4, X_1] = X_1, \quad [X_4, X_2] = \sqrt{2}X_2, \quad [X_4, X_3] = \sqrt{3}X_3.$$

Let $\ell := X_1^* + X_2^* + X_3^* \in \mathfrak{g}^*$. Then $\mathfrak{g}(\ell)$ is a two-dimensional subalgebra spanned by $-\sqrt{2}X_1 + X_2$ and $-\sqrt{3}X_1 + X_3$. It is easy to check that $[X_4, -\sqrt{2}X_1 + X_2] = -\sqrt{2}(X_1 - X_4)$ does not belong to $\mathfrak{g}(\ell)$, so that $\mathfrak{g}(\ell)$ is not an ideal of $\mathfrak{g}$. We take a polarisation $\mathfrak{b} := \mathbb{R}\text{-span}\{X_1, X_2, X_3\}$ satisfying the Pukanszky condition at ℓ and a unitary character χ_ℓ defined as in equation (1.2.3). Since $G = \exp(\mathbb{R}X_4) \ltimes B$, we realise $\pi_\ell = \operatorname{ind}_B^G \chi_\ell$ on $L^2(\mathbb{R})$ as

$$\begin{aligned}
&\pi_\ell(\exp(t_1X_1 + t_2X_2 + t_3X_3))\phi(t) \\
&\quad = \chi_\ell((\exp(tX_4))^{-1} \exp(t_1X_1 + t_2X_2 + t_3X_3) \exp(tX_4))\phi(t) \\
&\quad = e^{i(e^{-t}t_1 + e^{-\sqrt{2}t}t_2 + e^{-\sqrt{3}t}t_3)}\phi(t),
\end{aligned}$$

and

$$\pi_\ell(\exp(t_4X_4))\phi(t) = \phi(t - t_4) \qquad (\phi \in L^2(\mathbb{R})).$$

We thus obtain the following description of the infinitesimal representation $d\pi_\ell$ as

$$d\pi_\ell(X_1)\phi(t) = ie^{-t}\phi(t), \quad d\pi_\ell(X_2)\phi(t) = ie^{-\sqrt{2}t}\phi(t), \quad d\pi_\ell(X_3)\phi(t) = ie^{-\sqrt{3}t}\phi(t),$$
$$d\pi_\ell(X_4)\phi(t) = -\frac{d}{dt}\phi(t).$$

Thus for $A = \sum_{\alpha \in \mathbb{N}^4} c_\alpha X^\alpha \in \mathcal{U}(\mathfrak{g})$ $(X^\alpha = X_1^{\alpha_1} \cdots X_4^{\alpha_4})$, we see that $d\pi_\ell(A)$ equals the differential operator

$$\sum_{\alpha \in \mathbb{N}^4} c_\alpha (ie^{-t})^{\alpha_1} (ie^{-\sqrt{2}t})^{\alpha_2} (ie^{-\sqrt{3}t})^{\alpha_3} \left(-\frac{d}{dt}\right)^{\alpha_4}.$$

If $d\pi_\ell(A) = 0$, then for each $\alpha_4 \geq 0$, we have

$$\sum_{(\alpha_1, \alpha_2, \alpha_3) \in \mathbb{N}^3} i^{\alpha_1 + \alpha_2 + \alpha_3} c_\alpha e^{-(\alpha_1 + \alpha_2\sqrt{2} + \alpha_3\sqrt{3})t} = 0.$$

Since $1, \sqrt{2}, \sqrt{3}$ are linearly independent over $\mathbb{Q}$, the functions $t \mapsto e^{-(\alpha_1 + \alpha_2\sqrt{2} + \alpha_3\sqrt{3})t}$ are linearly independent over $\mathbb{C}$ for $(\alpha_1, \alpha_2, \alpha_3) \in \mathbb{N}^3$. Therefore we have $c_\alpha = 0$ for all $\alpha \in \mathbb{N}^4$, which means that $A = 0$. Namely, $\ker d\pi_\ell = \{0\}$, whereas Ω_ℓ is not open.

3.3.2 The setting of complex solvable Lie algebras

Let now $\mathfrak{s}$ be a complex solvable Lie algebra, and $\mathrm{Prim}(\mathcal{U}(\mathfrak{s}))$ the set of primitive ideals of $\mathcal{U}(\mathfrak{s})$. Dixmier [36] introduced a surjective map $\mathcal{I} : \mathfrak{s}^* \to \mathrm{Prim}(\mathcal{U}(\mathfrak{s}))$. Let $\mathcal{G}$ be the algebraic adjoint of $\mathfrak{s}$. Namely, $\mathcal{G}$ is the smallest algebraic subgroup of $GL(\mathfrak{s})$ whose Lie algebra contains $\mathrm{ad}(\mathfrak{s}) \subset \mathfrak{gl}(\mathfrak{s})$. The group $\mathcal{G}$ acts on $\mathfrak{s}^*$ by the contragredient representation, and the Dixmier map is invariant under this action, which means that $\mathcal{I}(g \cdot \ell) = \mathcal{I}(\ell)$ for $g \in \mathcal{G}$ and $\ell \in \mathfrak{s}^*$. Moreover, the Dixmier map $\mathcal{I}$ induces a bijection from the orbit space $\mathcal{G}\backslash\mathfrak{s}^*$ onto $\mathrm{Prim}(\mathcal{U}(\mathfrak{s}))$. Let $\mathcal{O}_\ell := \mathcal{G} \cdot \ell$ designate the $\mathcal{G}$-orbit through ℓ. As an insight on the relation between primitive ideals and algebraic adjoint orbits, our second main result reads as follows:

Theorem 3.3.4 *For a complex solvable Lie algebra $\mathfrak{s}$ and $\ell \in \mathfrak{s}^*$ with the notation above, the primitive ideal $\mathcal{I}(\ell)$ is zero if and only if the orbit $\mathcal{O}_\ell \subset \mathfrak{s}^*$ is open.*

Proof 3.14 *The proof is based on results by Pedersen [91] and Ooms [86]. In [91, Section 3.8], the orbit $\mathcal{O}_\ell$ is described by a family of algebraic equalities and algebraic inequalities. Each of the equalities corresponds to a non-zero element of $\mathcal{I}(\ell)$, whereas each of the inequalities corresponds to an element of $\mathcal{U}(\mathfrak{s})$ that does not belong to $\mathcal{I}(\ell)$. Actually, these properties give a characterisation of the ideal $\mathcal{I}(\ell)$ among all the primitive ideals of $\mathcal{U}(\mathfrak{s})$ ([91, Theorem 3.7.2]). More precisely, a layering of the dual vector space $\mathfrak{s}^* = \bigsqcup_{\varepsilon\in\mathcal{E}} \mathcal{L}_\varepsilon$ according to the $\mathcal{G}$-action is constructed in [91, Section 1.3], where $\mathcal{E}$ is a certain totally ordered finite set. Note that the set $\mathcal{L}_\varepsilon$ is denoted by Ω_ε in the original paper [91]. For each $\varepsilon \in \mathcal{E}$, a polynomial function Q_ε on $\mathfrak{s}^*$ is defined so that*

$$\mathcal{L}_\varepsilon = \{\ell \in \mathfrak{s}^* : Q_{\varepsilon'}(\ell) = 0 \text{ for all } \varepsilon' < \varepsilon \text{ and } Q_\varepsilon(\ell) \neq 0\}.$$

See [91, Proposition 1.3.2]. Besides, for any $\varepsilon \in \mathcal{E}$, there exist a finite set $E(\varepsilon)^+_{\mathrm{min}}$ and rational functions T^ε_α on $\mathcal{L}_\varepsilon$ parametrised by $\alpha \in E(\varepsilon)^+_{\mathrm{min}}$ such that, if $\ell \in \mathcal{L}_\varepsilon$, one has

$$\begin{aligned} \mathcal{O}_\ell &= \{\ell_0 \in \mathcal{L}_\varepsilon \ : \ T^\varepsilon_\alpha(\ell_0) = T^\varepsilon_\alpha(\ell) \textit{ for all } \alpha \in E(\varepsilon)^+_{\mathrm{min}}\} \\ &= \left\{ \ell_0 \in \mathfrak{s}^* \ : \ \begin{array}{c} Q_{\varepsilon'}(\ell_0) = 0 \text{ for all } \varepsilon' < \varepsilon \text{ and } Q_\varepsilon(\ell_0) \neq 0 \\ T^\varepsilon_\alpha(\ell_0) = T^\varepsilon_\alpha(\ell) \textit{ for all } \alpha \in E(\varepsilon)^+_{\mathrm{min}} \end{array} \right\}, \end{aligned} \tag{3.3.1}$$

see [91, Theorem 2.6.2]. On the other hand, a description of $\mathcal{L}_\varepsilon$ in terms of $\mathcal{U}(\mathfrak{s})$ was given as well. Let u_ε be the element of $\mathcal{U}(\mathfrak{s})$ corresponding to the polynomial function Q_ε by symmetrisation. Then as in [91, Proposition 3.4.2], we have

$$\mathcal{L}_\varepsilon = \{\ell \in \mathfrak{s}^* : u_{\varepsilon'} \in \mathcal{I}(\ell) \text{ for all } \varepsilon' < \varepsilon \text{ and } u_\varepsilon \notin \mathcal{I}(\ell)\}.$$

Besides, for any $\varepsilon \in \mathcal{E}$ and $\alpha \in E(\varepsilon)^+_{\mathrm{min}}$, there exist some elements v^ε_α and w^ε_α of $\mathcal{U}(\mathfrak{s})$ such that the following result holds: If $\ell \in \mathcal{L}_\varepsilon$, there exists one and only one primitive ideal $I \in \mathrm{Prim}(\mathcal{U}(\mathfrak{s}))$ such that:

(I1) $u_{\varepsilon'} \in I$ for all $\varepsilon' < \varepsilon$,

(I2) $u_\varepsilon \notin I$,

(I3) $v_\alpha^\varepsilon - T_\alpha^\varepsilon(\ell) w_\alpha^\varepsilon \in I$, *for all* $\alpha \in E(\varepsilon)^+_{\min}$.

This primitive ideal is $\mathcal{I}(\ell)$ *([91, Theorem 3.7.2]).*

If $\mathcal{O}_\ell$ *is not open, there exists at least one algebraic equality in the description of* $\mathcal{O}_\ell$ *at (3.3.1), so that* $\mathcal{I}(\ell)$ *is non-zero by (I1) and (I3). This implies the 'only if' part of the statement.*

Assume conversely that $\mathcal{O}_\ell$ *is open. Then [86, Theorem] tells us that* $\mathcal{U}(\mathfrak{s})$ *is a primitive algebra. In other words, the zero ideal* $\{0\}$ *is a primitive ideal of* $\mathcal{U}(\mathfrak{s})$*. On the other hand, since* $\mathcal{O}_\ell$ *is open, the description (3.3.1) of* $\mathcal{O}_\ell$ *contains no algebraic equality, which means that there are no* ε' *such that* $\varepsilon' < \varepsilon$*, and that the set* $E(\varepsilon)^+_{\min}$ *is empty. Thus the ideal* $\mathcal{I} = \{0\}$ *satisfies (I1) – (I3), so that we get* $\mathcal{I}(\ell) = \{0\}$ *by the characterisation theorem above. Hence the 'if part' is proved.*

Remark 3.3.5 Let $\mathfrak{g}$ be real exponential solvable and set $\mathfrak{s} := \mathfrak{g}_\mathbb{C}$ as in Theorem 3.3.4. We may then regard $\ell \in \mathfrak{g}^*$ as a complex linear form on $\mathfrak{s} = \mathfrak{g}_\mathbb{C}$. If Ω_ℓ is open, then the bilinear form $B_\ell(X, Y) = \ell([X, Y])$ is non-degenerate on $\mathfrak{g}$, so that B_ℓ is non-degenerate as a bilinear form on $\mathfrak{s}$, too. Thus $\mathcal{O}_\ell = \mathcal{G} \cdot \ell \subset \mathfrak{s}^*$ is open. Therefore, Theorem 3.3.4 tells us that $\mathcal{I}(\ell) = \{0\}$.

On the other hand, when the radical $\mathfrak{g}(\ell)$ is an ideal, our result gives more information about the primitive ideal $\mathcal{I}(\ell) = \ker d\pi_\ell$ than Pedersen's works [89, 91] because Theorem 3.2.2 gives a generator of $\ker d\pi_\ell$.

CHAPTER 4

Polynomial Conjectures in the exponential setting

The content of this chapter is a continuation of our previous study (cf. Chapters 2 and 3). It happens to more invest in the harmonic analysis on the solvable homogeneous spaces through monomial representations (which are induced from unitary characters of analytic subgroups), and restrictions of unitary representations to analytic subgroups. This concerns the extensions to the exponential setting of the polynomial conjectures, stating that the algebra $D_\tau(G/H)$ of the G-invariant differential operators on the fibre bundle associated to a data (H,χ), and also $D_\pi(G)^K$ are polynomial rings, where G is an exponential solvable Lie group, H,K are closed and connected subgroups of G, χ is a unitary character of H, $\pi\in\widehat{G}$, and τ a monomial representation of G. Our investigation focus also on Duflo's polynomial conjecture in both the settings, where the orbit method due to Kirillov turns out to be a crucial ingredient for our study.

4.1 THE MONOMIAL SETTING

4.1.1 Branching rules and the related Plancherel formula

Let $G=\exp\mathfrak{g}$ be an exponential solvable Lie group with Lie algebra $\mathfrak{g}$, $\widehat{G}$ the unitary dual of G, $\mathfrak{g}^*$ the dual vector space of $\mathfrak{g}$, $H=\exp\mathfrak{h}$ an analytic subgroup of G and χ a unitary character of H. The object of our study is the monomial representation $\tau=\mathrm{ind}_H^G\chi$ of G. It is well known that there exists $f\in\mathfrak{g}^*$ verifying $f([\mathfrak{h},\mathfrak{h}])=\{0\}$ and such that $\chi(\exp X)=\chi_f(\exp X)=e^{if(X)}$ for $X\in\mathfrak{h}$. Put $\Gamma_\tau=f+\mathfrak{h}^\perp\subset\mathfrak{g}^*$.

Let

$$\tau\simeq\int_{\widehat{G}}^{\oplus}\widetilde{\pi}\,d\mu(\pi),\tag{4.1.1}$$

be the factor disintegration of τ, where $\widetilde{\pi}$ is the factor representation which is a multiple of $\pi\in\widehat{G}$, and where μ is a Borel measure on $\widehat{G}$. More in detail, we take a finite measure $\widetilde{\mu}$ on Γ_τ equivalent to the Lebesgue measure. We regard $\widetilde{\mu}$ as a measure on $\mathfrak{g}^*$ and the measure μ is the image of $\widetilde{\mu}$ by the Kirillov-Bernat map $\theta_G:\mathfrak{g}^*\to\widehat{G}$. In what follows, the adjective generic signifies except a negligible set with respect to $\widetilde{\mu}$. Let $\mathcal{H}_\tau^{-\infty}$ be the anti-dual space of $\mathcal{H}_\tau^\infty$ and let us consider the Dirac measure

DOI: 10.1201/9781003685739-4

$\delta_\tau \in \mathcal{H}_\tau^{-\infty}$ defined by

$$\delta_\tau : \mathcal{H}_\tau^\infty \ni \varphi \mapsto \overline{\varphi(e)} \in \mathbb{C}.$$

From [92], the disintegration (4.1.1) brings that of δ_τ:

$$\delta_\tau = \int_{\widehat{G}}^{\oplus} a_{\tilde{\pi}} d\mu(\pi) \tag{4.1.2}$$

with certain $a_{\tilde{\pi}} \in \mathcal{H}_{\tilde{\pi}}^{-\infty}$.

Let $\mathcal{D}(G)$ be the space of the C^∞ functions on G with compact support. We denote by Δ_G the modular function of G and put $\Delta_{H,G} = \Delta_H/\Delta_G$. For $\varphi \in \mathcal{D}(G)$ we fabricate the element $\tau(\varphi)\delta_\tau = \varphi_H^f$ of $\mathcal{H}_\tau^\infty$:

$$\varphi_H^f(g) = \int_H \varphi(gh)\chi_f(h)\Delta_{H,G}^{-1/2}(h)dh \ (g \in G),$$

dh being the left invariant measure on H.

4.1.2 An algebra of differential operators

We now define a certain algebra of differential operators. Let

$$C^\infty(G,\tau) = \{\varphi \in C^\infty(G); \varphi(gh) = \chi(h)^{-1}\Delta_{H,G}^{1/2}(h)\varphi(g), g \in G, h \in H\}$$

and $Diff(G,\tau)$ the algebra of the C^∞ differential operators which leave $C^\infty(G,\tau)$ stable. Our object to study is $D_\tau(G/H)$ which consists of the restrictions $D|_{C^\infty(G,\tau)}$ of $D \in Diff(G,\tau)$ commuting with left translations of G on $C^\infty(G,\tau)$. In other words, $D_\tau(G/H)$ is nothing but the algebra of the G-invariant differential operators on the fibre bundle over G/H associated with data (G, H, χ). Let us describe this algebra in terms of the enveloping algebra $\mathcal{U}(\mathfrak{g})$ of $\mathfrak{g}_\mathbb{C}$. For $g \in G$ or $A \in \mathcal{U}(\mathfrak{g})$, we denote by R_g or $R(A)$ (resp. L_g or $L(A)$) the right (resp. left) action. Let $\mathcal{U}(\mathfrak{g},\tau)$ be the set of the $A \in \mathcal{U}(\mathfrak{g})$ such that $R(A)$ leaves $C^\infty(G,\tau)$ stable, $\mathfrak{a}_\tau$ the vector subspace of $\mathcal{U}(\mathfrak{g})$ spanned by

$$Y + if(Y) - \frac{1}{2}\mathrm{tr}\ \mathrm{ad}_{\mathfrak{g}/\mathfrak{h}}Y,$$

Y running over $\mathfrak{h}$, and $\mathcal{U}(\mathfrak{g})\mathfrak{a}_\tau$ the left ideal of $\mathcal{U}(\mathfrak{g})$ generated by $\mathfrak{a}_\tau$. Then, it comes that

$$\mathcal{U}(\mathfrak{g},\tau) = \{A \in \mathcal{U}(\mathfrak{g}); [A,Y] \in \mathcal{U}(\mathfrak{g})\mathfrak{a}_\tau, \ \forall Y \in \mathfrak{h}\}$$

and that the mapping $A \mapsto R(A)|_{C^\infty(G,\tau)}$ gives us a homomorphism from $\mathcal{U}(\mathfrak{g},\tau)$ onto $D_\tau(G/H)$ with the kernel $\mathcal{U}(\mathfrak{g})\mathfrak{a}_\tau$. In sum,

$$D_\tau(G/H) \cong \mathcal{U}(\mathfrak{g},\tau)/\mathcal{U}(\mathfrak{g})\mathfrak{a}_\tau.$$

4.2 MONOMIAL REPRESENTATIONS OF DISCRETE TYPE

Let $G = \exp \mathfrak{g}$ be an exponential solvable Lie group with Lie algebra $\mathfrak{g}$, $H = \exp \mathfrak{h}$ an analytic subgroup of G with Lie algebra $\mathfrak{h}$ and χ a unitary character of H. As a continuation of our study of the nilpotent case, we go to study the monomial representation $\tau = \mathrm{ind}_H^G \chi$ of G. The notion of monomial representations of discrete type is introduced in [16]. We say that τ is of discrete type if any connected component of $\Gamma_\tau \cap \Omega(\pi)$ is an H-orbit almost everywhere on Γ_τ with respect to μ, which is the case when τ has finite multiplicities. On the other hand, τ is of discrete type if and only if $\mathfrak{h} + \mathfrak{g}(\ell)$ is a Lagrangian subspace for B_ℓ almost everywhere with respect to $\widetilde{\mu}$ (cf. [48]). When G is completely solvable, τ is discrete type if and only if τ has finite multiplicities (cf. [20]).

Let now τ be of discrete type in the setting of exponential solvable Lie group. In this case, we have shown that the Plancherel formula due to Penney (cf. [16]) holds and is well described as follows. The measure μ in (4.1.1) can be chosen in such a manner that for any π in the spectre of τ, there exist distributions $a_\pi^k (1 \leq k \leq m(\pi))$ such that for all $\varphi \in C_c^\infty(G)$,

$$\varphi_H^f(e) = \int_{\hat{G}} \sum_{k=1}^{m(\pi)} \langle \pi(\varphi) a_\pi^k, a_\pi^k \rangle d\mu(\pi), \tag{4.2.1}$$

where

$$\varphi_H^f(g) = \int_H \varphi(gh)\chi_f(h)\Delta_{H,G}^{-1/2}(h)dh \quad (g \in G)$$

with $\Delta_{H,G} = \Delta_H/\Delta_G$ and dh denotes the Haar measure on H. Moreover, each distribution is explicitly written as an integral over a certain homogeneous space depending upon generic linear forms of Γ_τ.

4.2.1 A Penney distribution

Supposing that $\tau = \mathrm{ind}_H^G \chi_f$ is of discrete type, let us recall some results obtained in [16] (cf. [17], Chap. 7). Let $\mathfrak{n}$ be a nilpotent ideal of $\mathfrak{g}$ containing $[\mathfrak{g}, \mathfrak{g}]$ and

$$\mathcal{S} : \{0\} \subset \mathfrak{g}_1 \subset \cdots \subset \mathfrak{g}_n = \mathfrak{g}$$

a good sequence of subalgebras of $\mathfrak{g}$ which passes through $\mathfrak{n}$. For $\ell \in \Gamma_\tau$, we denote by $\mathfrak{b}[\ell]$ the Vergne polarisation at ℓ associated to $\mathcal{S}$. Then, μ-almost everywhere on Γ_τ, we find

$$\mathrm{Tr}\ \mathrm{ad}_{\mathfrak{h}/(\mathfrak{h}\cap\mathfrak{b}[\ell])} X + \mathrm{Tr}\ \mathrm{ad}_{\mathfrak{b}[\ell]/(\mathfrak{h}\cap\mathfrak{b}[\ell])} X = 0$$

for $X \in \mathfrak{h} \cap \mathfrak{b}[\ell]$. This allows us to consider the integral

$$\langle a_\ell, \varphi \rangle = \oint_{H/(H\cap B[\ell])} \overline{\varphi(h)\chi_\ell(h)} \Delta_{H,G}^{-1/2}(h) d\nu(h) \tag{4.2.2}$$

with $B[\ell] = \exp(\mathfrak{b}[\ell])$, where φ is a test function taken in the Hilbert space of the irreducible unitary representation $\pi_\ell = \mathrm{ind}_{B[\ell]}^G \chi_\ell$. We further see that this integral is convergent for the continuous test functions with compact support modulo $B[\ell]$ (cf. [17], Theorem 7.3.11). Let $W \in \mathcal{U}(\mathfrak{g}, \tau)$. We also proved that

$$\pi_\ell(\overline{W})a_\ell = \overline{P_W(\ell)}a_\ell \tag{4.2.3}$$

with a H-invariant function P_W defined almost everywhere on Γ_τ. The map

$$\Theta : \mathcal{U}(\mathfrak{g}, \tau) \ni W \mapsto P_W(\ell) \tag{4.2.4}$$

induces a homomorphism of the algebra $D_\tau(G/H)$ into the algebra of the H-invariant functions on Γ_τ. Further, the Plancherel formula for τ implies that this homomorphism has $\mathcal{U}(\mathfrak{g})\mathfrak{a}_\tau$ as its kernel. In this way, the algebra $D_\tau(G/H)$ is commutative (cf. Theorem 2.1.1). This is typically what happens in the nilpotent setting with the strong assertion that the commutativity of $D_\tau(G/H)$ is equivalent to the fact that the multiplicities are finite μ-almost everywhere (cf. Theorem 2.1.16).

4.2.2 The polynomial conjecture for discrete type monomial representations

We study here the algebra $D_\tau(G/H)$ of the G-invariant differential operators on the line bundle associated to the data (G, H, χ). We ask if the algebra $D_\tau(G/H)$ is commutative or even more is it isomorphic to the algebra $\mathbb{C}[\Gamma_\tau]^H$ of the H-invariant polynomial functions on Γ_τ as in the nilpotent case?

We shall discuss the polynomial conjecture for monomial representations of exponential solvable Lie groups. In this case, it is easy to find examples where τ has infinite multiplicities but the algebra $D_\tau(G/H)$ is commutative.

Example 4.2.1 Let $G = G_3(\alpha) = \exp(\mathfrak{g}_3(\alpha))$ with $\mathfrak{g}_3(\alpha) = \langle T, Y_1, Y_2\rangle_{\mathbb{R}} : [T, Y_1] = Y_1 - \alpha Y_2, [T, Y_2] = \alpha Y_1 + Y_2$, where α is a negative real number. Let $f = Y_1^*$ and $\mathfrak{h} = \mathbb{R}Y_1$. We have the parameterised expression of the orbit passing $\ell = (1, \lambda) \in \mathfrak{a}^* = \mathbb{R}Y_1^* + \mathbb{R}Y_2^*$. Using the (x, y)-coordinates,

$$\begin{aligned} x(t) &= e^t(\cos(\alpha t) - \lambda \sin(\alpha t)) \\ y(t) &= e^t(\sin(\alpha t) + \lambda \cos(\alpha t)). \end{aligned} \tag{4.2.5}$$

If the straight line $x = 1$ is tangent to the orbit $G \cdot \ell$, we see $(dx/dt)_{t=0} = 1 - \alpha\lambda = 0$, i.e. $\lambda = 1/\alpha$. We denote this value by λ_0. Let t^* be the first positive real number t verifying $e^t(\cos(\alpha t) - (1/\alpha)\sin(\alpha t)) = 1$, and λ_1 the y-coordinate of the intersection point. We take $\ell = (1, \lambda), \lambda_0 \leq \lambda < \lambda_1$ as the representative points of the orbits which encounter Γ_τ. These orbits are all saturated in the direction of T^*. Each of these orbit has the intersection with Γ_τ of disjoint straight lines in infinite number each of which is a H-orbit. Hence τ has infinite multiplicities, while it is easy to see that the algebra $D_\tau(G/H)$ is commutative.

Our objective in this section is to generalise Theorem 2.1.1, to encompass the setting of discrete type monomial representations of exponential solvable Lie groups. Our main result here is as follows:

Theorem 4.2.2 *Let $G = \exp \mathfrak{g}$ be an exponential solvable Lie group with Lie algebra $\mathfrak{g}$, $H = \exp \mathfrak{h}$ an analytic subgroup of G, χ a unitary character of H and $f \in \mathfrak{g}^*$ such that if $f|_{\mathfrak{h}} = d\chi$. Assume that the monomial representation $\tau = ind_H^G \chi$ is of discrete type. Set $\Gamma_\tau = f + \mathfrak{h}^\perp \subset \mathfrak{g}^*$. Then, the algebra $D_\tau(G/H)$ of G-invariant differential operators on the fibre bundle associated to data (H, χ) is isomorphic to the algebra $\mathbb{C}[\Gamma_\tau]^H$ of the H-invariant polynomial functions on Γ_τ.*

Proof 4.1 *We first prepare the following, which will be next of use in the proof (cf. [17], Corollary 7.6.2).*

Proposition 4.2.1 *Suppose that τ has multiplicities of discrete type. Let $\mathfrak{g}_0$ be an ideal of codimension 1 in $\mathfrak{g}$ such that the G-orbits are μ-almost everywhere saturated with respect to $\mathfrak{g}_0$. Then, the algebra $\mathcal{U}(\mathfrak{g}, \tau)$ is contained in $\mathcal{U}(\mathfrak{g}_0)$ modulo $\mathcal{U}(\mathfrak{g})\mathfrak{a}_\tau$.*

Further the following holds:

Proposition 4.2.2 *For $W \in \mathcal{U}(\mathfrak{g})$, $W \in \mathcal{U}(\mathfrak{g})\mathfrak{a}_\tau$ if and only if $\overline{W} \cdot a_\ell = 0$ for generic $\ell \in \Gamma_\tau$.*

Proof 4.2 *Let $W \in \mathcal{U}(\mathfrak{g})\mathfrak{a}_\tau$, it is evident that $\overline{W} \cdot a_\ell = 0$ for generic $\ell \in \Gamma_\tau$. Conversely, we suppose $\overline{W} \cdot a_\ell = 0$ for generic $\ell \in \Gamma_\tau$. Then, the Plancherel formula (4.2.1) for τ shows $W = 0$ under the right action in the space of τ. Namely, $W \in \mathcal{U}(\mathfrak{g})\mathfrak{a}_\tau$.*

We use an induction on the dimension of the group G. We propose to show that the homomorphism Θ defined in equation (4.2.4) induces as in the nilpotent case an isomorphism of algebras from $D_\tau(G/H)$ onto $\mathbb{C}[\Gamma_\tau]^H$. Clearly, we can suppose that $\mathfrak{h}$ contains the centre $\mathfrak{z}$ of $\mathfrak{g}$. If $\mathfrak{h} \cap \ker f$ contains an ideal $\mathfrak{a}$ of $\mathfrak{g}$, we can pass to the quotient $\mathfrak{g}/\mathfrak{a}$ to which applies the induction hypothesis. Hereafter we suppose that there exists no such ideal $\mathfrak{a}$. Hence, $\dim \mathfrak{z} \leq 1$ and $f|_{\mathfrak{z}} \neq 0$ if $\mathfrak{z} \neq \{0\}$.

Let $\mathfrak{a}$ be a minimal non-central ideal of $\mathfrak{g}$. This means that $\mathfrak{z} = \mathfrak{g}_1$ if it is not trivial and either $\mathfrak{a} = \mathfrak{g}_1, \mathfrak{g}_2$ or $\mathfrak{g}_3$ in the sequence $\mathcal{S}$. Put $A = \exp \mathfrak{a}$,

$$\tilde{\mathfrak{g}} = \mathfrak{a}^f = \{X \in \mathfrak{g}; f([X, \mathfrak{a}]) = \{0\}\}$$

and $\tilde{G} = \exp(\tilde{\mathfrak{g}})$. It becomes in our situation that $\tilde{G}$ is a proper subgroup of G and $A \cdot f = f + (\tilde{\mathfrak{g}})^\perp$.

Let as previously $\mathfrak{g}_0 = \{X \in \mathfrak{g}; [X, \mathfrak{a}] \subset \mathfrak{z}\}$, $G_0 = \exp(\mathfrak{g}_0)$, $\mathfrak{h}_0 = \mathfrak{h} \cap \mathfrak{g}_0$ and $H_0 = \exp(\mathfrak{h}_0)$. Let $W \in \mathcal{U}(\mathfrak{g}, \tau)$. We shall be using the structure details on minimal non-central ideals, as written in Subsection 1.1.2

Case 1: Assume first that $\mathfrak{h} \subset \tilde{\mathfrak{g}}$.

Case 1.1: If $\{0\} \neq \mathfrak{z}$ and $[\mathfrak{g}, \mathfrak{a}] = \mathfrak{a}$, we find in $\mathfrak{n}$ thanks to equations (1.1.11), (1.1.16) and (1.1.17), some co-exponential bases of $\mathfrak{g}$ to $\tilde{\mathfrak{g}}$ through the different situations. By means of these bases, we introduce coordinates in the space of π_ℓ almost everywhere in Γ_τ.

Note in this case that $\mathfrak{g}_0$ is a proper ideal of $\mathfrak{g}$ of codimension one. We shall discuss now the following two cases:

Case 1.1.1: Assume further that the representations π_ℓ for generic linear forms $\ell \in \Gamma_\tau$ correspond to saturated orbits with respect to $\mathfrak{g}_0$, according to the terminology of Subsection 2.3.4.3.

If $\mathfrak{h} \not\subset \mathfrak{g}_0$, it is evident that W belongs to $\mathcal{U}(\mathfrak{g}_0)$ modulo $\mathcal{U}(\mathfrak{g})\mathfrak{a}_\tau$. Write $\mathfrak{h} = \mathbb{R}X + \mathfrak{h}_0$ with $\mathfrak{h}_0 = \mathfrak{h} \cap \mathfrak{g}_0$, $H_0 = \exp(\mathfrak{h}_0)$, and let $\ell \in \Gamma_\tau$ be generic with a Pukanszky polarisation $\mathfrak{b}[\ell] \subset \mathfrak{g}_0$, and take in $\mathcal{H}_{\pi_\ell}$ a test function

$$\varphi(g) = \phi(x)\psi(g_0), \ g = \exp(xX)g_0 \ (x \in \mathbb{R}, \ g_0 \in G_0)$$

where ϕ is a continuous function with compact support on $\mathbb{R}$ and ψ is a test function of $\pi^0_{\ell_0} = ind_{B[\ell]}^{G_0}\chi_\ell$, $B[\ell] = \exp(\mathfrak{b}[\ell]) \subset G_0$. Then,

$$\langle W\cdot a_\ell, \varphi\rangle = \langle W\cdot a^0_{\ell_0}, \psi\rangle \int_{\mathbb{R}} \overline{\phi(t)} e^{-itf(X)} e^{-\frac{1}{2} Tr\ ad_{\mathfrak{g}/\mathfrak{h}} X} dt, \tag{4.2.6}$$

where $a^0_{\ell_0}$ designates the Penney's distribution for $\pi^0_{\ell_0}$. Indeed,

$$\begin{aligned}
\langle W\cdot a_\ell, \varphi\rangle = \langle a_\ell, W^*\cdot\varphi\rangle &= \oint_{H/(H\cap B[\ell])} \overline{W^* \cdot \varphi(h)\chi_\ell(h)} \Delta_{H,G}^{-1/2}(h) d\nu(h) \\
&= \int_{\mathbb{R}} \overline{\phi(t)} e^{-itf(X)} e^{-\frac{1}{2} Tr\ ad_{\mathfrak{g}/\mathfrak{h}} X} dt \\
&\times \oint_{H_0/(H_0\cap B[\ell])} \overline{W^* \cdot \psi(h_0)\chi_\ell(h_0)} \Delta_{H_0,G_0}^{-1/2}(h_0) d\nu(h_0) \\
&= \int_{\mathbb{R}} \overline{\phi(t)} e^{-itf(X)} e^{-\frac{1}{2} Tr\ ad_{\mathfrak{g}/\mathfrak{h}} X} dt \cdot \langle W\cdot a^0_{\ell_0}, \psi\rangle.
\end{aligned}$$

We now derive the desired result from the induction hypothesis applied to $\tau_0 = ind_{H_0}^{G_0}\chi_f$ and passing to the quotient by the action of H. Let us first see the:

Lemma 4.2.3 *τ_0 is of discrete type.*

Proof 4.3 *For generic $\ell \in \Gamma_\tau$, let us write $\mathfrak{h} + \mathfrak{g}(\ell) = (\mathfrak{h}_0 + \mathfrak{g}(\ell)) \oplus \mathbb{R}X$. On the other hand, $\mathfrak{g}_0(\ell_0) = \mathfrak{g}(\ell) \oplus \mathbb{R}X_0(\ell)$ for some $X_0(\ell) \in \mathfrak{g}_0$. Then obviously, $\ell([X, X_0(\ell)]) \neq 0$ generically and finally*

$$\mathfrak{h}_0 + \mathfrak{g}_0(\ell_0) = (\mathfrak{h}_0 + \mathfrak{g}(\ell)) \oplus \mathbb{R}X_0(\ell),$$

which means that $\mathfrak{h}_0 + \mathfrak{g}_0(\ell_0)$ is generically Lagrangian on $\Gamma_{\tau_0} = f|_{\mathfrak{g}_0} + (\mathfrak{h}_0)^\perp \subset \mathfrak{g}_0^*$.

By Lemma 4.2.3, the induction hypothesis applied to G_0, H_0 and τ_0 says that the function $P_W(\ell)$ is a polynomial function on Γ_{τ_0}. Moreover, we designate by $S(\mathfrak{g})$ the symmetric algebra of $\mathfrak{g}_\mathbb{C}$ and consider their elements as polynomial functions *on $i\mathfrak{g}, i = \sqrt{-1}$. Then, $\mathbb{C}[\Gamma_\tau]^H$ is identified with $(S(\mathfrak{g})/S(\mathfrak{g})\overline{\mathfrak{a}_\tau})^H$, the set of the H-invariant vectors under the adjoint action. This being in mind, the same induction hypothesis says also that the mapping*

$$\Theta_0 : \mathcal{U}(\mathfrak{g}_0, \tau_0) \ni W \mapsto P_W \in \mathbb{C}[\Gamma_0]^{H_0}$$

is surjective. Let $Q \in \mathbb{C}[\Gamma_\tau]^H$. Since Q belongs to $\mathbb{C}[\Gamma_{\tau_0}]^{H_0}$, there exists $W_0 \in \mathcal{U}(\mathfrak{g}_0, \tau_0)$ such that $\Theta_0(W_0) = Q$ by the induction hypothesis. Namely,

$$\overline{W_0}\cdot a'_{\ell'} = \overline{Q}(\ell')a'_{\ell'}$$

for generic $\ell' \in \Gamma_{\tau_0}$, where $a'_{\ell'}$ denotes the Penney distribution at ℓ' for τ_0. Further, Q being H-invariant, W_0 is $\mathfrak{h}$-invariant, namely $W_0 \in \mathcal{U}(\mathfrak{g}, \tau)$ and by (4.3.10),

$$\overline{W_0}\cdot a_\ell = \overline{Q}(\ell)a_\ell$$

for generic $\ell \in \Gamma_\tau$. In this way, $P_W = Q$ and the mapping $\Theta : \mathcal{U}(\mathfrak{g}, \tau) \ni W \mapsto P_W \in \mathbb{C}[\Gamma_\tau]^H$ is surjective.

When $\mathfrak{h} \subset \mathfrak{g}_0$, then by Proposition 4.2.1, W belongs to $\mathcal{U}(\mathfrak{g}_0)$ modulo $\mathcal{U}(\mathfrak{g})\mathfrak{a}_\tau$ and the same line of argumentation works out. Indeed, $\mathrm{ind}_H^{G_0}\chi_f$ is of discrete type as generically we have $X_0(\ell) \in \mathfrak{h} + \mathfrak{g}(\ell)$, where $X_0(\ell)$ is as in Lemma 4.2.3. Finally, $\mathfrak{h} + \mathfrak{g}_0(\ell_0) = \mathfrak{h} + \mathfrak{g}(\ell)$ generically as needed. That is, the correspondence (4.2.4) meets the following:

$$W \mapsto P_W \in \mathbb{C}[\Gamma^0_{\tau_0}]^H,$$

where $\Gamma^0_{\tau_0} = f + \mathfrak{h}^{\perp,\mathfrak{g}_0^*}$.

On the other hand, it is clear that $(S(\mathfrak{g})/S(\mathfrak{g})\overline{\mathfrak{a}_\tau})^H$ is identified to $(S(\mathfrak{g}_0)/S(\mathfrak{g}_0)\overline{\mathfrak{a}_\tau})^H$ and this finishes the proof in this case.

Case 1.1.2: Suppose now that π_ℓ, for generic ℓ, corresponds to the non-saturated orbit with respect to $\mathfrak{g}_0$.

This already implies that $[\mathfrak{g}, \mathfrak{a}] = \mathfrak{a}$. We first prove the following:

Lemma 4.2.4 *We can assume that $W \in \mathcal{U}(\tilde{\mathfrak{g}})$ modulo $\mathcal{U}(\mathfrak{g})\mathfrak{a}_\tau$.*

Proof 4.4 *The result is obvious when $\mathfrak{a} \subset \mathfrak{h}$. Suppose hereafter that $\mathfrak{a} \not\subset \mathfrak{h}$. Let us prove in a first step that W can be taken in $\mathcal{U}(\mathfrak{g}_0)$ modulo the kernel $\mathcal{U}(\mathfrak{g})\mathfrak{a}_\tau$. The result is already straight if $\mathfrak{h} \not\subset \mathfrak{g}_0$. Assume therefore that $\mathfrak{h} \subset \mathfrak{g}_0$ and, for generic $\ell \in \Gamma_\tau$, let*

$$\mathfrak{g} = \mathbb{R}X(\ell) + \mathfrak{g}_0$$

with $X(\ell) \in \mathfrak{g}(\ell)$, following the assumption that generic orbits are not saturated with respect to $\mathfrak{g}_0$. Since W acts from right in the space of τ, let

$$W = \sum_{j=0}^{m} w_j(\ell)X(\ell)^j, \; w_j(\ell) \in \mathcal{U}(\mathfrak{g}_0)(0 \le j \le m).$$

Then, acting from right in the space of π_ℓ,

$$W = \sum_{j=0}^{m} w_j(\ell)(-\sqrt{-1}\ell(X(\ell)) - \frac{1}{2} Tr\ ad_{\mathfrak{g}/\mathfrak{b}[\ell]} X(\ell))^j.$$

On the other hand, we can and do transfer the action of W *on* φ *in the expression* $\langle \overline{W} \cdot a_\ell, \varphi\rangle$ *from left to right using the* $\mathfrak{h}$*-invariance. Let* $\mathfrak{g} = \mathbb{R}X + \mathfrak{g}_0$. *Since* $X(\ell)$ *depends only on the restriction* $\ell|_{\mathfrak{g}_0}$, *we see that* $w_j \in \mathcal{U}(\mathfrak{g}, \tau)(0 \leq j \leq m)$ *by varying the value* $\ell(X)$ *over* $\mathbb{R}$. *The result then follows.*

The second step consists in showing that $W \in \mathcal{U}(\widetilde{\mathfrak{g}})$ *modulo* $\mathcal{U}(\mathfrak{g})\mathfrak{a}_\tau$. *The result follows from (1.1.12) and (1.1.15) because* W *does not contain partial derivatives relative to these external coordinates. Let us detail this point. Assume for a while that* $\dim \mathfrak{a} = 2$. *Then obviously the action of the representation on* X_1 *consists of a partial differential operator. Write next*

$$W = \sum_{j=0}^{m} w_j X_1^j,\ w_j \in \mathcal{U}(\mathfrak{g}_0)(0 \leq j \leq m)$$

and $w_m \notin \mathcal{U}(\mathfrak{g})\mathfrak{a}_\tau$.

In the case where $\mathfrak{h} \subset \mathfrak{g}_0$, *this leads to the fact that* $\mathfrak{h} \subset \mathfrak{m}$ *and this entails straightly that* $m = 0$, *as it is to be shown.*

Otherwise $\mathfrak{h} \not\subset \mathfrak{g}_0$, *and we can assume that* $X \in \mathfrak{h}$. *In this case, let*

$$\varphi(g) = \phi(x)\psi(\widetilde{g})\Delta^{1/2}_{\widetilde{G},G}(\widetilde{g}),\ \phi \in C_0(\mathbb{R}), g = \exp(xX_1)\widetilde{g},\ (x \in \mathbb{R}, \widetilde{g} \in \widetilde{G}),$$

be a test function of the space of π_ℓ, *where* ψ *denotes a test function of the space of* $\widetilde{\pi}_\ell = ind_{B[\ell]}^{\widetilde{G}}\chi_{\widetilde{\ell}}$. *Regarding the action of* W *on* φ, *we see from (4.2.3) that* $m = 0$ *since* X_1 *gives a partial derivative with respect to the coordinate* x *along it.*

When $\dim \mathfrak{a} = 3$, *then the actions of the representation on* X_1, X_2 *consist of a partial differential operator with analytic coefficients and no hurdles stand to write down similar proof. Then, our claim follows once again from the arguments above as* W *does not contain partial derivatives relative to these external coordinates.*

Put next $\widetilde{\tau} = ind_H^{\widetilde{G}}\chi_f$,

$$\widetilde{\Gamma}_{\widetilde{\tau}} = \{\widetilde{\ell} \in (\widetilde{\mathfrak{g}})^*; \widetilde{\ell}|_{\mathfrak{h}} = f|_{\mathfrak{h}}\}$$

and $\widetilde{\pi}_{\widetilde{\ell}} = ind_{B[\ell]}^{\widetilde{G}}\chi_\ell$ *for* $\ell \in \Gamma_\tau$. *Remark that for generic* $\ell \in \Gamma_\tau$, $\ell + (\mathfrak{a}^\ell)^\perp \subset H\cdot\ell$ *and hence* $\widetilde{\tau}$ *is of discrete type, as* $\mathfrak{h} + \mathfrak{g}(\ell) = \mathfrak{h} + \widetilde{\mathfrak{g}}(\ell)$ *is generically Lagrangian.*

Hence, the result comes from the induction hypothesis applied to $\widetilde{G}$ *concerning the fact that* $\mathcal{U}(\mathfrak{g}, \tau) \ni W \mapsto P_W \in \mathbb{C}[\Gamma_\tau]^H$. *Concerning the surjectivity, similar technics and steps as in the proof of Lemma 4.2.4 allow to conclude the identification*

$$(S(\mathfrak{g})/S(\mathfrak{g})\overline{\mathfrak{a}_\tau})^H \simeq (S(\widetilde{\mathfrak{g}})/S(\widetilde{\mathfrak{g}}))\overline{\mathfrak{a}_{\widetilde{\tau}}})^H.$$

Case 1.2: Let now either $\{0\} = \mathfrak{z}$, *or* $\{0\} \neq \mathfrak{z}$ *and* $[\mathfrak{g}, \mathfrak{a}] \subset \mathfrak{z}$. *Then either* $\mathfrak{g}_0 = \widetilde{\mathfrak{g}} = \mathfrak{k}$ *or* $\mathfrak{g}_0 = \mathfrak{g}$ *and* $\widetilde{\mathfrak{g}} = \mathfrak{k}$, *respectively. Here,* $\mathfrak{k}$ *is the centraliser of* $\mathfrak{a}$ *in* $\mathfrak{g}$. *Thus,* $\widetilde{\mathfrak{g}}$ *is an ideal of codimension 1 in* $\mathfrak{g}$ *in both cases and any element of* $\mathfrak{g} \setminus \widetilde{\mathfrak{g}}$ *is qualified as a co-exponential basis of* $\mathfrak{g}$ *to* $\widetilde{\mathfrak{g}}$*. In these situations, the representations* π_ℓ *for generic linear forms correspond to saturated orbits with respect to* $\mathfrak{k}$*. As* $W \in \mathcal{U}(\mathfrak{k})$ *modulo* $\mathcal{U}(\mathfrak{g})\mathfrak{a}_\tau$ *by Proposition 4.2.1, the induction procedure and arguments of Case 1.1 apply.*

Case 2. Suppose henceforth that $\mathfrak{h}$ *is not contained in* $\widetilde{\mathfrak{g}}$ *and therefore* $\mathfrak{a} \not\subset \mathfrak{h}$*. Put* $\widetilde{\mathfrak{h}} = \mathfrak{h} \cap \widetilde{\mathfrak{g}}$ *and* $\widetilde{H} = \exp(\widetilde{\mathfrak{h}})$*.*

Case 2.1: $\mathfrak{z} = \{0\}$*. We can and do repeat the arguments used above. In this situation, we have* $\mathfrak{g}_0 = \mathfrak{k} = \widetilde{\mathfrak{g}}$*. Then,* $\mathfrak{a}$ *is a minimal ideal of* $\mathfrak{g}$ *and* $\mathfrak{g}_0$ *is nothing but the centraliser of* $\mathfrak{a}$ *in* $\mathfrak{g}$ *which is an ideal of codimension 1 in* $\mathfrak{g}$*. In our situation, the orbit* $G{\cdot}\ell$ *of generic* $\ell \in \Gamma_\tau$ *is saturated with respect to* $\mathfrak{g}_0$ *and hence* $W \in \mathcal{U}(\mathfrak{g}_0) + \mathcal{U}(\mathfrak{g})\mathfrak{a}_\tau$*. Let* $\mathfrak{h} = \mathbb{R}X + \widetilde{\mathfrak{h}}$ *and let* $\ell \in \Gamma_\tau$ *be generic. Take in* $\mathcal{H}_{\pi_\ell}$ *a test function*

$$\varphi(g) = \phi(x)\psi(g_0),\ g = \exp(xX)g_0\ (x \in \mathbb{R},\ g_0 \in G_0),$$

where ϕ *is a continuous function with compact support on* $\mathbb{R}$*, i.e.* $\phi \in C_0(\mathbb{R})$*, and* ψ *is a test function of* $\widetilde{\pi}_{\widetilde{\ell}} = ind_{B[\ell]}^{G_0}\chi_\ell$*. Then, the same computation as in Case 1.1.1 shows that*

$$\langle W{\cdot}a_\ell, \varphi\rangle = \langle W{\cdot}\widetilde{a}_{\widetilde{\ell}}, \psi\rangle \int_{\mathbb{R}} \overline{\phi(t)} e^{-itf(X)} e^{-\frac{1}{2} Tr\ ad_{\mathfrak{g}/\mathfrak{h}} X} dt,$$

where $\widetilde{a}_{\widetilde{\ell}}$ *designates the Penney's distribution for* $\widetilde{\pi}_{\widetilde{\ell}}$*. This gives us the desired result from the induction hypothesis applied to* $\widetilde{\tau} = ind_{\widetilde{H}}^{\widetilde{G}}\chi_f$ *and passing to the quotient by the action of* H*. Remark that* $\widetilde{\tau}$ *is of discrete type by the same arguments as above.*

Case 2.2: $\mathfrak{z} \neq \{0\}$*. This means* $\dim \mathfrak{z} = 1$ *and it suffices for us to assume* $\mathfrak{z} \subset \mathfrak{a}$*. If* $[\mathfrak{g}, \mathfrak{a}] \subset \mathfrak{z}$*, then* $\mathfrak{g}_0 = \mathfrak{g}$ *and* $\widetilde{\mathfrak{g}} = \mathfrak{k}$*. In this case,* $W \in \mathcal{U}(\widetilde{\mathfrak{g}})$ *modulo* $\mathcal{U}(\mathfrak{g})\mathfrak{a}_\tau$ *and* $\widetilde{\tau} = ind_{\widetilde{H}}^{\widetilde{G}}\chi_f$ *is of discrete type for similar reasons as the cases above. Then no complications are expected to happen in this case (cf. Case 1.1.1).*

Assume then that $[\mathfrak{g}, \mathfrak{a}] = \mathfrak{a}$ *in the sequel.*

Case 2.2.1: Suppose first $\mathfrak{h} \not\subset \mathfrak{g}_0$*, what brings* $W \in \mathcal{U}(\mathfrak{g}_0)$ *modulo* $\mathcal{U}(\mathfrak{g})\mathfrak{a}_\tau$*. Then, we can assume the non-saturation of generic orbits with respect to* $\mathfrak{g}_0$*, otherwise we are back to Case 1.1.1. If* $\mathfrak{h}_0 \subset \widetilde{\mathfrak{g}}$*, then* $\mathfrak{h}_0$ *is contained in* $\mathfrak{k}$ *by (1.1.12), (1.1.15), and* P_W *belongs to* $S(\widetilde{\mathfrak{g}})$ *modulo* $S(\mathfrak{g})\overline{\mathfrak{a}_\tau}$*. Then, we deduce from these the result as before.*

If $\mathfrak{h}_0 \not\subset \widetilde{\mathfrak{g}}$*, then* $W \in \mathcal{U}(\widetilde{\mathfrak{g}})$ *modulo* $\mathcal{U}(\mathfrak{g})\mathfrak{a}_\tau$*. Further, by the same arguments of Lemma 4.2.3,* $\widetilde{\tau} = ind_{\widetilde{H}}^{\widetilde{G}}\chi_f$ *is of discrete type. On the other hand, thanks to equations (1.1.11), (1.1.16), and (1.1.17), there exists outside of* $\widetilde{\mathfrak{g}}$ *a subspace* V *of* $[\mathfrak{h}, \mathfrak{h}]$ *having the dimension* $\dim(\mathfrak{a}/\mathfrak{z})$*. This is well adapted to the integration process descending from* H *to* $\widetilde{H}$ *in the expression of the Penney distribution (4.2.2). Then, we can just argue as before, considering test functions* φ *in* $\mathcal{H}_{\pi_\ell}$*,* $\ell \in \Gamma_\tau$ *generic, verifying*

$$\varphi((\exp X)\widetilde{g}) = \phi(X)\psi(\widetilde{g})\Delta_{\widetilde{G},G}^{1/2}(\widetilde{g}),\ X \in V,\ \widetilde{g} \in \widetilde{G},$$

where ϕ *is a continuous function with compact support on* V *and* ψ *is a test function of* $\widetilde{\pi}_{\widetilde{\ell}} = ind_{B[\ell]}^{\widetilde{G}}\chi_\ell$*. The surjectivity of* Θ *is obvious in this case.*

Case 2.2.2: Suppose finally that $\mathfrak{h} \subset \mathfrak{g}_0$. *Then, we can assume once again the non-saturation of generic orbits with respect to* $\mathfrak{g}_0$. *Further,* $\widetilde{\tau}$ *is of discrete type by previous reasons.*

On the other hand, we can similarly as in Lemma 4.2.4 show that $W \in \mathcal{U}(\mathfrak{g}_0)$. *When* $\dim \mathfrak{a} = 2$, *we can assume that* $X_1 \in \mathfrak{h}$. *Otherwise* ($\dim \mathfrak{a} = 3$), X_1 *or* X_2 *or both belong to* $\mathfrak{h}$. *In this case, the same arguments above work out to show that* $W \in \mathcal{U}(\widetilde{\mathfrak{g}})$ *modulo* $\mathcal{U}(\mathfrak{g})\mathfrak{a}_\tau$, *as these generators correspond to independent partial differential operators of analytic coefficients.*

This being understood, $\mathfrak{h}/\widetilde{\mathfrak{h}}$ *is qualified as a part of a co-exponential basis of* $\mathfrak{g}$ *to* $B[\ell]$ *for generic* $\ell \in \Gamma_\tau$ *and* $Q \in \mathbb{C}[\Gamma_\tau]^H \simeq (S(\mathfrak{g})/S(\mathfrak{g})\overline{\mathfrak{a}_\tau})^H$ *belongs to* $S(\widetilde{\mathfrak{g}})$ *modulo* $S(\widetilde{\mathfrak{g}})\overline{\mathfrak{a}_\tau}$ *taking back the same arguments as before.*

4.2.3 Examples and a counterexample to Duflo's problem

We here treat some examples showing evidence of Theorem 4.2.2. These also allow us to construct generators of primitive ideals in the enveloping algebra in some specific situations. We also provide in Example 4.2.7 a second counterexample in the exponential context to a problem by M. Duflo in [40]. An example in the setting of completely solvable Lie groups was generated in ([16], Example 7).

Example 4.2.5 Let $G = G_3(\alpha) = \exp(\mathfrak{g}_3(\alpha)), \mathfrak{g}_3(\alpha) = \langle T, Y_1, Y_2 \rangle$:

$$[T, Y_1] = Y_1 - \alpha Y_2, [T, Y_2] = \alpha Y_1 + Y_2,$$

with a real number $\alpha \neq 0$. Let $f = Y_1^* \in (\mathfrak{g}_3(\alpha))^*$ and $\mathfrak{h} = \mathbb{R}Y_1$. Then, $\tau - \mathrm{ind}_H^G \chi_f, H = \exp \mathfrak{h}$, is of discrete type with uniform multiplicities equal to $+\infty$. Indeed, for generic $\ell = \gamma T^* + Y_1^* + \lambda Y_2^* \in \Gamma_\tau$, we have $\mathfrak{g}(\ell) = \mathbb{R}((\lambda+\alpha)Y_1 + (\lambda\alpha - 1)Y_2)$ so that $\mathfrak{h} + \mathfrak{g}(\ell)$ is a Lagrangian subspace of $\mathfrak{g}$ for the bilinear form B_ℓ. Here, the elements of $\mathcal{U}(\mathfrak{g}, \tau)$ are clearly the polynomials of Y_2 modulo the kernel and the same for $\mathbb{C}[\Gamma_\tau]^H$.

Example 4.2.6 Let $G_4 = \exp(\mathfrak{g}_4), \mathfrak{g}_4 = \langle T, X, Y, Z \rangle_{\mathbb{R}}$:

$$[T, X] = -X, [T, Y] = Y, [X, Y] = Z.$$

Let $f(\alpha, \beta) = \alpha T^* + \beta Z^* \in \mathfrak{g}_4^*$, $O(\alpha, \beta) = G{\cdot}f(\alpha, \beta)$ and $\pi(\alpha, \beta) \in \widehat{G}$ corresponds to $O(\alpha, \beta)$ for $\beta \neq 0$. Put $\mathfrak{h} = \mathbb{R}T, H = \exp \mathfrak{h}, \mathfrak{g}_1 = \mathfrak{z} = \mathbb{R}Z, \mathfrak{g}_2 = \mathfrak{a} = \mathfrak{z} \oplus \mathbb{R}Y$ and $\mathfrak{g}_3 = \mathfrak{a} + \mathbb{R}X$. Then,

$$\mathcal{S} : \{0\} \subset \mathfrak{g}_1 \subset \mathfrak{g}_2 \subset \mathfrak{g}_3 \subset \mathfrak{g}$$

is a Jordan-Hölder sequence of $\mathfrak{g}$ and $\mathfrak{b} = \mathbb{R}T + \mathfrak{a}$ is the Vergne polarisation at $f(\alpha, \beta)$ associated to $\mathcal{S}$. The Hilbert space $\mathcal{H}_{\pi(\alpha,\beta)}$ of $\pi(\alpha, \beta)$ is identified with $L^2(\mathbb{R})$ by the correspondence

$$\mathcal{H}_{\pi(\alpha,\beta)} \ni \varphi \leftrightarrow \Phi \in L^2(\mathbb{R}),\ \Phi(x) = \varphi(\exp(xX))(x \in \mathbb{R}).$$

Write simply π instead of $\pi(\alpha,\beta)$. This being done,

$$\begin{aligned}
(\pi(T)\Phi)(x) &= \frac{d}{dt}\varphi(\exp(-tT)\exp(xX))\big|_{t=0} \\
&= \frac{d}{dt}\varphi(\exp(xe^tX)\exp(-tT))\big|_{t=0} = \frac{d}{dt}\Phi(xe^t)e^{t/2}e^{i\alpha t}\big|_{t=0} \\
&= x\frac{d\Phi(x)}{dx} + (\frac{1}{2}+i\alpha)\Phi(x) \\
(\pi(X)\Phi)(x) &= \frac{d}{dt}\varphi(\exp(-tX)\exp(xX))\big|_{t=0} = -\frac{d\Phi(x)}{dx} \\
(\pi(Y)\Phi)(x) &= \frac{d}{dx}\varphi(\exp(-tY)\exp(xX))\big|_{t=0} = \frac{d}{dt}\varphi(\exp(xX)\exp(txZ))\big|_{t=0} \\
&= \frac{d}{dt}e^{-i\beta tx}\Phi(x)\big|_{t=0} = -i\beta x\Phi(x) \\
(\pi(Z)\Phi)(x) &= i\beta\Phi(x).
\end{aligned}$$

Thus,

$$W = \beta T + iYX - (\alpha - \frac{i}{2})Z \in \ker\pi.$$

In this case,

$$\tau = \mathrm{ind}_H^G\chi_{\alpha_0}, \chi_{\alpha_0}(\exp(tT)) = e^{i\alpha_0 t}(t\in\mathbb{R})$$

has finite multiplicities which are uniformly equal to 2 (cf. [48]). It is easy to see that $\mathcal{U}(\mathfrak{g},\tau)$ is generated by YX and Z modulo the kernel, and the same happens for $\mathbb{C}[\Gamma_\tau]^H$.

Example 4.2.7 Let $G_6 = \exp(\mathfrak{g}_6), \mathfrak{g}_6 = \langle T, X_1, X_2, Y_1, Y_2, Z\rangle_{\mathbb{R}}$:

$$[X_i, Y_j] = \delta_{ij}Z(1\le i,j\le 2), [T,X_1] = -X_1 - \alpha X_2, [T,X_2] = -X_2 + \alpha X_1,$$

$$[T,Y_1] = Y_1 - \alpha Y_2, [T,Y_2] = Y_2 + \alpha Y_1 (0\neq\alpha\in\mathbb{R}).$$

Put $f = \beta T^* + \gamma Z^* \in \mathfrak{g}_6^*(\gamma\neq 0)$, $\Omega = G{\cdot}f \subset \mathfrak{g}^*$ and let $\pi\in\widehat{G}$ be the irreducible unitary representation of G corresponding to Ω. Let $\mathfrak{g}_1 = \mathfrak{z} = \mathbb{R}Z, \mathfrak{g}_2 = \mathfrak{z}\oplus\mathbb{R}Y_2, \mathfrak{g}_3 = \mathfrak{a} = \mathfrak{g}_2\oplus\mathbb{R}Y_1, \mathfrak{g}_4 = \mathfrak{g}_3 + \mathbb{R}X_2, \mathfrak{g}_5 = \mathfrak{g}_4 + \mathbb{R}X_1$. Then,

$$\mathcal{S} : \{0\} \subset \mathfrak{g}_1 \subset \mathfrak{g}_2 \subset \mathfrak{g}_3 \subset \mathfrak{g}_4 \subset \mathfrak{g}_5 \subset \mathfrak{g}$$

is a good sequence of subalgebras of $\mathfrak{g}$ and $\mathfrak{b} = \mathfrak{a} + \mathbb{R}T$ is the Vergne polarisation at f associated to $\mathcal{S}$. Put $\mathfrak{h} = \mathbb{R}T$. The Hilbert space $\mathcal{H}_\pi$ of π is identified with $L^2(\mathbb{R}^2)$ by the correspondence

$$\mathcal{H}_\pi \ni \varphi \leftrightarrow \Phi \in L^2(\mathbb{R}^2),\ \Phi(x_1,x_2) = \varphi(\exp(x_1X_1 + x_2X_2))(x_1,x_2\in\mathbb{R}).$$

This being done, the following computations are straight:

$$\begin{aligned}(\pi(T)\Phi)(x_1,x_2) &= \frac{d}{dt}\varphi(\exp(-tT)\exp(x_1X_1+x_2X_2))\big|_{t=0}\\ &= \frac{d}{dt}\Big(\varphi(\exp(e^t(x_1\cos(\alpha t)-x_2\sin(\alpha t))X_1\\ &+e^t(x_1\sin(\alpha t)+x_2\cos(\alpha t))X_2)\exp(-tT))\Big)\big|_{t=0}\\ &= (x_1-\alpha x_2)\frac{\partial\Phi(x_1,x_2)}{\partial x_1}+(x_2+\alpha x_1)\frac{\partial\phi(x_1,x_2)}{\partial x_2}+(1+i\beta)\Phi(x_1,x_2)\end{aligned}$$

On the other hand,

$$\pi(X_1)\Phi = -\frac{\partial\Phi}{\partial x_1}$$

and

$$\pi(X_2)\Phi = -\frac{\partial\Phi}{\partial x_2}.$$

Further:

$$\begin{aligned}(\pi(Y_1)\Phi)(x_1,x_2) &= \frac{d}{dt}\varphi(\exp(-tY_1)\exp(x_1X_1+x_2X_2))\big|_{t=0}\\ &= \frac{d}{dt}\varphi(\exp(x_1X_1+x_2X_2)\exp(-tY_1)\exp(tx_1Z))\big|_{t=0}\\ &= \frac{d}{dt}e^{-i\gamma tx_1}\Phi(x_1,x_2)\big|_{t=0} = -i\gamma x_1\Phi(x_1,x_2).\end{aligned}$$

Likewise, $(\pi(Y_2)\Phi)(x_1,x_2) = -i\gamma x_2\Phi(x_1,x_2)$ and $\pi(Z)\Phi - i\gamma\Phi$. Thus, we find

$$W = \gamma T + i(Y_1-\alpha Y_2)X_1 + i(Y_2+\alpha Y_1)X_2 + (i-\beta)Z \in \ker\pi.$$

In this case, $\tau = \mathrm{ind}_H^G\chi_f, H = \exp\mathfrak{h}$, is not of discrete type. Nevertheless, we can see that $\mathcal{U}(\mathfrak{g},\tau)$ is generated by $Y_1X_1+Y_2X_2, Y_1X_2-Y_2X_1$ and Z modulo the kernel, the same for $\mathbb{C}[\Gamma_\tau]^H$. These elements complement each other from the above expression of $\ker\pi$. Indeed, let

$$W = \sum c_{p,q,r,s}Y_1^pY_2^qX_1^rX_2^s \in \mathcal{U}(\mathfrak{g},\tau),\ c_{p,q,r,s}\in\mathbb{C}.$$

Then, for any $m\in\mathbb{N}$,

$$\sum_{p+q+r+s=m} c_{p,q,r,s}Y_1^pY_2^qX_1^rX_2^s$$

is $\mathfrak{h}$-invariant and it is not difficult to get our claim. Since $Y_1X_1+Y_2X_2, Y_1X_2-Y_2X_1$ commute each other, the algebra $D_\tau(G/H)$ is commutative. This also provides a negative solution to Duflo's problem [40], stating that there is equivalence between the fact that $D_\tau(G/H)$ is commutative and the fact that the subspace $\mathfrak{h}+\mathfrak{g}(\ell)$ is generically Lagrangian on Γ_τ. Note that a first example in the completely solvable setting was given in ([16], Example 7).

4.3 RESTRICTIONS OF DISCRETE TYPE

The notion of monomial representations of discrete type is discussed in Subsection 4.1, where a full proof of the polynomial conjecture was provided (cf. Theorem 4.2.2). We here go to develop new techniques in the objective to study the analogy for the case of restrictions. Let $G = \exp \mathfrak{g}$ be an exponential solvable Lie group with Lie algebra $\mathfrak{g}$, $K = \exp \mathfrak{k}$ an analytic subgroup of G with Lie algebra $\mathfrak{k}$ and π an irreducible unitary representation of G. We assume that the coadjoint orbit Ω of G corresponding to π is of maximal dimension. As a continuation of our study of monomial representations, we go to define and study the restriction $\pi|_K$ of discrete type, especially the algebra $D_\pi(G)^K$ of K-invariant differential operators in the space of π defined in (Subsection 2.3.1, Chapter 2). We diagonalise these operators by means of Penney's distributions so that the algebra $D_\pi(G)^K$ turns out to be commutative when $\pi|_K$ is of discrete type, and the converse fails to hold. Furthermore, $D_\pi(G)^K$ is isomorphic to the algebra $\mathbb{C}[\Omega]^K$ of the K-invariant polynomial functions on Ω. We also provide a process to express some generators of the kernel $\ker \pi$ in the enveloping algebra of $\mathfrak{g}_{\mathbb{C}}$.

4.3.1 Discrete type restrictions of representations

We first single out a notion of discrete type for the restriction of a unitary representation. Let $G = \exp \mathfrak{g}$ be an exponential solvable Lie group with Lie algebra $\mathfrak{g}$, $K = \exp \mathfrak{k}$ an analytic subgroup of G and π an irreducible unitary representation of G whose coadjoint orbit of G is denoted by Ω. The restriction $\pi|_K$ of π to K is said to be of discrete type if, $\mathfrak{b}[\ell|_{\mathfrak{k}}] + \mathfrak{g}(\ell)$ is a Lagrangian subspace of $\mathfrak{g}$ for B_ℓ almost everywhere on Ω with respect to the canonical measure. In other words, each connected component of the intersection of Ω with the inverse image by the canonical projection $p : \mathfrak{g}^* \to \mathfrak{k}^*$ of a coadjoint orbit $\omega \subset \mathfrak{k}^*$ of K is a K-orbit almost everywhere. For discrete multiplicities restrictions of representations in different contexts, the readers could consult the references [60], [64] and [102].

4.3.2 Some trace formulae

Let $G = \exp \mathfrak{g}$ be an exponential solvable Lie group with Lie algebra $\mathfrak{g}$, $K = \exp \mathfrak{k}$ an analytic subgroup of G and π an irreducible unitary representation of G whose coadjoint orbit of G is denoted by Ω. We assume that Ω is of maximal dimension. This means (cf. [23], Chap II) that $\mathfrak{g}(\ell)$ is commutative for $\ell \in \Omega$. Let

$$\mathcal{S} : \{0\} = \mathfrak{g}_0 \subset \mathfrak{g}_1 \subset \cdots \subset \mathfrak{g}_{n-1} \subset \mathfrak{g}_n = \mathfrak{g} \tag{4.3.1}$$

be a good sequence of subalgebras of $\mathfrak{g}$ which passes $[\mathfrak{g}, \mathfrak{g}]$, namely the $\mathfrak{g}_j (0 \le j \le n)$ are subalgebras of $\mathfrak{g}$ verifying $\dim(\mathfrak{g}_j) = j$ such that $\mathfrak{g}_{j\pm 1}$ are ideals of $\mathfrak{g}$ if $\mathfrak{g}_j$ is not so, and the adjoint action of $\mathfrak{g}$ on $\mathfrak{g}_{j+1}/\mathfrak{g}_{j-1}$ is irreducible. Let

$$\mathcal{I} = \{1 \le i \le n; \mathfrak{k} \cap \mathfrak{g}_i \ne \mathfrak{k} \cap \mathfrak{g}_{i-1}\} = \{i_1 < \cdots < i_d\}$$

with $d = \dim \mathfrak{k}$ and $\mathfrak{k}_r = \mathfrak{k} \cap \mathfrak{g}_{i_r}$ for $1 \le r \le d$. Then,

$$\mathcal{S}|_{\mathfrak{k}} : \mathfrak{k}_0 = \{0\} \subset \mathfrak{k}_1 \subset \cdots \subset \mathfrak{k}_d = \mathfrak{k} \tag{4.3.2}$$

is a good sequence of subalgebras of $\mathfrak{k}$. Indeed, let $i_r \in \mathcal{I}$ be an index such that $\mathfrak{g}_{i_r}$ is not an ideal of $\mathfrak{g}$. If $[\mathfrak{k}, \mathfrak{k}_r] \subset \mathfrak{k}_{r-1}$, $\mathfrak{k}_r$ is an ideal of $\mathfrak{k}$. Otherwise, $\mathfrak{k}_{r\pm 1}$ are ideals of $\mathfrak{k}$ and the adjoint action of $\mathfrak{k}$ on $\mathfrak{k}_{r+1}/\mathfrak{k}_{r-1}$ is irreducible.

For $\ell \in \Omega$, we construct the Vergne polarisation $\mathfrak{b}[\ell]$ of $\mathfrak{g}$ by means of $\mathcal{S}$ and the Vergne polarisation $\mathfrak{b}[\ell|_{\mathfrak{k}}]$ of $\mathfrak{k}$ at $\ell|_{\mathfrak{k}} \in \mathfrak{k}^*$ by means of $\mathcal{S}|_{\mathfrak{k}}$. Clearly,

$$\mathfrak{k} \cap \mathfrak{b}[\ell] = \mathfrak{b}[\ell|_{\mathfrak{k}}] \cap \mathfrak{b}[\ell]$$

for $\ell \in \Omega$.

We now prove the following important upshot, which is crucial to construct the Penney type distributions.

Proposition 4.3.1 *Suppose that the restriction $\pi|_K$ is of discrete type. Then almost everywhere on Ω, we have*

$$Tr\ ad_{\mathfrak{b}[\ell|_{\mathfrak{k}}]/(\mathfrak{b}[\ell|_{\mathfrak{k}}]\cap\mathfrak{b}[\ell])}U + Tr\ ad_{\mathfrak{b}[\ell]/(\mathfrak{b}[\ell|_{\mathfrak{k}}]\cap\mathfrak{b}[\ell])}U = 0,$$

for $U \in \mathfrak{b}[\ell|_{\mathfrak{k}}] \cap \mathfrak{b}[\ell]$.

Proof 4.5 *Let $\mathfrak{z}$ be the centre of $\mathfrak{g}$. In what follows, we can suppose $\mathfrak{z} \subset \mathfrak{k}$. Further, if π vanishes on a non-zero ideal of $\mathfrak{g}$, we can suppose that $\mathfrak{k}$ contains this ideal and then we can pass to the quotient by this ideal. This being assumed, we can suppose* $\dim \mathfrak{z} \leq 1$.

Let $\mathfrak{a}$ be a minimal non-central ideal of $\mathfrak{g}$. If $\mathfrak{z} = \{0\}$, $\mathfrak{a}$ is a minimal ideal of $\mathfrak{g}$ and we can suppose that $\mathfrak{a}$ is nothing but $\mathfrak{g}_1$ or $\mathfrak{g}_2$ in the sequence $\mathcal{S}$. Let $\mathfrak{g}_0$ be the centraliser of $\mathfrak{a}$ in $\mathfrak{g}$ and $G_0 = \exp(\mathfrak{g}_0)$. Since π does not vanish on $\mathfrak{a}$, Ω is saturated with respect to $\mathfrak{g}_0$ which is an ideal of codimension 1 in $\mathfrak{g}$. Let

$$\pi|_{G_0} \simeq \int_{\mathbb{R}}^{\oplus} \pi'_t dt$$

be the canonical central disintegration of the restriction of π to G_0. Then, for generic $\ell \in \Omega$, $\mathfrak{b}[\ell] \subset \mathfrak{g}_0$.

The result is obvious if $\mathfrak{k} \subset \mathfrak{g}_0$. Assume that $\mathfrak{k} \not\subset \mathfrak{g}_0$ so that the vector X as in equation (1.1.7) belongs to $\mathfrak{k}$. We can and do assume that $X \in \mathfrak{b}[\ell|_{\mathfrak{k}}]$ as the situation where $\mathfrak{b}[\ell|_{\mathfrak{k}}] \subset \mathfrak{g}_0$ is easy to treat.

Put $\mathfrak{k}_0 = \mathfrak{k} \cap \mathfrak{g}_0$ and $K_0 = \exp(\mathfrak{k}_0)$. Then the restriction $\pi'_t|_{K_0}$ is discrete type for almost all $t \in \mathbb{R}$. Indeed, for generic $\ell \in \Omega$, we have

$$\mathfrak{g}(\ell) + \mathfrak{b}[\ell|_{\mathfrak{k}}] = \mathfrak{g}(\ell) + \mathfrak{b}[\ell|_{\mathfrak{k}_0}] + \mathbb{R}X, \tag{4.3.3}$$

and

$$\mathfrak{g}_0(\ell_0) + \mathfrak{b}[\ell|_{\mathfrak{k}_0}] = \mathfrak{g}(\ell) + \mathfrak{b}[\ell|_{\mathfrak{k}_0}] + \mathbb{R}Y. \tag{4.3.4}$$

Note that $Y \notin \mathfrak{g}(\ell) + \mathfrak{b}[\ell|_{\mathfrak{k}_0}]$ in (4.3.4). Finally, equations (4.3.3) and (4.3.4) allow to conclude.

This being in mind, we have

$$Tr\ ad_{\mathfrak{b}[\ell|_{\mathfrak{k}_0}]/(\mathfrak{b}[\ell|_{\mathfrak{k}_0}]\cap\mathfrak{b}[\ell])}U + Tr\ ad_{\mathfrak{b}[\ell]/(\mathfrak{b}[\ell|_{\mathfrak{k}_0}]\cap\mathfrak{b}[\ell])}U = 0$$

for every $U \in \mathfrak{b}[\ell|_{\mathfrak{k}_0}] \cap \mathfrak{b}[\ell]$ *from the induction hypothesis. Thus, we get the desired result.*

Now suppose $\dim \mathfrak{z} = 1$. *For generic* $\ell \in \Omega$, *put* $\mathfrak{g}' = \mathfrak{a}^\ell$ *as in definition (1.1.5). Then, if* $\mathfrak{a} \subset \mathfrak{k}$ *(which leads to* $\mathfrak{b}[\ell|_{\mathfrak{k}}] \subset \mathfrak{a}^\ell$*), the result follows immediately from the induction hypothesis applied to* $\mathfrak{g}'$. *Suppose* $\mathfrak{a} \not\subset \mathfrak{k}$. *Let* $\mathfrak{k}' = \mathfrak{k} \cap \mathfrak{g}'$, $K' = \exp(\mathfrak{k}')$, $G' = \exp(\mathfrak{g}')$ *and* $\pi' = ind_{B[\ell]}^{G'}\chi_\ell$. *Then as earlier, we easily remark that the restriction* $\pi'|_{K'}$ *is discrete type. Using the structure details of minimal non-central ideals, we get for any* $U \in \mathfrak{b}[\ell|_{\mathfrak{k}}] \cap \mathfrak{b}[\ell]$,

$$Tr\ ad_{\mathfrak{a}/(\mathfrak{a}\cap\mathfrak{b}[\ell|_{\mathfrak{k}}])}U + Tr\ ad_{\mathfrak{b}[\ell|_{\mathfrak{k}}]/(\mathfrak{b}[\ell|_{\mathfrak{k}}]\cap\mathfrak{a}^\ell)}U = 0.$$

The induction hypothesis applied to $\pi'|_{K'}$ *provides us with the desired result.*

The consideration of these examples rests valuable in the general case. Let $\ell \in \Omega$ be generic, $\mathfrak{g}' = \mathfrak{a}^\ell$, $G' = \exp(\mathfrak{g}')$, $\pi' = \mathrm{ind}_{B[\ell]}^{G'}\chi_\ell$ so that $\pi = \mathrm{ind}_{G'}^{G}\pi'$, $\mathfrak{k}' = \mathfrak{k} \cap \mathfrak{g}'$ and $K' = \exp(\mathfrak{k}')$ in what follows. Let $j_1 < j_2$ two consecutive indices such that $\mathfrak{g}_{j_1}, \mathfrak{g}_{j_2}$ are ideals of $\mathfrak{g}$ in the sequence $\mathcal{S}$, hence $j_2 - j_1 = 1$ or 2. If $\mathfrak{b}[\ell] \cap \mathfrak{g}_{j_2} \neq \mathfrak{b}[\ell] \cap \mathfrak{g}_{j_1}$, the part $\mathfrak{b}[\ell] \setminus ((\mathfrak{b}[\ell] \cap \mathfrak{b}[\ell|_{\mathfrak{k}}]) + \mathfrak{g}(\ell))$ which augments at this stage is found in the duality with the part of $\mathfrak{b}[\ell|_{\mathfrak{k}}] \setminus (\mathfrak{b}[\ell] \cap \mathfrak{b}[\ell|_{\mathfrak{k}}])$ which augments at the passage from $(\mathfrak{g}_{j_2})^\ell$ to $(\mathfrak{g}_{j_1})^\ell$. In this way, we have the following:

Lemma 4.3.1 *Under the same hypotheses, we have:*

$$Tr\ ad_{\mathfrak{b}[\ell|_{\mathfrak{k}}]/(\mathfrak{b}[\ell|_{\mathfrak{k}}]\cap\mathfrak{b}[\ell])}X + Tr\ ad_{\mathfrak{b}[\ell]/((\mathfrak{b}[\ell|_{\mathfrak{k}}]\cap\mathfrak{b}[\ell])+\mathfrak{g}(\ell))}X = 0$$

for $X \in \mathfrak{g}(\ell)$.

Remark 4.3.2 *This equality holds for* $X \in \mathfrak{g}$ *if* $B_\ell(X, [\mathfrak{g},\mathfrak{g}]) = \{0\}$.

4.3.3 Penney type distributions

At any generic point $\ell \in \Omega$, we realise π as $\pi_\ell = \mathrm{ind}_{B[\ell]}^{G}\chi_\ell$, where the unitary character χ_ℓ of $B[\ell]$ is defined by $\chi_\ell(\exp X) = e^{i\ell(X)} (X \in \mathfrak{b}[\ell])$. We denote by $\mathcal{H}_{\pi_\ell}$ the Hilbert space of π_ℓ. We define the Penney distribution $a_\ell := a_\ell^K$ by the formula

$$\mathcal{H}_{\pi_\ell} \ni \varphi \mapsto \langle a_\ell, \varphi\rangle = \oint_{B[\ell|_{\mathfrak{k}}]/(B[\ell|_{\mathfrak{k}}]\cap B[\ell])} \overline{\varphi(k)\chi_\ell(k)}\Delta_{B[\ell|_{\mathfrak{k}}],G}^{-1/2}(k)d\nu(k). \tag{4.3.5}$$

At least in a formal sense, we can consider this integral. Indeed, let

$$\Phi(k) = \overline{\varphi(k)\chi_\ell(k)}\Delta_{B[\ell|_{\mathfrak{k}}],G}^{-1/2}(k)\ (k \in B[\ell|_{\mathfrak{k}}]).$$

Then, for $k_0 \in B[\ell|_{\mathfrak{k}}] \cap B[\ell]$,

$$\begin{aligned}\Phi(kk_0) &= \overline{\varphi(kk_0)\chi_\ell(kk_0)}\Delta_{B[\ell|_{\mathfrak{k}}],G}^{-1/2}(kk_0)\\ &= \overline{\varphi(k)\chi_\ell(k)}\Delta_{B[\ell|_{\mathfrak{k}}],G}^{-1/2}(k)\Delta_{B[\ell],G}(k_0)^{1/2}\Delta_{B[\ell|_{\mathfrak{k}}],G}^{-1/2}(k_0)\\ &= \Phi(k)\Delta_{B[\ell],G}(k_0)^{1/2}\Delta_{B[\ell|_{\mathfrak{k}}],G}^{-1/2}(k_0)\\ &= \Phi(k)\Delta_{B[\ell|_{\mathfrak{k}}]\cap B[\ell],B[\ell|_{\mathfrak{k}}]}(k_0)\end{aligned}$$

from Proposition 4.3.1. This justifies that Φ satisfies the required covariance relation. It is now all about to prove that the integral formula (4.3.5) defining the distribution a_ℓ converges on a dense subspace of $\mathcal{H}_{\pi_\ell}$. One of the major hurdles we face up for our study, is the convergence of the Penney-type distributions for restrictions, which will be shown to diagonalise the differential operators, elements of $D_\pi(G)^K$.

We have the following:

Theorem 4.3.3 *Let G be an exponential solvable Lie group, K a closed connected subgroup of G and $\pi \in \widehat{G}$ such that the restriction $\pi|_K$ is of discrete type. Assume further that $\Omega(\pi)$ is of maximal dimension. Then generically on $\Omega(\pi)$, the integral $\langle a_\ell, \varphi\rangle$ as in formula (4.3.5) converges for $\varphi \in \mathcal{H}_\pi$ continuous with compact support modulo $B[\ell]$.*

Proof 4.6 *We show the convergence of $\langle a_\ell, \varphi\rangle$ for the test functions $\varphi \in \mathcal{H}_\pi$ with compact support modulo $B[\ell]$. We divide our proof into two cases, here is the first:*

Case 1. Suppose first that there exists an ideal $\mathfrak{g}_0$ of codimension 1 of $\mathfrak{g}$ such that Ω is saturated with respect to $\mathfrak{g}_0$. Let $G_0 = \exp(\mathfrak{g}_0)$, $\mathfrak{k}_0 = \mathfrak{k} \cap \mathfrak{g}_0$ and $K_0 = \exp(\mathfrak{k}_0)$. We can suppose that the sequence $\mathcal{S}$ passes through $\mathfrak{g}_0$. Let $\ell \in \Omega$ be generic. Then, $\pi \simeq ind_{G_0}^{G}\pi_0$ with $\pi_0 = ind_{B[\ell]}^{G_0}\chi_\ell$. Here, we prove the following:

Lemma 4.3.2 *The restriction $\pi_0|_{K_0}$ is of discrete type.*

Proof 4.7 *Remark that if $K \subset G_0$, everything is clear. Assume otherwise, if the K-orbits of $p(\Omega)$ are generically saturated with respect to $\mathfrak{k}_0$ in which case we have $\mathfrak{b}[\ell|_{\mathfrak{k}}] \subset \mathfrak{g}_0$, we are also done. In case of non-saturation, $\mathfrak{b}[\ell|_{\mathfrak{k}}] \cap \mathfrak{g}_0 = \mathfrak{b}[\ell|_{\mathfrak{k}_0}]$ stands for a polarisation of $\ell|_{\mathfrak{k}_0}$ generically on Ω. Write*

$$\mathfrak{b}[\ell|_{\mathfrak{k}}] = \mathbb{R}X(\ell) \oplus \mathfrak{b}[\ell|_{\mathfrak{k}_0}],$$

for some $X(\ell) \in \mathfrak{k} \setminus \mathfrak{k}_0$. In this case, we have:

$$\mathfrak{b}[\ell|_{\mathfrak{k}}] + \mathfrak{g}(\ell) = (\mathfrak{b}[\ell|_{\mathfrak{k}_0}] + \mathfrak{g}(\ell)) \oplus \mathbb{R}X(\ell),$$

and

$$\mathfrak{b}[\ell|_{\mathfrak{k}_0}] + \mathfrak{g}_0(\ell_0) = \mathfrak{b}[\ell|_{\mathfrak{k}_0}] + \mathfrak{g}(\ell) + \mathbb{R}Y_0(\ell),$$

for some $Y_0(\ell) \in \mathfrak{g}_0$ with the property that $\ell([X(\ell), Y_0(\ell)]) \neq 0$ generically on Ω. This of course shows that $Y_0(\ell) \notin \mathfrak{b}[\ell|_{\mathfrak{k}_0}] + \mathfrak{g}(\ell)$ and the restriction $\pi_0|_{K_0}$ is of discrete type for dimension reasons.

Hence, if $\mathfrak{b}[\ell|_{\mathfrak{k}}] \subset \mathfrak{g}_0$, the induction hypothesis assures that $\langle a_\ell, \varphi\rangle$ converges for the continuous test function $\varphi \in \mathcal{H}_\pi$ with compact support modulo $B[\ell|_{\mathfrak{k}}]$ thanks to Lemma 4.3.2. When $\mathfrak{b}[\ell|_{\mathfrak{k}}] \not\subset \mathfrak{g}_0$, let

$$\mathfrak{b}[\ell|_{\mathfrak{k}}] = \mathbb{R}X(\ell) + \mathfrak{b}[\ell|_{\mathfrak{k}}] \cap \mathfrak{g}_0.$$

We have, for the test function $\varphi \in \mathcal{H}_\pi$ verifying

$$\varphi(g) = \phi(x)\psi(g_0), g = \exp(xX(\ell))g_0 \ (x \in \mathbb{R}, g_0 \in G_0),$$

with $\phi \in C_0(\mathbb{R})$ and $\psi \in \mathcal{H}_{\pi_0}$ continuous with compact support modulo $B[\ell]$,

$$\langle a_\ell, \varphi\rangle = \int_{\mathbb{R}} \overline{\phi(x)} e^{-ix\ell(X(\ell))} \Delta^{-1/2}_{B[\ell|_{\mathfrak{k}}],G}(\exp(xX(\ell)))dx \tag{4.3.6}$$

$$\times \oint_{B[\ell|_{\mathfrak{k}_0}]/(B[\ell|_{\mathfrak{k}_0}]\cap B[\ell])} \overline{\psi(g_0)\chi_\ell(g_0)} \Delta^{-1/2}_{B[\ell|_{\mathfrak{k}_0}],G_0}(g_0)d\nu(g_0) \tag{4.3.7}$$

$$= \int_{\mathbb{R}} \overline{\phi(x)} e^{-ix\ell(X(\ell))} \Delta^{-1/2}_{B[\ell|_{\mathfrak{k}}],G}(\exp(xX(\ell)))dx \cdot \langle a^0_\ell, \psi\rangle, \tag{4.3.8}$$

where a^0_ℓ denotes the Penney distribution for the restriction $\pi_0|_{K_0}$. These being done, the convergence of $\langle a_\ell, \varphi\rangle$ follows from the induction hypothesis, thanks again to Lemma 4.3.2.

Case 2. Suppose hereafter that Ω is non-saturated with respect to any ideal $\mathfrak{g}_0$ of codimension 1 of $\mathfrak{g}$. Now as usual we pass to the quotient $\mathfrak{g}/\mathfrak{a}$ to which applies the induction hypothesis, if π vanishes on an ideal $\mathfrak{a} \neq \{0\}$ of $\mathfrak{g}$. This implies that the dimension of the centre $\mathfrak{z}$ of $\mathfrak{g}$ is equal to 1 and $\pi|_{\mathfrak{z}} \neq 0$. Let $\ell \in \Omega$ be generic and fix an ideal $\mathfrak{g}_0$ of codimension 1 of $\mathfrak{g}$. Let $G_0 = \exp(\mathfrak{g}_0)$, then

$$\pi|_{G_0} \simeq \pi_0,$$

the representation of G_0 corresponding to the linear form $\ell_0 = \ell|_{\mathfrak{g}_0}$ as in Lemma 1.5.8. If $\mathfrak{b}[\ell|_{\mathfrak{k}}] \subset \mathfrak{g}_0$, the subspace $\mathfrak{b}[\ell|_{\mathfrak{k}}] + \mathfrak{g}_0(\ell_0)$ is found to be Lagrangian as $\pi|_K$ is of discrete type. This of course entails that $\pi_0|_{K_0}$ is also of discrete type and the result follows from the induction hypothesis applied to $G_0 = \exp(\mathfrak{g}_0)$, where $K_0 = K \cap G_0$.

Let $\mathfrak{a}$ be a minimal non-central ideal of $\mathfrak{g}$. Then, $\mathfrak{a} \supset \mathfrak{z}$, $[\mathfrak{g}, \mathfrak{a}] = \mathfrak{a}$ as introduced in Subsection 1.1.2, and Ω is non-saturated with respect to $\mathfrak{g}_0 = \{U \in \mathfrak{g}; [U, \mathfrak{a}] \subset \mathfrak{z}\}$.

For simplicity, we treat here the situation where $\dim \mathfrak{a} = 2$, the case $\dim \mathfrak{a} = 3$ can be handled similarly. Let $\ell \in \Omega$ be generic. We can and do assume that as in Subsection 1.1.2, $T \in \mathfrak{b}[\ell], T \notin \mathfrak{g}_0, X \notin \mathfrak{g}' = \mathfrak{a}^\ell$ verifying $[T, X] \equiv -X$ modulo $\mathfrak{g}'$.

If $\mathfrak{b}[\ell|_{\mathfrak{k}}] \subset \mathfrak{g}'$, then no hurdles stand to apply the induction hypothesis as $\mathfrak{b}[\ell|_{\mathfrak{k}}] + \mathfrak{g}(\ell)$ is Lagrangian. Otherwise, we may assume that $T + \alpha X \in \mathfrak{b}[\ell|_{\mathfrak{k}}] \cap \ker \ell$ for some non-zero real number α. We can and do assume for simplicity that $\alpha = 1$. Indeed, we just have to ascertain about the fact that $\pi'|_{K'}$ is of discrete type, $K' = \exp \mathfrak{k}'$ with $\mathfrak{k}' = \mathfrak{k} \cap \mathfrak{g}'$. For that remark that for $\ell' = \ell|_{\mathfrak{g}'}$, we have

$$\mathfrak{b}[\ell|_{\mathfrak{k}}] + \mathfrak{g}(\ell) = \mathfrak{b}[\ell|_{\mathfrak{k}'}] + \mathfrak{g}(\ell) \oplus \mathbb{R}(X + T)$$

and

$$\mathfrak{b}[\ell|_{\mathfrak{k}'}] + \mathfrak{g}'(\ell') = \mathfrak{b}[\ell|_{\mathfrak{k}'}] + \mathfrak{g}(\ell) \oplus \mathbb{R}Y,$$

with the fact that $Y \notin \mathfrak{b}[\ell|_{\mathfrak{k}'}] + \mathfrak{g}(\ell)$, *which is enough to conclude.*

If $X \notin \mathfrak{b}[\ell] + [\mathfrak{g}, \mathfrak{g}]$, $X + T$ *is qualified as a member of a co-exponential basis of* $\mathfrak{g}$ *to* $\mathfrak{b}[\ell]$. *We have in this case that*

$$B[\ell|_{\mathfrak{k}}]/(B[\ell|_{\mathfrak{k}}] \cap B[\ell]) \simeq \mathbb{R}(X+T) \oplus B[\ell|_{\mathfrak{k}'}]/(B[\ell|_{\mathfrak{k}'}] \cap B[\ell]).$$

Hence, the integral $\langle a_\ell, \varphi \rangle$ *is convergent for continuous* φ *with compact support modulo* $B[\ell]$ *from the induction hypothesis applied to* $G' = \exp(\mathfrak{g}')$ *as in equations (4.3.6), (4.3.7), and (4.3.8).*

We now prove the following:

Lemma 4.3.3 *Suppose* $X \in \mathfrak{b}[\ell] + [\mathfrak{g}, \mathfrak{g}]$. *Then modifying* T *if necessary, we can suppose that* $X \in [\mathfrak{g}, \mathfrak{g}]$.

Proof 4.8 *Write* $X = X_1 + X_2$ *for some* $X_1 \in \mathfrak{b}[\ell]$ *and* $X_2 \in [\mathfrak{g}, \mathfrak{g}]$. *As* $[X, Y] = Z$, *then* $X_2 \notin \mathfrak{g}'$. *On the other hand,* $X + T = X_1 + T + X_2$ *and we may replace* T *by* $X_1 + T$ *so that* X_2 *replaces* X *as a vector of* $[\mathfrak{g}, \mathfrak{g}]$. *Now an easy computation of the bracket* $[[T, X], Y]$ *shows that we still have* $[T, X] = -X$ *modulo* $\mathfrak{g}'$ *regarding the new vectors.*

Now Lemma 4.3.3 allows to qualify X *as a member of a co-exponential basis of* $\mathfrak{g}$ *to* $\mathfrak{b}[\ell]$. *Let* $\varphi \in \mathcal{H}_\pi$ *such that*

$$\varphi(\exp(tX)g') = \phi(t)\psi(g') \ (t \in \mathbb{R}, g' \in G')$$

with $\phi \in C_0(\mathbb{R})$ *and* $\psi \in \mathcal{H}_{\pi'}$ *continuous with compact support modulo* $B[\ell]$. *For* $k' \in B[\ell|_{\mathfrak{k}'}]$, *we have*

$$\begin{aligned}
\varphi(\exp(t(X+T))k') =& \varphi\Big(\exp((1-e^{-t})X)\exp(tT)g_1(t)k'\Big) \\
=& \varphi\Big(\exp((1-e^{-t})X)\exp(tT)g_1(t)\exp(-tT)\exp(tT)k'\Big) \\
=& \varphi\Big(\exp((1-e^{-t})X)g'(t)\exp(tT)k'\Big), \\
& \text{(for } g'(t) = \exp(tT)g_1(t)\exp(-tT) \in G'\text{)} \\
=& \varphi\Big(\exp((1-e^{-t})X)g'(t)\exp(tT)k'\exp(-tT)\exp(tT)\Big) \\
=& \varphi\Big(\exp((1-e^{-t})X)g'(t)\exp(tT)k'\exp(-tT)\Big) \\
& \times \Delta^{1/2}_{B[\ell],G'}(\exp(tT))\Delta^{1/2}_{G',G}(\exp(tT)) \\
=& \phi(1-e^{-t})\psi\Big(g'(t)\exp(tT)k'\exp(-tT)\Big)\Delta^{1/2}_{B[\ell],G'}(\exp(tT))e^{-t/2}
\end{aligned}$$

with some element $g_1(t) \in G'$.

This leads to the following:

$$\begin{aligned}\langle a_\ell, \varphi\rangle = &\int_{\mathbb{R}} \oint_{B[\ell|_{\mathfrak{k}'}]/(B[\ell|_{\mathfrak{k}'}]\cap B[\ell])} \overline{\varphi\Big(\exp(t(X+T))k'\Big)\chi_\ell(k')} \\ &\times \Delta^{-1/2}_{B[\ell|_{\mathfrak{k}}],G}(\exp(t(X+T))k')d\nu(k')dt \\ = &\int_{\mathbb{R}} \overline{\phi(1-e^{-t})}\Big(\oint_{B[\ell|_{\mathfrak{k}'}]/(B[\ell|_{\mathfrak{k}'}]\cap B[\ell])} \overline{\psi\Big(g'(t)\exp(tT)k'\exp(-tT)\Big)\chi_\ell(k')} \\ &\times \Delta^{-1/2}_{B[\ell|_{\mathfrak{k}'}],G'}(\exp(t(X+T))k')\Delta^{1/2}_{B[\ell],G'}(\exp(tT))d\nu(k')\Big)e^{-t/2}dt \\ = &\int_{\mathbb{R}} \overline{\phi(1-e^{-t})}\Big(\oint_{B[\ell|_{\mathfrak{k}'}]/(B[\ell|_{\mathfrak{k}'}]\cap B[\ell])} \overline{\psi\Big(g'(t)\exp(tT)k'\exp(-tT)\Big)\chi_\ell(k')} \\ &\times \Delta^{-1/2}_{B[\ell|_{\mathfrak{k}'}],G'}(\exp(t(X+T)))\Delta^{1/2}_{B[\ell],G'}(\exp(tT)) \\ &\times \Delta^{-1/2}_{B[\ell|_{\mathfrak{k}'}],G'}(k')d\nu(k')\Big)e^{-t/2}dt.\end{aligned}$$

Now, as X is a nilpotent vector, we have that:

$$\begin{aligned}\Delta =&\Delta^{-1/2}_{B[\ell|_{\mathfrak{k}'}],G'}(\exp(t(X+T)))\Delta^{1/2}_{B[\ell],G'}(\exp(tT)) \\ =&\Delta^{-1/2}_{B[\ell|_{\mathfrak{k}'}],G'}(\exp(tT))\Delta^{1/2}_{B[\ell],G'}(\exp(tT)) \\ =&\exp\frac{t}{2}\Big(Tr\ ad_{\mathfrak{g}'/\mathfrak{b}[\ell|_{\mathfrak{k}'}]}T - Tr\ ad_{\mathfrak{g}'/\mathfrak{b}[\ell]}T\Big) \\ =&\exp\Big(-t\,Tr\ ad_{\mathfrak{b}[\ell|_{\mathfrak{k}'}]/(\mathfrak{b}[\ell|_{\mathfrak{k}'}]\cap\mathfrak{b}[\ell])}T\Big),\end{aligned}$$

by Lemma 4.3.1 and the fact that $Tr\ ad_{\mathfrak{g}(\ell)/(\mathfrak{b}[\ell|_{\mathfrak{k}'}]\cap\mathfrak{g}(\ell))}T = 0$, as $\mathfrak{g}(\ell)$ is commutative. In this way, Δ is nothing but the value of the Jacobian for the change of variables in ψ. We get therefore:

$$\begin{aligned}\langle a_\ell, \varphi\rangle = &\int_{\mathbb{R}} \overline{\phi(1-e^{-t})}\Big(\oint_{B[\ell|_{\mathfrak{k}'}]/(B[\ell|_{\mathfrak{k}'}]\cap B[\ell])} \overline{\psi\Big(g'(t)k'\Big)\chi_\ell(k')} \\ &\times \Delta^{-1/2}_{B[\ell|_{\mathfrak{k}'}],G'}(k')d\nu(k')\Big)e^{-t/2}dt.\end{aligned}$$

Up to this step, for almost all $t \in \mathbb{R}$, the integral

$$\widetilde{\psi}(t) = \oint_{B[\ell|_{\mathfrak{k}'}]/(B[\ell|_{\mathfrak{k}'}]\cap B[\ell])} \overline{\psi\Big(g'(t)k'\Big)\chi_\ell(k')} \times \Delta^{-1/2}_{B[\ell|_{\mathfrak{k}'}],G'}(k')d\nu(k')$$

converges by the induction hypothesis.

We consider the closure $\overline{B[\ell|_{\mathfrak{k}'}]B[\ell]}$ in G' of the simple product $B[\ell|'_{\mathfrak{k}}]B[\ell]$, which is right $B[\ell]$-invariant. Let C be a compact subset of G' such that the support of ψ is contained in $CB[\ell]$. If $\lim_{t\to\pm\infty} g'(t)$ belongs to the compact set $C \cap \overline{B[\ell|'_{\mathfrak{k}}]B[\ell]}$, say $\lim_{t\to\pm\infty} g'(t) = g_0$, then $\lim_{t\to\pm\infty} \widetilde{\psi}(t)$ converges to

$$\oint_{B[\ell|_{\mathfrak{k}'}]/(B[\ell|_{\mathfrak{k}'}]\cap B[\ell])} \overline{\psi_{g_0}(k')\chi_\ell(k')}\Delta^{-1/2}_{B[\ell|_{\mathfrak{k}'}],G'}(k')d\nu(k'),$$

where $\psi_{g_0}(g') = \psi(g_0 g')(g' \in G')$, *by the dominated convergence theorem of Lebesgue. Otherwise,* $\lim_{t\to\pm\infty} g'(t)$ *does not belong to* $C \cap \overline{B[\ell|_{\mathfrak{k}}]B[\ell]}$, *and* $\lim_{t\to\pm\infty} \widetilde{\psi}(t) = 0$.

These observations entail that the integral

$$\langle a_\ell, \varphi\rangle = \int_{\mathbb{R}} \overline{\phi(1-e^{-t})}\ \widetilde{\psi}(t)e^{-t/2}dt$$

is convergent by the dominated convergence theorem, and this closes the proof of the theorem.

4.3.4 Commutativity

We keep the same backgrounds and notations. Let $\ell \in \Omega$, taking there a real polarisation $\mathfrak{b}[\ell]$ of $\mathfrak{g}$, we realise π as $\pi = \mathrm{ind}_{B[\ell]}^{G}\chi_\ell$ with $B[\ell] = \exp(\mathfrak{b}[\ell])$ and χ_ℓ is the unitary character of $B[\ell]$ whose differential is $i\ell|_{\mathfrak{b}[\ell]}$. On the other hand, by means of the flag (4.3.2), we construct [23] the Vergne polarisation $\mathfrak{b}[\ell|_{\mathfrak{k}}]$ of $\mathfrak{k}$ at $\ell|_{\mathfrak{k}} \in \mathfrak{k}^*$. Put $B[\ell|_{\mathfrak{k}}] = \exp(\mathfrak{b}[\ell|_{\mathfrak{k}}])$ and consider the Penney type distribution a_ℓ as in formula (4.3.5). Suppose that $\pi|_K$ is of discrete type. This would say that $\mathfrak{b}[\ell|_{\mathfrak{k}}] + \mathfrak{g}(\ell)$ is a Lagrangian subspace of $\mathfrak{g}$ for B_ℓ at almost all $\ell \in \Omega$ with respect to the invariant measure.

Let us recall once again the algebra of differential operators on the space of the restriction of π to K defined in (Subsection 2.3.1, Chapter 2). We denote by $\mathcal{U}(\mathfrak{g})$ the enveloping algebra of $\mathfrak{g}_{\mathbb{C}}$ and let $\ker\pi$ be the primitive ideal of $\mathcal{U}(\mathfrak{g})$ associated to π. We introduce the algebra

$$\mathcal{U}_\pi(\mathfrak{g})^{\mathfrak{k}} = \{A \in \mathcal{U}(\mathfrak{g}); [A, \mathfrak{k}] \subset \ker\pi\}$$

and $D_\pi(G)^K$ the image of $\mathcal{U}_\pi(\mathfrak{g})^{\mathfrak{k}}$ under the homomorphism π. Namely,

$$D_\pi(G)^K \cong \mathcal{U}_\pi(\mathfrak{g})^{\mathfrak{k}}/\ker\pi \cong (\mathcal{U}(\mathfrak{g})/\ker\pi)^K,$$

where the last member designates the quotient algebra of the K-invariant elements. Then $D_\pi(G)^K$ is seen as the algebra of differential operators on the space of $C^{+\infty}$-vectors of π which commute with the action of K on that space.

Then, we shall show that μ_π-almost everywhere in Ω that a_ℓ is an eigen vector for all the elements of $D_\pi(G)^K$ acting on a_ℓ by continuity. This also means that for every $W \in \mathcal{U}_\pi(\mathfrak{g})^{\mathfrak{k}}$ we have

$$W \cdot a_\ell := \pi(W)a_\ell = P_\ell(W)a_\ell$$

with a certain scalar $P_\ell(W)$. Our main result here is the following:

Theorem 4.3.4 *Suppose that* $\pi|_K$ *is of discrete type, with* Ω *of maximal dimension. Let* $W \in \mathcal{U}_\pi(\mathfrak{g})^{\mathfrak{k}}$. *Then,*

$$W \cdot a_\ell = P_W(\ell)a_\ell, \tag{4.3.9}$$

for some K*-invariant function* $\ell \mapsto P_W(\ell) \in \mathbb{C}$, $\ell \in \Omega$ *being generic. In particular, the algebra* $D_\pi(G)^K$ *is commutative.*

Proof 4.9 *The proof basically uses the following proposition, which we state as follows:*

Proposition 4.3.5 *Suppose that $\pi|_K$ is of discrete type. Let $W \in \mathcal{U}(\mathfrak{g})$. If $W{\cdot}a_\ell = 0$ almost everywhere on Ω, then $W \in \ker\pi$.*

Proof 4.10 *Let us resume all the steps of Theorem 4.3.3. We then arrive to the case where we deal with a minimal non-central ideal $\mathfrak{a}$ of dimension 2. The case where $\dim\mathfrak{a} = 3$ is handled similarly.*

Case 1. Let us first suppose that $\mathfrak{b}[\ell|_\mathfrak{k}] \subset \mathfrak{g}' = \mathfrak{a}^\ell$ almost everywhere on Ω. Then we find $W \in \mathcal{U}(\mathfrak{g}')$. Indeed, suppose for example $\mathfrak{a} = \mathbb{R}Y \oplus \mathfrak{z}$. We immediately see that

$$Y{\cdot}a_\ell = i\ell(Y)a_\ell \tag{4.3.10}$$

and $Y{\cdot}(W{\cdot}a_\ell) = i\ell(Y)W{\cdot}a_\ell$. Hence, $[W,Y]{\cdot}a_\ell = 0$. This taken into account, it suffices to consider $W = uX + v$ with $u, v \in \mathcal{U}(\mathfrak{g}')$ verifying $u{\cdot}a_\ell \neq 0$ and $X \notin \mathfrak{g}'$. But,

$$[W,Y]{\cdot}a_\ell = u{\cdot}[X,Y]{\cdot}a_\ell = i\ell([X,Y])u{\cdot}a_\ell \neq 0,$$

which is absurd. Up to this step, $W{\cdot}a'_{\ell'} = 0$ for the Penney distribution $a'_{\ell'}, \ell' = \ell|_{\mathfrak{g}'}$. From the induction hypothesis,

$$W \in \ker\pi'_{\ell'}, \pi'_{\ell'} = ind_{B[\ell']}^{G'}\chi_{\ell'}, \tag{4.3.11}$$

where $G' = \exp(\mathfrak{g}')$.

We now argue similarly as in the proof of Proposition 3.2.5, to arrive to the fact that $W \in \ker\pi$ as equation (4.3.11) is taken for generic $\ell \in \Omega$.

Case 2. Now suppose that $\mathfrak{b}[\ell|_\mathfrak{k}] \not\subset \mathfrak{g}'$ almost everywhere on Ω. Let for example $W = uX + v, u, v \in \mathcal{U}(\mathfrak{g}'), X \in \mathfrak{b}[\ell|_\mathfrak{k}]$. Then,

$$X{\cdot}a_\ell = (-i\ell(X) + \frac{1}{2}Tr\ ad_{\mathfrak{g}/\mathfrak{b}[\ell|_\mathfrak{k}]}X)a_\ell = P_X(\ell)a_\ell. \tag{4.3.12}$$

Hence, $[\mathfrak{k}, X]{\cdot}a_\ell = 0$ and

$$[\mathfrak{k}, W]{\cdot}a_\ell = P_X(\ell)[\mathfrak{k}, u]{\cdot}a_\ell + [\mathfrak{k}, v]{\cdot}a_\ell = 0.$$

Taking the value $\ell(X)$ over the orbit $A{\cdot}\ell$ with $A = \exp\mathfrak{a}$, we conclude that

$$[\mathfrak{k}, u]{\cdot}a_\ell = 0, \ \ [\mathfrak{k}, v]{\cdot}a_\ell = 0.$$

Thus, $[\mathfrak{k}, u]$ and $[\mathfrak{k}, v]$ belong to $\ker\pi$. Taking all into account, we can suppose that $W \in \mathcal{U}(\mathfrak{g}')$. This being done, the induction hypothesis says that $W \in \ker\pi'_{\ell'}$ for generic $\ell \in \Omega$, what gives us the result.

Taking all our observations of the proof of Proposition 4.3.5 into account, we reach the proof of Theorem 4.3.4.

4.3.5 The polynomial conjecture for discrete type restrictions

Taking the nilpotent case in mind, we ask whether the function $P_W(\ell), W \in \mathcal{U}_\pi(\mathfrak{g})^{\mathfrak{k}}$ is a polynomial function on Ω. Recall once again the situation. Let $G = \exp \mathfrak{g}$ be an exponential solvable Lie group with Lie algebra $\mathfrak{g}$, $K = \exp \mathfrak{k}$ an analytic subgroup of G and $\pi \in \widehat{G}$ such that $\pi|_K$ is of discrete type. Let Ω be the coadjoint orbit of G, which is assumed to be of maximal dimension, corresponding to π, $W \in \mathcal{U}_\pi(\mathfrak{g})^{\mathfrak{k}}$ and $W \cdot a_\ell = P_W(\ell) a_\ell$ with Penney's distributions for generic $\ell \in \Omega$. Let $\mathbb{C}[\Omega]^K$ be the algebra of K-invariant polynomial functions on Ω. We show the following:

Theorem 4.3.6 *Let G be an exponential solvable Lie group with Lie algebra $\mathfrak{g}$, $K = \exp \mathfrak{k}$ an analytic subgroup of G and $\pi \in \widehat{G}$ such that $\pi|_K$ is discrete type. Let Ω be the maximal dimension coadjoint orbit of G corresponding to π, and $W \in \mathcal{U}_\pi(\mathfrak{g})^{\mathfrak{k}}$. Then, the mapping $W \mapsto P_W(\ell)$ gives an isomorphism from the algebra $D_\pi(G)^K$ onto the algebra $\mathbb{C}[\Omega]^K$.*

Proof 4.11 *We use an induction on the dimension of G. We can suppose that the centre $\mathfrak{z}$ of $\mathfrak{g}$ is contained in $\mathfrak{k}$ and, if $\dim \mathfrak{z} \geq 2$, we as usual can pass to the quotient $\mathfrak{g}/(\mathfrak{z} \cap \ker \pi)$ to which applies the induction hypothesis. Hence $\mathfrak{z} = \{0\}$ or $\mathfrak{z} = \mathbb{R}Z$, and $\pi(Z) = i$.*

Assume also that $\mathfrak{z} \neq \{0\}$. Let $\mathfrak{a}$ be a minimal non-central ideal of $\mathfrak{g}$. Then, $\mathfrak{a} \supset \mathfrak{z}$, and we can assume as in the proof of Theorem 4.3.3 that $[\mathfrak{g}, \mathfrak{a}] = \mathfrak{a}$, according to the notation of Subsection 1.1.2. In this case, $\mathfrak{g}_0 = \{U \in \mathfrak{g}; [U, \mathfrak{a}] \subset \mathfrak{z}\}$ is a proper ideal of $\mathfrak{g}$ of codimension 1. Note also that we shall concentrate on the case where $\dim \mathfrak{a} = 2$, the arguments in the case where $\dim \mathfrak{a} = 3$ run similarly.

In the case where $\mathfrak{z} = \{0\}$, then $\mathfrak{g}_0 = \widetilde{\mathfrak{g}}$ is nothing but the centraliser of $\mathfrak{a}$ which is an ideal of $\mathfrak{g}$ of codimension one, which saturates the orbit Ω. In this case, the arguments developed next, will also be sufficient to conclude.

Let now $\ell \in \Omega$ be generic and $\widetilde{\mathfrak{g}} = \mathfrak{a}^\ell$. In the case where $\mathfrak{b}[\ell|_{\mathfrak{k}}] \subset \widetilde{\mathfrak{g}}$ for generic $\ell \in \Omega$, $\mathcal{U}_\pi(\mathfrak{g})^{\mathfrak{k}} \subset \mathcal{U}(\widetilde{\mathfrak{g}})$ modulo $\ker \pi$, (cf. Case 1 in Proposition 4.3.5). This is also derived by equation (4.3.9), because there is no term of partial derivative with respect to the coordinate outside $\widetilde{\mathfrak{g}}$. When otherwise $\mathfrak{b}[\ell|_{\mathfrak{k}}] \not\subset \widetilde{\mathfrak{g}}$ generically, $W \in \mathcal{U}_\pi(\mathfrak{g})^{\mathfrak{k}}$ modulo $\ker \pi$ belongs essentially to $\mathcal{U}(\widetilde{\mathfrak{g}})$ as in Case 2 in the proof of Proposition 4.3.5.

Independently from the fact that whether K is included in $\widetilde{G} = \exp(\widetilde{\mathfrak{g}})$ or not, let $\widetilde{\ell} = \ell|_{\widetilde{\mathfrak{g}}}, \widetilde{\omega}_\ell = \widetilde{G} \cdot \widetilde{\ell}, \widetilde{K} = K \cap \widetilde{G}$ and $\widetilde{\pi}_\ell$ the irreducible unitary representation of $\widetilde{G}$ corresponding to the coadjoint orbit $\widetilde{\omega}_\ell$. Then, the restriction $\widetilde{\pi}_\ell|_{\widetilde{K}}$ is of discrete type by the argument in Case 2.1 in the proof of Theorem 4.3.3.

We now prove the following:

Lemma 4.3.7 *The function $\widetilde{\ell} \mapsto P_W(\widetilde{\ell})$ is polynomial on $\widetilde{\omega}_\ell$.*

Proof 4.12 *If $\mathfrak{b}[\ell|_{\mathfrak{k}}] \subset \widetilde{\mathfrak{g}}$, then $W \in \mathcal{U}(\widetilde{\mathfrak{g}}) + \ker\pi$ and the induction hypothesis allows to conclude in this case. Assume next that $\mathfrak{b}[\ell|_{\mathfrak{k}}] \not\subset \widetilde{\mathfrak{g}}$. The argument in Case 2 in the proof of Proposition 4.3.5 says that W belongs essentially to $\mathcal{U}(\widetilde{\mathfrak{g}})$. In this case, let $B[\ell|_{\mathfrak{k}}] = \exp(\mathbb{R}X(\ell))B[\ell|_{\widetilde{\mathfrak{k}}}]$, $\widetilde{\mathfrak{k}} = \mathfrak{k} \cap \widetilde{\mathfrak{g}}$. We take a test function*

$$\varphi(\exp(tX(\ell))g') = \phi(t)\psi(k')$$

with $\phi \in C_0(\mathbb{R})$ and $\psi \in \mathcal{H}^{\infty}_{\widetilde{\pi}_\ell}$ having compact support modulo $B[\ell]$. Then, since $W \in \mathcal{U}(\widetilde{\mathfrak{g}})$ is K-invariant,

$$\begin{aligned}
\langle W\cdot a_\ell, \varphi\rangle &= \int_{\mathbb{R}} \Big(\oint_{B[\widetilde{\ell}|_{\widetilde{\mathfrak{k}}}]/(B[\widetilde{\ell}|_{\widetilde{\mathfrak{k}}}]\cap B[\ell])} \overline{\varphi(W^* \exp(tX(\ell))k')\chi_\ell(\exp(tX(\ell)k'))} \\
&\quad \times \Delta^{-1/2}_{B[\ell|_{\mathfrak{k}}],G}(\exp(tX(\ell))k')d\nu(k')\Big)dt \\
&= \int_{\mathbb{R}} \overline{\phi(t)}e^{-t\ell(X(\ell))}\Delta^{-1/2}_{B[\widetilde{\ell}|_{\widetilde{\mathfrak{k}}}],G}(\exp(tX(\ell)))dt \oint_{B[\widetilde{\ell}|_{\widetilde{\mathfrak{k}}}]/(B[\widetilde{\ell}|_{\widetilde{\mathfrak{k}}}]\cap B[\ell])} \overline{\psi(W^* \exp(k'))\chi_\ell(k} \\
&\quad \times \Delta^{-1/2}_{B[\ell|_{\mathfrak{k}}],G}(k')d\nu(k') \\
&= \langle W\cdot a_{\widetilde{\ell}}, \psi\rangle \int_{\mathbb{R}} \overline{\phi(t)}e^{-t\ell(X(\ell))}\Delta^{-1/2}_{B[\ell|_{\mathfrak{k}}],G}(\exp(tX(\ell)))dt \\
&= P_W(\widetilde{\ell})\langle a_{\widetilde{\ell}}, \psi\rangle \int_{\mathbb{R}} \phi(t)e^{-t\ell(X(\ell))}\Delta^{-1/2}_{B[\ell|_{\mathfrak{k}}],G}(\exp(tX(\ell)))dt \\
&= P_W(\widetilde{\ell})\langle a_\ell, \varphi\rangle,
\end{aligned}$$

where $a_{\widetilde{\ell}}$ and $P_W(\widetilde{\ell})$ designate the objects corresponding to $\widetilde{\pi}_\ell$. In this way, the induction hypothesis says that $P_W(\ell) = P_W(\widetilde{\ell})$ is a polynomial function on $\widetilde{\omega}_\ell$.

Suppose first that Ω is saturated with respect to $\mathfrak{g}_0$. Let $G_0 = \exp(\mathfrak{g}_0)$, and

$$\pi|_{G_0} = \int_{\mathbb{R}}^{\oplus} \pi_t dt$$

as in formula (1.5.2). In this situation, $W \in \mathcal{U}(\mathfrak{g}_0)$ modulo $\ker\pi$, and $\pi_t|_{K_0}$ is of discrete type as similarly in Lemma 4.3.2 for almost all $t \in \mathbb{R}$. Here as above, $K_0 = \exp(\mathfrak{k}_0)$, and $\mathfrak{k}_0 = \mathfrak{k} \cap \mathfrak{g}_0$. Therefore the induction hypothesis says that the function P_W is a polynomial function on Ω_t, the corresponding orbit. Hence for an adequate base point of Ω, the function P_W is a polynomial function on Ω_0.

In the case where $\mathfrak{b}[\ell|_{\mathfrak{k}}] \subset \widetilde{\mathfrak{g}}$, $\widetilde{\ell} \mapsto P_W(\widetilde{\ell})$ is polynomial on $\widetilde{\omega}_\ell$ by Lemma 4.3.7. As it is polynomial on Ω_0 too, P_W is a polynomial function on Ω. Otherwise, using the equation (4.3.12) and the K-invariance, we can also extend P_W to a polynomial function on Ω. These facts together allow us to conclude that the function P_W is a polynomial function on Ω.

Suppose now that Ω is not saturated with respect to $\mathfrak{g}_0$. Let $G_0 = \exp(\mathfrak{g}_0)$. Then, there exists in $\ker\pi$ an element described in Proposition 3.2.5, which means that $\mathcal{U}_\pi(\mathfrak{g})^{\mathfrak{k}} \subset \mathcal{U}(\mathfrak{g}_0)$ modulo $\ker\pi$. Let $\Omega_0 = \Omega|_{\mathfrak{g}_0}$ and $\pi_0 = \pi|_{G_0}$. Then, π_0 is the irreducible unitary representation of G_0 corresponding to the coadjoint orbit Ω_0.

If $\mathfrak{b}[\ell|_{\mathfrak{k}}] \subset \mathfrak{g}_0$ *generically, then* $\pi_0|_{K_0}$ *is found to be of discrete type (cf. Case 2 in the proof of Theorem 4.3.3), and the induction hypothesis says that the function* $P_W(\ell)$ *is a polynomial function on* Ω_0 *and so on* Ω *too. Otherwise, we can assume that* $\mathfrak{b}[\ell|_{\mathfrak{k}}] \not\subset \widetilde{\mathfrak{g}}$ *generically. In this case, the* K*-invariance technique allows us to conclude.*

Now let $Q \in \mathbb{C}[\Omega]^K$ *and let us show the existence of* $W \in \mathcal{U}_\pi(\mathfrak{g})^{\mathfrak{k}}$ *satisfying* $Q = P_W$*. Suppose for instance* $\dim \mathfrak{a} = 2$*. First consider the situation where* Ω *is non-saturated with respect to* $\mathfrak{g}_0$*. Suppose* $\mathfrak{k} \not\subset \mathfrak{g}_0$ *and write* $\mathfrak{k} = \mathbb{R}T + \mathfrak{k}_0$ *with* $\mathfrak{k}_0 = \mathfrak{k} \cap \mathfrak{g}_0$*. We first verify that* $Q \in S(\widetilde{\mathfrak{g}})$*. Indeed, using Theorem 3.2.6 we cancel all the terms depending upon* T *from* Q *to get* $\widetilde{Q} \in S(\mathfrak{g}_0)$*. Remark here that the part* A *in* $W \in \ker\pi$ *of Remark 3.2.7 is taken in* $\mathcal{U}(\mathfrak{n})$ *with* $\mathfrak{n} = [\mathfrak{g}, \mathfrak{g}]$*, what is seen by applying the induction hypothesis to* $\widetilde{W} \in \mathcal{U}(\widetilde{\mathfrak{g}})$*. Let for example* $\widetilde{Q} = pX + q$ *with* $p, q \in S(\widetilde{\mathfrak{g}})$*. Then, from* $[T, \widetilde{Q}] = 0$*, we need* $[T, p] = p$*. This means*

$$[T, p] = -[YX + A, p] = p,$$

namely $p = 0$*. By induction hypothesis applied to* $\widetilde{\mathfrak{g}}$*, there exists* $\widetilde{V} \in \mathcal{U}_{\widetilde{\pi}_\ell}(\widetilde{\mathfrak{g}})^{\widetilde{\mathfrak{k}}}$ *with* $\widetilde{\mathfrak{k}} = \mathfrak{k} \cap \widetilde{\mathfrak{g}}$ *such that* $\widetilde{V} \cdot \widetilde{a}_\ell = \widetilde{Q}(\ell)\widetilde{a}_\ell$ *with Penney's distribution* $\widetilde{a}_\ell$ *for* $\widetilde{\pi}_\ell$*. Since* $\widetilde{Q}$ *is* K*-invariant,* $\widetilde{V}$ *is* $\mathfrak{k}$*-invariant too.*

Suppose now $\mathfrak{k} \subset \mathfrak{g}_0$*. Applying the induction hypothesis to* $(\widetilde{Q}, \pi_0)$ *with* $\pi_0 = \pi|_{G_0}$*, we immediately find the desired result.*

Finally, suppose that Ω *is saturated with respect to* $\mathfrak{g}_0$*. By the duality between* $\mathfrak{a}/\mathfrak{z}$ *and* $\mathfrak{g}/\mathfrak{a}^f$ *with generic* $f \in \Omega$*, there appears* $\mathfrak{k}$*-part at least one of them as* $\pi|_K$ *is of discrete type. Hence any* $Q \in \mathbb{C}[\Omega]^K$ *belongs to* $S(\mathfrak{a}^f)$ *if* $\mathfrak{k} \not\subset \mathfrak{g}_0$ *and we can apply the induction hypothesis.*

We are now in the case where $\mathfrak{k} \subset \mathfrak{g}_0$ *and* $f \in \Omega$ *generic. If* $\mathfrak{b}[f|_{\mathfrak{k}}] \not\subset \widetilde{\mathfrak{g}} = \mathfrak{a}^f$*, write* $\mathfrak{b}[f|_{\mathfrak{k}}] = \mathbb{R}X(f) + \mathfrak{b}[f|_{\mathfrak{k}}] \cap \widetilde{\mathfrak{g}}$*. When* $Q \notin S(\widetilde{\mathfrak{g}})$*, let for example* $Q = \beta X(\ell) + \gamma$ *with* $\beta, \gamma \in S(\widetilde{\mathfrak{g}})$*. Then, on the affine space* $A \cdot f = f + (\widetilde{\mathfrak{g}})^\perp$*,* $A = \exp \mathfrak{a}$*,* β, γ *are* K*-invariant. Extending them on* Ω *by using the* K*-invariance as developed in Case 2 of Proposition 4.3.5, we find that* Q *belongs essentially to* $S(\widetilde{\mathfrak{g}})$*. Then the induction hypothesis and formula (4.3.12) allow to conclude*

Now at last we suppose $\mathfrak{b}[f|_{\mathfrak{k}}] \subset \widetilde{\mathfrak{g}}$*. Since* $\pi|_K$ *is of discrete type,* $\mathfrak{a} \subset \mathfrak{b}[f|_{\mathfrak{k}}] + \mathfrak{g}(f)$ *and hence* $Q \in S(\widetilde{\mathfrak{g}})$*, what immediately gives us the desired result.*

4.3.6 Examples

Let us take back the following two examples discussed in Subsection 4.2.3.

Example 4.3.4 Let $G_4 = \exp(\mathfrak{g}_4), \mathfrak{g}_4 = \langle T, X, Y, Z \rangle_{\mathbb{R}}$:

$$[T, X] = -X, [T, Y] = Y, [X, Y] = Z.$$

Let $f(\alpha, \beta) = \alpha T^* + \beta Z^* \in \mathfrak{g}_4^*$, $O(\alpha, \beta) = G \cdot f(\alpha, \beta)$ and $\pi(\alpha, \beta) \in \widehat{G}$ corresponds to $O(\alpha, \beta)$ for $\beta \neq 0$. Put $\mathfrak{g}_1 = \mathfrak{z} = \mathbb{R}Z, \mathfrak{g}_2 = \mathfrak{a} = \mathfrak{z} \oplus \mathbb{R}Y$, and $\mathfrak{g}_3 = \mathfrak{a} + \mathbb{R}X$. Then,

$$\mathcal{S} : \{0\} \subset \mathfrak{g}_1 \subset \mathfrak{g}_2 \subset \mathfrak{g}_3 \subset \mathfrak{g}$$

is a Jordan-Hölder sequence of $\mathfrak{g}$ and $\mathfrak{b} = \mathbb{R}T + \mathfrak{a} = \mathbb{R}\text{-span}\{Z, Y, T\}$ is the Vergne polarisation at $f(\alpha,\beta)$ associated with $\mathcal{S}$. The Hilbert space $\mathcal{H}_{\pi(\alpha,\beta)}$ of $\pi(\alpha,\beta)$ is identified with $L^2(\mathbb{R})$ by the correspondence

$$\mathcal{H}_{\pi(\alpha,\beta)} \ni \varphi \leftrightarrow \Phi \in L^2(\mathbb{R}),\ \Phi(x) = \varphi(\exp(xX))(x \in \mathbb{R}).$$

Write simply π instead of $\pi(\alpha,\beta)$. This being done,

$$\begin{aligned}
(\pi(T)\Phi)(x) &= \frac{d}{dt}\varphi(\exp(-tT)\exp(xX))\big|_{t=0} \\
&= \frac{d}{dt}\varphi(\exp(xe^tX)\exp(-tT))\big|_{t=0} = \frac{d}{dt}\Phi(xe^t)e^{t/2}e^{i\alpha t}\big|_{t=0} \\
&= x\frac{d\Phi(x)}{dx} + (\frac{1}{2} + i\alpha)\Phi(x) \\
(\pi(X)\Phi)(x) &= \frac{d}{dt}\varphi(\exp(-tX)\exp(xX))\big|_{t=0} = -\frac{d\Phi(x)}{dx} \\
(\pi(Y)\Phi)(x) &= \frac{d}{dx}\varphi(\exp(-tY)\exp(xX))\big|_{t=0} = \frac{d}{dt}\varphi(\exp(xX)\exp(txZ))\big|_{t=0} \\
&= \frac{d}{dt}e^{-i\beta tx}\Phi(x)\big|_{t=0} = -i\beta x\Phi(x) \\
(\pi(Z)\Phi)(x) &= i\beta\Phi(x).
\end{aligned}$$

Hence,

$$W = \beta T + iYX - (\alpha - \frac{i}{2})Z \in \ker\pi,$$

confirming Theorem 3.2.6.

Let $\sigma : S(\mathfrak{g}) \to \mathcal{U}(\mathfrak{g})$ be the symmetrisation map and $f = f(\alpha,\beta)$. We find

$$\sigma^{-1}(W) = \beta T + iXY - \alpha Z.$$

Now, for

$$g = \exp(xX)\exp(yY)\exp(tT) \in G\ (x, y, t \in \mathbb{R}),$$

we get

$$g{\cdot}f = (\alpha - \beta xy)T^* + \beta yX^* - \beta xY^* + \beta Z^*.$$

Hence we confirm Theorem 3.2.7. Namely, $\sigma^{-1}(W)|_\Omega \equiv 0$.

Let now $\mathfrak{k} = \mathbb{R}(T + X)$. Then, the restriction $\pi(\alpha,\beta)|_K$ is discrete type. As

$$\exp(t(T + X)) = \exp((1 - e^{-t})X)\exp(tT)\ (t \in \mathbb{R}),$$

we see, for a test function of the space of $\pi(\alpha,\beta)$,

$$\begin{aligned}
\langle a_\ell, \varphi\rangle &= \int_{\mathbb{R}} \overline{\varphi(\exp((1 - e^{-t})X)\exp(tT))\chi_\ell(\exp(t(T + X)))} \\
&\quad \times \Delta^{-1/2}_{B[\ell|_{\mathfrak{k}}],G}(\exp(t(T + X)))dt \\
&= \int_{\mathbb{R}} \overline{\varphi(\exp((1 - e^{-t})X))}e^{-t/2}dt.
\end{aligned}$$

Therefore, if φ is continuous with compact support modulo $B[\ell]$, this integral is convergent because of the existence of the factor $e^{-t/2}$ when $t \to +\infty$, as in Theorem 4.3.3.

Example 4.3.5 Let $G_6 = \exp(\mathfrak{g}_6), \mathfrak{g}_6 = \langle T, X_1, X_2, Y_1, Y_2, Z\rangle_{\mathbb{R}} : [X_i, Y_j] = \delta_{ij} Z (1 \leq i, j \leq 2), [T, X_1] = -X_1 - \alpha X_2, [T, X_2] = -X_2 + \alpha X_1, [T, Y_1] = Y_1 - \alpha Y_2, [T, Y_2] = Y_2 + \alpha Y_1 (0 \neq \alpha \in \mathbb{R})$. Put $f = \beta T^* + \gamma Z^* \in \mathfrak{g}_6^* (\gamma \neq 0)$, $\Omega = G{\cdot}f \subset \mathfrak{g}^*$ and denote by $\pi \in \widehat{G}$ the irreducible unitary representation of G corresponding to Ω. Let $\mathfrak{g}_1 = \mathfrak{z} = \mathbb{R}Z, \mathfrak{g}_2 = \mathfrak{z} \oplus \mathbb{R}Y_2, \mathfrak{g}_3 = \mathfrak{a} = \mathfrak{g}_2 \oplus \mathbb{R}Y_1, \mathfrak{g}_4 = \mathfrak{g}_3 + \mathbb{R}X_2, \mathfrak{g}_5 = \mathfrak{g}_4 + \mathbb{R}X_1$. Then,

$$\mathcal{S} : \{0\} \subset \mathfrak{g}_1 \subset \mathfrak{g}_2 \subset \mathfrak{g}_3 \subset \mathfrak{g}_4 \subset \mathfrak{g}_5 \subset \mathfrak{g}$$

is a good sequence of subalgebras of $\mathfrak{g}$ and $\mathfrak{b} = \mathfrak{a} + \mathbb{R}T$ is the Vergne polarisation of $\mathfrak{g}$ at f associated with $\mathcal{S}$. The Hilbert space $\mathcal{H}_\pi$ of π is identified with $L^2(\mathbb{R}^2)$ by the correspondence

$$\mathcal{H}_\pi \ni \varphi \leftrightarrow \Phi \in L^2(\mathbb{R}^2),\ \Phi(x_1, x_2) = \varphi(\exp(x_1 X_1 + x_2 X_2))(x_1, x_2 \in \mathbb{R}).$$

This being done, the following computations are straight:

$$\begin{aligned}
(\pi(T)\Phi)(x_1, x_2) &= \frac{d}{dt}\varphi(\exp(-tT)\exp(x_1X_1 + x_2X_2))\big|_{t=0} \\
=&\frac{d}{dt}\varphi(\exp(e^t(x_1\cos(\alpha t) - x_2\sin(\alpha t)))X_1 \\
&+ e^t(x_1\sin(\alpha t) + x_2\cos(\alpha t))X_2)e^{(1+i\beta)t}\big|_{t=0} \\
=&(x_1 - \alpha x_2)\frac{\partial\Phi(x_1, x_2)}{\partial x_1} + (x_2 + \alpha x_1)\frac{\partial\phi(x_1, x_2)}{\partial x_2} + (1 + i\beta)\Phi(x_1, x_2)
\end{aligned}$$

On the other hand,

$$\pi(X_1)\Phi = -\frac{\partial\Phi}{\partial x_1}$$

and

$$\pi(X_2)\Phi = -\frac{\partial\Phi}{\partial x_2}.$$

.

Further:

$$\begin{aligned}
(\pi(Y_1)\Phi)(x_1, x_2) &= \frac{d}{dt}\varphi(\exp(-tY_1)\exp(x_1X_1 + x_2X_2))\big|_{t=0} \\
&= \frac{d}{dt}\varphi(\exp(x_1X_1 + x_2X_2)\exp(-tY_1)\exp(tx_1Z))\big|_{t=0} \\
&= \frac{d}{dt}e^{-i\gamma x_1 t}\Phi(x_1, x_2)\big|_{t=0} = -i\gamma x_1\Phi(x_1, x_2).
\end{aligned}$$

Likewise, $(\pi(Y_2)\Phi)(x_1,x_2) = -i\gamma x_2\Phi(x_1,x_2)$ and $\pi(Z)\Phi = i\gamma\Phi$. Thus, we find

$$W = \gamma T + i(Y_1 - \alpha Y_2)X_1 + i(Y_2 + \alpha Y_1)X_2 + (i - \beta)Z \in \ker \pi,$$

all as in Theorem 3.2.6.

Let $\sigma : S(\mathfrak{g}) \to \mathcal{U}(\mathfrak{g})$ be the symmetrisation map. We find

$$\sigma^{-1}(W) = \gamma T + iX_1Y_1 - i\alpha X_1Y_2 + iX_2Y_2 + i\alpha X_2Y_1 - \beta Z.$$

Now, for

$$g = \exp(x_1X_1)\exp(x_2X_2)\exp(y_1Y_1)\exp(y_2Y_2)\exp(tT) \in G \ (x_1, x_2, y_1, y_2, t \in \mathbb{R}),$$

we get

$$\begin{aligned} g{\cdot}f =& [\beta + \{y_1(\alpha x_2 - x_1) - y_2(\alpha x_1 + x_2)\}\gamma]T^* \\ &+ \gamma y_1 X_1^* + \gamma y_2 X_2^* - \gamma x_1 Y_1^* - \gamma x_2 Y_2^* + \gamma Z^*. \end{aligned}$$

Hence we confirm Theorem 3.2.7. Namely, $\sigma^{-1}(W)|_\Omega \equiv 0$.

We consider in this situation $\mathfrak{k} = \mathbb{R}(T + X)$ with $X \in \mathfrak{g}\backslash\mathfrak{a}^f$ verifying $X \in \ker f$ and $[X, \mathfrak{a}] \subset \mathfrak{z}$. Evidently, the restriction $\pi|_K$ is not of discrete type but nevertheless let us consider our distribution a_f:

$$\langle a_f, \varphi\rangle = \int_{\mathbb{R}} \overline{\varphi(\exp(t(T+X)))\chi_f(\exp(t(T+X)))}\Delta_{K,G}^{-1/2}(\exp(t(T+X)))dt.$$

Here, it matters of the complex number $\lambda = -(1 + i\alpha), i = \sqrt{-1}, \alpha \neq 0$. The factor corresponding to $1 - e^{\lambda t}$ becomes

$$\begin{aligned} &\begin{pmatrix} 1 & 0 \\ 0 & 1 \end{pmatrix} - e^{-t}\begin{pmatrix} \cos(\alpha t) & -sin(\alpha t) \\ sin(\alpha t) & \cos(\alpha t) \end{pmatrix} \\ =& \begin{pmatrix} 1 - e^{-t}\cos(\alpha t) & e^{-t}\sin(\alpha t) \\ -e^{-t}\sin(\alpha t) & 1 - e^{-t}\cos(\alpha t) \end{pmatrix}. \end{aligned}$$

This wants to say that, if $X = X_1$ for example, the distribution a_f becomes

$$\langle a_f, \varphi\rangle = \int_{\mathbb{R}} \overline{\varphi(\exp((1 - e^{-t}\cos(\alpha t))X_1 - e^{-t}\sin(\alpha t)X_2))}e^{-t}dt.$$

Hence everything as in the previous example, this integral is convergent if φ is continuous with compact support modulo $B[f]$, where $B[f] = \exp(\mathfrak{b}[f])$ with $\mathfrak{b}[f] = \mathbb{R}T + \mathfrak{a}$, as treated in Theorem 4.3.3.

This also provides a negative solution to the analogue question of Duflo's problem [40] in the case of induction (cf. [12] and ([16], Example 7)), stating that there is an equivalence between the fact that $D_\pi(G)^K$ is commutative and the fact that the subspace $\mathfrak{g}(\ell) + \mathfrak{b}[\ell\|_{\mathfrak{k}}]$ is generically Lagrangian on Ω (i.e. $\pi|_K$ is of discrete type).

4.4 DUFLO'S POLYNOMIAL CONJECTURES FOR DISCRETE TYPE REPRESENTATIONS

4.4.1 The monomial setting

Now we consider the exponential case where our monomial representation $\tau = \mathrm{ind}_H^G \chi_f$, supposed to be of discrete type (cf. Section 4.2).
Let again C_τ be the centre of $D_\tau(G/H)$, and consider the symmetric algebra $S(\mathfrak{g})$ of $\mathfrak{g}_\mathbb{C}$. As earlier, we regard an element $X \in \mathfrak{g}$ as a polynomial function on $\mathfrak{g}^*$ by the formula

$$\mathfrak{g}^* \ni \ell \mapsto i\ell(X)\ (i = \sqrt{-1}).$$

Let $\mathfrak{b}_\tau$ be the vector subspace of $S(\mathfrak{g})$ spanned by $Y - if(Y), Y \in \mathfrak{h}$. Then, the algebra $\mathbb{C}[\Gamma_\tau]^H$ of the H-invariant polynomial functions on Γ_τ is identified to the set $(S(\mathfrak{g})/S(\mathfrak{g})\mathfrak{b}_\tau)^H$ of the H-invariant elements which inherits the Poisson structure $\{\cdot,\cdot\}$ of $S(\mathfrak{g})$ defined by $\{X,Y\} = [X,Y]$ for $X, Y \in \mathfrak{g}$. We keep the notation Z_τ to designate its centre. We first have the following result:

Theorem 4.4.1 *If $\tau = ind_H^G \chi_f$ is discrete type, the Poisson algebra $\mathbb{C}[\Gamma_\tau]^H$ of the H-invariant polynomial functions on Γ_τ is commutative.*

Proof 4.13 *If $\mathfrak{h}$ contains a non-trivial ideal $\mathfrak{a}$ of $\mathfrak{g}$ on which f vanishes, we can pass to the quotient $\mathfrak{g}/\mathfrak{a}$ to which applies the induction hypothesis. Suppose hereafter that there is no such an ideal $\mathfrak{a}$. Let $\mathfrak{a}$ be a minimal non-central ideal of $\mathfrak{g}$. For generic $\ell \in \Gamma_\tau$, we put*

$$\widetilde{\mathfrak{g}} = \mathfrak{a}^\ell = \{X \in \mathfrak{g}; \ell([X,\mathfrak{a}]) = \{0\}\}$$

and $\widetilde{G} = \exp(\widetilde{\mathfrak{g}})$. Then, from our hypothesis $\widetilde{\mathfrak{g}}$ is a proper subalgebra of $\mathfrak{g}$. Let $A = \exp\mathfrak{a}$. Then,

$$A \cdot \ell = \ell + (\widetilde{\mathfrak{g}})^\perp \subset \mathfrak{g}^*.$$

When $\mathfrak{h} \subset \widetilde{\mathfrak{g}}$, $\mathfrak{a} \subset \mathfrak{h} + \mathfrak{g}(\ell)$ as τ is discrete type. Hence, setting $\widetilde{\tau} = ind_H^{\widetilde{G}} \chi_f$ which is discrete type and $\Gamma_{\widetilde{\tau}} = \Gamma_\tau|_{\widetilde{\mathfrak{g}}}$. Since

$$\mathbb{C}[\Gamma_\tau]^H = (S(\mathfrak{g})/S(\mathfrak{g})\mathfrak{b}_\tau)^H = (S(\widetilde{\mathfrak{g}})/S(\widetilde{\mathfrak{g}})\mathfrak{b}_\tau)^H = \mathbb{C}[\Gamma_{\widetilde{\tau}}]^H$$

which is Poisson commutative by induction hypothesis.

If $\mathfrak{h} \not\subset \widetilde{\mathfrak{g}}$, we can take the representatives of $(S(\mathfrak{g})/S(\mathfrak{g})\mathfrak{b}_\tau)^H$ in $S(\widetilde{\mathfrak{g}})$. We put $\widetilde{\mathfrak{h}} = \mathfrak{h} \cap \widetilde{\mathfrak{g}}$, $\widetilde{H} = \exp(\widetilde{\mathfrak{h}})$ and $\widetilde{\tau} = ind_{\widetilde{H}}^{\widetilde{G}} \chi_f$. Then, $\widetilde{\tau}$ is discrete type and therefore $(S(\widetilde{\mathfrak{g}})/S(\widetilde{\mathfrak{g}})\mathfrak{b}_{\widetilde{\tau}})^{\widetilde{H}}$, where $\mathfrak{b}_{\widetilde{\tau}} = \{Y - if(Y); Y \in (\widetilde{\mathfrak{h}})_\mathbb{C}\}$, is Poisson commutative by induction hypothesis. Since

$$\mathbb{C}[\Gamma_\tau]^H = (S(\mathfrak{g})/S(\mathfrak{g})\mathfrak{b}_\tau)^H \subset (S(\widetilde{\mathfrak{g}})/S(\widetilde{\mathfrak{g}})\mathfrak{b}_{\widetilde{\tau}})^{\widetilde{H}},$$

$\mathbb{C}[\Gamma_\tau]^H$ is Poisson commutative.

We now leave two questions which go back to Penney [92]. Recall its canonical central decomposition (1.5.1)

$$\tau \simeq \int_{\widehat{G}}^{\oplus} m(\pi)\pi d\mu(\pi).$$

Put $\Delta_{H,G} = \Delta_H / \Delta_G$ and

$$\varphi_H^f(g) = \int_H \varphi(gh)\chi_f(h)\Delta_{H,G}^{-1/2}(h)dh$$

for $\varphi \in \mathcal{D}(G)$, where dh denotes a left Haar measure on G. Then, when τ has finite multiplicities, Penney's Plancherel formula for τ is written as

$$\varphi_H^f(e) = \int_{\widehat{G}} \sum_{k=1}^{m(\pi)} \langle \pi(\varphi)a_\pi^k, a_\pi^k \rangle d\mu(\pi)$$

with some $a_\pi^k \in (\mathcal{H}_\pi^{-\infty})^{H,\chi_f\Delta_{H,G}^{1/2}}$. Here are our questions.

1. Reciprocity: Does the equality $m(\pi) = \dim\left((\mathcal{H}_\pi^{-\infty})^{H,\chi_f\Delta_{H,G}^{1/2}}\right)$ hold μ-almost everywhere?

2. Concrete Plancherel formula: Write explicitly generalised vectors a_π^k coming into the Plancherel formula.

Concerning the reciprocity, we have the results [52], [13] when G is nilpotent. For the Plancherel formula, see [16]. Here in what follows we shall see some examples.

Example 4.4.2 We first retake the example 4.2.1 and use all the notations introduced there. Through the parametrisation of the orbit and putting $e^t(cos(\alpha t) - \lambda \sin(\alpha t)) = 1$, we look for the intersection points with Γ_τ. Thus,

$$e^t(1+\lambda^2)^{1/2}\cos(\alpha t + \gamma) = 1$$

with γ satisfying

$$\sin\gamma = \lambda(1+\lambda^2)^{-1/2},\ \cos\gamma = (1+\lambda^2)^{-1/2}.$$

This brings

$$e^t(\sin(\alpha t) + \lambda\cos(\alpha t)) = (1+\lambda^2)^{1/2}e^t\sin(\alpha t+\gamma) = \tan(\alpha t + \gamma).$$

From these observations, we find the points $\ell_n = (1, \tan(\alpha t_n + \gamma)), \ell_0 = \ell$. Choose arbitrarily $g_n \in G$ such that $g_n \cdot \ell = \ell_n$ and fabricate the mapping

$$a_n : \phi \mapsto \oint_{H/(H\cap g_n B g_n^{-1})} \overline{\phi(hg_n)\chi_f(h)\Delta_{H,G}^{-1/2}(h)d\nu(h)},\ \phi \in \mathcal{H}_\pi^\infty, \tag{4.4.1}$$

which is nothing but the mapping $\phi \mapsto \overline{\phi(g_n)}$, where $B = \exp\mathfrak{b}$ associated to the polarisation $\mathfrak{b} = \mathbb{R}Y_1 \oplus \mathbb{R}Y_2$ and $\pi = \pi_\ell = \mathrm{ind}_B^G \chi_\ell$. It is easy to check $a_n \in (\mathcal{H}_\pi^{-\infty})^{H,\chi_f \Delta_{H,G}^{1/2}}$. Consequently, for $\psi \in \mathcal{D}(G)$, a computation shows

$$(\pi(\psi)a_n)(g) = \Delta_G(g_n)^{-1} \int_B \psi(gbg_n)\chi_\ell(b)db(g \in G).$$

And then

$$\langle \pi(\psi)a_n, a_n \rangle = \int_{\mathbb{R}} \psi_H^f(\exp(sY_2)) \exp(is \tan(\alpha t_n + \gamma))ds.$$

The formulas (4.2.5) lead, $y(t)$ being simply written y,

$$\begin{aligned}(1 - \alpha y)(dt/d\lambda) &= e^t sin(\alpha t)\\ (dy/d\lambda) - (y + \alpha)(dt/d\lambda) &= e^t \cos(\alpha t).\end{aligned} \tag{4.4.2}$$

On the other hand, from $e^{2t} = (1 + y^2)/(1 + \lambda^2)$,

$$[(1 + \lambda^2)(dt/d\lambda) + \lambda](1 + y^2)/(1 + \lambda^2) = y(dy/d\lambda). \tag{4.4.3}$$

The equalities 4.4.2 imply

$$\begin{aligned}\lambda(dy/d\lambda) + (1 - \alpha y - \lambda y - \alpha\lambda)(dt/d\lambda) &= y,\\ (dy/d\lambda) + (\alpha\lambda y - \lambda - y - \alpha)(dt/d\lambda) &= 1,\end{aligned}$$

hence

$$(\lambda - y)(dy/d\lambda) + (1 - \alpha\lambda)(1 + y^2)(dt/d\lambda) = 0.$$

Substituting (4.4.3),

$$(\lambda - y)(dy/d\lambda) + (1 - \alpha\lambda)[y(dy/d\lambda) - \lambda(1 + y^2)/(1 + \lambda^2)] = 0,$$

namely,

$$\lambda(1 - \alpha y)\frac{dy}{d\lambda} = \frac{\lambda(1 + \alpha\lambda)(1 + y^2)}{(1 + \lambda^2)}.$$

Consequently, for $\lambda \neq 0$,

$$\frac{dy}{d\lambda} = \frac{(1 - \alpha\lambda)(1 + y^2)}{(1 - \alpha y)(1 + \lambda^2)}.$$

$y_n = y_n(\lambda)$ denoting $\tan(\alpha t_n + \gamma)$, our formula to be seen becomes to

$$\begin{aligned}\psi_H^f(e) &= \int_{\lambda_0}^{\lambda_1} \xi(\lambda)d\lambda(\sum_{n=0}^{\infty} \kappa(n) \int_{\mathbb{R}} \psi_H^f(\exp(sY_2)) \exp(is \tan(\alpha t_n + \gamma))ds)\\ &= (2\pi)^{-1/2} \int_{\lambda_0}^{\lambda_1} ((\psi_H^f)^\wedge(y_n)\kappa(n))\xi(\lambda)d\lambda\end{aligned}$$

with certain measurable functions $\xi(\lambda)$ and $\kappa(n) \geq 0$, because if necessary we multiply a_n by an appropriate scalar, and $(\psi_H^f)^\wedge$ denoting the inverse Fourier transform of $\psi_H^f \circ \exp$.

In fact, let $\xi(\lambda) = (1-\alpha\lambda)/2\pi(1+\lambda^2)$ and $\kappa(n) = (1+y_n^2)/|1-\alpha y_n|$, i.e. we take $[(1+y_n^2)/|1-\alpha y_n|]^{1/2}a_n$. Then,

$$\begin{aligned}
&(2\pi)^{-1/2}\int_{\lambda_0}^{\lambda_1}((\psi_H^f)^\wedge(y_n)\kappa(n))\xi(\lambda)d\lambda\\
=&(2\pi)^{-1/2}\sum_{n=0}^{\infty}\int_{\lambda_0}^{\lambda_1}(\psi_H^f)^\wedge(y_n)\frac{(1-\alpha\lambda)(1+y_n^2)}{|1-\alpha y_n|(1+\lambda^2)}d\lambda\\
=&(2\pi)^{-1/2}\int_{\mathbb{R}}(\psi_H^f)^\wedge(s)ds = (\psi_H^f\circ\exp)(0) = \psi_H^f(e).
\end{aligned}$$

Example 4.4.3 Let $G = G_3(\alpha) = \exp(\mathfrak{g}_3(\alpha))$ as in the previous example. This time let $\mathfrak{h} = \mathbb{R}T$ and $f \in \mathfrak{g}^*$ arbitrary. Then,

$$\Gamma_\tau = f + \mathfrak{h}^\perp = f(T)T^* + \mathbb{R}Y_1^* + \mathbb{R}Y_2^*$$

and there the generic orbits have their representatives

$$\hat{\theta} = (\cos\theta)Y_1^* + (\sin\theta)Y_2^*$$

to which associate the irreducible representations $\pi_\theta = \mathrm{ind}_B^G\chi_{\hat{\theta}}$, where $B = \exp\mathfrak{b}$ with $\mathfrak{b} = \mathbb{R}Y_1 \oplus \mathbb{R}Y_2$.

Our formula (4.4.1) gives

$$a_\theta : \phi \mapsto \int_H \overline{\phi(h)\chi_f(h)}\Delta_{H,G}^{-1/2}(h)dh.$$

It is easy to see, for $\psi \in \mathcal{D}(G)$,

$$\begin{aligned}
(\pi_\theta(\psi)a_\theta)(g) &= \int_B \psi_H^f(gb)\chi_{\hat{\theta}}(b)db,\\
\langle\pi_\theta(\psi)a_\theta, a_\theta\rangle &= \int_{\mathbb{R}} e^{2t}dt\int_B \psi_H^f(b)\chi_\ell(b)db,
\end{aligned}$$

with the notation

$$\ell = h{\cdot}\hat{\theta} = e^t(\cos(\alpha t+\theta)Y_1^* + \sin(\alpha t+\theta)Y_2^*),\ h = \exp(tT).$$

Now, we have

$$\begin{aligned}
&(2\pi)^{-2}\int_0^{2\pi}\langle\pi_\theta(\psi)a_\theta, a_\theta\rangle d\theta\\
=&(2\pi)^{-2}\int_0^{2\pi}d\theta\int_{\mathbb{R}}e^{2t}dt\int_{\mathbb{R}^2}\psi_H^f(\exp(b_1Y_1+b_2Y_2))\\
&\times\exp(ie^t(b_1\cos(\alpha t+\theta)+b_2\sin(\alpha t+\theta)))db_1db_2.
\end{aligned}$$

Applying the change of variables

$$x = e^t\cos(\alpha t+\theta),\ y = e^t\sin(\alpha t+\theta),$$

whose Jacobian is

$$\frac{\partial(x,y)}{\partial(t,\theta)} = e^{2t}, \; dxdy = e^{2t}dtd\theta,$$

we deduce

$$(2\pi)^{-2}\int_0^{2\pi}\langle\pi_\theta(\psi)a_\theta, a_\theta\rangle d\theta = (2\pi)^{-2}\int_{\mathbb{R}^2} dxdy\int_{\mathbb{R}^2}(\psi_H^f)^*(b_1,b_2)$$
$$\times\exp(i(xb_1+yb_2))db_1db_2 = (\psi_H^f)^*(0,0) = \psi_H^f(e),$$

where $(\psi_H^f)^*(b_1,b_2) = (\psi_H^f\circ\exp)(b_1Y_1+b_2Y_2)$.

Example 4.4.4 Retake the example 1.5.11: $G_4 = \exp(\mathfrak{g}_4), \mathfrak{g}_4 = \langle T,X,Y,Z\rangle_{\mathbb{R}}$: $[T,X] = X, [T,Y] = -Y, [X,Y] = Z$ Let $f(\alpha,\beta) = \alpha T^* + \beta Z^* \in \mathfrak{g}_4^*$ and $O(\alpha,\beta) = G\cdot f(\alpha,\beta)$ for $\beta \neq 0$. Set $f = f(\alpha_0,\beta_0)$.

(i) Let first $\mathfrak{h} = \mathbb{R}T + \mathbb{R}X$. Then,

$$\tau = \mathrm{ind}_H^G\chi_f \simeq \int_{\mathbb{R}}\pi(\alpha_0,\beta)d\beta,$$

where $\pi(\alpha,\beta)\in\widehat{G}$ corresponds to $O(\alpha,\beta)$. By means of $\ell = f(\alpha_0,\beta)\in\Gamma_\tau\cap O(\alpha_0,\beta)$ and the polarisation $\mathfrak{b} = \langle T,X,Z\rangle_{\mathbb{R}}$ at ℓ, we realise $\pi(\alpha_0,\beta) = \mathrm{ind}_B^G\chi_\ell$, where $B = \exp\mathfrak{b}$. In this situation, the usual way proposes $a(\beta)$ by $\langle a(\beta),\phi\rangle = \overline{\phi(e)}$, which clearly satisfies the required conditions. Since $\mathfrak{b} = \mathfrak{h} + \mathbb{R}Z$ and Z is a central element,

$$(\mathcal{H}^{-\infty}_{\pi(\alpha_0,\beta)})^{H,\chi_f\Delta_{H,G}^{1/2}} = \mathbb{C}a(\beta).$$

Now for $\psi\in\mathcal{D}(G)$, $\pi(\alpha_0,\beta)(\psi)a(\beta) = \psi_B^\ell$, i.e.

$$(\pi(\alpha_0,\beta)(\psi)a(\beta))(g) = \int_B\psi(gb)\chi_\ell(b)\Delta_{B,G}^{-1/2}(b)db$$

and consequently

$$\langle\pi(\alpha_0,\beta)(\psi)a(\beta),a(\beta)\rangle = \int_B\psi(b)\chi_\ell(b)\Delta_{B,G}^{-1/2}(b)db.$$

Finally,

$$(2\pi)^{-1}\int_{\mathbb{R}}\langle\pi(\alpha_0,\beta)(\psi)a(\beta),a(\beta)\rangle d\beta$$
$$=(2\pi)^{-1}\int_{\mathbb{R}}d\beta\int_{\mathbb{R}}\psi_H^f(\exp(wZ))e^{i\beta w}dw = \psi_H^f(e).$$

(ii) Let second $\mathfrak{h} = \mathbb{R}T\oplus\mathbb{R}Z$. Put $\ell = f(\alpha,\beta_0) = \alpha T^* + \beta_0 Z^*$. Then,

$$(f(\alpha_0,\beta_0)+\mathfrak{h}^\perp)\cap O(\alpha,\beta_0) = \alpha_0T^* + xX^* + yY^* + \beta_0Z^*,$$
$$xy = \beta_0(\alpha-\alpha_0).$$

If the value

$$\begin{aligned}(\exp(aX)\exp(bY)\cdot\ell)(T) &= (\exp(bY)\cdot\ell)(T+aX)\\ &= \ell(T+aX-bY+abZ) = \alpha+ab\beta_0\end{aligned}$$

is equal to $f(T)$, it comes $\alpha+ab\beta_0=\alpha_0$, i.e. $ab\beta_0=\alpha_0-\alpha$. Modifying the basis by convenient scalar multiplications, we can suppose $\beta_0=1$ and the equality obtained above becomes $ab=\alpha_0-\alpha$. Consequently, for $f(\alpha,\beta_0)$ such that $\alpha\neq\alpha_0$, $g_j\cdot f(\alpha,\beta_0)\in\Gamma_\tau (j=1,2)$ with

$$g_1=\exp(X)\exp((\alpha_0-\alpha)Y),\ g_2=\exp(-X)\exp((\alpha-\alpha_0)Y).$$

Similarly to the case (i), we realise the representation $\pi(\alpha,\beta_0)=\mathrm{ind}_B^G\chi_\ell, \ell=f(\alpha,\beta_0)$, using the polarisation $\mathfrak{b}=\langle T,Y,Z\rangle_{\mathbb{R}}$. For $\phi\in\mathcal{H}^\infty_{\pi(\alpha,\beta_0)}$, we recall the usual formula

$$\begin{aligned}\langle a_1(\alpha),\phi\rangle &= \oint_{H/(H\cap g_1Bg_1^{-1})}\overline{\phi(hg_1)\chi_f(h)}\Delta_{H,G}^{-1/2}(h)d\nu(h)\\ &=\int_{\mathbb{R}}\overline{\phi(\exp(tT)\exp(X))}e^{-i\alpha_0 t}dt=\int_{\mathbb{R}}\overline{\phi(\exp(e^tX))}e^{i(\alpha-\alpha_0)t+t/2}dt\\ &=\int_0^\infty\overline{\phi(\exp(sX))}s^{i(\alpha-\alpha_0)-1/2}ds.\end{aligned}$$

The last integral is well convergent and immediately

$$a_1(\alpha)\in(\mathcal{H}^{-\infty}_{\pi(\alpha,\beta_0)})^{H,\chi_f\Delta_{H,G}^{1/2}}.$$

Similarly, the formula

$$\begin{aligned}\langle a_2(\alpha),\phi\rangle &= \oint_{H/(H\cap g_2Bg_2^{-1})}\overline{\phi(hg_2)\chi_f(h)}\Delta_{H,G}^{-1/2}(h)d\nu(h)\\ &=\int_0^\infty\overline{\phi(\exp(-sX))}s^{i(\alpha-\alpha_0)-1/2}ds\end{aligned}$$

defines a non-zero element

$$a_2(\alpha)\in(\mathcal{H}^{-\infty}_{\pi(\alpha,\beta_0)})^{H,\chi_f\Delta_{H,G}^{1/2}}.$$

Let a be any element of $(\mathcal{H}^{-\infty}_{\pi(\alpha,\beta_0)})^{H,\chi_f\Delta_{H,G}^{1/2}}$. The semi-invariance of a with respect to $h=\exp(tT), t\in\mathbb{R}$, gives us

$$\langle e^{it(\alpha_0-\alpha)-1/2}a,\phi(\exp(xX))\rangle=\langle a,\phi(\exp(e^txX))\rangle.$$

This tells us

$$\langle a,\phi\rangle=c_1\int_{\mathbb{R}_+}\overline{\phi(\exp(xX))}x^{i(\alpha-\alpha_0)-1/2}dx\ (c_1:\text{constant})$$

if the support of ϕ is contained in $\mathbb{R}_+=\{s\in\mathbb{R}; s>0\}$, namely $a=c_1a_1(\alpha)$ (c_1 : constant) on $\mathbb{R}_+$. Similarly, $a=c_2a_2(\alpha)$ (c_2 : constant) on $\mathbb{R}_-=\{s\in\mathbb{R}; s<0\}$.

Now suppose that the support of a is contained in B and write $a = \sum_{j=0}^{m} \lambda_j D_j (\lambda_j \in \mathbb{C}, 0 \leq j \leq m)$, where

$$\langle D_j, \phi \rangle = \overline{\frac{d^j \hat{\phi}}{dx^j}}, \ \hat{\phi}(x) = \phi(\exp(xX)).$$

Then the semi-invariance of a for $h = \exp(tT)$ gives

$$e^{i\alpha_0 t} \sum_{j=0}^{m} \overline{\frac{d^j \hat{\phi}}{dx^j}(0)} = e^{i\alpha t} \sum_{j=0}^{m} \lambda_j \overline{\frac{d^j \hat{\phi}}{dx^j}(0)} e^{(j+1/2)t}.$$

If we choose ϕ verifying $\frac{d^j \hat{\phi}}{dx^j}(0) = \delta_{jm}$, it follows that

$$\lambda_m e^{i\alpha_0 t} = \lambda_m e^{(m+1/2)t} e^{i\alpha t} (t \in \mathbb{R}).$$

This leads $\lambda_m = 0$, what means $a = 0$. In sum,

$$(\mathcal{H}^{-\infty}_{\pi(\alpha,\beta_0)})^{H, \chi_f \Delta^{1/2}_{H,G}} = \mathbb{C} a_1(\alpha) + \mathbb{C} a_2(\alpha).$$

We pass to the concrete Plancherel formula for the monomial representation $\tau = \mathrm{ind}_H^G \chi_f, f = f(\alpha_0, \beta_0)$. To lighten the notation, $\pi(\alpha, \beta_0)$, examined in detail for the moment, will be simply written π. Let $\psi \in \mathcal{D}(G)$ and $\phi \in \mathcal{H}^{\infty}_{\pi}$ with compact support modulo B. We calculate: writing $a_j(\alpha)(j = 1, 2)$ simply a_j,

$$\begin{aligned} \langle \pi(\psi) a_1, \phi \rangle = & \oint_{H/(H \cap g_1 B g_1^{-1})} d\nu(h) \oint_{G/B} d\nu(g) \int_B \psi(g b g_1^{-1} h^{-1}) \\ & \times \overline{\phi(g)} \Delta_{B,G}^{-1/2}(h) \chi_\ell(b) \overline{\chi_f(h)} \Delta_G^{-1}(h) \Delta_{H,G}^{-1/2}(h) db. \end{aligned} \tag{4.4.4}$$

We shall see in the sequel that the order of the first two integrals in the right member may be interchanged. First of all $\Delta_G(h) = \Delta_{H,G}(h) = 1$ and in the expression (4.4.4) $g = \exp(xX)$, x running in a certain finite interval J. We denote by $\Xi(h, g, b)$ the integrand in (4.4.4) and write

$$h = \exp(tT), \ b = \exp(sT) \exp(yY) \exp(wZ).$$

Then,

$$\begin{aligned} & \int_B \Xi(h, g, b) db = \overline{\hat{\phi}(x)} \int_{\mathbb{R}^3} \psi(\exp(xX) \exp(yY) \exp(wZ) \exp((\alpha - \alpha_0)Y) \\ & \times \exp(-X) \exp(-tT)) e^{-s/2} e^{i(w + \alpha s - \alpha_0)} ds dy dw. \end{aligned}$$

We find

$$\begin{aligned} & \exp(xX) \exp(sT) \exp(yY) \exp(wZ) \exp((\alpha - \alpha_0)Y) \exp(-X) \exp(-tT) \\ = & \exp((w + e^{-s} x (y + \alpha - \alpha_0))Z) \\ & \times \exp(e^{-s}(y + \alpha - \alpha_0)Y) \exp((x - e^s)X) \exp((s - t)T). \end{aligned}$$

Taking this into account,

$$\begin{aligned} |\int_B \Xi(h,g,b)db| \leq &|\hat{\phi}(x)| \int_{\mathbb{R}^3} |\psi(\exp(wZ)\exp(e^{-s}yY) \\ &\times \exp((x-e^s)X)\exp((s-t)T))|e^{-s/2}dsdydw \\ =&|\hat{\phi}(x)| \int_{\mathbb{R}^3} |\psi(\exp(wZ)\exp(yY) \\ &\times \exp((x-e^{s+t})X)\exp(sT))|e^{(s+t)/2}dsdydw. \end{aligned} \tag{4.4.5}$$

Because x runs in the finite interval J, (4.4.5) is integrable with respect to $dtdx$ and we can interchange the order of the first two integrals in (4.4.4).

We arrive in this way to

$$\begin{aligned} \langle \pi(\psi)a_1, \phi\rangle = &\oint_{G/B} d\nu(g) \oint_{H/(H\cap g_1 B g_1^{-1})} d\nu(h) \int_B \psi(gbg_1^{-1}h^{-1}) \\ &\times \overline{\phi(g)} \Delta_{B,G}^{-1/2}(b)\chi_\ell(b)\overline{\chi_f(h)}\Delta_G^{-1}(h)\Delta_{H,G}^{-1/2}(h)db. \end{aligned}$$

Hence,

$$\begin{aligned} (\pi(\psi)a_1)(g) = &\oint_{H/(H\cap g_1 B g_1^{-1})} d\nu(h) \int_B \psi(gbg_1^{-1}h^{-1}) \\ &\times \Delta_{B,G}^{-1/2}(b)\chi_\ell(b)\overline{\chi_f(h)}\Delta_G^{-1}(h)\Delta_{H,G}^{-1/2}(h)db \\ = &\oint_{H/(H\cap g_1 B g_1^{-1})} d\nu(h) \oint_{B/(B\cap g_1^{-1} H g_1)} \overline{\chi_f(h)}\chi_\ell(b)\Delta_G^{-1}(h)\Delta_{H,G}^{-1/2}(b)d\nu(b) \\ &\times \int_{B\cap g_1^{-1}Hg_1} \psi(gbb_0^{-1}g_1^{-1}h^{-1})\Delta_{B,G}^{1/2}(b_0)\overline{\chi_\ell(b_0)} \\ &\times \Delta_{B\cap g_1^{-1}Hg_1,B}(b_0)\Delta_{B\cap g_1^{-1}Hg_1}(b_0)db_0. \end{aligned}$$

Arguments analogous to those employed above confirm that the first two integrals may be interchanged. Finally,

$$\begin{aligned} (\pi(\psi)a_1)(g) = &\oint_{B/(B\cap g_1^{-1} H g_1)} \chi_\ell(b)\Delta_{B,G}^{-1/2}(b)d\nu(b) \\ &\times \oint_{H/(H\cap g_1 B g_1^{-1})} \overline{\chi_f(h)}\Delta_G^{-1}(h)\Delta_{H,G}^{-1/2}(h)d\nu(h) \\ &\times \int_{B\cap g_1^{-1}Hg_1} \psi(gbb_0^{-1}g_1^{-1}h^{-1})\overline{\chi_\ell(b_0)} \\ &\times\Delta_{B,G}^{1/2}(b_0)\Delta_{B\cap g_1^{-1}Hg_1,B}(b_0)\Delta_{B\cap g_1^{-1}Hg_1}^{-1}(b_0)db_0, \end{aligned}$$

whose last integral is equal to

$$\begin{aligned} &\int_{H\cap g_1 B g_1^{-1}} \psi(gbg_1^{-1}b'^{-1}h^{-1})\chi_{g_1\cdot\ell}(b') \\ &\times\Delta_{g_1 B g_1^{-1},G}^{1/2}(b')\Delta_{g_1 B g_1^{-1}\cap H, g_1 B g_1}(b')\Delta_{g_1 B g_1^{-1}\cap H}^{-1}(b')db'. \end{aligned}$$

As

$$\Delta_H^{1/2}(b')\Delta_{g_1Bg_1^{-1}\cap H}(b') = \Delta_{g_1^{-1}Bg_1}(b')(b' \in g_1Bg_1^{-1} \cap H),$$

we have

$$\begin{aligned}&\Delta^{1/2}_{g_1Bg_1^{-1},G}(b')\Delta_{g_1Bg_1^{-1}\cap H, g_1Bg_1^{-1}}(b')\Delta^{-1}_{g_1Bg_1^{-1}\cap H}(b')\\ =&\Delta_H^{-1}(b')\Delta_{H,G}^{1/2}(b')\Delta^{-1}_{g_1Bg_1^{-1}\cap H,H}(b').\end{aligned}$$

We at last come to

$$\begin{aligned}(\pi(\psi)a_1)(g) =& \oint_{B/(B\cap g_1^{-1}Hg_1)} \Delta_{B,G}^{-1/2}(b)\chi_\ell(b)d\nu(b)\\ &\times \int_H \psi(gbg_1^{-1}h)\chi_f(h)\Delta_{H,G}^{-1/2}(h)dh\\ =& \oint_{B/(B\cap g_1^{-1}Hg_1)} \psi_H^f(gbg_1^{-1})\chi_\ell(b)\Delta_{B,G}^{-1/2}(b)d\nu(b).\end{aligned}$$

By the same manner,

$$(\pi(\psi)a_2)(g) = \oint_{B/(B\cap g_2^{-1}Hg_2)} \psi_H^f(gbg_2^{-1})\chi_\ell(b)\Delta_{B,G}^{-1/2}(b)d\nu(b).$$

From what we have seen until now,

$$\begin{aligned}\langle\pi(\psi)a_1, a_1\rangle =& \oint_{H/(H\cap g_1Bg_1^{-1})} \chi_\ell(h)\Delta_{H,G}^{-1/2}(h)d\nu(h)\\ \times& \oint_{B/(B\cap g_1^{-1}Hg_1)} \psi_H^f(hg_1bg_1^{-1})\chi_\ell(b)\Delta_{B,G}^{-1/2}(b)d\nu(b)\\ =& \int_{\mathbb{R}} e^{it(\alpha_0-\alpha)+t/2}dt\\ \times& \int_{\mathbb{R}} \psi_H^f(\exp(e^tX)\exp(xT)\exp(yY)\exp(-X))e^{i\alpha x-x/2}dxdy\\ =& \int_{\mathbb{R}} e^{i(t-x)(\alpha_0-\alpha)+(t+x)/2}dt\\ \times& \int_{\mathbb{R}^2} \psi_H^f(\exp((e^t-e^x)X))\exp(yY)e^{-iye^x}dxdy.\end{aligned}$$

Making the change of variables $t \mapsto t+x, e^x = s$, we obtain

$$\begin{aligned}&\langle\pi(\psi)a_1, a_1\rangle\\ =& \int_{\mathbb{R}} e^{it(\alpha_0-\alpha)+t/2}dt \int_{\mathbb{R}\times\mathbb{R}_+} \psi_H^f(\exp(s(e^t-1)X)\exp(yY))e^{-iys}dyds.\end{aligned}$$

By a similar way,

$$\begin{aligned}\langle\pi(\psi)a_2, a_2\rangle =& \oint_{H/(H\cap g_2Bg_2^{-1})} \chi_f(h)\Delta_{H,G}^{-1/2}(h)d\nu(h)\\ \times& \oint_{B/(B\cap g_2^{-1}Hg_2)} \psi_H^f(hg_2bg_2^{-1})\chi_\ell(b)\Delta_{B,G}^{-1/2}(b)d\nu(b)\end{aligned}$$

$$= \int_{\mathbb{R}} e^{it(\alpha_0-\alpha)+t/2}dt \int_{\mathbb{R}\times\mathbb{R}_+} \psi(\exp(s(1-e^t)X)\exp(yY))e^{iys}dyds.$$

These calculations finish in

$$\langle \pi(\psi)a_1, a_1\rangle + \langle \pi(\psi)a_2, a_2\rangle = \int_{\mathbb{R}} e^{it(\alpha_0-\alpha)+t/2}dt$$
$$\times \int_{\mathbb{R}^2} \psi_H^f(\exp(s(1-t)X)\exp(yY))e^{iys}dyds.$$

If we put

$$\Psi(t) = \int_{\mathbb{R}^2} \psi_H^f(\exp(s(1-e^t)X)\exp(yY))e^{iys}dyds,$$

this function is C^∞ for $t \neq 0$, because in this case the integral is carried out on a compact set. Besides, the function $\mathbb{R}^2 \ni (s,y) \mapsto \psi_H^f(\exp(sX)\exp(yY))$ belonging to $\mathcal{D}(\mathbb{R}^2)$, we make its L^1-approximations by functions of the form $\sum_j \xi_j(s)\eta_j(y)(\xi_j, \eta_j \in \mathcal{D}(\mathbb{R}))$. For arbitrary positive real number ϵ, we can choose ξ_j, η_j in such a manner that

$$|\Psi(t) - (2\pi)^{1/2}\sum_j \int_{\mathbb{R}} \xi_j(s(1-e^t))(\eta_j)^\wedge(s)ds| < \epsilon$$

for any $t \in \mathbb{R}$, where $(\eta_j)^\wedge$ denotes the inverse Fourier transform of η_j. Taking into account the fact that the limit, when t tends to zero, of

$$\int_{\mathbb{R}} \xi_j(s(1-e^t))(\eta_j)^\wedge(s)ds$$

is equal to $(2\pi)^{1/2}\xi_j(0)\eta_j(0)$, $\Psi(t)$ is continuous even at $t = 0$.

We consider at last the function $e^{t/2}\Psi(t)$, which rapidly decrease when $t \mapsto -\infty$ because of the factor $e^{t/2}$. Let us examine its behaviour when t tends to $+\infty$. A change of variables shows

$$e^{t/2}\Psi(t) = e^{t/2}(1-e^t)^{-1}\int_{\mathbb{R}^2} \psi_H^f(\exp(sX)\exp(yY))e^{isy(1-e^t)^{-1}}dsdy$$

and

$$|e^{t/2}\Psi(t)| \leq e^{t/2}(1-e^t)^{-1}\int_{\mathbb{R}^2} |\psi_H^f(\exp(sX)\exp(yY))|dsdy,$$

which shows that $e^{t/2}\Psi(t)$ rapidly decreases, when t tends to $+\infty$.

It follows that the Fourier inversion formula brings us to the concrete Plancherel formula for the monomial representation $\tau = \mathrm{ind}_H^G \chi_f$:

$$(2\pi)^{-2}\int_{\mathbb{R}} (\langle \pi(\psi)a_1(\alpha), a_1(\alpha)\rangle + \langle \pi(\psi)a_2(\alpha), a_2(\alpha)\rangle)d\alpha$$
$$=(2\pi)^{-1}\int_{\mathbb{R}} \psi_H^f(\exp(yY))e^{isy}dsdy = \psi_H^f(e).$$

Let us consider the monomial representation $\tau = \mathrm{ind}_H^G \chi_f$ which has infinite multiplicities. When G is exponential, there are two types of infinity of the multiplicities: discrete or continuous. This difference does not occur when G is nilpotent. We showed [16] that the algebra $D_\tau(G/H)$ is commutative if τ is of discrete type. But the inverse is not true.

Example 4.4.5 When $G = \exp \mathfrak{g}$ is no longer nilpotent, the algebra $D_\tau(G/H)$ can be very small. Let

$$\mathfrak{g} = \mathbb{R} - \mathrm{span}(T, X, Y, Z) : [T, X] = X, [T, Y] = Y, [T, Z] = 2Z, [X, Y] = Z.$$

Therefore,

$$G = \left\{ \begin{pmatrix} a^2 & x & z \\ 0 & a & y \\ 0 & 0 & 1 \end{pmatrix} ; a > 0,\ x, y, z \in \mathbb{R} \right\}$$

is completely solvable. Take $f = 0 \in \mathfrak{g}^*$ and $\mathfrak{h} = \mathbb{R}T$. Then, it follows immediately that $\mathfrak{g}(\ell) = \{0\}$ for $\ell \in \mathfrak{g}^*$ verifying $\ell(Z) \neq 0$. It results that $\tau = \mathrm{ind}_H^G 1$ has infinite multiplicities of continuous type. However, we see that the algebra $D_\tau(G/H)$ is trivial. Indeed, each monomial $X^p Y^q Z^r$ $(p, q, r \in \mathbb{N})$ is an eigenvector for the action ad T with the eigenvalue $p+q+2r$, what immediately leads our assertion. In this example, we see $D_\tau(G/H) \cong \mathbb{C} \cong \mathbb{C}[\Gamma_\tau]^H$.

4.4.2 The case of restrictions

Let $G = \exp \mathfrak{g}$ be an exponential solvable Lie group with Lie algebra $\mathfrak{g}$, $\pi \in \widehat{G}$ and $K = \exp \mathfrak{k}$ an analytic subgroup of G. We assume that the restriction $\pi|_K$ is discrete type and keep all the notations introduced before in the section of the polynomial conjecture for the restriction $\pi|_K$. The algebra $\mathbb{C}[\Omega]^K$ of the K-invariant polynomials on Ω would be constituted of the elements which is written as polynomial of $\mathfrak{k}^\ell$ at almost all the points ℓ of Ω. If this would be true, this algebra could be Poisson commutative when $\pi|_K$ is discrete type, because the Poisson product is given by $\{x, y\}(\ell) = B_\ell(x, y)$ for $x, y \in \mathfrak{g}$ and $\ell \in \Omega$ and we know that $\mathfrak{k}^\ell$ is isotropic for B_ℓ almost everywhere on Ω.

Theorem 4.4.6 *Assume that Ω is of maximal dimension. If $\pi|_K$ is discrete type, then $\mathbb{C}[\Omega]^K$ is Poisson commutative. Thus, the mapping $W \mapsto P_W \in \mathbb{C}[\Omega]^K$ is an algebra isomorphism from $D_\pi(G)^K$ onto the Poisson algebra $\mathbb{C}[\Omega]^K$.*

Proof 4.14 *We as usual employ the induction on* $\dim G$ *and assume that $\mathfrak{k}$ contains the centre $\mathfrak{z}$ of $\mathfrak{g}$. The Poisson structure in the symmetric algebra $S(\mathfrak{g})$ is defined by $\{x, y\} = [x, y](x, y \in \mathfrak{g})$ and $\mathbb{C}[\Omega]^K$ inherits this structure. If* $\ker \pi$ *contains a non-trivial ideal $\mathfrak{a}$ of $\mathfrak{g}$, we pass to the quotient $\mathfrak{g}/\mathfrak{a}$ to which applies the induction hypothesis. So, we can assume* $\dim \mathfrak{z} \leq 1$ *and $\pi|_{\mathfrak{z}} \neq 0$ if* $\dim \mathfrak{z} = 1$. *We take as before a minimal non-central ideal $\mathfrak{a}$ of $\mathfrak{g}$ and keep all the notations introduced before.*

We refer the proof of Theorem 4.3.6. Suppose first that Ω is non-saturated with respect to $\mathfrak{g}_0 = \{X \in \mathfrak{g}; [X, \mathfrak{a}] \subset \mathfrak{z}\}$. This implies $\dim \mathfrak{z} = 1$. *Let $\mathfrak{z} = \mathbb{R}Z$ with $\pi(Z) = i = \sqrt{-1}$. Here the restriction π_0 of π to $G_0 =$* $\exp(\mathfrak{g}_0)$ *is irreducible. Let W be the element of* $\ker \pi$ *introduced by Theorem 3.2.6. Since $\sigma^{-1}(W) \equiv 0$ on Ω, we can think that $\mathbb{C}[\Omega]^K$ is contained in $S(\mathfrak{g}_0)$. Then, if $\mathfrak{k} \subset \mathfrak{g}_0$, $\pi_0|_K$ is discrete type and the induction hypothesis says that $\mathbb{C}[\Omega]^K = \mathbb{C}[\omega]^K$, where $\omega = \Omega|_{\mathfrak{g}_0}$, is*

Poisson commutative. Suppose $\mathfrak{k} \not\subset \mathfrak{g}_0$. *Let* $f \in \Omega$ *be generic,* $\widetilde{\mathfrak{g}} = \mathfrak{a}^f, \widetilde{G} = \exp\widetilde{\mathfrak{g}}, \widetilde{f} = f|_{\widetilde{\mathfrak{g}}}, \widetilde{\omega} = \widetilde{G}\cdot\widetilde{f}, \widetilde{\pi} = ind_{B[f]}^{\widetilde{G}}\chi_f, \widetilde{\mathfrak{k}} = \mathfrak{k}\cap\widetilde{\mathfrak{g}}$ *and* $\widetilde{K} = \exp\widetilde{\mathfrak{k}}$. *Then,* $\widetilde{\pi}|_{\widetilde{K}}$ *is discrete type and* $\mathbb{C}[\Omega]^K \subset S(\widetilde{\mathfrak{g}})$ *(cf. the proof of Theorem 4.3.6). The induction hypothesis says that* $\mathbb{C}[\widetilde{\omega}]^{\widetilde{K}}$ *is Poisson commutative and hence* $\mathbb{C}[\Omega]^K$ *too because of* $\mathbb{C}[\Omega]^K \subset \mathbb{C}[\widetilde{\omega}]^{\widetilde{K}}$.

Suppose now that Ω *is saturated with respect to* $\mathfrak{g}_0$. *Let* $f \in \Omega$ *be generic,* $f_0 = f|_{\mathfrak{g}_0}, G_0 = \exp(\mathfrak{g}_0), \omega = G_0 \cdot f_0$ *and* $\pi_0 = ind_{B[f]}^{G}\chi_f$. *In this case,* $\pi_0|_{K_0}$, *where* $K_0 = \exp(\mathfrak{k}_0)$ *with* $\mathfrak{k}_0 = \mathfrak{k}\cap\mathfrak{g}_0$ *is discrete type. If* $\mathfrak{b}[f|_{\mathfrak{k}}] \subset \mathfrak{g}_0$, $\mathbb{C}[\Omega]^K \subset \mathbb{C}[\omega]^{K_0}$ *which is Poisson commutative by induction hypothesis. Finally, if* $\mathfrak{b}[f|_{\mathfrak{k}}] \not\subset \mathfrak{g}_0$, *we saw before that* $\mathfrak{k} = \mathbb{R}x + \mathfrak{k}_0$, *where* $[\mathfrak{k}, x] \subset \ker\pi$. *Therefore,*

$$\mathbb{C}[\Omega]^K = \mathbb{C}x + (\mathbb{C}[\Omega]^K \cap S(\widetilde{\mathfrak{g}})^K)$$

and x *is Poisson commutative with* $S(\widetilde{\mathfrak{g}})^K$. *Since* $\widetilde{\pi}|_{\widetilde{K}}$ *is discrete type, the induction hypothesis says that* $\mathbb{C}[\widetilde{\omega}]^{\widetilde{K}}$ *is Poisson commutative. As* $\mathbb{C}[\Omega]^K \cap S(\widetilde{\mathfrak{g}})^K$ *is contained in* $\mathbb{C}[\widetilde{\omega}]^{\widetilde{K}}$, $\mathbb{C}[\Omega]^K$ *is also Poisson commutative.*

CHAPTER 5

Intertwining operators for irreducible representations of exponential groups

Our main focus in this chapter is how to construct an intertwining operator for two irreducible unitary representations of exponential solvable Lie groups. Let $G = \exp \mathfrak{g}$ be an exponential solvable Lie group with Lie algebra $\mathfrak{g}$, $f \in \mathfrak{g}^*$ and $\mathfrak{h}_1, \mathfrak{h}_2$ two (real) polarisations of $\mathfrak{g}$ at f satisfying the Pukanszky condition. Put $H_j = \exp(\mathfrak{h}_j)$ $(j = 1, 2)$. Then, it is well known that the two irreducible monomial representations $\pi_j = \mathrm{ind}_{H_j}^G \chi_f$ $(j = 1, 2)$ are mutually equivalent and we even have a natural candidate of the intertwining operator between them. In order to verify that it is a true intertwining operator, the principal obstacle is the convergence of the integral which intervenes (cf. [14]). A property which assures this convergence is the closedness of the simple product set H_2H_1 in G. Here we shall establish this property (cf. [8]).

5.1 POSITION OF THE PROBLEM

Let us consider then $f \in \mathfrak{g}^*$ and two (real) polarisations $\mathfrak{h}_1, \mathfrak{h}_2$ of $\mathfrak{g}$ at f which satisfy the Pukanszky condition. This means that $f + \mathfrak{h}_j^{\perp} = H_j \cdot f$ with $H_j = \exp(\mathfrak{h}_j)(1 \leq j \leq 2)$. Starting from the unitary character χ_f of H_j defined by $\chi_f(\exp X) = e^{if(X)}$ $(X \in \mathfrak{h}_j)$, we induce a monomial representation $\pi_j = \mathrm{ind}_{H_j}^G \chi_f$ of G. Thus, it comes [23] that π_1, π_2 are irreducible and mutually equivalent. For $j = 1, 2$, we denote by $\mathcal{H}_{\pi_j}$ the Hilbert space of π_j. We propose to construct explicitly an intertwining operator between them. Our first step in [46] was to establish the relation

$$\mathrm{Tr}\ \mathrm{ad}_{\mathfrak{h}_1/(\mathfrak{h}_1 \cap \mathfrak{h}_2)} X + \mathrm{Tr}\ \mathrm{ad}_{\mathfrak{h}_2/(\mathfrak{h}_1 \cap \mathfrak{h}_2)} X = 0$$

for all $X \in \mathfrak{h}_1 \cap \mathfrak{h}_2$. This leads to

$$\Delta_{H_1,G}(h) = \Delta_{H_2,G}(h) \Delta^2_{H_1 \cap H_2, H_2}(h) \quad (h \in H_1 \cap H_2).$$

This being set, for $\varphi \in \mathcal{H}_{\pi_1}$ and $g \in G$, the function Φ_g on H_2 defined by

$$\Phi_g(h) = \varphi(gh)\chi_f(h)\Delta^{-1/2}_{H_2,G}(h)$$

DOI: 10.1201/9781003685739-5

verifies the relation

$$\Phi_g(hx) = \Delta_{H_1 \cap H_2, H_2}(x)\Phi_g(h) \quad (h \in H_2, x \in H_1 \cap H_2).$$

Hence we are able to describe formally the integral

$$\left(T_{\mathfrak{h}_2\mathfrak{h}_1}\varphi\right)(g) = \oint_{H_2/(H_2\cap H_1)} \varphi(gh)\chi_f(h)\Delta_{H_2,G}^{-1/2}(h)d\nu(h) \quad (g \in G). \tag{5.1.1}$$

At least on formal level, it is clear that $T_{\mathfrak{h}_2\mathfrak{h}_1}\varphi$ verifies the covariance condition required to belong to the space $\mathcal{H}_{\pi_2}$ and that $T_{\mathfrak{h}_2\mathfrak{h}_1}$ commutes with the action of G by left translations. In fact, it is not exaggerated [14] to say that one of the principal problems is the question of the convergence of the integral. We here consider the following claim:

Claim 5.1.1 *The product H_2H_1 is closed in G.*

It turns out from the Pukanszky condition that the simple product set H_2H_1 is locally closed in G and hence that the space H_2H_1/H_1 is homeomorphic to homogeneous space $H_2/(H_1 \cap H_2)$. Therefore, if Claim 5.1.1 holds, the integral (5.1.1) is convergent for all continuous function φ with compact support modulo H_1 of $\mathcal{H}_{\pi_1}$.

Note that Claim 5.1.1 holds, in the setting when G is nilpotent or at least one of the polarisations $\mathfrak{h}_j$ is of Vergne type (cf. [3]), and also in the case where $\mathfrak{h}_1 + \mathfrak{h}_2$ is a subalgebra of $\mathfrak{g}$ (cf. [17]). The aim of this chapter is to provide a proof in the general setting of exponential solvable Lie groups.

5.2 ORIENTATION OF THE STUDY

We begin with groping our way to find an orientation.

Example 5.1 Let $\mathfrak{g} = \langle x, y, z, a\rangle_{\mathbb{R}} : [x, y] = z, [x, a] = a$. Then, the derived algebra $[\mathfrak{g}, \mathfrak{g}]$ is commutative. At the point $f = z^*$ in $\mathfrak{g}^*$, the subalgebra $\mathfrak{h} = \langle x, a, z\rangle_{\mathbb{R}}$ is a polarisation satisfying the Pukanszky condition and $\mathfrak{h}$ is not contained in the nilpotent radical of $\mathfrak{g}$.

5.2.1 Indices and notations

Let

$$\mathcal{S} : \{0\} = \mathfrak{g}_0 \subset \mathfrak{g}_1 \subset \cdots \subset \mathfrak{g}_{n-1} \subset \mathfrak{g}_n = \mathfrak{g}, \dim \mathfrak{g}_j = j \ (0 \leq j \leq n)$$

be a good sequence of subalgebras of $\mathfrak{g}$ which passes a nilpotent ideal $\mathfrak{n} = \mathfrak{g}_{j_0}$ containing $[\mathfrak{g}, \mathfrak{g}]$. What is to say, if $\mathfrak{g}_j$ is not an ideal of $\mathfrak{g}$, then $\mathfrak{g}_{j\pm1}$ are ideals of $\mathfrak{g}$ and the adjoint representation of $\mathfrak{g}$ on $\mathfrak{g}_{j+1}/\mathfrak{g}_{j-1}$ is irreducible. If $\mathfrak{g}_j$ and $\mathfrak{g}_{j-1}$ are ideals of $\mathfrak{g}$, we obtain a root $\alpha_j : \mathfrak{g} \to \mathbb{R}$ of $\mathfrak{g}$:

$$\mathrm{ad}(X)(X_j) - \alpha_j(X)X_j \in \mathfrak{g}_{j-1}, \ X \in \mathfrak{g}, \ X_j \in \mathfrak{g}_j \setminus \mathfrak{g}_{j-1}.$$

If $\mathfrak{g}_j$ is not an ideal of $\mathfrak{g}$, then we take a subspace $\mathfrak{v}_j$ of dimension 2 of $\mathfrak{g}_{j+1}$ such that $\mathfrak{g}_{j+1} = \mathfrak{v}_j + \mathfrak{g}_{j-1}$ and there exists a homomorphism α_j of $\mathfrak{g}$ to the algebra of the endomorphisms of $\mathfrak{v}_j$ such that

$$\mathrm{ad}(X)v - \alpha_j(X)v \in \mathfrak{g}_{j-1},\ X \in \mathfrak{g},\ v \in \mathfrak{v}_j.$$

We define for a subspace $\mathfrak{v}$ of $\mathfrak{g}$ the index set $I^{\mathfrak{v}} \subset \{1, 2, \ldots, n\}$ by

$$\begin{aligned} I^{\mathfrak{v}} &= \{1 \le j \le n; \mathfrak{v} + \mathfrak{g}_j = \mathfrak{v} + \mathfrak{g}_{j-1}\} \\ &= \{1 \le j \le n; \mathfrak{v} \cap \mathfrak{g}_j \neq \mathfrak{v} \cap \mathfrak{g}_{j-1}\}. \end{aligned}$$

Choose for $j \in I^{\mathfrak{h}_2} \backslash I^{\mathfrak{h}_1 \cap \mathfrak{h}_2}$ an element $U_j \in \mathfrak{h}_2 \cap \mathfrak{g}_j$ outside of $\mathfrak{g}_{j-1}$ and determine the integer $k = k(j) \le j$ as the smallest index such that

$$(U_j + (\mathfrak{h}_2 \cap \mathfrak{g}_{j-1}) + \mathfrak{h}_1) \cap \mathfrak{g}_k \neq \emptyset.$$

Then, $k \ge 1$. This gives us, retaking U_j again if necessary, an element $S_j \in \mathfrak{h}_1$ and an element $V_j \in \mathfrak{g}_{k(j)} \setminus \mathfrak{g}_{k(j)-1}$ such that

$$U_j = V_j + S_j.$$

We see that $j \in I^{\mathfrak{h}_2} \backslash I^{\mathfrak{h}_1}$ if and only if $k(j) = j$. In this case, we can take $S_j = 0$ and $V_j = U_j$. Because of the minimality of $k(j)$, we know that $k(j) \notin I^{\mathfrak{h}_1}$ and that the mapping

$$I^{\mathfrak{h}_2} \cap I^{\mathfrak{h}_1} \backslash I^{\mathfrak{h}_1 \cap \mathfrak{h}_2} \ni j \mapsto k(j) \in I^{\mathfrak{h}_2 + \mathfrak{h}_1} \backslash (I^{\mathfrak{h}_1} \cup I^{\mathfrak{h}_2})$$

is injective.

We set

$$\widehat{I}^{\mathfrak{h}_2} = \{j \in I^{\mathfrak{h}_2} \backslash I^{\mathfrak{h}_1 \cap \mathfrak{h}_2}; k(j) \le j_0\}.$$

For $0 \le j \le j_0$, we put $\mathfrak{m}_j = \mathfrak{g}_j + \mathfrak{h}_1$ and

$$I^{\mathfrak{h}_2}_{\mathfrak{m}} = \{1 \le j \le j_0; \mathfrak{h}_2 \cap \mathfrak{m}_j \neq \mathfrak{h}_2 \cap \mathfrak{m}_{j-1}\}.$$

Let $j \in I^{\mathfrak{h}_2}_{\mathfrak{m}}$. Let us show that there exists X_j in $\mathfrak{h}_2 \cap \mathfrak{m}_j$ outside of $\mathfrak{h}_2 \cap \mathfrak{m}_{j-1}$ such that we have $[X_j, \mathfrak{g}_j] \subset \mathfrak{g}_{j-1}$. When we can take X_j in $\mathfrak{n}$, our assertion is trivial.

5.2.2 A key result

The following proposition stands to be a key result for the proof of Claim 5.1.1. We have the following:

Proposition 5.2.1 *Let $G = \exp \mathfrak{g}$ be an exponential solvable Lie group with Lie algebra $\mathfrak{g}$ such that there exists an abelian ideal $\mathfrak{n}$ between the nilpotent radical of $\mathfrak{g}$ and $[\mathfrak{g}, \mathfrak{g}]$. Let $f \in \mathfrak{g}^*$, $\mathfrak{h}_1, \mathfrak{h}_2$ two polarisations at f satisfying the Pukanszky condition and $H_j = \exp(\mathfrak{h}_j)$ for $j = 1, 2$. Then, Claim 5.1.1 holds.*

Proof 5.1 *Let us proceed by induction on $\dim G$. If there exists a non-trivial ideal on which f vanishes, then this ideal is contained in $\mathfrak{h}_1 \cap \mathfrak{h}_2$ and we can descend to the quotient by this ideal and apply there the induction hypothesis, because the image of $\mathfrak{n}$*

in the quotient is found between the nilpotent radical and the derived algebra. Suppose hereafter that there is no such an ideal, what brings that the dimension of the centre $\mathfrak{z}$ *of* $\mathfrak{g}$ *is inferior or equal to 1. If there exists a minimal ideal* $\mathfrak{a} \neq \{0\}$*, then, since* f *does not vanish on* $\mathfrak{a}$*, the subalgebra*

$$\widetilde{\mathfrak{g}} = \{X \in \mathfrak{g}; f([X, \mathfrak{a}]) = \{0\}\}$$

turns out to be proper in $\mathfrak{g}$*. Since the Pukanszky condition requires that* $\mathfrak{h}_1, \mathfrak{h}_2$ *are contained in* $\widetilde{\mathfrak{g}}$*, the induction hypothesis suffices for us. These observations apply to the general setting without any assumption and we can assume in what follows* $\mathfrak{z} = \mathbb{R}Z$ *with* $f(Z) = 1$*.*

In this way, we are led to the following situation: $\mathfrak{g}_1 = \mathfrak{z}$*,* f *does not vanish on* $\mathfrak{g}_1$ *and* $\mathfrak{a} = \mathfrak{g}_2$ *or* $\mathfrak{a} = \mathfrak{g}_3$ *is a minimal non-central ideal of* $\mathfrak{g}$*. In order to simplify the notation, designate* $\widetilde{\mathfrak{g}}$ *by* $\mathfrak{k}$*, which is a proper subalgebra of* $\mathfrak{g}$*. Remark that* $\mathfrak{g} = \mathfrak{k} + [\mathfrak{g}, \mathfrak{g}]$ *except the case where* $\mathfrak{a} = \mathfrak{g}_2$ *and where* $[\mathfrak{g}, \mathfrak{a}] = \mathfrak{z}$*, namely that* $\alpha_2 = 0$*. In this last case,* $\mathfrak{k}$ *is an ideal of* $\mathfrak{g}$ *and our hypothesis on the existence of* $\mathfrak{n}$ *regulate this eventuality. All taken into account, we obtain a good sequence of subalgebras of* $\mathfrak{k}$ *by taking the intersection* $\mathcal{S}'$ *of* $\mathcal{S}$ *with* $\mathfrak{k}$ *and* $\mathfrak{k} \cap \mathfrak{n}$ *is found between the nilpotent radical of* $\mathfrak{k}$ *and* $[\mathfrak{k}, \mathfrak{k}]$*.*

This taken in mind, if $\mathfrak{a} \subset \mathfrak{h}_1 \cap \mathfrak{h}_2$*, it suffices for us to apply the induction hypothesis to the subgroup* $K = \exp \mathfrak{k}$*. For* $i = 1, 2$*, put* $\mathfrak{h}_j^0 = \mathfrak{h}_j \cap \mathfrak{k}$ *and* $\mathfrak{h}_j' = \mathfrak{h}_j^0 + \mathfrak{a}$*. When* $\mathfrak{h}_2$ *contains* $\mathfrak{a}$ *and* $\mathfrak{h}_1$ *does not contain* $\mathfrak{a}$*, it suffices for us to apply the induction hypothesis to the data* $(\mathfrak{k}, \mathfrak{h}_1', \mathfrak{h}_2, \mathcal{S}')$*.*

Suppose that $\mathfrak{h}_1$ *contains* $\mathfrak{a}$ *and so does not* $\mathfrak{h}_2$*. Since* $\mathfrak{g}_{j_0}$ *is commutative and* $\mathfrak{g}_{j_0}$ *contains* $\mathfrak{a}$*,* $\mathfrak{g}_{j_0}$ *is contained in* $\mathfrak{k}$*. It follows that* $\mathfrak{m}_{j_0} \cap \mathfrak{h}_2 \subset \mathfrak{h}_2^0$*. Therefore, our assertion results from that already established for* $\mathfrak{h}_1$ *and* $\mathfrak{h}_2'$*.*

Suppose from now on that neither $\mathfrak{h}_1$ *nor* $\mathfrak{h}_2$ *contains* $\mathfrak{a}$*. If* $\mathfrak{h}_1 \cap \mathfrak{h}_2$ *is not contained in* $\mathfrak{k}$*, our assertion follows from that already established for* $\mathfrak{h}_1'$ *and* $\mathfrak{h}_2'$*. It remains to examine the situation where* $\mathfrak{a} \subset \mathfrak{h}_1 + \mathfrak{h}_2$*. Recall that* $\mathfrak{g}_1 = \mathfrak{z}$ *and* $\mathfrak{a} = \mathfrak{g}_2$*. So, let* $\mathfrak{z} = \mathbb{R}Z$*,* $f(Z) = 1$*, and* $\mathfrak{g}_2 = \mathbb{R}Y + \mathbb{R}Z$ *with* $f(Y) = 0$*.*

First, suppose that $\mathfrak{a} \subset \mathfrak{h}_1^0 + \mathfrak{h}_2^0$*. Write* $Y = X_2 - X_1$ *with* $X_k \in \mathfrak{h}_k^0 (1 \leq k \leq 2)$*. Let* i_0 *be the smallest index in* $I_{\mathfrak{m}}^{\mathfrak{h}_2}$ *such that* $\mathfrak{h}_2 \cap \mathfrak{m}_{i_0} \not\subset \mathfrak{h}_2^0$*. If there does not exist such an index* i_0*, we have nothing to do. It comes that* $\mathfrak{h}_2^0 \cap \mathfrak{m}_{i_0} \subset \mathfrak{m}_{i_0 - 1}$ *and then that there exists* $V \in \mathfrak{g}_{i_0} \setminus \mathfrak{g}_{i_0 - 1}$ *which is written as* $V = T_2 - T_1$ *with* $T_k \in \mathfrak{h}_k \setminus \mathfrak{h}_k^0$ *verifying* $[T_k, Y] = Z$ $(1 \leq k \leq 2)$ *(cf. Lemma 5.3.2 in the sequel). Modifying the elements* X_2, X_1 *by elements of* $\mathfrak{h}_1 \cap \mathfrak{h}_2$*, we can choose* X_2, X_1 *in such a manner that* $B_f(X_k, \mathfrak{g}_{i_0 - 1}) = \{0\}$*. Write* $[T_2, X_2] = \beta X_2 + w_2$ *with* $\beta \in \mathbb{R}$ *and* $w_2 \in \mathfrak{h}_2^0$ *such that* $B_f(T_1, w_2) = 0$*. As* $\mathfrak{n} = \mathfrak{g}_{j_0}$ *is commutative, we have*

$$\begin{aligned} 0 = f([V, [T_2, X_2]]) &= B_f(V, [T_2, X_2]) = -B_f(T_1, [T_2, X_2]) \\ &= -\beta B_f(T_1, X_2) = -\beta B_f(T_1, Y) = -\beta. \end{aligned}$$

Hence, $[T_2, X_2] = w_2$*. Similarly,* $[T_1, X_1] = w_1$ *with* $w_1 \in \mathfrak{h}_1^0$ *verifying* $B_f(T_2, w_1) = 0$*. Then, we have*

$$[V, X_2] = [T_2 - T_1, X_2] = [T_2, X_2] - [T_1, X_1 + Y] = w_2 - w_1 - Z$$

and consequently

$$\begin{aligned} 0 &= B_f(T_2, [V, X_2]) = B_f([T_2, V], X_2) + B_f(V, [T_2, X_2]) \\ &= B_f([T_2, V], X_2), \end{aligned}$$

what says that $[T_2, V]$ *belongs to* $\mathfrak{g}_{i_0-1}$.

Second, suppose that $\mathfrak{a} \not\subset \mathfrak{h}_1^0 + \mathfrak{h}_2^0$. *Writing* $Y = T_2 - T_1$ *with* $T_k \in \mathfrak{h}_k \setminus \mathfrak{h}_k^0$ $(1 \leq k \leq 2)$, *we know that the Pukanszky condition allows us to choose* T_k *in such a manner that* $[T_k, Y]$ *belongs to* $\mathfrak{g}_1$ *(cf. Lemma 5.3.2).*

We can treat similarly the case where $\dim \mathfrak{a} = 3$. *Let* i_0 *be the smallest index in* $I_{\mathfrak{m}}^{\mathfrak{h}_2}$ *such that* $\mathfrak{h}_2 \cap \mathfrak{m}_{i_0} \not\subset \mathfrak{h}_2^0$. *Then, there exists* $V \in \mathfrak{g}_{i_0} \setminus \mathfrak{g}_{i_0-1}$ *which is written as* $V = T_2 - T_1$ *with* $T_k \in \mathfrak{h}_k \setminus \mathfrak{h}_k^0$ *verifying* $[T_k, Y] = Z$ $(1 \leq k \leq 2)$ *because of Lemma 5.3.2. Following the same way as above, we find* $[T_k, V] \in \mathfrak{g}_{i_0-1}$.

Now, putting $M_j = \exp(\mathfrak{m}_j)$ *for* $1 \leq j \leq j_0$, *let us show by induction on* j *that* $H_2H_1 \cap M_j$ *is a closed subset of* M_j. *When* $j = 1$, *it is trivial. Suppose that* $H_2H_1 \cap M_{j-1}$ *is closed in* M_{j-1}. *If* $j \notin I_{\mathfrak{m}}^{\mathfrak{h}_2}$, *it turns out that* $H_2H_1 \cap M_j = H_2H_1 \cap M_{j-1}$. *Suppose that* $j \in I_{\mathfrak{m}}^{\mathfrak{h}_2}$. *Take* X_j *in* $\mathfrak{h}_2 \cap \mathfrak{m}_j$ *outside of* $\mathfrak{m}_{j-1}$ *in the way that we have* $[X_j, \mathfrak{g}_j] \subset \mathfrak{g}_{j-1}$. *Let* $X_j = V_j + X'_j$ *with* $V_j \in \mathfrak{g}_j \setminus \mathfrak{g}_{j-1}$ *and* $X'_j \in \mathfrak{h}_1$. *Let now* $\{x_i\}_{i=1}^{\infty}$ *be a sequence in* $H_2H_1 \cap M_j$, *convergent in* G. *Write*

$$x_i = \exp(t_i X_j) \cdot g_i \ (t_i \in \mathbb{R},\ g_i \in H_2H_1 \cap M_{j-1}).$$

Because

$$x_i = \exp(t_i(V_j + X'_j)) \cdot g_i = \exp(t_i V_j) \cdot g'_i$$

with $g'_i \in M_{j-1}$, *the sequence* $\{t_i\}_{i=1}^{\infty}$ *is convergent and consequently the sequence* $g_i = \exp(-t_i X_j) \cdot x_i$ *too. Say,* $\lim_{i\to\infty} t_i = t_0 \in \mathbb{R}$ *and* $\lim_{i\to\infty} g_i = g_0 \in M_{j-1}$. *Since* $H_2H_1 \cap M_{j-1}$ *is closed in* M_{j-1}, *we confirm that* $g_0 \in H_2H_1$. *After these observations,* $\lim_{i\to\infty} x_i = \exp(t_i X_j) \cdot g_i$ *is found in* H_2H_1. *We see in this way that* $H_2H_1 \cap M_j$ *is a closed subset of* M_j.

Finally, $H_2H_1 \cap M_{j_0}$ *is closed in* M_{j_0}. *By adding to the subalgebra* $\mathfrak{m}_{j_0}$ *a co-exponential basis* $\{W_1, \ldots, W_p\}$ *in* $\mathfrak{h}_2$ *to* $\mathfrak{h}_2 \cap \mathfrak{m}_{j_0}$, *we conclude that*

$$H_2H_1 = \exp(\mathbb{R}W_1) \cdots \exp(\mathbb{R}W_p)(H_2H_1 \cap M_{j_0})$$

is a closed subset in G.

5.3 CASE OF COMPLETELY SOLVABLE LIE GROUPS

Suppose here that $\mathfrak{g}$ is completely solvable, and we shall provide a proof of Claim 5.1.1. Our main result in this section is the following:

Theorem 5.3.1 *Let* $G = \exp \mathfrak{g}$ *be a completely solvable Lie group with Lie algebra* $\mathfrak{g}$, $f \in \mathfrak{g}^*$ *and* $\mathfrak{h}_j \in S(f, \mathfrak{g})(j = 1, 2)$ *two Pukanszky (real) polarisations of* $\mathfrak{g}$ *at* $f \in \mathfrak{g}^*$. *Put* $H_j = \exp(\mathfrak{h}_j)(j = 1, 2)$. *Then, the product* H_2H_1 *is closed in* G.

5.3.1 First preparations

Let $\mathfrak{n}$ be a nilpotent ideal of $\mathfrak{g}$ containing $[\mathfrak{g}, \mathfrak{g}]$. Put $\mathfrak{n}_0 = \{0\}$ and let $\mathfrak{n}_1$ be an ideal of $\mathfrak{g}$ contained in $\mathfrak{n}$ and having the dimension 1. We denote by $c(\mathfrak{n}_1, \mathfrak{n}_0)$ the centraliser of $\mathfrak{n}_1$ in $\mathfrak{n}$, what evidently leads the equality $c(\mathfrak{n}_1, \mathfrak{n}_0) = \mathfrak{n}$. Take now an ideal $\mathfrak{n}_2$ of $\mathfrak{g}$ included in $\mathfrak{n}$, containing $\mathfrak{n}_1$ and having the dimension 2. If we put

$$c(\mathfrak{n}_2, \mathfrak{n}_0) = \{X \in \mathfrak{n}; [X, \mathfrak{n}_2] \subset \mathfrak{n}_0\},$$

then,

$$\{0\} = \mathfrak{n}_0 \subsetneq \mathfrak{n}_1 \subsetneq \mathfrak{n}_2 \subset c(\mathfrak{n}_2, \mathfrak{n}_0) \subset \mathfrak{n}, \quad \dim\left(\mathfrak{n}/c(\mathfrak{n}_2, \mathfrak{n}_0)\right) \leq 1.$$

If $\mathfrak{n}_2 \neq c(\mathfrak{n}_2, \mathfrak{n}_0)$, we take an ideal $\mathfrak{n}_3$ of $\mathfrak{g}$ such that

$$\mathfrak{n}_2 \subset \mathfrak{n}_3 \subset c(\mathfrak{n}_2, \mathfrak{n}_0), \quad \dim(\mathfrak{n}_3/\mathfrak{n}_2) = 1.$$

Next we consider the ideal

$$c(\mathfrak{n}_3, \mathfrak{n}_1) = \{X \in c(\mathfrak{n}_2, \mathfrak{n}_0) : [X, \mathfrak{n}_3] \subset \mathfrak{n}_1\}.$$

of $\mathfrak{g}$. Then,

$$\mathfrak{n}_1 \subsetneq \mathfrak{n}_2 \subsetneq \mathfrak{n}_3 \subset c(\mathfrak{n}_3, \mathfrak{n}_1) \subset c(\mathfrak{n}_2, \mathfrak{n}_0), \quad \dim\left(c(\mathfrak{n}_2, \mathfrak{n}_0)/c(\mathfrak{n}_3, \mathfrak{n}_1)\right) \leq 1.$$

If $\mathfrak{n}_3 \neq c(\mathfrak{n}_3, \mathfrak{n}_1)$, we take an ideal $\mathfrak{n}_4$ of $\mathfrak{g}$ such that

$$\mathfrak{n}_3 \subset \mathfrak{n}_4 \subset c(\mathfrak{n}_3, \mathfrak{n}_1), \quad \dim(\mathfrak{n}_4/\mathfrak{n}_3) = 1.$$

Next we consider the ideal

$$c(\mathfrak{n}_4, \mathfrak{n}_2) = \{X \in c(\mathfrak{n}_3, \mathfrak{n}_1); [X, \mathfrak{n}_4] \subset \mathfrak{n}_2\}$$

of $\mathfrak{g}$. Then,

$$\mathfrak{n}_1 \subsetneq \mathfrak{n}_2 \subsetneq \mathfrak{n}_3 \subsetneq \mathfrak{n}_4 \subset c(\mathfrak{n}_4, \mathfrak{n}_2) \subset c(\mathfrak{n}_3, \mathfrak{n}_1) \subset c(\mathfrak{n}_2, \mathfrak{n}_0) \subset \mathfrak{n}$$

and $\dim\left(c(\mathfrak{n}_3, \mathfrak{n}_1)/c(\mathfrak{n}_4, \mathfrak{n}_2)\right) \leq 1$.

We continue the same process until we have an index k_0 such that

$$\mathfrak{n}_{k_0} = c(\mathfrak{n}_{k_0}, \mathfrak{n}_{k_0-2}).$$

By this way, we arrive to a sequence of ideals of $\mathfrak{g}$:

$$\begin{aligned}\{0\} = \mathfrak{n}_0 \subsetneq \mathfrak{n}_1 \subsetneq \mathfrak{n}_2 \subsetneq \cdots \subsetneq \mathfrak{n}_{k_0} &= c(\mathfrak{n}_{k_0}, \mathfrak{n}_{k_0-2}) \\ &\subset c(\mathfrak{n}_{k_0-1}, \mathfrak{n}_{k_0-3}) \subset \cdots \subset c(\mathfrak{n}_2, \mathfrak{n}_0) \subset \mathfrak{n}.\end{aligned}$$

Denote this sequence made by different terms by $\mathcal{S}$:

$$\{0\} = \mathfrak{n}_0 \subset \mathfrak{n}_1 \subset \cdots \subset \mathfrak{n}_{m-1} \subset \mathfrak{n}_m = \mathfrak{n}, \ \dim(\mathfrak{n}_j/\mathfrak{n}_{j-1}) = 1 \ (1 \leq j \leq m).$$

For $1 \leq j \leq m$, we set $\mathfrak{m}_j = \mathfrak{n}_j + \mathfrak{h}_1$ and

$$I^{\mathfrak{h}_2} = \{1 \leq j \leq m; \mathfrak{h}_2 \cap \mathfrak{m}_j \neq \mathfrak{h}_2 \cap \mathfrak{m}_{j-1}\}.$$

Now, we propose the following claim:

Claim 5.3.1 *Let $j \in I^{\mathfrak{h}_2}$. There exists X_j in $\mathfrak{h}_2 \cap \mathfrak{m}_j$ outside of $\mathfrak{h}_2 \cap \mathfrak{m}_{j-1}$ such that we have $[X_j, \mathfrak{n}_j] \subset \mathfrak{n}_{j-1}$.*

Then the following is immediate from the proof of Proposition 5.2.1:

Proposition 5.3.2 *If Claim 5.3.1 holds, then so does Claim 5.1.1, so a proof of Theorem 5.3.1.*

5.3.2 Proof of the closedness property

We shall provide a proof of Theorem 5.3.1 through different steps, here is the first one:

5.3.3 An induction procedure

Step 1:

We proceed by induction on $\dim \mathfrak{g}$. If $\mathfrak{h}_2 = \mathfrak{h}_1 = \mathfrak{g}(f) = \mathfrak{g}$, $H_2H_1 = H_1 = H_2 = G$. If f vanishes on an ideal $\mathfrak{a} \neq \{0\}$ of $\mathfrak{g}$, $\mathfrak{a} \subset \mathfrak{g}(f) \subset \mathfrak{h}_1 \cap \mathfrak{h}_2$ and we can pass to the quotient $\mathfrak{g}/\mathfrak{a}$ to which applies the induction hypothesis. Hence, we can suppose that the centre $\mathfrak{z}$ satisfies $\dim \mathfrak{z} \leq 1$, and that f does not vanish on $\mathfrak{z}$ when $\dim \mathfrak{z} = 1$. Let $\mathfrak{a}$ be a minimal non-central ideal of $\mathfrak{g}$ and $\widetilde{\mathfrak{g}} = \{X \in \mathfrak{g}; B_f(X, \mathfrak{a}) = \{0\}\}$ which is a proper subalgebra of $\mathfrak{g}$. If $\mathfrak{a}$ is a minimal ideal of $\mathfrak{g}$, then the Pukanszky condition implies $\mathfrak{h}_j (j = 1, 2)$ are contained in $\widetilde{\mathfrak{g}}$ and the result comes immediately from the induction hypothesis applied to $\widetilde{G} = \exp(\widetilde{\mathfrak{g}})$.

Suppose hereafter that $\mathfrak{a}$ is not a minimal ideal of $\mathfrak{g}$. This means $\mathfrak{n}_1 = \mathfrak{z} = 1, \dim \mathfrak{a} = 2$ and $\mathfrak{z} \subset \mathfrak{a}$. Let $\mathfrak{z} = \mathbb{R}Z, \mathfrak{a} = \mathbb{R}Y \oplus \mathfrak{z}$ such that $f(Z) = 1$ and $f(Y) = 0$. Note that the intersection of $\widetilde{\mathfrak{g}}$ with $\mathfrak{n}$ is nothing but $c(\mathfrak{n}_2, \mathfrak{n}_0) - c(\mathfrak{n}_2, \{0\})$.

Now let $\mathfrak{a} \not\subset \mathfrak{h}_1 \cap \mathfrak{h}_2$ and put $\mathfrak{h}_i^0 = \mathfrak{h}_i \cap \widetilde{\mathfrak{g}}$, $\mathfrak{h}_i' = \mathfrak{h}_i^0 + \mathfrak{a}$ for $1 \leq i \leq 2$. It is well known that $\mathfrak{h}_1'$ is a polarisation at f verifying the Pukanszky condition (cf. [23]). Let us introduce

$$\mathfrak{g}_0 = \{X \in \mathfrak{g}; [X, \mathfrak{a}] \subset \mathfrak{n}_1\}.$$

Case 1. $\mathfrak{h}_1 \subset \widetilde{\mathfrak{g}}$ and $\mathfrak{h}_2 \not\subset \widetilde{\mathfrak{g}}$. It matters only that $j = m$, otherwise we can apply the induction hypothesis to $\mathfrak{h}_1, \mathfrak{h}_2'$ contained in $\widetilde{\mathfrak{g}}$. If $\mathfrak{h}_2 \subset \mathfrak{g}_0$, then $[\mathfrak{h}_2, \mathfrak{n}] \subset \mathfrak{n} \cap \widetilde{\mathfrak{g}}$ and hence $[\mathfrak{h}_2, \mathfrak{n}_j] \subset \mathfrak{n}_{j-1}$, from which comes the result. If $\mathfrak{h}_2 \not\subset \mathfrak{g}_0$, the Pukanszky condition implies that $\mathfrak{h}_2^0 \not\subset \mathfrak{g}_0$ and hence there is a nilpotent element in $\mathfrak{h}_2 \setminus \mathfrak{h}_2^0$ which serves as a co-exponential basis in $\mathfrak{g}$ to $\widetilde{\mathfrak{g}}$.

Case 2. $\mathfrak{h}_1 \not\subset \widetilde{\mathfrak{g}}$ and $\mathfrak{h}_2 \subset \widetilde{\mathfrak{g}}$. Just as in the case 1, we are led to the situation where $j = m$. If $\mathfrak{h}_1 \subset \mathfrak{g}_0$, $[\mathfrak{h}_1 + \mathfrak{n}, \mathfrak{n}] \subset \mathfrak{n}_{j-1}$ and hence $[\mathfrak{m}_j, \mathfrak{n}_j] \subset \mathfrak{n}_{j-1}$, further $[\mathfrak{h}_2 \cap \mathfrak{m}_j, \mathfrak{n}_j] \subset \mathfrak{n}_{j-1}$. If $\mathfrak{h}_1 \not\subset \mathfrak{g}_0$, the Pukanszky condition demands just as in Case 1 that $\mathfrak{h}_1 \cap \mathfrak{n}_j \neq \mathfrak{h}_1 \cap \mathfrak{n}_{j-1}$. Namely, $\mathfrak{m}_j = \mathfrak{m}_{j-1}$ and hence $j \notin I^{\mathfrak{h}_2}$.

Case 3. $\mathfrak{h}_1 \not\subset \widetilde{\mathfrak{g}}$ and $\mathfrak{h}_2 \not\subset \widetilde{\mathfrak{g}}$. First, let $\mathfrak{a} \not\subset \mathfrak{h}_1 + \mathfrak{h}_2$. Then, there exists $0 \neq X \in \mathfrak{h}_1 \cap \mathfrak{h}_2$ outside of $\widetilde{\mathfrak{g}}$. If $j < m$, modifying by X the concerned elements of $\mathfrak{h}_1, \mathfrak{h}_2$, we immediately see that

$$\mathfrak{h}_2' \cap (\mathfrak{n}_j + \mathfrak{h}_1') \neq \mathfrak{h}_2' \cap (\mathfrak{n}_{j-1} + \mathfrak{h}_1').$$

If $j = m$, we can proceed as in the previous cases. Indeed, if $\mathfrak{h}_2 \not\subset \mathfrak{g}_0$, we can take in $[\mathfrak{h}_2, X]$ a co-exponential basis in $\mathfrak{g}$ to $\tilde{\mathfrak{g}}$. Let $\mathfrak{h}_2 \subset \mathfrak{g}_0$. If $\mathfrak{h}_1 \subset \mathfrak{g}_0$, we can apply the induction hypothesis to $\mathfrak{g}_0$. If $\mathfrak{h}_1 \not\subset \mathfrak{g}_0$, we have that $\mathfrak{h}_1 \cap \mathfrak{n}_m \neq \mathfrak{h}_1 \cap \mathfrak{n}_{m-1}$. Namely, $\mathfrak{m}_m = \mathfrak{m}_{m-1}$ and hence $m \notin I^{\mathfrak{h}_2}$.

Suppose in what follows that $\mathfrak{a} \subset \mathfrak{h}_1 + \mathfrak{h}_2$. First of all let $\mathfrak{a} \not\subset \mathfrak{h}_1^0 + \mathfrak{h}_2^0$. Write $Y = T_2 - T_1$ with $T_i \in \mathfrak{h}_i \backslash \mathfrak{h}_i^0$ for $1 \leq i \leq 2$. We know that $T_i \in \mathfrak{g}_0$ $(1 \leq i \leq 2)$ from the Pukanszky condition (cf. Lemma 5.3.2 below). By replacing X by T_1, T_2 we can make the same argument as in the precedent situation.

Finally, let us treat the essential case where $\mathfrak{a} \subset \mathfrak{h}_1^0 + \mathfrak{h}_2^0$. Let j_0 be the smallest index such that there exists $V \in \mathfrak{n}_{j_0} \backslash \mathfrak{n}_{j_0 - 1}$ which is written $V = T_2 - T_1$ with $T_i \in \mathfrak{h}_i \backslash \mathfrak{h}_i^0$ $(1 \leq i \leq 2)$. Then $(\mathfrak{h}_1^0 + \mathfrak{h}_2^0) \cap \mathfrak{n}_{j_0} \subset \mathfrak{n}_{j_0-1}$. If $j_0 = m$, we can repeat the arguments utilised in the case 1 or 2. Thus, let $j_0 < m$, namely $V \in \tilde{\mathfrak{g}}$. When we write $Y = X_2 - X_1$ with $X_i \in \mathfrak{h}_i^0$ $(1 \leq i \leq 2)$, we have $\mathfrak{h}_1' \cap \mathfrak{h}_2 = \mathbb{R}X_2 + (\mathfrak{h}_1 \cap \mathfrak{h}_2)$ and $\mathfrak{h}_1 \cap \mathfrak{h}_2' = \mathbb{R}X_1 + (\mathfrak{h}_1 \cap \mathfrak{h}_2)$. Since $(\mathfrak{h}_1 + \mathfrak{h}_2) \cap \mathfrak{n}_{j_0-1} \subset \mathfrak{h}_1^0 + \mathfrak{h}_2^0$, we ascertain

$$B_f(X_2, \mathfrak{n}_{j_0-1}) = B_f(X_1, \mathfrak{n}_{j_0-1}) = \{0\}$$

by modifying X_i by the elements of $\mathfrak{h}_1 \cap \mathfrak{h}_2$ if necessary. In fact, let $\mathfrak{n}_k = \mathbb{R}V_k + \mathfrak{n}_{k-1}$ with $k \leq j_0 - 1$. If $V_k \notin \mathfrak{h}_1 + \mathfrak{h}_2$, there exists $x_k \in \mathfrak{h}_1 \cap \mathfrak{h}_2$ such that $B_f(x_k, V_k) \neq 0$. Modifying X_j by x_k, we can assume $B_f(X_j, V_k) = 0$. If $V_k \in \mathfrak{h}_1 + \mathfrak{h}_2$, then $V_k = x_2 - x_1$ with $x_j \in \mathfrak{h}_j^0$. This means $B_f(X_1, x_2) = B_f(X_2, x_1) = 0$ and hence $B_f(X_2, V_k) = B_f(X_1, V_k) = 0$. Moreover, we can suppose that $f(X_1) = f(X_2) = 0$ by modifying X_i by Z if necessary. Let us normalise T_i or Y so that $B_f(T_i, Y) = 1$. This leads to

$$B_f(T_1, X_2) = 1,\ B_f(T_2, X_1) = B_f(V, X_1) = B_f(V, X_2) = -1.$$

Lemma 5.3.2 *Let us assume more generally that $\mathfrak{g}$ is exponential here. We can take T_j in $\mathfrak{g}_0$, for $1 \leq j \leq 2$.*

Proof 5.2 *Suppose for a while that $\mathfrak{g}$ is completely solvable and that $T_j \notin \mathfrak{g}_0$. Then, from the Pukanszky condition there exist (cf. [23]) $T_j' \in \mathfrak{h}_j^0$ outside $\mathfrak{g}_0$ and hence two elements $0 \neq v_j = [T_j', T_j] \in (\mathfrak{h}_j \cap \mathfrak{n}) \backslash \mathfrak{h}_j^0$ $(j = 1, 2)$ and let k_0 be the smallest index satisfying $a = v_2 - v_1 \in \mathfrak{n}_{k_0}$ by the use of such nilpotent elements v_1, v_2 in $\mathfrak{h}_j \setminus \mathfrak{h}_j^0$. This brings that $j_0 \leq k_0 \leq m - 1$. If $k_0 = j_0$, we can choose $T_j = v_j \in \mathfrak{n} \subset \mathfrak{g}_0$. Suppose hence $j_0 < k_0$. This being done, from the induction hypothesis there exists $S_j \in \mathfrak{h}_j^0$ $(j = 1, 2)$ such that*

$$b = S_2 - S_1 \in \mathfrak{n}_{k_0} \backslash \mathfrak{n}_{k_0-1},\quad [S_j, \mathfrak{n}_{k_0}] \subset \mathfrak{n}_{k_0-1}.$$

This says that

$$[S_2, a] = [S_2, v_2] - [S_2, v_1] \in \mathfrak{n}_{k_0-1},$$

namely,

$$\begin{aligned} [S_2, a] &= [S_2, v_2] - [S_1 + b, v_1] \\ &= [S_2, v_2] - [S_1, v_1] - [b, v_1] \in \mathfrak{n}_{k_0-1}. \end{aligned}$$

Consequently,

$$[S_2, v_2] - [S_1, v_1] \in \mathfrak{n}_{k_0-1},$$

which contradicts the choice of k_0 unless $S_j \in \mathfrak{g}_0$ $(j = 1, 2)$.

By considering an appropriate multiple of S_j, we can suppose that

$$b + a = (S_2 + v_2) - (S_1 + v_1) \in \mathfrak{n}_{k_0-1}.$$

Then,

$$\begin{aligned}\mathfrak{n}_{k_0-1} \ni [T_2, b + a] &= [T_2, S_2 + v_2] - [T_2, S_1 + v_1] \\ &= [T_2, S_2 + v_2] - [T_1 + V, S_1 + v_1] \\ &= [T_2, S_2 + v_2] - [T_1, S_1 + v_1] - [V, S_1 + v_1].\end{aligned}$$

Since $[V, S_1 + v_1] \in \mathfrak{n}_{j_0}$ and $j_0 \leq k_0 - 1$, we have

$$[T_2, S_2 + v_2] - [T_1, S_1 + v_1] \in \mathfrak{n}_{k_0-1}.$$

All taken into account, if $T_j \notin \mathfrak{g}_0$, this contradicts the choice of k_0 because $[T_j, S_j + v_j] \notin \mathfrak{h}_j^0$ $(j = 1, 2)$. In sum, $T_j \in \mathfrak{g}_0$.

Now suppose that $\mathfrak{g}$ is exponential. Let $k_0 = j_0$. If $[T_j, \mathfrak{n}_{j_0}] \not\subset \mathfrak{n}_{j_0-1}$, the equation

$$[T_2, a] = [T_2, v_2] - [T_1, v_1] - [V, v_1]$$

shows that we can take T_j in $\mathfrak{h}_j \cap \mathfrak{n}$. Thus, we can assume $j_0 < k_0$ and we can repeat the above arguments done for the completely solvable case.

Remark 5.3.3 In the proof of Lemma 5.3.2, if $j_0 = k_0$, we establish our assertion by taking $T_j = v_j (j = 1, 2)$. Besides, if $j_0 < k_0$, it is obligatory that $T_j \in \mathfrak{g}_0 (j = 1, 2)$. So, when we write an element A of $\mathfrak{n}_i, i \leq j_0$, as $A = a_2 - a_1$ with $a_i \in \mathfrak{h}_i (i = 1, 2)$, we can always suppose that $a_i \in \mathfrak{g}_0 (i = 1, 2)$. For example, this is the case for X_i introduced above. (See the proof of the next lemma.)

Taking into account $B_f(T_1, X_2) = 1$ and $B_f(T_2, X_1) = -1$, write

$$[T_1, X_1] = \beta X_1 + w_1, \ [T_2, X_2] = \gamma X_2 + w_2$$

with $w_j \in \mathfrak{h}_j^0 (j = 1, 2)$ verifying

$$B_f(T_1, w_2) = B_f(T_2, w_1) = 0.$$

Then, it is immediate that

$$\begin{aligned}[V, X_1] = [V, X_2] &= [T_2 - T_1, X_2] = [T_2, X_2] - [T_1, X_2] \\ &= [T_2, X_2] - [T_1, X_1 + Y] = \gamma X_2 + w_2 - (\beta X_1 + w_1) - Z \\ &= (\gamma - \beta) X_2 + w_2 - w_1 + \beta Y - Z \\ &= (\gamma - \beta) X_1 + w_2 - w_1 + \gamma Y - Z.\end{aligned} \tag{5.3.1}$$

Let us calculate further

$$B_f(T_2, [V, X_2]) = f([T_2, [V, X_2]]) = (\gamma - \beta)f([T_2, X_2]) + \beta f(Z) = \beta.$$

On the other hand,

$$\begin{aligned} B_f(T_2, [V, X_2]) &= B_f([T_2, V], X_2) + B_f(V, [T_2, X_2]) \\ =&B_f([T_2, V], X_2) + B_f(V, \gamma X_2 + w_2) = B_f([T_2, V], X_2) - \gamma. \end{aligned}$$

Consequently,

$$[T_1, V] = [T_2, V] = -(\beta + \gamma)V + W \tag{5.3.2}$$

with a certain element $W \in \mathfrak{n}_{j_0-1}$. So, the task consists in showing that $\beta + \gamma = 0$, which will close the proof thanks to Proposition 5.3.2. Note that this aim is immediately attained if both $\mathfrak{h}_1$ and $\mathfrak{h}_2$ are abelian.

5.3.4 Extended computations

Step 2: Towards $\beta + \gamma = 0$

Let us first compute:

$$\begin{aligned} [T_2, [V, X_2]] &= (\gamma - \beta)[T_2, X_2] + [T_2, w_2] - [T_2, w_1] + \beta Z \\ &= -(\beta + \gamma)[V, X_2] + [W, X_2] + \gamma[V, X_2] + [V, w_2] \\ &= -\beta\{(\gamma - \beta)X_2 + w_2 - w_1 + \beta Y - Z\} + [W, X_2] + [V, w_2]. \end{aligned}$$

Hence,

$$\begin{aligned} &(\gamma - \beta)[T_2, X_2] + [T_2, w_2] - [T_2, w_1] \\ &= -\beta(\gamma - \beta)X_2 - \beta w_2 + \beta w_1 - \beta^2 Y + [W, X_2] + [V, w_2]. \end{aligned} \tag{5.3.3}$$

Thus, in particular $f([W, X_2]) = 0$.

The three equations (5.3.1)-(5.3.3) are our poor tools.

As we may replace V by any element of $\mathfrak{g}_{j_0}$ which is found in $\mathfrak{h}_1 + \mathfrak{h}_2$ but outside of $\mathfrak{h}_1^0 + \mathfrak{h}_2^0$, think for a while about the possibility to modify V so that we would have $\gamma = 0$ for instance. Let $U = R_2 - R_1 \in \mathfrak{n}_{j_0-1}$ with $R_j \in \mathfrak{h}_j^0$ $(j = 1, 2)$. Let us compute γ', the new γ, for $V + \alpha U$. We find that

$$\gamma' = B_f(T_1 + \alpha R_1, [T_2 + \alpha R_2, X_2]).$$

Namely,

$$\begin{aligned} \gamma' =&B_f(R_1, [R_2, X_2])\alpha^2 + (B_f(R_1, [T_2, X_2]) \\ &+ B_f(T_1, [R_2, X_2]))\alpha + \gamma \end{aligned} \tag{5.3.4}$$

and similarly

$$\begin{aligned} -\beta' =&B_f(R_2, [R_1, X_1])\alpha^2 + (B_f(R_2, [T_1, X_1]) \\ &+ B_f(T_2, [R_1, X_1]))\alpha - \beta. \end{aligned} \tag{5.3.5}$$

Thus,

$$\beta' + \gamma' = \beta + \gamma$$
$$+ (B_f(R_1, w_2) + B_f(T_1, [R_2, X_2]) - B_f(R_2, w_1) - B_f(T_2, [R_1, X_1]))\alpha$$

and we may choose $\alpha \in \mathbb{R}$ in such a manner that

$$\beta' + \gamma' = 0$$

provided that

$$B_f(R_1, w_2) + B_f(T_1, [R_2, X_2]) - B_f(R_2, w_1) - B_f(T_2, [R_1, X_1]) \neq 0.$$

We can hence suppose that

$$B_f(R_1, w_2) + B_f(T_1, [R_2, X_2]) - B_f(R_2, w_1) - B_f(T_2, [R_1, X_1]) = 0.$$

When we make the changes

$$T_j \mapsto kT_j, \ X_j \mapsto k^{-1}X_j \ (k \in \mathbb{R} \setminus \{0\}, \ j = 1, 2)$$

in the expressions of (5.3.4) and (5.3.5), we check that

$$B_f(R_1, [R_2, X_2]) = B_f(R_2, [R_1, X_1]) = 0,$$

otherwise we can choose the sign of $B_f(R_1, [R_2, X_2])$ by multiplying it by k^{-1} so that we would have a real α such that $\gamma' = 0$ for example. Consequently,

$$B_f(R_1, [T_2, X_2]) + B_f(T_1, [R_2, X_2]) = B_f(R_2, [T_1, X_1]) + B_f(T_2, [R_1, X_1]) = 0, \tag{5.3.6}$$

otherwise we may arrive to the situation $\beta + \gamma = 0$ by a suitable choice of α. The last equation (5.3.6) is written again

$$B_f(R_1, w_2) + B_f(T_1, [R_2, X_2]) = B_f(R_2, w_1) + B_f(T_2, [R_1, X_1]) = 0.$$

Next, as

$$B_f(T_2, [R_1, X_1]) = B_f([T_2, R_1], X_1) + B_f(R_1, [T_2, X_1]) = B_f(R_1, [V, X_1]) = B_f(R_1, w_2),$$

we have

$$B_f(U, w_1 - w_2) = B_f(R_2 - R_1, w_1 - w_2) = B_f(R_1, w_2) + B_f(R_2, w_1) = 0.$$

Lemma 5.3.3 $R_j \in \mathfrak{g}_0$ *for* $1 \leq j \leq 2$.

Proof 5.3 *If $R_j \notin \mathfrak{g}_0$, we find*

$$[R_2, T_2] - [R_1, T_1] = [R_2, V] - [T_1, U]$$

in $\mathfrak{n}_{j_0}$. We can take this relation instead of $T_2 - T_1 = V$, if $[R_2, V] \notin \mathfrak{n}_{j_0-1}$. Otherwise, $[R_2, V] \in \mathfrak{n}_{j_0-1}$ and

$$[R_j, T_j] \notin \mathfrak{h}_j^0 \ (j = 1, 2)$$

if $R_j \notin \mathfrak{g}_0$, the above relation contradicts the choice of j_0.

Now let us start the following computations. Put

$$\begin{aligned} B_f(T_1, [T_1, w_2]) = p, \ B_f(T_2, [T_2, w_1]) = q, \\ B_f(T_1, [T_2, w_1]) = r, \ B_f(T_2, [T_1, w_2]) = s. \end{aligned} \tag{5.3.7}$$

and let us begin our calculations. From (5.3.3),

$$[T_1, w_2] - [T_2, w_1] = (\beta^2 - \gamma^2)X_2 + \beta w_1 - \gamma w_2 - \beta^2 Y + [W, X_2]. \tag{5.3.8}$$

Since

$$\begin{aligned} B_f(T_2, [W, X_2]) = B_f(W, [T_2, X_2]) = B_f(W, w_2) \\ B_f(T_1, [W, X_1]) = B_f(W, [T_1, X_1]) = B_f(W, w_1), \end{aligned}$$

it follows from (5.3.8) that

$$p - r = B_f(W, w_1) - \gamma^2, \ q - s = \beta^2 - B_f(W, w_2). \tag{5.3.9}$$

Furthermore,

$$\begin{aligned} r = B_f(T_2, [T_1, w_1]) + B_f(W, w_1), \\ s = B_f(T_1, [T_2, w_2]) - B_f(W, w_2). \end{aligned} \tag{5.3.10}$$

We emphasise

$$B_f(W, X_1) = B_f(W, X_2) = 0. \tag{5.3.11}$$

Suppose first $r \neq 0$, $s \neq 0$ and put

$$\mathfrak{m}_i = \{X \in \mathfrak{h}_i; B_f(T_j, X) = B_f(T_i, [T_j, X]) = 0\}$$

for $1 \leq i \neq j \leq 2$. Then, as vector spaces,

$$\mathfrak{h}_i = \mathbb{R}X_i \oplus \mathbb{R}w_i \oplus \mathfrak{m}_i \ (1 \leq i \leq 2).$$

Let

$$\begin{aligned} [T_1, w_1] = (B_f(W, w_1) - r)X_1 + bw_1 + u, \ u \in \mathfrak{m}_1, \\ [T_2, w_2] = (s + B_f(W, w_2))X_2 + dw_2 + v, \ v \in \mathfrak{m}_2. \end{aligned} \tag{5.3.12}$$

Further, let us consider the subspaces

$$\begin{aligned}\mathfrak{w}_i &= \{X \in \mathfrak{g}; B_f(T_1, X) = B_f(T_2, X) \\ &= B_f(T_i, [T_j, X]) = B_f(T_i, [T_i, X]) = 0\}\end{aligned}$$

for $1 \leq i \neq j \leq 2$. Then, supposing first $pq \neq 0$, we have

$$\mathfrak{g} = \mathbb{R}X_2 \oplus \mathbb{R}X_1 \oplus \mathbb{R}w_1 \oplus \mathbb{R}w_2 \oplus \mathfrak{w}_i.$$

We immediately see that $\{X_1, X_2, w_1, w_2\}$ are linearly independent if $pq \neq 0$, what will turn out little important in the future situation where $sr = 0$ or $pq = 0$. In fact, if $r = 0$ or $s = 0$, we can take as $\mathfrak{m}_i$ or $\mathfrak{w}_i (i = 1, 2)$ a subspace of $\mathfrak{h}_i$ or $\mathfrak{g}$ complementary to $\mathbb{R}X_i \oplus \mathbb{R}w_i$ or $\mathbb{R}X_1 \oplus \mathbb{R}X_2 \oplus \mathbb{R}w_1 \oplus \mathbb{R}w_2$. Let

$$\begin{aligned}[T_1, w_2] &= pX_2 + aw_2 - sX_1 + ew_1 + m, \ m \in \mathfrak{w}_1, \\ [T_2, w_1] &= rX_2 + cw_2 - qX_1 + hw_1 + n, \ n \in \mathfrak{w}_2.\end{aligned} \tag{5.3.13}$$

From these, the relations (5.3.9), (5.3.13) give us

$$\begin{aligned}&[W, X_1] = [W, X_2] \\ =&[T_1, w_2] - [T_2, w_1] + (\gamma^2 - \beta^2)X_2 + \gamma w_2 - \beta w_1 + \beta^2 Y \\ =&(p - r + \gamma^2)X_2 + (a - c + \gamma)w_2 + (q - s - \beta^2)X_1 \\ &+ (e - h - \beta)w_1 + m - n \\ =&(a - c + \gamma)w_2 + (e - h - \beta)w_1 + B_f(W, w_2)Y \\ &+ (B_f(W, w_1) - B_f(W, w_2))X_2 + m - n.\end{aligned} \tag{5.3.14}$$

We further calculate:

$$[T_2, [T_1, w_1]] = (B_f(W, w_1) - r)[T_2, X_1] + b[T_2, w_1] + [T_2, u]. \tag{5.3.15}$$

Taking the value $B_f(T_1, \cdot)$ of both members, we have

$$B_f(T_1, [T_2, [T_1, w_1]]) = (B_f(W, w_1) - r)\gamma + br(+B_f(T_1, [T_2, u]) \text{ if } r = 0).$$

On the other hand,

$$\begin{aligned}B_f(T_1, [T_2, [T_1, w_1]]) =&B_f(T_1, [(\beta + \gamma)V - W, w_1]) \\ &+ B_f(T_1, [T_1, rX_2 + cw_2 - qX_1 + hw_1 + n]) \\ =&(\beta + \gamma)r - B_f(T_1, [W, w_1]) + cp + B_f(T_1, [T_1, n]).\end{aligned}$$

Hence

$$\begin{aligned}B_f(T_1, [W, w_1]) = (r - B_f(W, w_1))\gamma + (\beta + \gamma - b)r \\ +cp + B_f(T_1, [T_1, n])(-B_f(T_1, [T_2, u]) \text{ if } r = 0).\end{aligned} \tag{5.3.16}$$

Taking the value $B_f(T_2, \cdot)$ of both members of equation (5.3.15), we have

$$B_f(T_2, [T_2, [T_1, w_1]]) = bq + B_f(T_2, [T_2, u]).$$

On the other hand,

$$\begin{aligned}
&B_f(T_2, [T_2, [T_1, w_1]]) = B_f(T_2, [(\beta + \gamma)V - W, w_1]) \\
&+B_f(T_2, [T_1, rX_2 + cw_2 - qX_1 + hw_1 + n]) \\
&=(\beta + \gamma)(q + B_f(W, w_1) - r) - B_f(T_2, [W, w_1]) \\
&+\beta(q - r) + cs + h(r - B_f(W, w_1))(+B_f(T_2, [T_1, n]) \text{ if } s = 0).
\end{aligned}$$

Hence,

$$\begin{aligned}
&B_f(T_2, [W, w_1]) = cs + (2\beta + \gamma - b)q + (h - 2\beta - \gamma)r \\
&+(\beta + \gamma - h)B_f(W, w_1) - B_f(T_2, [T_2, u]) \\
&(+B_f(T_2, [T_1, n]) \text{ if } s = 0).
\end{aligned} \tag{5.3.17}$$

Now,

$$[T_1, [T_2, w_2]] = (s + B_f(W, w_2))[T_1, X_2] + d[T_1, w_2] + [T_1, v]. \tag{5.3.18}$$

Taking the value $B_f(T_1, \cdot)$ of both members of this relation, we see

$$B_f(T_1, [T_1, [T_2, w_2]]) = dp + B_f(T_1, [T_1, v]).$$

On the other hand,

$$\begin{aligned}
&B_f(T_1, [T_1, [T_2, w_2]]) = B_f(T_1, [-(\beta + \gamma)V + W, w_2]) \\
&+B_f(T_1, [T_2, pX_2 + aw_2 - sX_1 + ew_1 + m]) \\
&= -(\beta + \gamma)(s + B_f(W, w_2) - p) + B_f(T_1, [W, w_2]) \\
&+\gamma(p - s) + a(s + B_f(W, w_2)) + er(+B_f(T_1, [T_2, m]) \text{ if } r = 0).
\end{aligned}$$

Thus,

$$\begin{aligned}
&B_f(T_1, [W, w_2]) = (d - \beta - 2\gamma)p + (\beta + 2\gamma - a)s - er \\
&+(\beta + \gamma - a)B_f(W, w_2) + B_f(T_1, [T_1, v]) \\
&(-B_f(T_1, [T_2, m]) \text{ if } r = 0).
\end{aligned} \tag{5.3.19}$$

Next, we take the value $B_f(T_2, \cdot)$ of both members of the equation (5.3.18). On one hand,

$$B_f(T_2, [T_1, [T_2, w_2]]) = -\beta(s + B_f(W, w_2)) + ds(+B_f(T_2, [T_1, v]) \text{ if } s = 0).$$

On the other hand,

$$\begin{aligned}
B_f(T_2, [T_1, [T_2, w_2]]) =& B_f(T_2, [-(\beta + \gamma)V + W, w_2]) \\
&+ B_f(T_2, [T_2, pX_2 + aw_2 - sX_1 + ew_1 + m]) \\
=&(\beta + \gamma)s + B_f(T_2, [W, w_2]) + eq + B_f(T_2, [T_2, m]).
\end{aligned}$$

Therefore,

$$\begin{aligned}B_f(T_2,[W,w_2]) =&(d-2\beta-\gamma)s-eq-\beta B_f(W,w_2)-B_f(T_2,[T_2,m])\\&(+B_f(T_2,[T_1,v]) \text{ if } s=0).\end{aligned} \tag{5.3.20}$$

Now,

$$\begin{aligned}[V,[V,X_2]] =&(\gamma-\beta)\{(\gamma-\beta)X_2+w_2-w_1+\beta Y-Z\}\\&+[V,w_2-w_1].\end{aligned} \tag{5.3.21}$$

Hence, on one hand

$$\begin{aligned}&B_f(T_2,[T_2,[V,[V,X_2]]])\\&=B_f(T_2,[T_2,(\gamma-\beta)\{(\gamma-\beta)X_2+w_2-w_1\}+[V,w_2-w_1]])\\&=(\beta-\gamma)B_f(T_2,[T_2,w_1])+B_f(T_2,[T_2,[V,w_2-w_1]])\\&=(\beta-\gamma)q+B_f(T_2,[T_2,sX_1-ew_1-m+qX_1-hw_1-n\\&+(B_f(W,w_1)-r)X_1+bw_1+u])\\&=(\beta-\gamma)q+(b-e-h)q-B_f(T_2,[T_2,m+n])+B_f(T_2,[T_2,u])\\&=(\beta-\gamma+b-e-h)q-B_f(T_2,[T_2,m+n])+B_f(T_2,[T_2,u]).\end{aligned} \tag{5.3.22}$$

On the other hand,

$$\begin{aligned}&B_f(T_2,[T_2,[V,[V,X_2]]])\\&=B_f(T_2,[-(\beta+\gamma)V+W,(\gamma-\beta)X_2+w_2-w_1])\\&+B_f(T_2,[V,[-(\beta+\gamma)V+W,X_2]])\\&+B_f(T_2,[V,[V,\gamma X_2+w_2]])\\&=(\beta^2-\gamma^2)\beta-(\beta+\gamma)B_f(T_2,[V,w_2-w_1])\\&+(\gamma-\beta)B_f(W,w_2)+B_f(T_2,[W,w_2-w_1])\\&-(\beta+\gamma)B_f(T_2,[V,(\gamma-\beta)X_2+w_2-w_1])\\&+B_f(T_2,[V,(a-c+\gamma)w_2+(e-h-\beta)w_1\\&+(B_f(W.w_1)-B_f(W,w_2))X_2+m-n])\\&+\gamma B_f(T_2,[V,(\gamma-\beta)X_2+w_2-w_1])\\&+B_f(T_2,[V,(s+B_f(W,w_2))X_2+dw_2+v\\&-pX_2-aw_2+sX_1-ew_1-m])\\&=\beta(\beta^2-\gamma^2)+(\beta+\gamma)(s+q+B_f(W,w_1)-r)\\&+(\gamma-\beta)B_f(W,w_2)+B_f(T_2,[W,w_2-w_1])-(\gamma^2-\beta^2)\beta\\&-(\beta+\gamma)B_f(T_2,[V,w_2-w_1])-(a-c+\gamma)s\\&+(e-h-\beta)(q+B_f(W,w_1)-r)\\&+\beta(B_f(W,w_1)-B_f(W,w_2))\\&+B_f(T_2,[V,m-n])+\beta\gamma(\gamma-\beta)+\gamma B_f(T_2,[V,w_2-w_1])\\&+\beta(s+B_f(W,w_2))-ds+B_f(T_2,[V,v])+\beta(s-p)+as\end{aligned}$$

$$
\begin{aligned}
&-e(q+B_f(W,w_1)-r)-B_f(T_2,[V,m])\\
&=2\beta(\beta^2-\gamma^2)+(2\beta+\gamma)(s+q-r+B_f(W,w_1))\\
&+(\gamma-\beta)B_f(W,w_2)+B_f(T_2,[W,w_2-w_1])-(a-c+\gamma)s\\
&+(e-h-\beta)(q+B_f(W,w_1)-r)\\
&+\beta(B_f(W,w_1)-B_f(W,w_2))\\
&+B_f(T_2,[V,m-n])+\beta\gamma(\gamma-\beta)+\beta(s+B_f(W,w_2))\\
&+(a-d)s+B_f(T_2,[V,v])+\beta(s-p)\\
&+e(r-B_f(W,w_1)-q)-B_f(T_2,[V,m]).
\end{aligned}
\tag{5.3.23}
$$

Step 3: $(2\beta+3\gamma)q=0$.

Putting in equality the results of the two last calculations, we have

$$
\begin{aligned}
&B_f(T_2,[W,w_2-w_1])\\
&=2\beta(\gamma^2-\beta^2)+(2\beta+\gamma)(r-s-q-B_f(W,w_1))\\
&+(\beta-\gamma)B_f(W,w_2)+(a-c+\gamma)s\\
&+(\beta+h-e)(q+B_f(W,w_1)-r)-\beta B_f(W,w_1)\\
&-B_f(T_2,[V,m-n])+\beta\gamma(\beta-\gamma)\\
&+(d-a-\beta)s-B_f(T_2,[V,v])+\beta(p-s)\\
&+e(q+B_f(W,w_1)-r)+B_f(T_2,[V,m])\\
&+(\beta+b-\gamma-e-h)q-B_f(T_2,[T_2,m+n])+B_f(T_2,[T_2,u]).
\end{aligned}
\tag{5.3.24}
$$

On the other hand, equations (5.3.17) and (5.3.20) become

$$
\begin{aligned}
&B_f(T_2,[W,w_2-w_1])\\
&=(d-2\beta-\gamma)s-\beta B_f(W,w_2)-eq+B_f(T_2,[T_1,v])\\
&-B_f(T_2,[T_2,m])-cs-(2\beta+\gamma-b)q-(h-2\beta-\gamma)r\\
&-(\beta+\gamma-h)B_f(W,w_1)+B_f(T_2,[T_2,u])-B_f(T_2,[T_1,n]).
\end{aligned}
$$

Now, we have from (5.3.21):

$$
\begin{aligned}
&B_f(T_1,[T_2,[V,[V,X_2]]])=B_f(-(\beta+\gamma)V+W,\\
&(\gamma-\beta)((\gamma-\beta)X_2+w_2-w_1)+[V,w_2-w_1])\\
&+B_f(T_2,[-(\beta+\gamma)V+W,(\gamma-\beta)X_2+w_2-w_1])\\
&+B_f(T_2,[V,[-(\beta+\gamma)V+W,X_2]])+B_f(T_2,[V,[V,\beta X_1+w_1]])\\
&=(\beta+\gamma)(\beta-\gamma)^2-(\beta+\gamma)B_f(V,[V,w_2-w_1])\\
&+(\gamma-\beta)B_f(W,w_2-w_1)+B_f(W,[V,w_2-w_1])-\beta(\gamma^2-\beta^2)\\
&-(\beta+\gamma)B_f(T_2,[V,w_2-w_1])+(\gamma-\beta)B_f(T_2,[W,X_2])\\
&+B_f(T_2,[W,w_2-w_1])-(\beta+\gamma)B_f(T_2,[V,(\gamma-\beta)X_2+w_2-w_1])\\
&+B_f(T_2,[V,[W,X_2]])+\beta B_f(T_2,[V,(\gamma-\beta)X_2+w_2-w_1])\\
&+B_f(T_2,[V,[V,w_1]])
\end{aligned}
$$

$$
\begin{aligned}
=&(\beta+\gamma)(\gamma-\beta)^2-(\beta+\gamma)B_f(V,[V,w_2-w_1])\\
&+(\gamma-\beta)B_f(W,w_2-w_1)+B_f(W,[V,w_2-w_1])+\beta(\beta^2-\gamma^2)\\
&-(\beta+2\gamma)B_f(T_2,[V,w_2-w_1])+(\gamma-\beta)B_f(W,w_2)\\
&+B_f(T_2,[W,w_2-w_1])+\beta(\beta^2-\gamma^2)+B_f(T_2,[V,(a-c+\gamma)w_2\\
&+(e-h-\beta)w_1+(B_f(W,w_1)-B_f(W,w_2))X_2+m-n])\\
&+\beta^2(\gamma-\beta)+B_f(T_2,[V,rX_2+cw_2-qX_1+hw_1+n\\
&-(B_f(W,w_1)-r)X_1-bw_1-u])\\
=&(\beta+\gamma)(\gamma-\beta)^2+(\beta+\gamma)B_f(T_1,[V,w_2-w_1])\\
&-(2\beta+3\gamma)B_f(T_2,[V,w_2-w_1])+(\gamma-\beta)B_f(W,w_2)\\
&+(\beta-\gamma)B_f(W,w_1)+B_f(W,[T_2,w_2-w_1])-B_f(W,[T_1,w_2-w_1])\\
&+\beta(\beta^2-\gamma^2)+(\gamma-\beta)B_f(W,w_2)+B_f(T_2,[W,w_2-w_1])\\
&+\beta(\beta^2-\gamma^2)-(a-c+\gamma)s+(e-h-\beta)(q+B_f(W,w_1)-r)\\
&+\beta(B_f(W,w_1)-B_f(W,w_2))+B_f(T_2,[V,m-n])+\beta^2(\gamma-\beta)\\
&+\beta(r-q+r-B_f(W,w_1))-cs+(h-b)(q+B_f(W,w_1)-r)\\
&+B_f(T_2,[V,n-u])\\
=&(\beta+\gamma)(\gamma-\beta)^2+(\beta+\gamma)(s+B_f(W,w_2)-p-r)\\
&-(2\beta+3\gamma)(q-s+r-B_f(W,w_1))+(\gamma-\beta)B_f(W,w_2)\\
&+(\beta-\gamma)B_f(W,w_1)+B_f(W,dw_2+v-cw_2-hw_1-n)\\
&-B_f(W,aw_2+ew_1+m-bw_1-u)+\beta(\beta^2-\gamma^2)\\
&+(\gamma-\beta)B_f(W,w_2)+B_f(T_2,[W,w_1-w_2])+\beta(\beta^2-\gamma^2)\\
&-(a-c+\gamma)s+(e-h-\beta)(q+B_f(W,w_1)-r)\\
&+\beta(B_f(W,w_1)-B_f(W,w_2))+B_f(T_2,[V,m-n])+\beta^2(\gamma-\beta)\\
&+\beta(2r-q-B_f(W,w_1))-cs+(h-b)(q+B_f(W,w_1)-r)\\
&+B_f(T_2,[V,n-u]).
\end{aligned}
$$

From (5.3.21), we also get

$$
\begin{aligned}
&B_f(T_1,[T_2,[V,[V,X_2]]])\\
=&B_f(T_1,[T_2,(\gamma-\beta)\{(\gamma-\beta)X_2+w_2-w_1\}+[V,w_2-w_1]])\\
=&(\gamma-\beta)^2\gamma+(\gamma-\beta)B_f(T_1,[T_2,w_2-w_1])\\
&+B_f(T_1,[T_2,[V,w_2-w_1]])\\
=&(\gamma-\beta)^2\gamma+(\gamma-\beta)(s+B_f(W,w_2)-r)\\
&+B_f(T_1,[T_2,(s+B_f(W,w_2))X_2+dw_2+v-pX_2\\
&-aw_2+sX_1-ew_1-m-rX_2-cw_2+qX_1-hw_1-n\\
&+(B_f(W,w_1)-r)X_1+bw_1+u])\\
=&(\gamma-\beta)^2\gamma+(\gamma-\beta)(s+B_f(W,w_2)-r)\\
&+(s+B_f(W,w_2))(\gamma+d-a-c)\\
&+(B_f(W,w_1)-r+s-p-r+q)\gamma\\
&+(b-e-h)r+B_f(T_1,[T_2,u-m-n+v]).
\end{aligned}
\tag{5.3.25}
$$

Comparing these two computations, we find

$$\begin{aligned}
&B_f(T_2,[W,w_2-w_1]) = \beta\gamma^2-\beta(\gamma-\beta)^2-\beta^2\gamma\\
&-\beta(\beta^2-\gamma^2)+(\beta+\gamma)(p+r-s-B_f(W,w_2))\\
&+(2\beta+3\gamma)(q-s+r-B_f(W,w_1))+(\gamma-\beta)B_f(W,w_1)\\
&-(\gamma-2\beta)B_f(W,w_2)-B_f(W,dw_2+v-n)\\
&+B_f(W,ew_1+m-u)+(h+\beta-e)(q+B_f(W,w_1)-r)\\
&+\beta(q-2r)+(\gamma-c)s+(b-h)q-B_f(T_2,[V,m])\\
&+B_f(T_2,[V,u])+(\gamma-\beta)(s-r)\\
&+(s+B_f(W,w_2))(\gamma+d)+(B_f(W,w_1)-2r-p+s+q)\gamma\\
&-er+B_f(T_1,[T_2,u-m-n+v])\\
&=\beta(\gamma^2+\beta\gamma-2\beta^2)-(4\beta+c-d)s+\beta p\\
&+(\beta+\gamma-h)r+(4\beta+4\gamma+b-e)q+(\beta-\gamma)B_f(W,w_2)\\
&-(2\beta+\gamma-h)B_f(W,w_1)-B_f(T_2,[T_2,m])+B_f(T_2,[T_2,u])\\
&+B_f(T_2,[T_1,v])-B_f(T_2,[T_1,n]).
\end{aligned} \tag{5.3.26}$$

Combining this with (5.3.24), we find

$$\beta(\gamma^2+\beta\gamma-2\beta^2)+(4\beta+6\gamma)q=2\beta(\gamma^2-\beta^2)+\beta\gamma(\beta-\gamma).$$

In sum,

$$(2\beta+3\gamma)q=0. \tag{5.3.27}$$

Step 4: $\gamma^2(\gamma-\beta)=0$.

From (5.3.21) we get:

$$\begin{aligned}
&B_f(T_1,[T_2,[V,[V,X_2]]])\\
&=B_f(T_1,[T_2,(\gamma-\beta)\{(\gamma-\beta)X_2+w_2-w_1\}+[V,w_2-w_1]])\\
&=(\gamma-\beta)^2\gamma+(\gamma-\beta)B_f(T_1,[T_2,w_2-w_1])\\
&+B_f(T_1,[T_2,[V,w_2-w_1]])\\
&=(\gamma-\beta)^2\gamma+(\gamma-\beta)(s+B_f(W,w_2)-r)\\
&+B_f(T_1,[T_2,(s+B_f(W,w_2))X_2+dw_2+v-pX_2\\
&-aw_2+sX_1-ew_1-m-rX_2-cw_2+qX_1-hw_1-n\\
&+(B_f(W,w_1)-r)X_1+bw_1+u])\\
&=(\gamma-\beta)^2\gamma+(\gamma-\beta)(s+B_f(W,w_2)-r)\\
&+(s+B_f(W,w_2))(\gamma+d-a-c)\\
&+(B_f(W,w_1)-r+s-p-r+q)\gamma\\
&+(b-e-h)r+B_f(T_1,[T_2,u-m-n+v]).
\end{aligned} \tag{5.3.28}$$

On the other hand,

$$B_f(T_1,[T_2,[V,[V,X_2]]])$$

$$
\begin{aligned}
&= B_f(T_1, [-(\beta+\gamma)V + W, (\gamma-\beta)X_2 + w_2 - w_1]) \\
&+ B_f(T_1, [V, [-(\beta+\gamma)V + W, X_2]]) + B_f(T_1, [V, [V, \gamma X_2 + w_2]]) \\
&= \gamma(\beta^2 - \gamma^2) - (\beta+\gamma)B_f(T_1, [V, w_2 - w_1]) + (\gamma-\beta)B_f(W, w_1) \\
&+ B_f(T_1, [W, w_2 - w_1]) - (\beta+\gamma)B_f(T_1, [V, (\gamma-\beta)X_2 + w_2 - w_1]) \\
&+ B_f(T_1, [V, (a - c + \gamma)w_2 + (e - h - \beta)w_1 \\
&+ (B_f(W, w_1) - B_f(W, w_2))X_2 + m - n]) \\
&+ \gamma B_f(T_1, [V, (\gamma-\beta)X_2 + w_2 - w_1]) \\
&+ B_f(T_1, [V, (s + B_f(W, w_2))X_2 + dw_2 + v \\
&- pX_2 - aw_2 + sX_1 - ew_1 - m]) \\
&= (\beta^2 - \gamma^2)\gamma + (\gamma-\beta)B_f(W, w_1) - 2(\beta+\gamma)B_f(T_1, [V, w_2 - w_1]) \\
&+ B_f(T_1, [W, w_2 - w_1]) - \gamma(\gamma^2 - \beta^2) \\
&+ (a - c + \gamma)(s + B_f(W, w_2) - p) + (e - h - \beta)r + \gamma(B_f(W, w_1) \\
&- B_f(W, w_2)) + B_f(T_1, [V, m - n]) + \gamma^2(\gamma - \beta) \\
&+ \gamma(s + B_f(W, w_2) - p - r) + \gamma(s + B_f(W, w_2)) \\
&+ \gamma(s - p) + (d - a)(s + B_f(W, w_2) - p) - er + B_f(T_1, [V, v - m]) \\
&= (\beta^2 - \gamma^2)\gamma + (\gamma-\beta)B_f(W, w_1) \\
&- (2\beta+\gamma)(s + B_f(W, w_2) - p - r) + B_f(T_1, [W, w_2 - w_1]) \\
&- \gamma(\gamma^2 - \beta^2) + (c - d)(p - s - B_f(W, w_2)) + 2\gamma(s - p) \\
&- (\beta + h)r + \gamma B_f(W, w_1) + B_f(T_1, [V, m - n]) + \gamma^2(\gamma - \beta) \\
&+ \gamma(s + B_f(W, w_2)) + B_f(T_1, [V, v - m]).
\end{aligned}
$$

Equalling these two expressions, we have

$$
\begin{aligned}
&B_f(T_1, [W, w_2 - w_1]) = \gamma(\gamma^2 - \beta^2) + (\beta - \gamma)B_f(W, w_1) \\
&+ (2\beta + \gamma)(s + B_f(W, w_2) - p - r) + \gamma(\gamma^2 - \beta^2) - cp \\
&+ \beta r - \gamma B_f(W, w_1) + B_f(T_1, [V, n]) \\
&+ \gamma^2(\beta - \gamma) + dp + B_f(T_1, [T_1, v]) + (\gamma - \beta)^2\gamma \\
&+ (\gamma - \beta)(s + B_f(W, w_2) - r) - a(s + B_f(W, w_2)) - \gamma(s - p) \\
&+ (B_f(W, w_1) - 2r + q)\gamma + (b - e)r + B_f(T_1, [T_2, u - m - n]).
\end{aligned}
\tag{5.3.29}
$$

On the other side, the equations (5.3.16) and (5.3.19) give

$$
\begin{aligned}
&B_f(T_1, [W, w_2]) - B_f(T_1, [W, w_1]) \\
&= (\beta + 2\gamma)(s - p) + (\beta + \gamma - a)B_f(W, w_2) \\
&- er + dp + B_f(T_1, [T_1, v]) - B_f(T_1, [T_2, m]) - as - (\beta + 2\gamma)r \\
&+ \gamma B_f(W, w_1) + br - cp - B_f(T_1, [T_1, n]) + B_f(T_1, [T_2, u]).
\end{aligned}
\tag{5.3.30}
$$

Equalling once again these two expressions, we have

$$
\begin{aligned}
&2\gamma(\gamma^2 - \beta^2) + (\beta - 2\gamma)B_f(W, w_1) - \beta p - 2\gamma r \\
&+ \gamma(\gamma - \beta)^2 + \gamma B_f(W, w_2) + \gamma q = -2rp - \beta r + \gamma s
\end{aligned}
$$

Taking into account the equations (5.3.9), we get

$$\gamma^2(\gamma - \beta) = 0. \tag{5.3.31}$$

Step 5: $\beta(2s + p + \beta\gamma) = 0.$

We need to continue our calculations:

$$\begin{aligned}
&B_f(T_2, [T_1, [V, [V, X_2]]]) \\
=&B_f(T_2, [T_1, (\gamma - \beta)\{(\gamma - \beta)X_2 + w_2 - w_1\}]) \\
&+B_f(T_2, [T_1, [V, w_2 - w_1]]) \\
=&-\beta(\gamma - \beta)^2 + (\gamma - \beta)(s + B_f(W, w_1) - r) \\
&+B_f(T_2, [T_1, (s + B_f(W, w_2))X_2 + dw_2 + v \\
&-pX_2 - aw_2 + sX_1 - ew_1 - m - rX_2 - cw_2 \\
&+qX_1 - hw_1 - n + (B_f(W, w_1) - r)X_1 + bw_1 + u]) \\
=&-\beta(\gamma - \beta)^2 + (\gamma - \beta)(s + B_f(W, w_1) - r) \\
&-\beta(s + B_f(W, w_2) - p + s - r + q + B_f(W, w_1) - r) \\
&+(d - a - c)s + (b - e - h)(r - B_f(W, w_1)) \\
&+B_f(T_2, [T_1, v - m - n + u]).
\end{aligned} \tag{5.3.32}$$

On the other hand,

$$\begin{aligned}
&B_f(T_2, [T_1, [V, [V, X_2]]]) \\
=&B_f(T_2, [-(\beta + \gamma)V + W, (\gamma - \beta)X_2 + w_2 - w_1]) \\
&+B_f(T_2, [V, [-(\beta + \gamma)V + W, X_2]]) + B_f(T_2, [V, [V, \beta X_1 + w_1]]) \\
=&\beta(\beta^2 - \gamma^2) - (\beta + \gamma)B_f(T_2, [V, w_2 - w_1]) + (\gamma - \beta)B_f(W, w_2) \\
&+B_f(T_2, [W, w_2 - w_1]) - (\beta + \gamma)B_f(T_2, [V, (\gamma - \beta)X_2 + w_2 - w_1]) \\
&+B_f(T_2, [V, (a - c + \gamma)w_2 + (e - h - \beta)w_1 \\
&+(B_f(W, w_1) - B_f(W, w_2))X_2 + m - n]) \\
&+\beta B_f(T_2, [V, (\gamma - \beta)X_2 + w_2 - w_1]) \\
&+B_f(T_2, [V, rX_2 + cw_2 - qX_1 + hw_1 + n \\
&-(B_f(W, w_1) - r)X_1 - bw_1 - u]) \\
=&\beta(\beta^2 - \gamma^2) + (\beta + 2\gamma)(s + q + B_f(W, w_1) - r) \\
&+(\gamma - \beta)B_f(W, w_2) + B_f(T_2, [W, w_2 - w_1]) - \beta(\gamma^2 - \beta^2) \\
&-(a - c + \gamma)s + (e - h - \beta)(q + B_f(W, w_1) - r) \\
&+\beta(B_f(W, w_1) - B_f(W, w_2)) + B_f(T_2, [V, m - n]) + \beta^2(\gamma - \beta) \\
&-\beta(s + q + B_f(W, w_1) - r) - cs + (h - b)(q + B_f(W, w_1) - r) \\
&+B_f(T_2, [V, n - u]).
\end{aligned}$$

Everything as before we find from these calculations

$$B_f(T_2, [W, w_2 - w_1])$$

$$
\begin{aligned}
=&2\beta(\gamma^2-\beta^2)-\gamma(q-r)-(\beta+\gamma)(s+B_f(W,w_1))\\
&+\gamma(r-s-q-B_f(W,w_1))-B_f(T_2,[V,m-n])+\gamma s\\
&-eq+\beta^2(\beta-\gamma)+bq-B_f(T_2,[V,n-u])-\beta(\gamma-\beta)^2\\
&+(\gamma-\beta)(s-r)-\beta(s-p-r)+(d-c)s\\
&+h(B_f(W,w_1)-r)+B_f(T_2,[T_1,v-m-n+u])\\
&+(\gamma-\beta)(B_f(W,w_1)-B_f(W,w_2)).
\end{aligned} \tag{5.3.33}
$$

While we have

$$
\begin{aligned}
&B_f(T_2,[W,w_2-w_1])\\
=&-(2\beta+\gamma)s-\beta B_f(W,w_2)+ds-eq+B_f(T_2,[T_1,v])\\
&-B_f(T_2,[T_2,m])-cs-(2\beta+\gamma-b)q-(h-2\beta-\gamma)r\\
&-(\beta+\gamma-h)B_f(W,w_1)+B_f(T_2,[T_2,u])-B_f(T_2,[T_1,n]).
\end{aligned}
$$

The equality of these two expressions gives us

$$
\begin{aligned}
&\beta(\gamma^2+\beta\gamma-2\beta^2)\\
&+(2\beta-\gamma)(q+B_f(W,w_2)-s)+\beta s+\beta(p-B_f(W,w_1))=0.
\end{aligned}
$$

Taking into account the equations (5.3.9),

$$
\beta(s+r)=0. \tag{5.3.34}
$$

Further we compute

$$
\begin{aligned}
&B_f(T_2,[T_1,[V,[V,X_2]]])\\
=&B_f((\beta+\gamma)V-W,(\gamma-\beta)((\gamma-\beta)X_2+w_2-w_1)+[V,w_2-w_1])\\
&+B_f(T_1,[-(\beta+\gamma)V+W,(\gamma-\beta)X_2+w_2-w_1])\\
&+B_f(T_1,[V,[-(\beta+\gamma)V+W,X_2]])+B_f(T_1,[V,[V,\gamma X_2+w_2]])\\
=&-(\gamma-\beta)^2(\beta+\gamma)+(\beta+\gamma)B_f(V,[V,w_2-w_1])+(\beta-\gamma)B_f(W,w_2)\\
&+(\gamma-\beta)B_f(W,w_1)-B_f(W,[V,w_2-w_1])+(\beta^2-\gamma^2)\gamma\\
&-(\beta+\gamma)B_f(T_1,[V,w_2-w_1])+(\gamma-\beta)B_f(T_1,[W,X_2])\\
&+B_f(T_1,[W,w_2-w_1])-(\beta+\gamma)B_f(T_1,[V,(\gamma-\beta)X_2+w_2-w_1])\\
&+B_f(T_1,[V,[W,X_2]])+\gamma B_f(T_1,[V,(\gamma-\beta)X_2+w_2-w_1])\\
&+B_f(T_1,[V,[V,w_2]])\\
=&-(\beta+\gamma)(\gamma-\beta)^2-(3\beta+2\gamma)B_f(T_1,[V,w_2-w_1])\\
&+(\beta+\gamma)B_f(T_2,[V,w_2-w_1])+(\beta-\gamma)B_f(W,w_2)+(\gamma-\beta)B_f(W,w_1)\\
&-B_f(W,[V,w_2-w_1])+(\beta^2-\gamma^2)\gamma+(\gamma-\beta)B_f(W,w_1)\\
&+B_f(T_1,[W,w_2-w_1])-\beta\gamma(\gamma-\beta)+B_f(T_1,[V,[W,X_2]])\\
&+B_f(T_1,[V,[V,w_2]])
\end{aligned}
$$

$$
\begin{aligned}
&= -(\beta+\gamma)(\gamma-\beta)^2 - (3\beta+2\gamma)(s+B_f(W,w_2)-p-r)\\
&+(\beta+\gamma)B_f(T_2,[V,w_2-w_1]) + (\beta-\gamma)B_f(W,w_2) + 2(\gamma-\beta)B_f(W,w_1)\\
&-B_f(W,dw_2+v-aw_2-ew_1-cw_2-hw_1-m-n+bw_1+u)\\
&+(\beta^2-\gamma^2)\gamma + B_f(T_1,[W,w_2-w_1]) - \beta\gamma(\gamma-\beta)\\
&+(a-c+\gamma)(s+B_f(W,w_2)-p) + (e-h-\beta)r\\
&+\gamma(B_f(W,w_1)-B_f(W,w_2)) + B_f(T_1,[V,m-n])\\
&+\gamma(s+B_f(W,w_2)-p+s) + (d-a)(s+B_f(W,w_2)-p)\\
&-er + B_f(T_1,[V,v-m]).
\end{aligned}
$$

Comparing this with (5.3.32),

$$
\begin{aligned}
&B_f(T_1,[W,w_2-w_1]) = (\gamma-\beta)(2\beta^2+2\gamma^2+3\beta\gamma)\\
&+(\beta+\gamma-a)s + (d-c-2\beta)p + (b-e-4\gamma)r + \gamma q\\
&+(\beta+2\gamma-a)B_f(W,w_2) + (\beta-\gamma)B_f(W,w_1) + B_f(T_1,[T_1,v])\\
&-B_f(T_1,[T_2,m]) - B_f(T_1,[T_1,n]) + B_f(T_1,[T_2,u]).
\end{aligned}
\tag{5.3.35}
$$

Now the equations (5.3.29), (5.3.35) give us

$$
\begin{aligned}
&(\gamma-\beta)(2\beta^2+2\gamma^2+3\beta\gamma) + (\beta+\gamma-a)s + (d-c-2\beta)p\\
&+(b-e-4\gamma)r + \gamma q + (\beta+2\gamma-a)B_f(W,w_2) + (\beta-\gamma)B_f(W,w_1)\\
&+B_f(T_1,[T_1,v]) - B_f(T_1,[T_2,m]) - B_f(T_1,[T_1,n]) + B_f(T_1,[T_2,u])\\
&=(d-\beta-2\gamma)p + (\beta+2\gamma-a)s - er + (\beta+\gamma-a)B_f(W,w_2)\\
&+B_f(T_1,[T_1,v]) - B_f(T_1,[T_2,m]) - (\gamma-B_f(W,w_1))\gamma - (\beta+\gamma-b)r\\
&-cp - B_f(T_1,[T_1,n]) + B_f(T_1,[T_2,u]).
\end{aligned}
$$

Namely,

$$
\begin{aligned}
&(\gamma-\beta)(2\beta^2+2\gamma^2+3\beta\gamma) + (\beta-2\gamma)(r-p+B_f(W,w_1))\\
&+\gamma(q-s+B_f(W,w_2)) = 0.
\end{aligned}
$$

From (5.3.9) this becomes

$$2\beta(\gamma-\beta)(\beta+\gamma) = 0.$$

Hence, if $\beta+\gamma \neq 0$,

$$\beta(\gamma-\beta) = 0. \tag{5.3.36}$$

Here, substituting (5.3.33) into (5.3.23) and comparing the result with (5.3.22), we get

$$
\begin{aligned}
&2\beta(\beta^2-\gamma^2) - (2\beta+\gamma)(r-B_f(W,w_1)-s-q) + (\gamma-\beta)B_f(W,w_2)\\
&+2\beta(\gamma^2-\beta^2) - \gamma(q-r) - (\beta+\gamma)(s+B_f(W,w_1))\\
&+\gamma(r-s-q-B_f(W,w_1)) - B_f(T_2,[V,m-n]) + \gamma s - eq + \beta^2(\beta-\gamma)\\
&+bq - B_f(T_2,[V,n-u]) - \beta(\gamma-\beta)^2 + (\gamma-\beta)(s-r) - \beta(s-p-r)
\end{aligned}
$$

$$
\begin{aligned}
&+(d-c)s+h(B_f(W,w_1)-r)+B_f(T_2,[T_1,v-m-n+u])\\
&+(\gamma-\beta)(B_f(W,w_1)-B_f(W,w_2))+(c-a-\gamma)s\\
&+(e-h-\beta)(q+B_f(W,w_1)-r)+\beta(B_f(W,w_1)-B_f(W,w_2))\\
&+B_f(T_2,[V,m-n])+\beta\gamma(\gamma-\beta)+\beta(B_f(W,w_2)+s)\\
&-ds+B_f(T_2,[V,v])+\beta(s-p)+as\\
&-e(q+B_f(W,w_1)-r)-B_f(T_2,[V,m])\\
&=(\beta-\gamma+b-e-h)q-B_f(T_2,[T_2,m+n])+B_f(T_2,[T_2,u]).
\end{aligned}
$$

Thus,

$$2\beta s+\beta(r+B_f(W,w_1))+\beta^2(\beta-\gamma)-\beta(\gamma-\beta)^2=0.$$

Taking into account the equations (5.3.9) again,

$$2\beta s+\beta(p+\gamma^2)-\beta\gamma^2+\beta^2\gamma=0.$$

In sum,

$$\beta(2s+p+\beta\gamma)=0. \tag{5.3.37}$$

Step 6: $\beta+\gamma=0$.

We add further two calculations.

$$
\begin{aligned}
&B_f(T_1,[T_1,[V,[V,X_2]]])\\
=&B_f(T_1,[-(\beta+\gamma)V+W,(\gamma-\beta)X_2+w_2-w_1])\\
&+B_f(T_1,[V,[T_1,(\gamma-\beta)X_2+w_2-w_1]])\\
&--(\beta+\gamma)B_f(-(\beta+\gamma)V+W,(\gamma-\beta)X_2+w_2-w_1)\\
&-(\beta+\gamma)B_f(V,[T_1,(\gamma-\beta)X_2+w_2-w_1])\\
&+B_f(T_1,[W,(\gamma-\beta)X_2+w_2-w_1])+(\gamma-\beta)B_f(T_1,[V,\beta X_1+w_1])\\
&+B_f(T_1,[V,pX_2+aw_2-sX_1+ew_1+m\\
&-(B_f(W,w_1)-r)X_1-bw_1-u])\\
=&-(\beta+\gamma)(\gamma^2-\beta^2)-(\beta+\gamma)(B_f(W,w_2)-B_f(W,w_1))\\
&-\beta(\beta^2-\gamma^2)-(\beta+\gamma)(s-p+B_f(W,w_1)-r)+(\gamma-\beta)B_f(W,w_1)\\
&+B_f(T_1,[W,w_2-w_1])+\beta\gamma(\gamma-\beta)-(\gamma-\beta)r\\
&+\gamma(p-s+r-B_f(W,w_1))+a(s+B_f(W,w_2)-p)+(e-b)r\\
&+B_f(T_1,[V,m])-B_f(T_1,[T_2,u])\\
=&(\gamma-\beta)p+dp+B_f(T_1,[T_1,v])-(a+c)p\\
&-B_f(T_1,[T_1,m])-B_f(T_1,[T_1,n]).
\end{aligned}
$$

Therefore,

$$
\begin{aligned}
&B_f(T_1,[W,w_2-w_1])\\
=&\gamma(\gamma^2-\beta^2)+(\beta+\gamma)B_f(W,w_2)-(\beta+\gamma)(p+r)
\end{aligned}
$$

$$\begin{aligned}&+(\gamma-\beta)p+dp-(a+c)p+B_f(T_1,[T_1,v])-B_f(T_1,[T_1,m])\\&-B_f(T_1,[T_1,n])+\beta\gamma(\beta-\gamma)+(\gamma-\beta)r\\&+\gamma(s+B_f(W,w_1)-p-r)+a(p-s-B_f(W,w_2))+(b-e)r\\&-B_f(T_1,[V,m])+B_f(T_1,[T_2,u]).\end{aligned} \tag{5.3.38}$$

The equality of two equations (5.3.29) and (5.3.38) brings

$$\begin{aligned}&\beta\gamma(\beta-\gamma)-\beta r+\gamma(s+B_f(W,w_1))\\=&(\beta-\gamma)B_f(W,w_1)+(2\beta+\gamma)s+\gamma(\gamma^2-\beta^2)+\beta r\\&+\gamma^2(\beta-\gamma)+(\gamma-\beta)^2\gamma+(\gamma-\beta)s-\gamma(s-p)+(q-2r)\gamma\end{aligned}$$

By means of (5.3.9), this reduces to

$$\beta^2\gamma-\beta\gamma^2+\gamma^3=\beta(s+p+r)+\gamma(q-p). \tag{5.3.39}$$

Finally,

$$\begin{aligned}&B_f(T_1,[T_2,[V,[V,X_2]]])\\=&B_f(T_1,[-(\beta+\gamma)V+W,(\gamma-\beta)X_2+w_2-w_1])\\&+B_f(T_1,[V,[T_2,(\gamma-\beta)X_2+w_2-w_1]])\\=&-(\beta+\gamma)B_f(T_1,[V,(\gamma-\beta)X_2+w_2-w_1])\\&+B_f(T_1,[W,(\gamma-\beta)X_2+w_2-w_1])\\&+(\gamma-\beta)B_f(T_1,[V,\gamma X_2+w_2])+B_f(T_1,[V,[T_2,w_2-w_1]])\\=&-(\beta+\gamma)B_f(-(\beta+\gamma)V+W,(\gamma-\beta)X_2+w_2-w_1)\\&-(\beta+\gamma)B_f(V,[T_1,(\gamma-\beta)X_2+w_2-w_1])+(\gamma-\beta)B_f(W,w_1)\\&+B_f(T_1,[W,w_2-w_1])+\gamma^2(\gamma-\beta)+(\gamma-\beta)(s+B_f(W,w_2)-p)\\&+B_f(T_1,[V,(s+B_f(W,w_2))X_2+dw_2+v\\&-rX_2-cw_2+qX_1-hw_1-n])\\=&(\beta+\gamma)(\beta^2-\gamma^2)-(\beta+\gamma)(B_f(W,w_2)-B_f(W,w_1))+\beta(\gamma^2-\beta^2)\\&-(\beta+\gamma)(s-p+B_f(W,w_1)-r)+(\gamma-\beta)B_f(W,w_1)\\&+B_f(T_1,[W,w_2-w_1])+\gamma^2(\gamma-\beta)+(\gamma-\beta)(s+B_f(W,w_2)-p)\\&+\gamma(s+B_f(W,w_2)-r+q)+(d-c)(s+B_f(W,w_2)-p)-hr\\&-B_f(T_1,[V,n])+B_f(T_1,[V,v])\\=&(\gamma-\beta)^2\gamma+(\gamma-\beta)(s+B_f(W,w_2)-p)\\&+(s+B_f(W,w_2))(\gamma+d-a-c)+(B_f(W,w_1)-r+s-r+q)\gamma\\&+(b-e-h)r+B_f(T_1,[T_2,u-m-n+v]),\end{aligned}$$

taking (5.3.23) into account. Thus,

$$\begin{aligned}&B_f(T_1,[W,w_2-w_1])=(\beta+\gamma)(\gamma^2-\beta^2)\\&+(\beta+\gamma)(B_f(W,w_2)-B_f(W,w_1))+\beta(\beta^2-\gamma^2)\end{aligned}$$

$$
\begin{aligned}
&+(\beta+\gamma)(s-p+B_f(W,w_1)-r)+(\beta-\gamma)B_f(W,w_1)\\
&-\gamma^2(\gamma-\beta)+(\beta-\gamma)(s+B_f(W,w_2)-p)\\
&-\gamma(s+B_f(W,w_2)-r+q)+hr+B_f(T_1,[V,n]) \qquad (5.3.40)\\
&+(c-d)(s+B_f(W,w_2)-p)-B_f(T_1,[V,v])+(\gamma-\beta)^2\gamma\\
&+(\gamma-\beta)(s+B_f(W,w_2)-r)+(s+B_f(W,w_2))(\gamma+d-a-c)\\
&+(B_f(W,w_1)-2r+s-p+q)\gamma+(b-e-h)r\\
&+B_f(T_1,[T_2,u-m-n+v]).
\end{aligned}
$$

If $(\beta,\gamma)\neq(0,0)$, then equations (5.3.31), (5.3.36) imply $\beta=\gamma$. Then, equation (5.3.27) implies $q=0$ and finally (5.3.34), (5.3.39) say $\beta=\gamma=0$. In this way, we arrive to the desired result $\beta+\gamma=0$. From the result $[T_2,V]\in\mathfrak{n}_{j_0-1}$, we can argue just as in the proof of Proposition 5.2.1 to show that the limit point of a convergent sequence $\{x_i\}_{i=1}^{\infty}$ in $H_2H_1\cap M_j$ belongs to H_2H_1. □

5.4 THE GENERAL CASE

We now proceed in parallel for the general exponential case. Employing the same induction procedure, we are led to the case where the centre $\mathfrak{z}$ of $\mathfrak{g}$ is of dimension 1 and f does not vanish there. This being assumed, let $\mathfrak{n}$ be a nilpotent ideal of $\mathfrak{g}$ containing $[\mathfrak{g},\mathfrak{g}]$ and $\mathfrak{a}$ a minimal non-central ideal of $\mathfrak{g}$ contained in $\mathfrak{n}$. Then,

$$\{0\}=\mathfrak{g}_0\subset\mathfrak{a}\subset\mathfrak{n}\cap\mathfrak{a}^f\subset\mathfrak{n}\subset\mathfrak{g}$$

is a sequence of ideals of $\mathfrak{g}$. Cutting this sequence, we obtain a good sequence

$$\{0\}=\mathfrak{g}_0\subset\mathfrak{g}_1\subset\cdots\subset\mathfrak{g}_{n-1}\subset\mathfrak{g}_n=\mathfrak{g},\ \dim(\mathfrak{g}_j/\mathfrak{g}_{j-1})=1(1\leq j\leq n),$$

where $\mathfrak{n}=\mathfrak{g}_{j_0}$, of subalgebras of $\mathfrak{g}$. Just as Proposition 5.2.1, in order to pass from M_{j-1} to M_j for $j\leq j_0$ the essential case is one where there exist $Y\in\mathfrak{a}$ outside of $\mathfrak{a}\cap\mathfrak{z}$ which is written $Y=X_2-X_1$ with $X_i\in\mathfrak{h}_i^0(i=1,2)$ and $V\in\mathfrak{g}_j\backslash\mathfrak{g}_{j-1}$ which is written $V=T_2-T_1$ with $T_i\in\mathfrak{h}_i(i=1,2)$ where $[T_2,Y]=[T_1,Y]=Z,\mathfrak{z}=\mathbb{R}Z,f(Z)=1$ and $f(Y)=0$ as before. We choose such an index j minimal. In these circumstances, we can see that the root of the adjoint action of T_i on V vanishes even if $\mathfrak{g}_j$ or $\mathfrak{g}_{j-1}$ is not an ideal of $\mathfrak{g}$. Indeed, let us place ourselves in this situation and put

$$[T_j,W]\equiv\lambda W+\delta V$$

modulo the precedent ideal of the Jordan-Hölder sequence of $\mathfrak{g}$ in question. We are obliged to follow the route made above. We go to notice the necessary changes in the calculations already done above in the completely solvable case.

We have

$$[T_1,w_2]-[T_2,w_1]=(\beta^2-\gamma^2)X_2+\beta w_1-\gamma w_2-\beta^2Y+[W,X_2].$$

and

$$
\begin{aligned}
&[W,X_1]=[W,X_2]=(a-c+\gamma)w_2+(e-h-\beta)w_1\\
&+(B_f(W,w_2)-\delta)Y+(B_f(W,w_1)-B_f(W,w_2))X_2+m-n.
\end{aligned}
$$

Since

$$B_f(T_j,[W,X_j]) = B_f([T_j,W],X_j) + B_f(W,[T_j,X_j]) = -\delta + B_f(W,w_j)$$

for $j = 1,2$, it follows that

$$p - r = B_f(W,w_1) - \gamma^2 - \delta, q - s = \beta^2 + \delta - B_f(W,w_2).$$

Repeating with these changes all the calculations done in the completely solvable case, we remark that equations (5.3.27)–(5.3.39) all hold without change. So, we arrive to the result $\beta + \gamma = 0$ even in the exponential case.

Our algebra $\mathfrak{g}$ being exponential, the result $\beta+\gamma = 0$ signifies that the two roots of the action $\mathrm{ad}(T_i)(i = 1,2)$ at the level of $\mathfrak{g}_j$ are zeros and this allows us to apply the same reasoning as before about a convergent sequence in H_2H_1. There is a possibility that we would have the same situation at other level, but we can treat it in parallel. Namely, when $\mathfrak{a} = \mathbb{R}Y + \mathbb{R}Y' + \mathfrak{z}$, we repeat the same arguments for Y' and at other levels we apply the induction hypothesis applied to $(\mathfrak{h}_1', \mathfrak{h}_2')$ contained in the proper subalgebra $\widetilde{\mathfrak{g}}$.

It is in this way that we achieve the proof of Claim 5.1.1, and verify the following:

Theorem 5.4.1 *Let $G = \exp\mathfrak{g}$ be an exponential solvable Lie group with Lie algebra $\mathfrak{g}$. Let $f \in \mathfrak{g}^*$ and $\mathfrak{h}_j$ $(j = 1,2)$ two polarisations verifying the Pukanszky condition of $\mathfrak{g}$ at f. Put $H_j = \exp(\mathfrak{h}_j)$ for $j = 1,2$. Then the product set H_2H_1 is a closed set of G.*

As explained in the introduction, the followings are straightforward:

Corollary 5.4.1 *(cf. [14]) Let $G = \exp\mathfrak{g}$ be an exponential solvable Lie group with Lie algebra $\mathfrak{g}$. Let $f \in \mathfrak{g}^*$ and let $\mathfrak{h}_j$ $(j = 1,2)$ be two polarisations verifying the Pukanszky condition of $\mathfrak{g}$ at f. Put $H_j = \exp(\mathfrak{h}_j)$ and $\pi_j = \mathrm{ind}_{H_j}^G\chi_f$ for $j = 1,2$, where χ_f denotes the unitary character of H_j defined by $\chi_f(\exp X) = e^{if(X)}$ $(X \in \mathfrak{h}_j)$. Let $\mathcal{H}_{\pi_1}$ be the Hilbert space of π_1. Then the integral T_{21} as defined in formula (5.1.1) by*

$$(T_{21}\varphi)(g) := (T_{\mathfrak{h}_2\mathfrak{h}_1}\varphi)(g) = \oint_{H_2/(H_2\cap H_1)} \varphi(gh)\chi_f(h)\Delta_{H_2,G}^{-1/2}(h)d\nu(h),$$

converges for continuous functions $\varphi \in \mathcal{H}_{\pi_1}$ with compact supports modulo H_1 and extends into an intertwining operator from π_1 to π_2.

For given three Lagrangian subspaces for the bilinear form B_f, let $\tau(\mathfrak{h}_3,\mathfrak{h}_2,\mathfrak{h}_1)$ designates the Maslov index for the triple $(\mathfrak{h}_3,\mathfrak{h}_2,\mathfrak{h}_1)$ (cf. [14]). The following is a direct consequence from Corollary 5.4.1 and [14]. For more details, the readers could consult the reference [73] for the treatment of the nilpotent contexts, and also [74], [75].

Corollary 5.4.2 *(cf. [3]) Let $G = \exp\mathfrak{g}$ be an exponential solvable Lie group with Lie algebra $\mathfrak{g}$. Let $f \in \mathfrak{g}^*$ and let $\mathfrak{h}_j$ $(j = 1,2,3)$ be three polarisations verifying the*

Pukanszky condition of $\mathfrak{g}$ *at* f. *Put* $H_j = \exp(\mathfrak{h}_j)$ *and* $\pi_j = ind_{H_j}^G \chi_f$ *for* $j = 1, 2, 3$, *where* χ_f *denotes the unitary character of* H_j *defined by* $\chi_f(\exp X) = e^{if(X)}$ $(X \in \mathfrak{h}_j)$. *For* $1 \leq i, j \leq 3$, *we normalise the intertwining operators* T_{ij} *as defined in Corollary 5.4.1, in order to obtain an isometry* I_{ij} *which intertwines* π_j *and* π_i. *Then, we have the composition formula*

$$I_{13} \circ I_{32} \circ I_{21} = e^{\frac{i\pi\tau(\mathfrak{h}_3, \mathfrak{h}_2, \mathfrak{h}_1)}{4}} Id.$$

CHAPTER 6

C^*-algebras

We develop in this chapter a new approach to operator-valued Fourier transforms of C^*-algebras of solvable Lie groups, blending complete positivity techniques and existence of suitable global canonical symplectic coordinates on coadjoint orbits of solvable Lie groups. Instead of the traditional way of constructing Lie group representations as induced representations, we use the Lie theoretic method of constructing firstly the representations of the corresponding Lie algebras via canonical coordinates on coadjoint orbits [90], and then we integrate the Lie algebra representations to Lie group representations. This leads to a realisation of the operator-valued Fourier transform, which is studied with the help of some basic properties of C^*-algebra extensions and of completely positive maps. We shall describe the C^*-algebras of some basic solvable Lie groups: the Heisenberg group, the thread-like nilpotent Lie group, the six dimensional normal-j-group, and the $(ax+b)$-like group. We first explain the general techniques of the Fourier transform of liminary C^*-algebras and we introduce the notion of norm controlled dual limits.

6.1 BASIC NOTATIONS AND DEFINITIONS

In this section, we recall basic notations and results on C^*-algebras, and we record some tools for proving norm-continuity of operator fields.

6.1.1 C^*-algebras and notations

We denote by $\mathcal{K}(\mathcal{H})$ the C^*-algebra of all compact operators on some complex separable infinite-dimensional Hilbert space $\mathcal{H}$. We denote by $\mathfrak{S}_p(\mathcal{H})$ the pth Schatten ideal, for $1 \leq p \leq \infty$. For any C^*-algebra $\mathcal{A}$ and locally compact Hausdorff space X we denote by $C_0(X, \mathcal{A})$ the C^*-algebra of continuous $\mathcal{A}$-valued functions on X which vanish at infinity. We also denote by $\widehat{\mathcal{A}}$ the dual space of $\mathcal{A}$ endowed with its Jacobson topology.

Any Lie group G has a unique left invariant Borel measure called Haar measure. We then provide the Banach space $L^1(G)$ of all integrable functions on G with respect

DOI: 10.1201/9781003685739-6

to the Haar measure with the convolution product $*$. That is,

$$f * g(y) := \int_G f(x)g(x^{-1}y)dx \quad \text{for } f, g \in L^1(G) \text{ and } y \in G.$$

Then $(L^1(G), *)$ is a Banach algebra which has an isometric involution defined by

$$f^*(x) := \Delta_G(x)^{-1}\overline{f(x^{-1})} \quad \text{for } f \in L^1(G) \text{ and } \ x \in G.$$

Here Δ_G denotes the modular function defined in Subsection 1.2.1. Every irreducible representation $(\pi, \mathcal{H}_\pi)$ of G integrates to an irreducible unitary representation (also denoted by π) of $L^1(G)$ in the following sense. The integral

$$\pi(F) = \int_G F(x)\pi(x)dx, \ F \in L^1(G),$$

converges in the Banach space $\mathcal{B}(\mathcal{H}_\pi)$, the space of all bounded linear operators on the Hilbert space $\mathcal{H}_\pi$ equipped with the operator norm $\|\cdot\|_{\mathrm{op}}$. Furthermore, for every irreducible unitary representation $\overline{\pi}$ of the algebra $L^1(G)$ on a Hilbert space $\mathcal{H}$, there exists a unique irreducible representation $(\pi, \mathcal{H})$ of G such that $\overline{\pi}(F) = \pi(F)$ for $F \in L^1(G)$. The C^*-algebra of G is defined to be the completion of the involutive Banach algebra $L^1(G)$ with respect to the C^*-norm

$$\|F\|_{C^*} = \sup_{\pi \in \widehat{G}} \|\pi(F)\|_{\mathrm{op}}. \tag{6.1.1}$$

The unitary dual space $\widehat{C^*(G)}$ of the C^*-algebra $C^*(G)$ can be identified with the unitary dual $\widehat{G}$ of G. Since connected Lie groups are second countable, the algebra $C^*(G)$ and its dual space $\widehat{G}$ are then separable topological spaces. This allows us to work in $\widehat{G}$ with sequences instead of nets.

If G is amenable, then the left regular representation $(\lambda, L^2(G))$ of $C^*(G)$ defined by $\lambda(F)\xi := F * \xi$ for $F \in L^1(G)$ and $\xi \in L^2(G)$ is injective. So we can also identify $C^*(G)$ with an algebra of convolution operators on the Hilbert space $L^2(G)$.

Definition 6.1 A *topological* $\mathbb{R}$*-space* is a topological space X endowed with a continuous map $\mathbb{R} \times X \to X$, $(t, x) \mapsto t \cdot x$, and with a distinguished point $x_0 \in X$ satisfying the following conditions:

1. For every $x \in X$ and $t \in \mathbb{R}$ one has $0 \cdot x = t \cdot x_0$.

2. For all $t, s \in \mathbb{R}$ and $x \in X$ one has $t \cdot (s \cdot x) = ts \cdot x$.

3. For every $x \in X \setminus \{x_0\}$ the map $\mathbb{R} \to X$, $t \mapsto t \cdot x$ is a homeomorphism onto its image.

An $\mathbb{R}$*-subspace* of the topological $\mathbb{R}$-space X is any subset $\Gamma \subseteq X$ such that $\mathbb{R} \cdot \Gamma \subseteq \Gamma \cup \{x_0\}$. If this is the case, then $\Gamma \cup \{x_0\}$ is a topological $\mathbb{R}$-space on its own.

In the above framework, a function $\varphi\colon X \to \mathbb{R}$ is called *homogeneous* if there exists $r \in [0, \infty)$ such that $\varphi(t \cdot x) = t^r \varphi(x)$ for all $t \in \mathbb{R}$ and $x \in X$.

Example 6.1 Every finite-dimensional real vector space is a topological $\mathbb{R}$-space. Moreover, if $\varphi_1, \dots, \varphi_{n_1}, \psi_1, \dots, \psi_{n_2} \colon \mathbb{R}^m \to \mathbb{R}$ are any homogeneous polynomials, then the semi-algebraic cone

$$\Gamma := \{x \in \mathbb{R}^m \mid \varphi_{j_1}(x) = 0 \neq \psi_{j_2}(x) \text{ for } 1 \leq j_1 \leq n_1 \text{ and } 1 \leq j_2 \leq n_2\}$$

is an $\mathbb{R}$-subspace of $\mathbb{R}^m$ in the sense of Definition 6.1.

As another type of examples, if G is any nilpotent Lie group, then the topological $\mathbb{R}$-space structure of $\mathfrak{g}^*$ gives rise to a topological $\mathbb{R}$-space structure of the orbit space $\mathfrak{g}^*/G$ (hence also of the dual space $\widehat{G}$ via Kirillov's correspondence), and the vector space of characters $[\mathfrak{g}, \mathfrak{g}]^\perp$, viewed as the set of singleton orbits, is an $\mathbb{R}$-subspace of $\mathfrak{g}^*/G$. More specifically, the $\mathbb{R}$-space structure of $\mathfrak{g}^*/G$ is the map

$$\mathbb{R} \times (\mathfrak{g}^*/G) \to \mathfrak{g}^*/G, \quad (t, \mathcal{O}_\xi) \mapsto \mathcal{O}_{t\xi}$$

where we denote by $\mathcal{O}_\xi$ the coadjoint orbit of every $\xi \in \mathfrak{g}^*$.

We close this subsection by the following definition:

Definition 6.2 *Let X be a topological space.*

i) A point $\gamma \in X$ is said to be separated in X if for every $\gamma' \in X$ that is not in the closure of the set $\{\gamma\}$, there exist open subsets V, $V' \subset X$ such that $\gamma \in V$, $\gamma' \in V'$ and $V \cap V' = \emptyset$.

ii) We denote by $\operatorname{Sep}(X)$ *the collection of subsets $Y \subset X$ such that all points in Y are closed and separated in X.*

6.1.2 First results

Let us first prove the following:

Lemma 6.1.1 *Let X be any topological space and for $j = 1, 2$ let V_j be any open subset of X that is homeomorphic to an open subset of $\mathbb{R}^{r_j}$, where $r_j \geq 1$ is some integer. If $V_1 \cap V_2 \neq \emptyset$, then $r_1 = r_2$.*

Proof 6.1 *The non-empty open set $V := V_1 \cap V_2$ is homeomorphic to some open subsets of $\mathbb{R}^{r_1}$ and of $\mathbb{R}^{r_2}$, hence $r_1 = r_2$ by Brouwer's theorem on the invariance of domain.*

The next lemma is well known, but we record it in order to fix some notation.

Lemma 6.1.2 *Let $\mathcal{A}$ be any C^*-algebra. For any closed two-sided ideal $\mathcal{J}$ of $\mathcal{A}$ denote*

$$\widehat{\mathcal{A}}_{\mathcal{J}} := \{[\pi] \in \widehat{\mathcal{A}} \mid \mathcal{J} \subseteq \operatorname{Ker} \pi\} \simeq \widehat{\mathcal{A}/\mathcal{J}},$$
$$\widehat{\mathcal{A}}^{\mathcal{J}} := \{[\pi] \in \widehat{\mathcal{A}} \mid \mathcal{J} \not\subseteq \operatorname{Ker} \pi\} \simeq \widehat{\mathcal{J}}.$$

Then the following assertions hold:

1. *$\mathcal{J} \mapsto \widehat{\mathcal{J}}$ is an increasing bijection between the closed two-sided ideals of $\mathcal{A}$ and the open subsets of $\widehat{\mathcal{A}}$.*

2. *$\mathcal{J} \mapsto \widehat{\mathcal{A}/\mathcal{J}}$ is a decreasing bijection between the closed two-sided ideals of $\mathcal{A}$ and the closed subsets of $\widehat{\mathcal{A}}$.*

Proof 6.2 *See [37, Props. 2.11.2, 3.2.2].*

Remark 6.1.3 In the present section, a key role is played by the well-known fact that for any C^*-algebra $\mathcal{A}$ the locally closed subsets of $\widehat{\mathcal{A}}$ are precisely the spectra of sub-quotients of $\mathcal{A}$.

In fact, recall that locally closed subset means any set of the form $F \cap D$, where F is any closed subset of $\widehat{\mathcal{A}}$ while D is any open subset of $\widehat{\mathcal{A}}$. Moreover, a sub-quotient of $\mathcal{A}$ is any C^*-algebra of the form $\mathcal{J}_2/\mathcal{J}_1$, where $\mathcal{J}_1 \subseteq \mathcal{J}_2$ are any closed two-sided ideals of $\mathcal{A}$. For any such a pair of ideals, it follows by Lemma 6.1.2 that $D := \widehat{\mathcal{J}_1}$ is an open subset of $\widehat{\mathcal{A}}$ and $\widehat{\mathcal{J}_2/\mathcal{J}_1}$ is a closed subset of the open set $\widehat{\mathcal{J}_2}$, hence it is easily checked that the disjoint union $F := \widehat{\mathcal{J}_2/\mathcal{J}_1} \cup (\widehat{\mathcal{A}} \setminus \widehat{\mathcal{J}_1})$ is a closed subset of $\widehat{\mathcal{A}}$ and $F \cap D = \widehat{\mathcal{J}_2/\mathcal{J}_1}$ is locally closed. Conversely, for any locally closed subset $F \cap D \subseteq \widehat{\mathcal{A}}$ we have $F \cap D = D \setminus (D \setminus F)$, where $D \setminus F = D \cap (\widehat{\mathcal{A}} \setminus F)$ is an open subset of D. Hence by Lemma 6.1.2 there exist uniquely determined two-sided closed ideals $\mathcal{J}_1 \subseteq \mathcal{J}_2$ of $\mathcal{A}$ with $\widehat{\mathcal{J}_1} = D \setminus F$ and $\widehat{\mathcal{J}_2} = D$. Moreover, since $\mathcal{J}_1$ is in particular an ideal of $\mathcal{J}_2$, it follows by Lemma 6.1.2 again that $\widehat{\mathcal{J}_2/\mathcal{J}_1} = D \setminus (D \setminus F) = D \cap F$. See [93, Lemma 7.3.5] for the fact that the $*$-isomorphism class of the C^*-algebra $\mathcal{J}_2/\mathcal{J}_1$ depends only on the locally closed set $\widehat{\mathcal{J}_2} \setminus \widehat{\mathcal{J}_1}$. □

6.1.3 The Fourier transform

The classical Fourier transform gives a $*$-isomorphism $C^*(\mathcal{V}) \simeq C_0(\mathcal{V}^*)$ for any finite-dimensional real vector space $\mathcal{V}$ regarded as an abelian Lie group. Some of the main results of this chapter give a result of this type when the abelian group $(\mathcal{V}, +)$ is replaced by an arbitrary connected, simply connected solvable Lie group G. (See Theorem 6.4.6 and Corollary 6.3.3 below). If G is non-commutative, then its unitary dual space $\widehat{G}$ is not Hausdorff and moreover $C^*(G)$ is non-commutative, hence one must replace the classical Fourier transform by a suitable operator-valued Fourier transform which realises $C^*(G)$ as a C^*-algebra of operator fields on $\widehat{G}$.

The problem of describing the image of the operator-valued Fourier transform on a solvable Lie group is notoriously difficult and some of its aspects were nicely discussed in [79].

We shall prove that for every connected, simply connected nilpotent Lie group G its C^*-algebra admits a finite sequence of closed two-sided ideals

$$\{0\} = \mathcal{J}_0 \subseteq \mathcal{J}_1 \subseteq \cdots \subseteq \mathcal{J}_n = C^*(G) \tag{6.1.2}$$

with $*$-isomorphisms $\mathcal{J}_j/\mathcal{J}_{j-1} \simeq C_0(\Gamma_j, \mathcal{K}(\mathcal{H}_j))$ for suitable locally compact spaces Γ_j that are homeomorphic to real semi-algebraic cones, where Γ_1 is a Zariski open subset of $\mathbb{R}^k$, denoting by k the codimension of generic coadjoint orbits of G. Here $\mathcal{H}_j$ for $j = 1, \dots, n$ are complex separable Hilbert spaces with $\dim \mathcal{H}_1 = \cdots = \dim \mathcal{H}_{n-1} = \infty$, and $\dim \mathcal{H}_n = 1$. Using a direct sum of topological spaces, one obtains a continuous bijection

$$\Gamma_1 \sqcup \cdots \sqcup \Gamma_n \to \widehat{G}. \tag{6.1.3}$$

whose restriction to Γ_j is a homeomorphism onto its image for $j = 1, \dots, n$. The above map is never a homeomorphism if G is non-abelian, since for instance it turns out that Γ_j is a relatively dense open subset of $\Gamma_j \cup \Gamma_{j+1} \cup \cdots \cup \Gamma_n$ for $j = 1, \dots, n-1$ (see Theorem 6.2.6 and Definition 6.4 below). We describe the image of the operator-valued Fourier transform of $C^*(G)$ as a C^*-algebra of operator fields on $\widehat{G}$, determined via the boundary behaviour of the restrictions of operator fields to Γ_j for $j = 1, \dots, n$ (see Theorem 6.4.6 and Corollary 6.3.3 below). The intricate topological nature of the C^*-algebra extensions involved in this picture can already be seen in the case of the Heisenberg group; see [103].

For nilpotent Lie groups of step two and dimension ≤ 6 (and a solvable Lie group), composition series like those in (6.1.2) were constructed in [42, Ch. 6] using techniques of twisted crossed products. The results thus obtained are optimal in terms of the length of the composition sequence, for the case of the free two-step nilpotent Lie group with tree generators, see [42, Ex. 6.3.5].

In the particular cases of the Heisenberg, thread-like Lie groups, and all nilpotent Lie groups of dimension ≤ 6, explicit calculations have been used to provide descriptions of the image of the operator-valued Fourier transform in [83], [81] and [95].

Our results can also be regarded as a sharpening of the results of [88], where one proved that there exists a composition series as in (6.1.2), where the successive quotients are however C^*-algebras with continuous trace.

6.1.4 Special solvable C^*-algebras

We first recall a notion introduced in [41] (see also [59, Sect. 2] for several completions).

Definition 6.3 A C^*-algebra $\mathcal{A}$ is called *solvable* if it has a *solving series*, that is, a finite sequence of ideals $\{0\} = \mathcal{J}_0 \subsetneq \mathcal{J}_1 \subsetneq \cdots \subsetneq \mathcal{J}_n = \mathcal{A}$ with $*$-isomorphisms $\mathcal{J}_j/\mathcal{J}_{j-1} \simeq C_0(\Gamma_j, \mathcal{K}(\mathcal{H}_j))$ for suitable locally compact spaces Γ_j and complex Hilbert spaces $\mathcal{H}_j$ for $j = 1, \dots, n$, with $\dim \mathcal{H}_1 \geq \cdots \geq \dim \mathcal{H}_n$. If $n \geq 1$ is the least integer for which there exists a sequence of ideals as above, then $n-1$ is called the *length* of the C^*-algebra $\mathcal{A}$.

Definition 6.4 We say that $\mathcal{A}$ is a *special solvable C^*-algebra* if it is separable and it has a *special solving series*, that is, a solving series as in Definition 6.3 with the following additional properties:

1. $\widehat{\mathcal{A}}$ has the structure of a topological $\mathbb{R}$-space and $\Gamma_j \subseteq \widehat{\mathcal{A}}$ is an $\mathbb{R}$-subspace for $j = 1, \dots, n$.

2. One has $\dim \mathcal{H}_n = 1$ and Γ_n is isomorphic as a topological $\mathbb{R}$-space to a finite-dimensional vector space.

3. For $j = 1, \dots, n-1$ one has $\dim \mathcal{H}_j = \infty$, the set Γ_{j+1} is dense in $\widehat{\mathcal{A}} \setminus \widehat{\mathcal{J}_j}$, and the points of Γ_{j+1} are closed and separated in $\widehat{\mathcal{A}} \setminus \widehat{\mathcal{J}_j}$.

4. For $j = 1, \dots, n$, Γ_j is isomorphic as a topological $\mathbb{R}$-space to a semi-algebraic cone in a finite-dimensional vector space. In addition, Γ_1 is assumed to be a Zariski open set, and the dimension of the corresponding ambient vector space is called the *index of* $\mathcal{A}$ and is denoted by $\operatorname{ind} \mathcal{A}$.

5. For $j = 1, \dots, n$, there exists a homogeneous function $\varphi_j \colon \widehat{\mathcal{A}} \to \mathbb{R}$ such that $\varphi_j|_{\Gamma_1}$ is a polynomial function (via the above homeomorphism) and

$$\Gamma_j = \{\gamma \in \widehat{\mathcal{A}} \mid \varphi_j(\gamma) \neq 0 \text{ and } \varphi_i(\gamma) = 0 \text{ if } i < j\}.$$

Since Γ_1 is open and dense in $\widehat{\mathcal{A}}$, it follows by Lemma 6.1.1 that $\operatorname{ind} \mathcal{A}$ does not depend on the choice of the solving series of $\mathcal{A}$.

Remark 6.1.4 In Definition 6.4, the distinguished point of the topological $\mathbb{R}$-space $\widehat{\mathcal{A}}$ is the origin $0 \in \Gamma_n$ of the vector space Γ_n. Therefore, for $j = 1, \dots, n$ and every $\gamma \in \Gamma_j$, one has $\overline{\mathbb{R}^* \cdot \gamma} \cap \mathbb{R} \cdot \gamma = \{0\} \subset \Gamma_n$. □

Remark 6.1.5 In Definition 6.4, since $\dim \mathcal{H}_n = 1 < \dim \mathcal{H}_j$ for $1 \leq j < n$, the set Γ_n is precisely the set of characters of $\mathcal{A}$, that is, the non-zero $*$-homomorphisms $\chi \colon \mathcal{A} \to \mathbb{C}$. Denoting by $\operatorname{Comm}(\mathcal{A})$ the closed two-sided ideal of $\mathcal{A}$ generated by the set $\{ab - ba \mid a, b \in \mathcal{A}\}$, it follows that $\mathcal{A}/\operatorname{Comm}(\mathcal{A})$ is a commutative C^*-algebra and for every $\chi \in \Gamma_n$ one has $\operatorname{Comm}(\mathcal{A}) \subseteq \operatorname{Ker} \chi$, hence χ can be identified with an element in the spectrum of $\mathcal{A}/\operatorname{Comm}(\mathcal{A})$. Thus the Gelfand representation provides a $*$-isomorphism $\mathcal{A}/\operatorname{Comm}(\mathcal{A}) \simeq C(\Gamma_n)$, and the condition in Definition 6.4(2) implies that the spectrum of the commutative C^*-algebra $\mathcal{A}/\operatorname{Comm}(\mathcal{A})$ is homeomorphic to a finite-dimensional real vector space, whose dimension is uniquely determined because of Brouwer's theorem on the invariance of domain. □

6.2 CONTINUITY OF OPERATOR FIELDS

6.2.1 First results

We begin this section with some results. The following two lemmas go back to [96, Proposition 2.2 and Theorem 2.3].

Lemma 6.2.1 *Let $\mathcal{A}$ be any C^*-algebra with some subset of its spectrum $\Gamma \subseteq \widehat{\mathcal{A}}$ such that the relative topology of Γ is Hausdorff. Assume $\mathcal{H}$ is a complex Hilbert space and in every $\gamma \in \Gamma$ we have picked $\pi_\gamma \colon \mathcal{A} \to \mathcal{B}(\mathcal{H})$. Also let $\mathcal{V}_1$ and $\mathcal{V}_2$ be any total subsets* *of $\mathcal{H}$. If $a \in \mathcal{A}$ has the property that for every $v_1 \in \mathcal{V}_1$ and $v_2 \in \mathcal{V}_2$, the function $\Gamma \to \mathbb{C}$, $\gamma \mapsto \langle \pi_\gamma(a)v_1, v_2\rangle$, is continuous, then for every $R \in \mathfrak{S}_1(\mathcal{H})$ the function*

$$f_R\colon \Gamma \to \mathbb{C}, \quad f_R(\gamma) = \mathrm{Tr}\,(\pi_\gamma(a)R)$$

is continuous and bounded.

Proof 6.3 *For every $R \in \mathfrak{S}_1(\mathcal{H})$ and $\gamma \in \Gamma$ we have $|f_R(\gamma)| \le \|R\|_1\|a\|$, hence $f_R \in \ell^\infty(\Gamma)$ and $\|f_R\|_\infty \le \|R\|_1\|a\|$. Since the limit of any uniformly convergent sequence of continuous functions is in turn continuous, it then easily follows that the set $\{R \in \mathfrak{S}_1(\mathcal{H}) \mid f_R \in C(\Gamma)\}$ is a closed linear subspace of $\mathfrak{S}_1(\mathcal{H})$. As that closed linear subspace has a dense linear subspace consisting of rank-one operators by the hypothesis on a, the conclusion follows.*

Lemma 6.2.2 *In the setting of Lemma 6.2.1, assume $\mathcal{S}$ is a dense $*$-subalgebra of $\mathcal{A}$ for which every element $a \in \mathcal{S}$ satisfies the condition of Lemma 6.2.1, and moreover for all $\gamma \in \Gamma$ we have $\pi_\gamma(a) \in \mathfrak{S}_1(\mathcal{H})$ and the function $\Gamma \to \mathbb{C}$, $\gamma \mapsto \mathrm{Tr}\,\pi_\gamma(a)$ is continuous.*

Then for every $a \in \mathcal{A}$ the map $\Pi_a\colon \Gamma \to \mathcal{B}(\mathcal{H})$, $\gamma \mapsto \pi_\gamma(a)$, is norm continuous.

Proof 6.4 *For arbitrary $a \in \mathcal{A}$ and $\gamma \in \Gamma$ we have $\|\Pi_a(\gamma)\| = \|\pi_\gamma(a)\| \le \|a\|$, hence just as in the proof of Lemma 6.2.1 we can see that the set $\{a \in \mathcal{A} \mid \Pi_a \text{ is continuous}\}$ is a closed linear subspace of $\mathcal{A}$. Therefore, it suffices to check that Π_a is continuous for $a \in \mathcal{S}$.*

If $a \in \mathcal{S}$ and $\{\gamma_j\}_{j\in J}$ is any net in Γ which is convergent to some $\gamma \in \Gamma$, then for all $j \in J$ we have

$$\begin{aligned}\|\Pi_a(\gamma_j) - \Pi_a(\gamma)\|^2 \le& \|\Pi_a(\gamma_j) - \Pi_a(\gamma)\|_2^2\\ =&\|\pi_{\gamma_j}(a) - \pi_\gamma(a)\|_2^2\\ =&\mathrm{Tr}\,\pi_{\gamma_j}(a^*a) - 2\mathrm{Re}\,\mathrm{Tr}\,(\pi_{\gamma_j}(a)\pi_\gamma(a)^*) + \mathrm{Tr}\,\pi_\gamma(a^*a).\end{aligned}$$

Since $a^*a \in \mathcal{S}$, it follows by hypothesis that

$$\lim_{j\in J} \mathrm{Tr}\,\pi_{\gamma_j}(a^*a) = \mathrm{Tr}\,\pi_\gamma(a^*a),$$

and on the other hand Lemma 6.2.1 implies

$$\lim_{j\in J} \mathrm{Tr}\,(\pi_{\gamma_j}(a)\pi_\gamma(a)^*) = \mathrm{Tr}\,(\pi_\gamma(a)\pi_\gamma(a)^*) = \mathrm{Tr}\,\pi_\gamma(a^*a),$$

Hence, by the above estimate we obtain $\lim\limits_{\gamma\in\Gamma} \pi_{\gamma_j}(a) = \pi_\gamma(a)$ in $\mathcal{B}(\mathcal{H})$, which concludes the proof.

Lemma 6.2.3 *Let $\mathcal{A}$ be any liminary separable C^*-algebra and consider any locally closed subset $\Gamma \subseteq \widehat{\mathcal{A}}$ for which the relative topology of Γ is Hausdorff, and let $\mathcal{J}_1 \subseteq \mathcal{J}_2$ be two-sided closed ideals of $\mathcal{A}$ with $\widehat{\mathcal{J}_2/\mathcal{J}_1} = \Gamma$ (see Remark 6.1.3). Then the following properties are equivalent:*

1. *The canonical field of elementary C^*-algebras defined by $\mathcal{J}_2/\mathcal{J}_1$ on Γ is trivial.*

2. *There exist a complex Hilbert space $\mathcal{H}$ and a complete system of distinct representatives $\{\pi_\gamma\colon \mathcal{A} \to \mathcal{B}(\mathcal{H})\}_{\gamma\in\Gamma}$ of the equivalence classes of representations corresponding to the elements of Γ such that for every $a \in \mathcal{J}_2$ the mapping $\Gamma \to \mathcal{B}(\mathcal{H})$, $\gamma \mapsto \pi_\gamma(a)$ is continuous with respect to the norm operator topology of $\mathcal{B}(\mathcal{H})$.*

Proof 6.5 *The implication (1)⇒(2) is clear, so we are left to proving only the converse implication.*

By Lemma 6.1.2, we may replace $\mathcal{A}$ by its sub-quotient $\mathcal{J}_2/\mathcal{J}_1$. Thus we may assume $\mathcal{J}_1 = \{0\}$ and $\mathcal{J}_2 = \mathcal{A}$, hence $\Gamma = \widehat{\mathcal{A}}$. Then the hypothesis (2) shows that the continuous sections of the canonical field of elementary C^*-algebras defined by $\mathcal{A}$ (which is a continuous field of C^*-algebras by [37, 10.5.1] since $\Gamma = \widehat{\mathcal{A}}$ is Hausdorff) are also continuous sections of the trivial field with the fibre $\mathcal{K}(\mathcal{H})$ over $\widehat{\mathcal{A}}$. Now [37, Prop. 10.2.4] ensures that the two aforementioned continuous fields of C^*-algebras are isomorphic, and in particular the canonical field of elementary C^*-algebras defined by $\mathcal{A}$ is trivial.

6.2.2 Norm controls for the boundary values of C^*-Fourier transforms

The present subsection proves our main abstract results on the image of the operator-valued Fourier transform of liminary C^*-algebras that have a finite composition series such that the spectra of the corresponding successive quotients are Hausdorff sets in the relative topology, and the canonical fields of elementary C^*-algebras defined by the sub-quotients are trivial.

6.2.2.1 Boundary values of Fourier transforms

Proposition 6.2.4 *Assume the following:*

- *$\mathcal{A}$ is any separable nuclear C^*-algebra.*
- *$T \in \mathrm{Sep}(\widehat{\mathcal{A}})$ is an open dense subset of $\widehat{\mathcal{A}}$, such that its corresponding ideal of $\mathcal{A}$ is $*$-isomorphic to $\mathcal{C}_0(T,\mathcal{K})$. Denote by $\mathcal{F}_T\colon \mathcal{A} \to \mathcal{C}_b(T,\mathcal{K})$ the Fourier transform of $\mathcal{A}$ restricted to T.*
- *$0 \to \mathcal{C}_0(T,\mathcal{K}) \hookrightarrow \mathcal{A} \xrightarrow{q} \mathcal{B} \to 0$ is an exact sequence of C^*-algebras.*

Then the following assertions hold:

1. *The map*
$$\Phi\colon \mathcal{A} \to \mathcal{C}_b(T,\mathcal{K}) \oplus \mathcal{B}, \quad \Phi(a) = \mathcal{F}_T(a) \oplus q(a)$$
is an isometric $*$-homomorphism.

2. *There exists a linear map* $\nu\colon \mathcal{B} \to C_b(T,\mathcal{K})$ *which is completely positive, completely isometric, almost* $*$*-homomorphism, and*

$$\operatorname{Ran}\Phi = \Big\{ f \oplus b \in C_b(T,\mathcal{K}) \oplus \mathcal{B} \mid \lim_{t\to\infty} f(t) - (\nu(b))(t) = 0 \Big\}.$$

3. *There exists a completely isometric cross section of* q.

Proof 6.6 *Since* T *is the spectrum of the ideal* $\mathcal{J} = C_0(T,\mathcal{K})$ *of* $\mathcal{A}$, *and the points of* T *are closed and separated in* $\widehat{\mathcal{A}}$ *sense of [35, Def. II.6], it follows by [35, Prop. IV.1.3] that the Busby invariant* γ *of the extension of* C^**-algebras*

$$0 \to \mathcal{J} \hookrightarrow \mathcal{A} \xrightarrow{q} \mathcal{B} \to 0$$

has the range contained into $L(\mathcal{J})/\mathcal{J}$, *where* $L(\mathcal{J})$ *is the largest two-sided closed ideal of the multiplier algebra of* $\mathcal{J}$ *which contains* $\mathcal{J}$ *and for which* $\widehat{\mathcal{J}} \in \operatorname{Sep}(\widehat{L(\mathcal{J})})$, *that is, the points of* $T = \widehat{\mathcal{J}}$ *are closed and separated in* $\widehat{L(\mathcal{J})}$ *([35, Lemme III.2.4]). On the other hand, since* $\mathcal{J} = C_0(T,\mathcal{K})$, *we have* $L(\mathcal{J}) = C_b(T,\mathcal{K})$ *by [35, Prop. III.4.1]. Consequently the Busby invariant of the extension of* C^**-algebras from the statement is a* $*$*-homomorphism*

$$\beta\colon \mathcal{B} \to C_b(T,\mathcal{K})/C_0(T,\mathcal{K}).$$

Moreover, since T *was assumed to be dense in* $\widehat{\mathcal{A}}$, *it follows by [27, Cor. 6.4] that the* $*$*-homomorphism* β *is injective. Therefore* β *is a complete isometry and it is moreover completely positive.*

On the other hand, since the C^**-algebra* $\mathcal{A}$ *is nuclear, it follows by [29, Cor. 4] that also its quotient* $\mathcal{B}$ *is nuclear. Also* $\mathcal{B}$ *is separable since* $\mathcal{A}$ *is. It then follows by the completely positive lifting theorem [28, Th. 3.10] that there exists a completely positive contraction* $\nu\colon \mathcal{B} \to C_b(T,\mathcal{K})$ *with the property* $\beta(b) = \nu(b) + C_0(T,\mathcal{K})$ *for all* $b \in \mathcal{B}$. *The unital extension of* ν *to the unitisations of the* C^**-algebras* $\mathcal{B}$ *and* $C_b(T,\mathcal{K})$ *is completely positive by [28, Lemma 3.9], hence completely contractive by [24, 1.3.3]. Therefore* ν *is both completely positive and completely contractive.*

Now note that the canonical map

$$\widetilde{q}\colon C_b(T,\mathcal{K}) \to C_b(T,\mathcal{K})/C_0(T,\mathcal{K}), \quad f \mapsto f + C_0(T,\mathcal{K}),$$

is a $*$*-homomorphism, hence completely contractive, and moreover* $\widetilde{q}\circ\nu = \beta$ *is completely isometric. Since we have proved above that* ν *is completely contractive, it then follows that* ν *is actually completely isometric. Moreover, since* $\widetilde{q}\circ\nu = \beta$ *is a* $*$*-homomorphism, it follows that* ν *is an almost* $*$*-homomorphism, which means that for all* $b_1, b_2 \in \mathcal{B}$ *we have* $\lim\limits_{t\to\infty} (\nu(b_1 b_2) - \nu(b_1)\nu(b_2))(t) = 0$ *and* $\nu(b_1^*) = \nu(b_1)^*$.

For proving the assertion on $\operatorname{Ran}\Phi$, *note that we have the commutative diagram*

$$\begin{array}{ccccccccc}
0 & \longrightarrow & \mathcal{J} & \longrightarrow & \mathcal{A} & \xrightarrow{\;q\;} & \mathcal{B} & \longrightarrow & 0 \\
 & & \big\downarrow{\scriptstyle \mathcal{F}|_{\mathcal{J}}} & & & \swarrow{\scriptstyle \nu} & \big\downarrow{\scriptstyle \beta} & & \\
0 & \longrightarrow & C_0(T,\mathcal{K}) & \longrightarrow & C_b(T,\mathcal{K}) & \xrightarrow{\;\widetilde{q}\;} & C_b(T,\mathcal{K})/C_0(T,\mathcal{K}) & \longrightarrow & 0
\end{array}$$

and the fact that β is the Busby invariant of the extension from the upper row of the above diagram implies by [27, Proof of Prop. 4.2, Th. 4.3] that we have the $*$-isomorphism of C^*-algebras

$$\mathcal{A} \simeq \{f \oplus b \in C_b(T, \mathcal{K}) \oplus \mathcal{B} \mid \tilde{q}(f) = \beta(b)\}, \quad a \mapsto \mathcal{F}_T(a) \oplus q(a). \tag{6.2.1}$$

Note that for all $f \oplus b \in C_b(T, \mathcal{K}) \oplus \mathcal{B}$ we have

$$\tilde{q}(f) = \beta(b) \iff f + C_0(T, \mathcal{K}) = \nu(b) + C_0(T, \mathcal{K}) \iff f - \nu(b) \in C_0(T, \mathcal{K})$$

and for every $f \in C_b(T, \mathcal{K})$ there exists at most one $b \in \mathcal{B}$ satisfying the above condition, since if we have $f - \nu(b_1), f - \nu(b_2) \in C_0(T, \mathcal{K})$, then

$$\beta(b_1) = \nu(b_1) + C_0(T, \mathcal{K}) = \nu(b_2) + C_0(T, \mathcal{K}) = \beta(b_2)$$

hence $b_1 = b_2$.

Finally, since ν is completely isometric, it follows that the map

$$\iota \colon \mathcal{B} \to C_b(T, \mathcal{K}) \oplus \mathcal{B}, \quad b \mapsto \nu(b) \oplus b$$

is completely isometric and its image is contained in the image of $\mathcal{A}$ by the $*$-isomorphism (6.2.1)*. By composing the inverse of that $*$-isomorphism with ι we obtain a completely isometric cross section of $q \colon \mathcal{A} \to \mathcal{B}$, and this completes the proof.*

Remark 6.2.5 If we have a short exact sequence $0 \to \mathcal{J} \to \mathcal{A} \to \mathcal{B} \to 0$ where $\mathcal{A}$ is any C^*-algebra of type I, then it follows that also its quotient $\mathcal{B}$ is of type I, and both $\mathcal{A}$ and $\mathcal{B}$ are nuclear by [99, Prop. 1.6, Ch. XV]. □

Definition 6.5 *Let $\mathcal{A}$ be a C^*-algebra with spectrum $\widehat{\mathcal{A}}$. We choose for every $\gamma \in \widehat{\mathcal{A}}$ a representation $(\pi_\gamma, \mathcal{H}_\gamma)$ in the equivalence class γ. Let $\ell^\infty(\widehat{\mathcal{A}})$ be the algebra of all bounded operator fields defined over $\widehat{\mathcal{A}}$ by*

$$\ell^\infty(\widehat{\mathcal{A}}) := \left\{ \phi = (\phi(\pi_\gamma) \in \mathcal{B}(\mathcal{H}_\gamma))_{\gamma \in \widehat{\mathcal{A}}} \mid \|\phi\|_\infty := \sup_\gamma \|\phi(\pi_\gamma)\|_{\mathcal{B}(\mathcal{H}_\gamma)} < \infty \right\}.$$

We define for $a \in \mathcal{A}$ its Fourier transform $\mathcal{F}_\mathcal{A}(a) = \hat{a}$ by

$$\mathcal{F}_\mathcal{A}(a)(\gamma) = \hat{a}(\gamma) = \pi_\gamma(a), \quad \gamma \in \hat{\mathcal{A}}.$$

Then $\mathcal{F}_\mathcal{A}(a)$ is a bounded field of operators over $\widehat{\mathcal{A}}$, and the mapping

$$\mathcal{F}_\mathcal{A} \colon \mathcal{A} \to \ell^\infty(\widehat{\mathcal{A}}), \quad a \mapsto \mathcal{F}_\mathcal{A}(a)$$

is an isometric $*$-homomorphism.

Theorem 6.2.6 *Let $\mathcal{A}$ be any separable liminary C^*-algebra with an increasing sequence of open subsets of its spectrum*

$$\emptyset = V_0 \subseteq V_1 \subseteq \cdots \subseteq V_n = \widehat{\mathcal{A}}$$

with the corresponding sequence of closed two-sided ideals of $\mathcal{A}$

$$\{0\} = \mathcal{J}_0 \subseteq \mathcal{J}_1 \subseteq \cdots \subseteq \mathcal{J}_n = \mathcal{A}$$

with $\widehat{\mathcal{J}_\ell} = V_\ell$ satisfying the following conditions for $\ell = 1, \ldots, n$:

1. *The set $\Gamma_\ell := V_\ell \setminus V_{\ell-1}$ is dense in $\widehat{\mathcal{A}} \setminus V_{\ell-1}$.*
2. *There exist a complex Hilbert space $\mathcal{H}_\ell$ and a complete system of distinct representatives $\{\pi_\gamma \colon \mathcal{A} \to \mathcal{B}(\mathcal{H}_\ell)\}_{\gamma \in \Gamma_\ell}$ of the equivalence classes of representations corresponding to the elements of Γ_ℓ such that for every $a \in \mathcal{A}$ the mapping $\Gamma_\ell \to \mathcal{K}(\mathcal{H}_\ell)$, $\gamma \mapsto \pi_\gamma(a)$ is continuous with respect to the norm operator topology of $\mathcal{K}(\mathcal{H}_\ell)$.*

For $\ell = 0, \ldots, n$ we define

$$\mathcal{L}_\ell := \{f \colon \widehat{\mathcal{A}} \setminus V_\ell \to \mathcal{K}(\bigoplus_{j=\ell+1}^{n} \mathcal{H}_j) \mid (\forall j \in \{\ell+1, \ldots, n\})(\forall \gamma \in \Gamma_j) \quad f(\gamma) \in \mathcal{K}(\mathcal{H}_j)\}$$

and

$$\mathcal{F}_{\mathcal{A}/\mathcal{J}_\ell} \colon \mathcal{A}/\mathcal{J}_\ell \to \mathcal{L}_\ell, \quad (\mathcal{F}_{\mathcal{A}/\mathcal{J}_\ell}(a + \mathcal{J}_\ell))(\gamma) := \pi_\gamma(a).$$

Then there is a family of linear maps $\{\nu_\ell \colon \mathcal{F}_{\mathcal{A}/\mathcal{J}_\ell}(\mathcal{A}/\mathcal{J}_\ell) \to C_b(\Gamma_\ell, \mathcal{K}(\mathcal{H}_\ell))\}_{1 \le \ell \le n}$, which are completely positive, completely isometric, almost $*$-homomorphisms, and for which the image $\mathcal{F}_{\mathcal{A}}(\mathcal{A})$ is precisely the set of all $f \in \mathcal{L}_0$ such that

$$f|_{\Gamma_\ell} - \nu_\ell(f|_{\widehat{\mathcal{A}} \setminus V_\ell}) \in C_0(\Gamma_\ell, \mathcal{K}(\mathcal{H}_\ell)) \tag{6.2.2}$$

for all $\ell \in \{1, \ldots, n-1\}$.

Remark 6.2.7 Under the assumptions in Theorem 6.2.6, for all $a \in \mathcal{A}$ we have

$$(\mathcal{F}_{\mathcal{A}/\mathcal{J}_{\ell-1}}(a + \mathcal{J}_{\ell-1}))(\gamma) = \pi_\gamma(a) = \mathcal{F}_{\mathcal{A}}(a)(\gamma) \in \mathcal{K}(\mathcal{H}_\ell)$$

for all $\gamma \in \hat{\mathcal{A}} \setminus V_{l-1}$. This follows from [37, Prop. 2.10.4] and the fact that

$$\hat{\mathcal{A}} \setminus V_{\ell-1} = \widehat{\mathcal{A}/\mathcal{J}_{\ell-1}} = \{[\pi] \in \hat{\mathcal{A}} \mid \mathcal{J}_{\ell-1} \subseteq \operatorname{Ker} \pi\}.$$

□

Proof 6.7 (Proof of Theorem 6.2.6) *Recall from Lemma 6.1.2 and Remark 6.1.3 that* $\widehat{\mathcal{J}_\ell/\mathcal{J}_{\ell-1}} = V_\ell \setminus V_{\ell-1}$*, hence Lemma 6.2.3 implies that the canonical field of elementary* C^**-algebras defined by* $\mathcal{J}_\ell/\mathcal{J}_{\ell-1}$ *is trivial. Thus we obtain an isometric* $*$*-isomorphism* $\mathcal{J}_\ell/\mathcal{J}_{\ell-1} \simeq C_0(V_\ell \setminus V_{\ell-1}, \mathcal{K}(\mathcal{H}_\ell))$ *by the Fourier transform. Then the canonical short exact sequence*

$$0 \to \mathcal{J}_\ell/\mathcal{J}_{\ell-1} \to \mathcal{A}/\mathcal{J}_{\ell-1} \to \mathcal{A}/\mathcal{J}_\ell \to 0$$

leads to a short exact sequence

$$0 \to C_0(V_\ell \setminus V_{\ell-1}, \mathcal{K}(\mathcal{H}_\ell)) \to \operatorname{Ran} \mathcal{F}_{\mathcal{A}}/\mathcal{J}_{\ell-1} \to \operatorname{Ran} \mathcal{F}_{\mathcal{A}/\mathcal{J}_\ell} \to 0.$$

Here the second arrow is simply the inclusion, while the third is the restriction to $\hat{\mathcal{A}} \setminus V_\ell$.

Note that the set $\Gamma_\ell := V_\ell \setminus V_{\ell-1}$ *is in* $\operatorname{Sep}(\widehat{\mathcal{A}} \setminus V_{\ell-1})$*. Indeed, all points of* $\widehat{\mathcal{A}}$ *are closed since* $\mathcal{A}$ *is liminary. In addition, the function* $[\pi] \mapsto \|\pi(a)\|$ *is continuous on* Γ_ℓ *for every* $a \in \mathcal{A}$ *by (2). This implies that every point of the open set* Γ_ℓ *separated in* $\widehat{\mathcal{A}}$ *(see [35, p. 116, (iii)] or the proof of [43, Th. 2.1]). Then by Proposition 6.2.4 we obtain a completely positive, completely contractive, almost* $*$*-homomorphism* $\nu_\ell\colon \operatorname{Ran} \mathcal{F}_{\mathcal{A}/\mathcal{J}_\ell} \to C_b(\Gamma_\ell, \mathcal{K}(\mathcal{H}_\ell))$ *with*

$$\begin{aligned}\operatorname{Ran} \mathcal{F}_{\mathcal{A}/\mathcal{J}_{\ell-1}} &= \\ &\Big\{ f \oplus b \in C_b(\Gamma_\ell, \mathcal{K}(\mathcal{H}_\ell)) \oplus \operatorname{Ran} \mathcal{F}_{\mathcal{A}/\mathcal{J}_\ell} \mid f - \nu_\ell(b) \in C_0(\Gamma_\ell, \mathcal{K}(\mathcal{H}_\ell)) \Big\}.\end{aligned}$$

We have that

$$f \oplus b|_{\hat{\mathcal{A}} \setminus V_\ell} = b, \quad f \oplus b|_{\Gamma_\ell} = f,$$

hence when $a \in \mathcal{A}$

$$\mathcal{F}_{\mathcal{A}/\mathcal{J}_{\ell-1}}(a + \mathcal{J}_{\ell-1})|_{\Gamma_\ell} - \nu_l(\mathcal{F}_{\mathcal{A}/\mathcal{J}_{\ell-1}}(a + \mathcal{J}_{\ell-1})|_{\hat{\mathcal{A}} \setminus V_\ell}) \in C_0(\Gamma_\ell, \mathcal{K}(\mathcal{H}_\ell)).$$

On the other hand, for all $\gamma \in \Gamma_\ell$ *and* $a \in \mathcal{A}$ *we have*

$$(\mathcal{F}_{\mathcal{A}/\mathcal{J}_{\ell-1}}(a + \mathcal{J}_{\ell-1}))(\gamma) = \pi_\gamma(a) = \mathcal{F}_{\mathcal{A}}(a)(\gamma) \in \mathcal{K}(\mathcal{H}_\ell).$$

Using now the above description of $\operatorname{Ran} \mathcal{F}_{\mathcal{A}/\mathcal{J}_{\ell-1}}$*, we see that* $\mathcal{F}_{\mathcal{A}}(a)$ *satisfies* (6.2.2).

Then, by successively considering the cases $\ell = n, n-1, \ldots, 1$*, we obtain the description of* $\operatorname{Ran} \mathcal{F}_{\mathcal{A}/\mathcal{J}_{\ell-1}}$*, that for* $\ell = 1$ *concludes the proof.*

6.3 C^*-ALGEBRAS WITH NORM CONTROLLED DUAL LIMITS

6.3.1 Basic notions

Definition 6.6 (see also [81]) *i) Let* S *be a topological space. We say that* S *is locally compact of step* $\leq d$ *if there exists a finite increasing family* $\emptyset \neq S_d \subset S_{d-1} \subset$

$\cdots \subset S_0 = S$ of closed subsets of S, such that the subsets $\Gamma_d = S_d$ and $\Gamma_i := S_{i-1} \setminus S_i$, $i = 1, \ldots, d$, are locally compact and Hausdorff in their relative topologies.

ii) Let S be locally compact of step $\leq d$, and let $\{\mathcal{H}_i\}_{i=1,\ldots,d}$ be Hilbert spaces. For a closed subset $M \subset S$, denote by $CB(M)$ the unital C^*-algebra of all uniformly bounded operator fields $(\psi(\gamma) \in \mathcal{B}(\mathcal{H}_i))_{\gamma \in S \cap \Gamma_i, i=1,\ldots,d}$, which are operator norm continuous on the subsets $\Gamma_i \cap M$ for every $i \in \{1, \ldots, d\}$ with $\Gamma_i \cap M \neq \emptyset$. We provide the algebra $CB(M)$ with the infinity-norm

$$\|\varphi\|_M = \sup\left\{\|\varphi(\gamma)\|_{\mathcal{B}(\mathcal{H}_i)} \mid M \cap \Gamma_i \neq \emptyset, \gamma \in M \cap \Gamma_i\right\}.$$

Definition 6.7 (see also [81]) *Let $\mathcal{A}$ be a separable liminary C^*-algebra. We assume that the spectrum $\widehat{\mathcal{A}}$ of $\mathcal{A}$ is a locally compact space of step $\leq d$,*

$$\emptyset = S_{d+1} \subset S_d \subset S_{d-1} \subset \cdots \subset S_0 = \widehat{\mathcal{A}},$$

and that for $0 \leq i \leq d$ there is a Hilbert space $\mathcal{H}_i$, and for every $\gamma \in \Gamma_i$ a concrete realisation $(\pi_\gamma, \mathcal{H}_i)$ of γ on the Hilbert space $\mathcal{H}_i$. The set S_d is the collection of all characters of $\mathcal{A}$.

We say that the C^*-algebra $\mathcal{A}$ has norm controlled dual limits if for every $a \in \mathcal{A}$ one has

1. *The mappings $\gamma \to \mathcal{F}(a)(\gamma)$ are norm continuous on the difference sets $\Gamma_i = S_{i-1} \setminus S_i$.*

2. *For any $i = 0, \ldots, d+1$ and for any converging sequence contained in Γ_i with limit set outside Γ_i, hence in S_i, there exists a properly converging sub-sequence $\overline{\gamma} = (\gamma_k)_{k\in\mathbb{N}}$, a constant $C > 0$ and for every $k \in \mathbb{N}$ an involutive linear mapping $\tilde{\sigma}_{\overline{\gamma},k} : CB(S_i) \to \mathcal{B}(\mathcal{H}_i)$, that is bounded by $C\|\cdot\|_{S_i}$, such that*

$$(\forall a \in \mathcal{A}) \quad \lim_{k\to\infty} \|\mathcal{F}(a)(\gamma_k) - \tilde{\sigma}_{\overline{\gamma},k}(\mathcal{F}(a)|_{S_i})\|_{\mathcal{B}(\mathcal{H}_i)} = 0.$$

Theorem 6.3.1 *Assume the C^*-algebra $\mathcal{A}$ satisfies the conditions Theorem 6.2.6. Then $\mathcal{A}$ has norm controlled dual limits. Specifically, with the notations of Theorem 6.2.6, let $1 \leq \ell \leq n$ be fixed, and $\bar{\gamma} = (\gamma_k)_k \in \Gamma_\ell$ be a properly convergent sequence with limit set outside Γ_ℓ. Then there exists a sequence $(\sigma_{\bar{\gamma},k})_k$ of completely positive and completely contractive maps $\sigma_{\bar{\gamma},k} \colon CB(\hat{\mathcal{A}} \setminus V_\ell) \to \mathcal{B}(\mathcal{H}_\ell)$ such that*

$$(\forall a \in \mathcal{A}) \quad \lim_{k\to\infty} \|\mathcal{F}(a)(\gamma_k) - \sigma_{\bar{\gamma},k}(\mathcal{F}(a)|_{\hat{\mathcal{A}}\setminus V_\ell})\|_{\mathcal{B}(\mathcal{H}_\ell)} = 0.$$

Proof 6.8 *From Theorem 6.2.6, we see that there is a linear map*

$$\nu_\ell \colon \operatorname{Ran} \mathcal{F}_{\mathcal{A}/\mathcal{J}_\ell} \to C_b(\Gamma_\ell, \mathcal{K}(\mathcal{H}_\ell))$$

that is completely positive, completely isometric and almost $*$-homomorphism, and such that

$$(\forall a \in \mathcal{A}) \quad \mathcal{F}(a)|_{\Gamma_\ell} - \nu_\ell(\mathcal{F}(a)|_{\hat{\mathcal{A}}\setminus V_\ell}) \in C_0(\Gamma_\ell, \mathcal{K}(\mathcal{H}_\ell)). \tag{6.3.1}$$

Define now $\tilde{\sigma}_{\overline{\gamma},k}\colon \mathcal{F}_{\mathcal{A}/\mathcal{J}_\ell}(\mathcal{A}/\mathcal{J}_\ell) \to \mathcal{B}(\mathcal{H}_\ell)$,

$$\tilde{\sigma}_{\overline{\gamma},k}(\cdot) := (\nu_\ell(\cdot))(\gamma_k),$$

for every $k \in \mathbb{N}$. *These maps are completely positive, and by [4, Thm. 1.2.3] and [28, Lemma 3.9], they extend from the* C^**-subalgebra* $\operatorname{Ran}\mathcal{F}_{\mathcal{A}/\mathcal{J}_\ell} \subset CB(\hat{\mathcal{A}} \setminus V_\ell)$ *to completely positive and completely contractive linear maps*

$$\sigma_{\overline{\gamma},k}\colon CB(\hat{\mathcal{A}} \setminus V_\ell) \to \mathcal{B}(\mathcal{H}_\ell).$$

Due to the properties of ν_ℓ*, and using* (6.3.1) *via [45, rem. II, p. 474] which implies* $\gamma_k \to \infty$ *in* Γ_ℓ*, the sequence* $(\sigma_{\overline{\gamma},k})_k$ *satisfies all the properties in the statement.*

6.3.2 Fourier transforms of C^*-algebras with norm controlled dual limits

Definition 6.8 (see also [81]) *Let* $\emptyset = S_{d+1} \subset S_d \subset \cdots \subset S_0 = S$ *be a locally compact topological space of step* $\leq d$. *Choose for every* $i = 1, \ldots, d$ *a Hilbert space* $\mathcal{H}_i$ *and assume that* $\mathcal{H}_d = \mathbb{C}$.

Let $B^*(S)$ *be the set of all operator fields* φ *defined over* S *such that*

1. $\varphi(\gamma) \in \mathcal{K}(\mathcal{H}_i)$ *for every* $\gamma \in \Gamma_i = S_{i-1} \setminus S_i$, $i = 1, \ldots, d$.
2. *The field* φ *is uniformly bounded, that is, we have that*

$$\|\varphi\| = \sup\Big\{\|\varphi(\gamma)\|_{\mathcal{B}(\mathcal{H}_i)} \mid \gamma \in \Gamma_i,\, i = 1, \ldots, d\Big\} < \infty.$$

3. *The mappings* $\gamma \to \varphi(\gamma)$ *are norm continuous on the difference sets* Γ_i.
4. *We have for any sequence* $(\gamma_k)_{k\in\mathbb{N}} \subset S$ *going to infinity, that*

$$\lim_{k\to\infty} \|\varphi(\gamma_k)\|_{op} = 0.$$

5. *For any* $i = 1, \ldots, d+1$, *and for any converging sequence contained in* $\Gamma_i = S_{i-1} \backslash S_i$ *with limit set outside* Γ_i*, there exists a properly converging sub-sequence* $\overline{\gamma} = (\gamma_k)_{k\in\mathbb{N}}$*, a constant* $C > 0$ *and for every* $k \in \mathbb{N}$ *an involutive linear mapping* $\tilde{\sigma}_{\overline{\gamma},k}\colon CB(S_i) \to \mathcal{B}(\mathcal{H}_i)$*, which is bounded by* $C\|.\|_{S_i}$*, such that*

$$\lim_{k\to\infty} \|\varphi(\gamma_k) - \tilde{\sigma}_{\overline{\gamma},k}(\varphi|_{S_i})\|_{\mathcal{B}(\mathcal{H}_i)} = 0.$$

Theorem 6.3.2 (see also [81]) *Let* S *be a locally compact topological space of step* $\leq d$. *Then the set* $B^*(S)$ *of Definition 6.8 is a closed involutive subspace of* $\ell^\infty(S)$. *Furthermore* $B^*(S)$*is a* C^**-subalgebra of* $\ell^\infty(S)$ *with spectrum* S *if and only if all the mappings* $\tilde{\sigma}_{\overline{\gamma},k}$ *are almost homomorphisms, i.e.*

$$\lim_{k\to\infty} \|\tilde{\sigma}_{\overline{\gamma},k}(\varphi \cdot \psi) - \tilde{\sigma}_{\overline{\gamma},k}(\varphi) \cdot \tilde{\sigma}_{\overline{\gamma},k}(\psi)\|_{\mathcal{B}(\mathcal{H}_i)} = 0, \quad \varphi, \psi \in B^*(S),$$

and the restrictions $B^*(S)|_{S_{i-1}}$ *contain the spaces* $C_0(\Gamma_i, \mathcal{H}_i)$, $i = 1, \ldots, d+1$.

Proof 6.9 *We easily see that conditions (1)–(4) in Definition 6.8 imply that $B^*(S)$ is a closed involution-invariant subspace of $\ell^\infty(S)$.*

For $i = 1, \dots, d+1$, let B_i^* be the set of all operator fields defined over S_{i-1}, satisfying conditions (1)–(5) on the sets S_j, $j = d, \dots, i-1$. Then obviously for every i the restriction $B^*(S)|_{S_{i-1}}$ of the space $B^*(S)$ to S_{i-1} is contained in B_i^*, and $B^*(S)$ is also an algebra and hence a C^*-subalgebra of $\ell^\infty(S)$.

Let us show that the spectrum of $B^*(S)$ can be identified with the space S. Since $C_0(\Gamma_i, \mathcal{H}_i)$ is contained in B_i^* for every i, it follows that the representations $\pi_s\colon \varphi \to \varphi(s) \in \mathcal{K}(\mathcal{H}_i)$ of $B^*(S)$ are irreducible. It follows from the choice of $\mathcal{H}_d$ and the properties of $B^*(S)$, that $B_d^* = B^*(S)|_{S_d} = C_0(S_d)$.

Suppose that for some $0 \le i < d$ the spectrum of the algebra B_{i+1}^* is the space S_i. Let $\pi \in \widehat{B_i^*}$. Consider the kernel K_{i+1} of the restriction mapping R_{i+1} from B_i^* into $l^\infty(S_i)$. If $\pi(K_{i+1}) = \{0\}$, then we can consider π as being a representation of the quotient algebra B_i^*/K_{i+1}. But the image $\mathcal{B}^*(S)|_{S_i}$ of R_{i+1} is a C^*-subalgebra of B_{i+1}^*, the spectrum of B_{i+1}^* is by assumption the set S_i and S_i is also contained in the spectrum of the subalgebra $\mathcal{B}^*(S)|_{S_i}$. Hence by the Stone-Weierstrass theorem [37, Th. 11.1.8] the algebras B_{i+1}^* and $B^*(S)|_{S_i}$ coincide. Hence π is an evaluation at a point in S_i.

If $\pi(K_{i+1}) \neq \{0\}$, then we look at the restriction of π to this ideal. The elements in K_{i+1} are operator fields defined on S_{i-1} which are norm continuous, which go to 0 at infinity and by condition (5) in Definition 6.8, for any properly converging sequence $\overline{\gamma} \subset \Gamma_i$ with limit outside Γ_i, for every $\varphi \in K_{i+1}$, we have that

$$\lim_{k\to\infty} \|\varphi(\gamma_k)\|_{\mathcal{B}(\mathcal{H}_i)} = \lim_{k\to\infty} \|\varphi(\gamma_k) - \tilde{\sigma}_{\overline{\gamma},k}(\varphi|_{S_i})\|_{\mathcal{B}(\mathcal{H}_i)} = 0.$$

This shows that $K_{i+1} \subset C_0(\Gamma_i, \mathcal{K}(\mathcal{H}_i))$. Since by condition (1) in Definition 6.8 we know that $C_0(\Gamma_i, \mathcal{K}(\mathcal{H}_i)) \subset K_{i+1}$ it follows that $C_0(\Gamma_i, \mathcal{K}(\mathcal{H}_i)) = K_{i+1}$ and then π is an evaluation at an element $s \in \Gamma_i$. Finally the spectrum of B_i^* is the set S_{i-1}. Therefore again by the aforementioned Stone-Weierstrass theorem the algebras B_i^* and $B^*(S)|_{S_{i-1}}$ coincide.

If $B^*(S)$ is a C^*-algebra, then obviously the conditions of Definition 6.8 are fulfilled.

Corollary 6.3.3 *Let $\mathcal{A}$ be a C^*-algebra with norm controlled dual limits. Then the Fourier transform of $\mathcal{A}$ is the C^*-algebra $B^*(\widehat{\mathcal{A}})$.*

Proof 6.10 *Use Theorem 6.3.2.*

6.3.3 Complements on properly convergent sequences

Here we discuss some uniqueness properties of the boundary value mappings that occur in Definition 6.7 (see Proposition 6.3.7 below).

Definition 6.9 A sequence $\bar{x} := \{x_k\}_{k\in\mathbb{N}}$ in a topological space X is said to be *properly convergent* if its set of cluster points

$$L(\bar{x}) := \bigcap_{k\in\mathbb{N}} \overline{\{x_i \mid i \geq k\}}$$

has the property that for every point $y \in L(\bar{x})$ and every subsequence $\{x_{k_j}\}_{j\in\mathbb{N}}$ one has $x_{k_j} \to y$ as $j \to \infty$.

Lemma 6.3.4 *For every C^*-algebra $\mathcal{A}$ and any closed sets $S_1, S_2 \subseteq \widehat{\mathcal{A}}$ one has*

$$S_1 \subseteq S_2 \iff (\forall a \in \mathcal{A}) \sup_{[\rho]\in S_1} \|\rho(a)\| \leq \sup_{[\rho]\in S_2} \|\rho(a)\|.$$

Proof 6.11 *"$\Rightarrow$" is obvious, and the converse follows by [43, Lemma 2.1].*

Lemma 6.3.5 *Let $\mathcal{A}$ be any C^*-algebra, $\bar{\pi} := \{[\pi_k]\}_{k\in\mathbb{N}}$ be any properly convergent sequence in $\widehat{\mathcal{A}}$. Then for any closed set $S \subseteq \widehat{\mathcal{A}}$ one has*

$$L(\bar{\pi}) = S \iff (\forall a \in \mathcal{A}) \lim_{k\to\infty} \|\pi_k(a)\| = \sup_{[\rho]\in S} \|\rho(a)\|;$$
$$L(\bar{\pi}) \subseteq S \iff (\forall a \in \mathcal{A}) \lim_{k\to\infty} \|\pi_k(a)\| \leq \sup_{[\rho]\in S} \|\rho(a)\|;$$
$$L(\bar{\pi}) \supseteq S \iff (\forall a \in \mathcal{A}) \lim_{k\to\infty} \|\pi_k(a)\| \geq \sup_{[\rho]\in S} \|\rho(a)\|.$$

Proof 6.12 *We prove only the first equivalence in the statement, as this proof contains the arguments needed for the second and third equivalences.*

"$\Rightarrow$" Use [43, Theorem 2.1].

"$\Leftarrow$" Let $S \subseteq \widehat{\mathcal{A}}$ be any closed set with the property

$$(\forall a \in \mathcal{A}) \quad \lim_{k\to\infty} \|\pi_k(a)\| = \sup_{[\rho]\in S} \|\rho(a)\|.$$

Since $\bar{\pi} := \{[\pi_k]\}_{k\in\mathbb{N}}$ is a properly convergent sequence, it follows by "$\Rightarrow$" that

$$(\forall a \in \mathcal{A}) \quad \lim_{k\to\infty} \|\pi_k(a)\| = \sup_{[\rho]\in L(\bar{\pi})} \|\rho(a)\|$$

hence

$$(\forall a \in \mathcal{A}) \quad \sup_{[\rho]\in S} \|\rho(a)\| = \sup_{[\rho]\in L(\bar{\pi})} \|\rho(a)\|.$$

Then, as both $L(\bar{\pi})$ and S are closed subsets of $\widehat{\mathcal{A}}$, it follows by Lemma 6.9 that $L(\bar{\pi}) \subseteq S$ and $L(\bar{\pi}) \supseteq S$, hence $L(\bar{\pi}) = S$, and this completes the proof.

Lemma 6.3.6 *Let $\mathcal{A}$ be any C^*-algebra, and $\bar{\pi} := \{[\pi_k]\}_{k\in\mathbb{N}}$ and $\bar{\tau} := \{[\tau_k]\}_{k\in\mathbb{N}}$ be any properly convergent sequences in $\widehat{\mathcal{A}}$. Then one has*

$$L(\bar{\pi}) \supseteq L(\bar{\tau}) \iff (\forall a \in \mathcal{A}) \ \lim_{k\to\infty} (\|\pi_k(a)\| - \|\tau_k(a)\|) \geq 0 \tag{6.3.2}$$

and

$$L(\bar{\pi}) = L(\bar{\tau}) \iff (\forall a \in \mathcal{A}) \ \lim_{k\to\infty} |\|\pi_k(a)\| - \|\tau_k(a)\|| = 0. \tag{6.3.3}$$

Proof 6.13 *It follows by Lemma 6.3.5 that*

$$(\forall a \in \mathcal{A}) \quad \lim_{k\to\infty} \|\pi_k(a)\| = \sup_{\rho\in L(\bar{\pi})} \|\rho(a)\|$$

and

$$(\forall a \in \mathcal{A}) \quad \lim_{k\to\infty} \|\tau_k(a)\| = \sup_{\rho\in L(\bar{\tau})} \|\rho(a)\|$$

and then the implications "$\Rightarrow$" in both (6.3.2) and (6.3.3) follow at once. For (6.3.2) one also needs the elementary remark that if $\{t_k\}_{k\in\mathbb{N}}$ and $\{s_k\}_{k\in\mathbb{N}}$ are any convergent sequences of real numbers, then

$$\lim_{k\to\infty} t_k = \lim_{k\to\infty} s_k \iff \lim_{k\to\infty} |t_k - s_k| = 0.$$

Furthermore, if the condition in the right-hand side of (6.3.2) is satisfied, it follows by the above displayed equalities that

$$(\forall a \in \mathcal{A}) \quad \lim_{k\to\infty} |\|\pi_k(a)\| \geq \sup_{[\rho]\in L(\bar{\tau})} \|\rho(a)\|$$

and then $L(\bar{\pi}) \supseteq L(\bar{\tau})$ by the last equivalence in Lemma 6.3.5. This completes the proof of (6.3.2). The implication "$\Leftarrow$" in (6.3.3) can be proved similarly.

Now we can establish a kind of uniqueness property of some mappings that occur in Definition 6.7, which also shows how these mappings depend on the limit set of the properly convergent sequence to which they are associated. It is worth pointing out that no linearity properties of these mappings are needed in the following proposition.

Proposition 6.3.7 *In the setting of Definition 6.7, fix $i \in \{0, \ldots, d+1\}$. Let $\bar{\pi} := \{[\pi_k]\}_{k\in\mathbb{N}}$ and $\bar{\tau} := \{[\tau_k]\}_{k\in\mathbb{N}}$ be any properly convergent sequences contained in Γ_i with $L(\bar{\pi}) \cup L(\bar{\tau}) \subseteq S_i$. Assume that for every $k \in \mathbb{N}$ one has some mappings $\tilde{\sigma}_{\bar{\pi},k} : CB(S_i) \to \mathcal{B}(\mathcal{H}_i)$ and $\tilde{\sigma}_{\bar{\tau},k} : CB(S_i) \to \mathcal{B}(\mathcal{H}_i)$ such that*

$$\lim_{k\to\infty} \|\mathcal{F}(a)(\pi_k) - \tilde{\sigma}_{\bar{\pi},k}(\mathcal{F}(a)|_{S_i})\|_{\mathcal{B}(\mathcal{H}_i)} = \lim_{k\to\infty} \|\mathcal{F}(a)(\tau_k) - \tilde{\sigma}_{\bar{\tau},k}(\mathcal{F}(a)|_{S_i})\|_{\mathcal{B}(\mathcal{H}_i)} = 0$$

for every $a \in \mathcal{A}$. Then the following assertions hold:

1. *If* $L(\bar{\pi}) \supseteq L(\bar{\tau})$, *then*

$$(\forall a \in \mathcal{A}) \quad \lim_{k\to\infty} (\|\tilde{\sigma}_{\bar{\pi},k}(\mathcal{F}(a)|_{S_i})\|_{\mathcal{B}(\mathcal{H}_i)} - \|\tilde{\sigma}_{\bar{\tau},k}(\mathcal{F}(a)|_{S_i})\|_{\mathcal{B}(\mathcal{H}_i)}) \geq 0.$$

2. *If* $L(\bar{\pi}) = L(\bar{\tau})$, *then*

$$(\forall a \in \mathcal{A}) \quad \lim_{k\to\infty} \big|\|\tilde{\sigma}_{\bar{\pi},k}(\mathcal{F}(a)|_{S_i})\|_{\mathcal{B}(\mathcal{H}_i)} - \|\tilde{\sigma}_{\bar{\tau},k}(\mathcal{F}(a)|_{S_i})\|_{\mathcal{B}(\mathcal{H}_i)}\big| = 0.$$

Proof 6.14 *The hypothesis can be written as*

$$\lim_{k\to\infty} \|\pi_k(a) - \tilde{\sigma}_{\bar{\pi},k}(\mathcal{F}(a)|_{S_i})\|_{\mathcal{B}(\mathcal{H}_i)} = \lim_{k\to\infty} \|\tau_k(a) - \tilde{\sigma}_{\bar{\tau},k}(\mathcal{F}(a)|_{S_i})\|_{\mathcal{B}(\mathcal{H}_i)} = 0$$

and this implies

$$\lim_{k\to\infty} \big|\|\pi_k(a)\|_{\mathcal{B}(\mathcal{H}_i)} - \|\tilde{\sigma}_{\bar{\pi},k}(\mathcal{F}(a)|_{S_i})\|_{\mathcal{B}(\mathcal{H}_i)}\big| = 0$$

and

$$\lim_{k\to\infty} \big|\|\tau_k(a)\|_{\mathcal{B}(\mathcal{H}_i)} - \|\tilde{\sigma}_{\bar{\tau},k}(\mathcal{F}(a)|_{S_i})\|_{\mathcal{B}(\mathcal{H}_i)}\big| = 0.$$

Now the assertions follow by Lemma 6.3.6.

6.4 C^*-ALGEBRAS OF NILPOTENT LIE GROUPS

We prove in this section that the C^*-algebras of nilpotent Lie groups are special solvable, and this allows us to use the results of the previous section to describe the image of their operator-valued Fourier transform.

Throughout this section we denote by $\mathfrak{g}$ any nilpotent Lie algebra with its corresponding Lie group $G = (\mathfrak{g}, \cdot)$ and with a fixed Jordan-Hölder sequence

$$\{0\} = \mathfrak{g}_0 \subseteq \cdots \subseteq \mathfrak{g}_m = \mathfrak{g}$$

and we pick $X_j \in \mathfrak{g}_j \setminus \mathfrak{g}_{j-1}$ for $j = 1, \dots, m$. We denote by

$$\langle \cdot, \cdot \rangle \colon \mathfrak{g}^* \times \mathfrak{g} \to \mathbb{R}$$

the duality pairing between $\mathfrak{g}$ and its linear dual space $\mathfrak{g}^*$, and for every subalgebra $\mathfrak{h} \subseteq \mathfrak{g}$ we define for $\xi \in \mathfrak{g}^*$:

$$\mathfrak{h}(\xi) := \{X \in \mathfrak{h} \mid (\forall Y \in \mathfrak{h}) \ \langle \xi, [X, Y] \rangle = 0\}.$$

We denote by $\mathcal{E}$ the set of all subsets of $\{1, \dots, m\}$ endowed with the total ordering defined for all $e_1 \neq e_2$ in $\mathcal{E}$ by

$$e_1 \prec e_2 \iff \min(e_1 \setminus e_2) < \min(e_2 \setminus e_1)$$

where we use the convention $\min \emptyset = \infty$, so in particular $\max \mathcal{E} = \emptyset$. This is precisely the ordering introduced in [88, subsect. 3.4, p. 454]. We also endow the m-th Cartesian power $\mathcal{E}^m$ with the total ordering obtained from the above ordering $\prec$ as in [91, Def. 1.2.5].

6.4.1 Coarse stratification of $\mathfrak{g}^*$ and continuity of trace on $\widehat{G}$

Let us define the jump indices for $\xi \in \mathfrak{g}^*$:

$$J_\xi := \{j \in \{1,\dots,m\} \mid \mathfrak{g}_j \not\subset \mathfrak{g}(\xi) + \mathfrak{g}_{j-1}\}$$

and

$$(\forall e \in \mathcal{E}) \quad \Omega_e := \{\xi \in \mathfrak{g}^* \mid J_\xi = e\}.$$

The *coarse stratification of* $\mathfrak{g}^*$ is the family $\{\Omega_e\}_{e\in\mathcal{E}}$, which is a finite partition of $\mathfrak{g}^*$ consisting of G-invariant sets. For every coadjoint G-orbit $\mathcal{O} \in \mathfrak{g}^*/G$ we define $J_{\mathcal{O}} := J_\xi$ for any $\xi \in \mathcal{O}$ and then

$$(\forall e \in \mathcal{E}) \quad \Xi_e := \{\mathcal{O} \in \mathfrak{g}^*/G \mid J_{\mathcal{O}} = e\}. \tag{6.4.1}$$

Lemma 6.4.1 *Assume the above setting and for every $\mathcal{O} \in \mathfrak{g}^*/G$ pick any unitary irreducible representation $\pi_{\mathcal{O}}\colon G \to \mathcal{B}(\mathcal{H}_{\mathcal{O}})$ which is associated with $\mathcal{O}$ via Kirillov's correspondence. If we endow the space $\mathfrak{g}^*/G \simeq \widehat{G}$ with its canonical topology, then for every index set $e \in \mathcal{E}$ the following assertions hold:*

1. *The relative topology of $\Xi_e \subseteq \mathfrak{g}^*/G$ is Hausdorff.*
2. *For every test function $\phi \in C_0^\infty(G)$ the function*

$$\Xi_e \to \mathbb{C}, \quad \mathcal{O} \mapsto \operatorname{Tr}(\pi_{\mathcal{O}}(\phi))$$

is well defined and continuous.

Proof 6.15 *The first assertion follows as an application of [31, Theorem 3.1.14(iv)] for the coadjoint action of G, which provides a homeomorphism of Ξ_e onto a certain algebraic subset of the vector space* $\operatorname{span}\{X_j \mid j \in \{1,\dots,m\}\setminus e\}$.

The second assertion is just [88, Lemma 4.4.4].

6.4.2 Fine stratification of $\mathfrak{g}^*$ and continuity of operator fields on $\widehat{G}$

We define for $\xi \in \mathfrak{g}^*$:

$$(\forall k = 1,\dots,m) \quad J_\xi^k := \{j \in \{1,\dots,k\} \mid \mathfrak{g}_j \not\subset \mathfrak{g}_k(\xi) + \mathfrak{g}_{j-1}\}$$

and

$$(\forall \varepsilon \in \mathcal{E}^m) \quad \Omega_\varepsilon := \{\xi \in \mathfrak{g}^* \mid (J_\xi^1,\dots,J_\xi^m) = \varepsilon\}.$$

The *fine stratification of* $\mathfrak{g}^*$ is the family $\{\Omega_\varepsilon\}_{\varepsilon\in\mathcal{E}^m}$, which is again a finite partition of $\mathfrak{g}^*$ consisting of G-invariant sets. For every coadjoint G-orbit $\mathcal{O} \in \mathfrak{g}^*/G$ we define $J_{\mathcal{O}}^k := J_\xi^k$ for any $\xi \in \mathcal{O}$ and $k = 1,\dots,m$, and then

$$(\forall \varepsilon \in \mathcal{E}^m) \quad \Xi_\varepsilon := \{\mathcal{O} \in \mathfrak{g}^*/G \mid (J_{\mathcal{O}}^1,\dots,J_{\mathcal{O}}^m) = \varepsilon\}.$$

For every $\varepsilon \in \mathcal{E}^m$ we also define $\Gamma_\varepsilon \subseteq \widehat{G}$ as the image of Ξ_ε through Kirillov's correspondence $\mathfrak{g}^*/G \simeq \widehat{G}$, which is actually a homeomorphism.

Remark 6.4.2 Every stratum of the coarse stratification of $\mathfrak{g}^*$ is the disjoint union of a few strata of the fine stratification of $\mathfrak{g}^*$. More precisely, since $\mathfrak{g}_m = \mathfrak{g}$, we have

$$(\forall e \in \mathcal{E}) \quad \Omega_e = \{\xi \in \mathfrak{g}^* \mid J_\xi^m = e\} = \bigsqcup_{\varepsilon \in \mathcal{E}^{m-1}\times\{e\}} \Omega_\varepsilon.$$

Lemma 6.4.3 *For every $\varepsilon \in \mathcal{E}^m$ there exists a Hilbert space $\mathcal{H}_\varepsilon$ and every equivalence class of representations $\gamma \in \Gamma_\varepsilon$ has a representative $\pi_\gamma \colon G \to \mathcal{B}(\mathcal{H}_\varepsilon)$ with the property that for every $a \in C^*(G)$ the map $\Pi_a \colon \Gamma_\varepsilon \to \mathcal{B}(\mathcal{H}_\varepsilon)$, $\gamma \mapsto \pi_\gamma(a)$, is norm continuous.*

Proof 6.16 *By using [79, Corollary 2.15], we obtain a Hilbert space $\mathcal{H}_\varepsilon = L^2(\mathbb{R}^d)$ and for every $\gamma \in \Gamma_\varepsilon$ a representative $\pi_\gamma \colon G \to \mathcal{B}(\mathcal{H}_\varepsilon)$ with the property that for all $\phi \in C_0^\infty(G)$, $f_1 \in C_0^\infty(\mathbb{R}^d)$, and $f_2 \in \mathcal{S}(\mathbb{R}^d)$ with $\widehat{f_2} \in C_0^\infty(\mathbb{R}^d)$ the function $\Gamma_\varepsilon \to \mathbb{C}$, $\gamma \mapsto \langle \pi_\gamma(\phi) f_1, f_2\rangle$ is continuous.*

On the other hand, it follows by Lemma 6.4.1(2) that for every $\phi \in C_0^\infty(G)$ the function $\Gamma_\varepsilon \to \mathbb{C}$, $\gamma \mapsto \operatorname{Tr} \pi_\gamma(\phi)$, is continuous. Hence we may use Lemma 6.2.2 with $\mathcal{S} = C_0^\infty(G)$ to obtain the conclusion.

Lemma 6.4.4 *For every $\varepsilon \in \mathcal{E}^m$ the set Ξ_ε is dense and open in $(\mathfrak{g}^*/G) \setminus \bigcup_{\mathcal{E}^m \ni \delta \prec \varepsilon} \Xi_\delta$.*

Proof 6.17 *If $e \in \mathcal{E}$ is the set for which $\varepsilon \in \mathcal{E}^{m-1} \times \{e\}$, then Remark 6.4.2 implies in particular that $\Xi_\varepsilon \subseteq \Xi_e$. It then follows by Lemma 6.4.1(1) that the relative topology of Ξ_ε is Hausdorff.*

On the other hand, it follows by [91, Prop. 1.3.2] that Ω_ε is a Zariski-open subset of $\mathfrak{g}^* \setminus \bigcup_{\mathcal{E}^m \ni \delta \prec \varepsilon} \Omega_\delta$. Since the quotient map $\mathfrak{g}^* \to \mathfrak{g}^*/G$ is an open map (see [26, Ch. 1, no. 2, Ex. 1]), it then follows that Ξ_ε is a dense open subset of $(\mathfrak{g}^*/G) \setminus \bigcup_{\mathcal{E}^m \ni \delta \prec \varepsilon} \Xi_\delta$.

6.4.3 Image of the Fourier transform of $C^*(G)$

Proposition 6.4.5 *Define the sequence*

$$\emptyset = V_0 \subsetneqq V_1 \subsetneqq \cdots \subsetneqq V_n = \widehat{G}$$

by the conditions

- $\{\emptyset\} \cup \{V_j \setminus V_{j-1} \mid 1 \le j \le n\} = \{\Gamma_\varepsilon \mid \varepsilon \in \mathcal{E}^m\}$;
- *if $j_1, j_2 \in \{1, \dots, n\}$ and $\varepsilon_1, \varepsilon_2 \in \mathcal{E}^m$ with $V_{j_1} \setminus V_{j_1-1} = \Gamma_{\varepsilon_1}$ and $V_{j_2} \setminus V_{j_2-1} = \Gamma_{\varepsilon_2}$, then we have $j_1 < j_2$ if and only if $\varepsilon_1 \prec \varepsilon_2$.*

Then V_j is an open subset of $\widehat{G}$ for all $j = 1, \dots, n$, and both hypotheses (1)–(2) of Theorem 6.2.6 are satisfied.

Proof 6.18 *If $j = 1, \dots, n$, then V_j is an open subset of $\widehat{G}$ as a consequence of Lemma 6.4.4, since Kirillov's correspondence $\mathfrak{g}^*/G \simeq \widehat{G}$ is a homeomorphism. Moreover, to see that the other hypotheses of Theorem 6.2.6 are satisfied, we just apply Lemma 6.4.4 again (for hypothesis (1)), and Lemma 6.4.3 (for hypothesis (2)). This completes the proof.*

We have thus obtained the following result:

Theorem 6.4.6 *Let G be connected simply connected nilpotent Lie group. Then the C^*-algebra $C^*(G)$ has norm controlled dual limits.*

Proof 6.19 *The theorem follows from Proposition 6.4.5, Theorem 6.2.6, and Theorem 6.3.1.*

6.5 THE C^*-ALGEBRA OF THE HEISENBERG GROUP

6.5.1 Introduction and notations

Let H_n be the Heisenberg group of dimension $2n+1$ as defined in Subsection 1.3.1. We refer to that section for all the notations and the backgrounds.

It has been known for a long time that the C^*-algebra, $C^*(H_n)$, of H_n is an extension of an ideal J isomorphic to $C_0(\mathbb{R}^*, \mathcal{K})$ with the quotient algebra isomorphic to $C^*(\mathbb{R}^{2n})$, where $\mathcal{K}$ is the C^*-algebra of compact operators on a separable Hilbert space, $\mathbb{R}^* = \mathbb{R}\backslash\{0\}$ and $C_0(\mathbb{R}^*, \mathcal{K})$ is the C^*-algebra of continuous functions vanishing at infinity from $\mathbb{R}^*$ to $\mathcal{K}$.

We obtain an exact characterisation of this extension giving a linear mapping from $C^*(\mathbb{R}^{2n})$ to $C^*(H_n)/J$ which is a cross section of the quotient mapping $i : C^*(H_n) \to C^*(H_n)/J$. More precisely, realising $C^*(H_n)$ as a C^*-subalgebra of the C^*-algebra $\mathcal{F}_n$ of all operator fields $(F = F(\lambda))_{\lambda\in\mathbb{R}}$ taking values in $\mathcal{K}$ for $\lambda \in \mathbb{R}^*$ and in $C^*(\mathbb{R}^{2n})$ for $\lambda = 0$, norm continuous on $\mathbb{R}^*$ and vanishing as $\lambda \to \infty$, we construct a linear map ν from $C^*(\mathbb{R}^{2n})$ to $\mathcal{F}_n$, such that the C^*-subalgebra is isomorphic to the C^*-algebra $D_\nu(H_n)$ of all $(F = F(\lambda))_{\lambda\in\mathbb{R}} \in \mathcal{F}_n$ such that

$$\|F(\lambda) - \nu(F(0))\|_{\text{op}} \to 0,$$

where $\|\cdot\|_{\text{op}}$ is the operator norm on $\mathcal{K}$. The constructed mapping ν is an almost homomorphism in the sense that

$$\lim_{\lambda\to 0} \|\nu(f \cdot h)(\lambda) - \nu(f)(\lambda) \circ \nu(h)(\lambda)\|_{\text{op}} = 0.$$

Moreover, any such almost homomorphism $\tau : C^*(\mathbb{R}^2) \to \mathcal{F}_n$ defines a C^*-algebra, $D_\tau(H_n)$, which is an extension of $C_0(\mathbb{R}^*, \mathcal{K})$ by $C^*(\mathbb{R}^{2n})$. A question we left unanswered : what mappings τ give the C^*-algebras which are isomorphic to $C^*(H_n)$. We note that the condition

$$\lim_{\lambda\to 0} \|\tau(h)(\lambda)\|_{\text{op}} = \|h\|_{C^*(\mathbb{R}^{2n})}, \text{ for all } h \in C^*(\mathbb{R}^{2n}),$$

which is equivalent to the condition that the topologies of $D_\tau(H_n)$ and that of $C^*(H_n)$ agree, is not the right condition: there are examples of splitting extensions of type $D_\tau(H_n)$ with the same spectrum as $C^*(H_n)$ (see [34] and Example 6.2) while it is known that $C^*(H_n)$ is a non-splitting extension.

We note that another characterisation of $C^*(H_n)$ as a C^*-algebra of operator fields is given without proof in a short paper by Gorbachev [57].

We shall use the following notation: $L^p(\mathbb{R}^n)$ denotes the space of (almost everywhere equivalence classes) p-integrable functions for $p = 1, 2$ with norm $\|\cdot\|_p$. By $\|f\|_\infty$ we denote the supremum norm $\sup_{x\in\Omega} |f(x)|$ of a continuous function f vanishing at infinity from a locally compact space Ω to $\mathbb{C}$. $\mathcal{D}(\mathbb{R}^n)$ is the space of complex-valued C^∞ functions with compact support and $\mathcal{S}(\mathbb{R}^n)$ is the space of Schwartz functions, i.e. rapidly decreasing complex-valued C^∞ functions on $\mathbb{R}^n$. The space of Schwartz functions on the groups H_n and G_N (see [31]) will be denoted by $\mathcal{S}(H_n)$ and $\mathcal{S}(G_N)$, respectively. We use the usual notation $B(H)$ for the space of all linear bounded operators on a Hilbert space H with the operator norm $\|\cdot\|_{\text{op}}$.

6.5.2 The topology of $\widehat{C^*(H_n)}$

Let $C^*(H_n)$ denote the full C^*-algebra of H_n. We recall that $C^*(H_n)$ is obtained by the completion of $L^1(H_n)$ with respect to the norm

$$\|f\|_{C^*(H_n)} = \sup \| \int f(x, y, t)\pi(x, y, t) dx dy dt\|_{\text{op}},$$

where the supremum is taken over all unitary representations π of H_n.

Definition 6.10 Let

$$\rho = \operatorname{ind}_{\mathcal{Z}}^{H_n} 1$$

be the left regular representation of H_n on the Hilbert space $L^2(H_n/\mathcal{Z})$. Then the image $\rho(C^*(H_n))$ is just the C^*-algebra of $\mathbb{R}^{2n}$ considered as an algebra of convolution operators on $L^2(\mathbb{R}^{2n})$ and $\rho(C^*(H_n))$ is isomorphic to the algebra $C_0(\mathbb{R}^{2n})$ of continuous functions vanishing at infinity on $\mathbb{R}^{2n}$ via the Fourier transform. For $f \in L^1(H_n)$ we have $\widehat{\rho(f)}(a, b) = \hat{f}(a, b, 0)$, a, $b \in \mathbb{R}^n$. Here, $\hat{f}(a, b, 0)$ is defined as in equation (1.3.1.3).

Definition 6.11 Define for $C^*(H_n)$ the *Fourier transform* $F(c)$ of c by

$$F(c)(\lambda) := \pi_\lambda(c) \in B(L^2(\mathbb{R}^n)), \lambda \in \mathbb{R}^*$$

and

$$F(c)(0) := \rho(c) \in C^*(\mathbb{R}^{2n}).$$

6.5.2.1 *Behaviour on* $\mathbb{R}^*$:

As for the topology of the dual space, it is well known that $[\pi_\lambda]$ tends to $[\pi_\nu]$ in $\widehat{H}_n$ if and only if λ tends to ν in $\mathbb{R}^*$, where $[\pi]$ denotes the unitary equivalence class of the unitary representation π. Furthermore, if λ tends to 0, then the representations π_λ converge in the dual space topology to all the characters $\chi_{a,b}$, $a, b \in \mathbb{R}^n$.

Let us compute for $f \in L^1(H_n)$ the operator $\pi_\lambda(f)$. We have for $\xi \in L^2(\mathbb{R}^n)$ and $s \in \mathbb{R}^n$ that

$$\pi_\lambda(f)\xi(s) = \int_{\mathbb{R}^n\times\mathbb{R}^n\times\mathbb{R}} f(x,y,t)\pi_\lambda(x,y,t)\xi(s)dxdydt \tag{6.5.1}$$

$$\begin{aligned} &= \int_{\mathbb{R}^n\times\mathbb{R}^n\times\mathbb{R}} f(x,y,t)e^{-2\pi i\lambda t-\frac{2\pi i\lambda}{2}x\cdot y+2\pi i\lambda s\cdot y}\xi(s-x)dxdydt \\ &= \int_{\mathbb{R}^n\times\mathbb{R}^n\times\mathbb{R}} f(s-x,y,t)e^{-2\pi i\lambda t-\frac{2\pi i\lambda}{2}(s-x)\cdot y+2\pi i\lambda s\cdot y}\xi(x)dxdydt \\ &= \int_{\mathbb{R}^n} \hat{f}^{2,3}(s-x,-\frac{\lambda}{2}(s+x),\lambda)\xi(x)dx. \end{aligned}$$

Here

$$\hat{f}^{2,3}(s,u,\lambda) = \int_{\mathbb{R}^n\times\mathbb{R}} f(s,y,t)e^{-2\pi i(y\cdot u+\lambda t)}dydt, \ (s,u,\lambda) \in \mathbb{R}^n\times\mathbb{R}^n\times\mathbb{R}$$

denotes the partial Fourier transform of f in the variables y and t.

Hence $\pi_\lambda(f)$ is a kernel operator with kernel

$$f_\lambda(s,x) := \hat{f}^{2,3}(s-x,-\frac{\lambda}{2}(s+x),\lambda), \ s,x \in \mathbb{R}^n. \tag{6.5.2}$$

If we take now a Schwartz-functions $f \in \mathcal{S}(H_n)$, then the operator $\pi_\lambda(f)$ is Hilbert-Schmidt and its Hilbert-Schmidt norm $\|\pi_\lambda(f)\|_{\text{H.S.}}$ is given by

$$\|\pi_\lambda(f)\|^2_{\text{H.S.}} = \int_{\mathbb{R}^2} |f_\lambda(s,x)|^2dxds = \int_{\mathbb{R}^2} |\hat{f}^{2,3}(s,\lambda x,\lambda)|^2dsdx < \infty. \tag{6.5.3}$$

Proposition 6.5.1 *For any $c \in C^*(H_n)$ and $\lambda \in \mathbb{R}^*$, the operator $\pi_\lambda(c)$ is compact , the mapping $\mathbb{R}^* \to B(L^2(\mathbb{R}^n)) : \lambda \mapsto \pi_\lambda(c)$ is norm continuous and tending to 0 for λ going to infinity.*

Proof 6.20 *Indeed, for $f \in \mathcal{S}(H_n)$, the compactness of the operator $\pi_\lambda(f)$ is a consequence of (6.5.3) and by (6.5.1) we have the estimate:*

$$\begin{aligned} \|\pi_\lambda(f)-\pi_\nu(f)\|^2_{H.S} &= \int_{\mathbb{R}^n\times\mathbb{R}^n} |\hat{f}^{2,3}(s-x,-\frac{\lambda}{2}(s+x),\lambda) \\ &\quad -\hat{f}^{2,3}(s-x,-\frac{\nu}{2}(s+x),\nu)|^2dsdx \\ &= \int_{\mathbb{R}^n\times\mathbb{R}^n} |\hat{f}^{2,3}(s,\lambda x,\lambda)-\hat{f}^{2,3}(s,\nu x,\nu)|^2dsdx \end{aligned}$$

Hence, since f is a Schwartz function, this expression goes to 0 if λ tends to ν by Lebesgue's theorem of dominated convergence. Therefore the mapping $\lambda \mapsto \pi_\lambda(f)$ is norm continuous. Furthermore, the Hilbert-Schmidt norms of the operators $\pi_\lambda(f)$ go to 0, when λ tends to infinity. The proposition follows from the density of $\mathcal{S}(H_n)$ in $C^*(H_n)$.

6.5.2.2 Behaviour at 0:

Let us now see the behaviour of $\pi_\lambda(f)$ for Schwartz functions $f \in \mathcal{S}(H_n)$, as λ tends to 0. Choose a Schwartz-function η in $\mathcal{S}(\mathbb{R}^n)$ with L^2-norm equal to 1. For $u = (a, b)$ in $\mathbb{R}^n \times \mathbb{R}^n$, $\lambda \in \mathbb{R}^*$, we define the function $\eta(\lambda, a, b)$ by

$$\eta(\lambda, a, b)(s) := |\lambda|^{n/4} e^{2\pi i a\cdot s} \eta(|\lambda|^{1/2}(s + \frac{b}{\lambda}))\ s \in \mathbb{R}^n. \tag{6.5.4}$$

and let $\eta_\lambda(s) = |\lambda|^{n/4}\eta(|\lambda|^{1/2}s),\ s \in \mathbb{R}^n$.

Let us compute

$$\begin{aligned}
c_{\lambda,u,u'}(x, y, t) &= \langle \pi_\lambda(x, y, t)\eta(\lambda, u), \eta(\lambda, u')\rangle \\
&= \int_{\mathbb{R}^n} e^{-2\pi i\lambda t - 2\pi i(\lambda/2)x\cdot y} e^{2\pi i\lambda s\cdot y} \eta(\lambda, u)(s - x)\overline{\eta(\lambda, u')(s)} ds \\
&= |\lambda|^{n/2} e^{-2\pi i\lambda t - 2\pi i(\lambda/2)x\cdot y} \int_{\mathbb{R}^n} e^{2\pi i\lambda s\cdot y - 2\pi i a\cdot x} e^{2\pi i(a-a')\cdot s} \\
&\quad \eta(|\lambda|^{1/2}(s - x + \frac{b}{\lambda}))\overline{\eta(|\lambda|^{1/2}(s + \frac{b'}{\lambda}))} ds \\
&= |\lambda|^{n/2} e^{-2\pi i\lambda t - 2\pi i(\lambda/2)x\cdot y} e^{-2\pi i b\cdot y} e^{-2\pi i a\cdot x} \int_{\mathbb{R}^n} e^{2\pi i\lambda s\cdot y} e^{2\pi i(a-a')\cdot(s - \frac{b}{\lambda})} \\
&\quad \eta(|\lambda|^{1/2}(s - x)\overline{\eta(|\lambda|^{1/2}(s + \frac{b' - b}{\lambda}))} ds.
\end{aligned}$$

Hence for $u = u'$ we get

$$\begin{aligned}
c_{\lambda,u,u}(x, y, t) &= e^{-2\pi i\lambda t - 2\pi i\frac{\lambda}{2}x\cdot y} e^{-2\pi i a\cdot x - 2\pi i b\cdot y} \int_{\mathbb{R}^n} e^{2\pi i(\text{sign}\lambda)|\lambda|^{1/2}s\cdot y} \eta(s - |\lambda|^{1/2}x)\overline{\eta(s)} ds \\
&\to e^{-2\pi i a\cdot x - 2\pi i b\cdot y} \int_{\mathbb{R}^n} \eta(s)\overline{\eta(s)} ds = e^{-2\pi i a\cdot x - 2\pi i b\cdot y}.
\end{aligned}$$

It follows also that the convergence of the coefficients $c_{\lambda,u,u}$ to the characters $\chi_{a,b}$ is uniform in u and uniform on compacta in (x, y, t) since

$$\begin{aligned}
|c_{\lambda,u,u}(x, y, t) - \chi_{a,b}(x, y, t)| &= |\int_{\mathbb{R}^n} (e^{-2\pi i\lambda t - 2\pi i\frac{\lambda}{2}x\cdot y} e^{2\pi i(\text{sign}\lambda)|\lambda|^{1/2}s\cdot y} \\
&\quad \eta(s - |\lambda|^{1/2}x)\overline{\eta(s)} - |\eta(s)|^2) ds| \to 0 \\
&\quad \text{as } \lambda \to 0.
\end{aligned}$$

Proposition 6.5.2 *For every $u = (a, b) \in \mathbb{R}^n \times \mathbb{R}^n, c \in C^*(H_n)$, we have that*

$$\lim_{\lambda\to 0} \|F(c)(\lambda)\eta(\lambda, u) - \widehat{F(c)(0)}(u)\eta(\lambda, u)\|_2 = 0$$

uniformly in (a, b).

Proof 6.21 *For $c \in C^*(H_n)$ we have that*

$$\begin{aligned}
&\|F(c)(\lambda)\eta(\lambda, u) - \widehat{F(c)(0)}(u)\eta(\lambda, u)\|_2^2 = \|\pi_\lambda(c)\eta(\lambda, u) - \chi_{a,b}(c)\eta(\lambda, u)\|_2^2 = \\
&= \langle \pi_\lambda(c)\eta(\lambda, u) - \chi_{a,b}(c)\eta(\lambda, u), \pi_\lambda(c)\eta(\lambda, u) - \chi_{a,b}(c)\eta(\lambda, u)\rangle \\
&= \langle \pi_\lambda(c^* * c)\eta(\lambda, u), \eta(\lambda, u)\rangle - \overline{\chi_{a,b}(c)}\langle \pi_\lambda(c)\eta(\lambda, u), \eta(\lambda, u)\rangle \\
&- \chi_{a,b}(c)\overline{\langle \pi_\lambda(c)\eta(\lambda, u), \eta(\lambda, u)\rangle} + |\chi_{a,b}(c)|^2 \\
&\to |\chi_{a,b}(c)|^2 - |\chi_{a,b}(c)|^2 - |\chi_{a,b}(c)|^2 + |\chi_{a,b}(c)|^2 = 0.
\end{aligned}$$

6.5.3 A C^*-condition

The aim of this section is to obtain a characterisation of the C^*-algebra $C^*(H_n)$ as a C^*-algebra of operator fields ([69, 52]). Let us first define a larger C^*-algebra $\mathcal{F}_n$.

Definition 6.12 Let $\mathcal{F}_n$ be the family consisting of all operator fields $(F = F(\lambda))_{\lambda\in\mathbb{R}}$ satisfying the following conditions:

1. $F(\lambda)$ is a compact operator on $L^2(\mathbb{R}^n)$ for every $\lambda \in \mathbb{R}^*$,
2. $F(0) \in C^*(\mathbb{R}^{2n})$,
3. the mapping $\mathbb{R}^* \to B(L^2(\mathbb{R}^n)) : \lambda \mapsto F(\lambda)$ is norm continuous,
4. $\lim_{\lambda\to\infty} \|F(\lambda)\|_{\text{op}} = 0$.

The proof of the following proposition is straight forward:

Proposition 6.5.3 *$\mathcal{F}_n$ is a C^*-algebra.*

Next, we prove the following:

Proposition 6.5.4 *The Fourier transform $F : C^*(H_n) \to \mathcal{F}_n$ is an injective homomorphism.*

Proof 6.22 *It is clear from the definition of F and Proposition 6.5.1 that F is a homomorphism with values in $\mathcal{F}_n$. If $F(c) = 0$, then for each irreducible representation π of $C^*(H_n)$, $\pi(c) = 0$. Hence $c = 0$.*

We also need to show the following:

Lemma 6.5.5 *Let $\xi \in \mathcal{S}(\mathbb{R}^n)$. Then, for any $\lambda \in \mathbb{R}^*$,*

$$\xi = \frac{1}{|\lambda|^n} \int_{\mathbb{R}^n\times\mathbb{R}^n} \langle \xi, \eta(\lambda, u)\rangle \eta(\lambda, u) du,$$

where $\eta(\lambda, u)$ is given by (6.5.4), the integral converging in $L^2(\mathbb{R}^n)$.

Proof 6.23 *Let $\xi \in \mathcal{S}(\mathbb{R}^n)$. Then*

$$\begin{aligned}
&\int_{\mathbb{R}^n\times\mathbb{R}^n} \langle \xi, \eta(\lambda, a, b)\rangle \eta(\lambda, a, b)(x) dadb \\
&= \int_{\mathbb{R}^n\times\mathbb{R}^n} \left(\int_{\mathbb{R}^n} \xi(s) e^{-2\pi i a\cdot s} \overline{\eta_\lambda(s + \frac{b}{\lambda})} ds \right) e^{2\pi i a\cdot x} \eta_\lambda(x + \frac{b}{\lambda}) dadb \\
&\text{(by Fourier's inversion formula)} \\
&= \int_{\mathbb{R}^n\times\mathbb{R}^n} \xi(x) \overline{\eta_\lambda(x + \frac{b}{\lambda})} \eta_\lambda(x + \frac{b}{\lambda}) db = |\lambda|^n \xi(x)
\end{aligned}$$

giving $\xi = \frac{1}{|\lambda|^n} \int_{\mathbb{R}^n\times\mathbb{R}^n} \langle \xi, \eta(\lambda, a, b)\rangle \eta(\lambda, a, b) dadb$.

Furthermore, since ξ is a Schwartz function, it follows that the mapping

$$(a,b) \to \langle \xi, \eta(\lambda,a,b)\rangle = |\lambda|^{n/4} \int_{\mathbb{R}^n} \xi(s) e^{-2\pi i a\cdot s} \overline{\eta(|\lambda|^{1/2}(s+\frac{b}{\lambda}))} ds$$

is also a Schwartz function in the variables a, b. Hence the integral

$$\int_{\mathbb{R}^n\times\mathbb{R}^n} \langle \xi, \eta(\lambda,a,b)\rangle \eta(\lambda,a,b) da db$$

converges in $\mathcal{S}(\mathbb{R}^n)$ and hence also in $L^2(\mathbb{R}^n)$.

Remark 6.5.6 By Lemma 6.5.5,

$$\pi_\lambda(f)\xi = \frac{1}{|\lambda|^n}\int_{\mathbb{R}^{2n}} \pi_\lambda(f)\eta(\lambda,u)\langle \xi, \eta(\lambda,u)\rangle du = \frac{1}{|\lambda|^n}\int_{\mathbb{R}^{2n}} \pi_\lambda(f)\circ P_{\eta(\lambda,u)}\xi du$$

for any $f \in C^*(H_n)$, where $P_{\eta(\lambda,u)}$ is the orthogonal projection onto the one dimensional subspace $\mathbb{C}\eta(\lambda,u)$.

Definition 6.13 For a vector $0 \neq \eta \in L^2(\mathbb{R}^n)$, we let P_η be the orthogonal projection onto the one dimensional subspace $\mathbb{C}\eta$.

Define for $\lambda \in \mathbb{R}^*$ and $h \in C^*(\mathbb{R}^{2n})$ the linear operator

$$\nu_\lambda(h) := \int_{\mathbb{R}^{2n}} \hat{h}(u) P_{\eta(\lambda,u)} \frac{du}{|\lambda|^n}. \tag{6.5.5}$$

Proposition 6.5.7 1. *For every $\lambda \in \mathbb{R}^*$ and $h \in \mathcal{S}(\mathbb{R}^{2n})$ the integral (6.5.5) converges in operator norm.*

2. *$\nu_\lambda(h)$ is compact and $\|\nu_\lambda(h)\|_{op} \leq \|h\|_{C^*(\mathbb{R}^{2n})}$.*

3. *The mapping $\nu_\lambda : C^*(\mathbb{R}^{2n}) \to \mathcal{F}_n$ is involutive, i.e. $\nu_\lambda(h^*) = \nu_\lambda(h)^*, h \in C^*(\mathbb{R}^{2n})$, where by ν_λ we denote also the extension of ν_λ to $C^*(\mathbb{R}^{2n})$.*

Proof 6.24 *Since $\|P_{\eta(\lambda,u)}\|_{\text{op}} = \|\eta(\lambda,u)\|_2^2 = 1$, we have that*

$$\|\nu_\lambda(h)\|_{\text{op}} = \|\int_{\mathbb{R}^{2n}} \hat{h}(u) P_{\eta(\lambda,u)} \frac{du}{|\lambda|^n}\|_{\text{op}} \leq \int_{\mathbb{R}^{2n}} |\hat{h}(u)| \frac{du}{|\lambda|^n} = \frac{\|\hat{h}\|_1}{|\lambda|^n}.$$

Hence the integral

$$\int_{\mathbb{R}^{2n}} \hat{h}(u) P_{\eta(\lambda,u)} \frac{du}{|\lambda|^n}$$

converges in operator norm for $h \in \mathcal{S}(\mathbb{R}^{2n})$.

We compute $\nu_\lambda(h)$ applied to a Schwartz function $\xi \in \mathcal{S}(\mathbb{R}^n)$:

$$\begin{aligned}
\nu_\lambda(h)\xi(x) &= \int_{\mathbb{R}^{2n}} \hat{h}(u)\langle \xi, \eta(\lambda, u)\rangle \eta(\lambda, u)(x)\frac{du}{|\lambda|^n} \\
&= \int_{\mathbb{R}^{2n}} \hat{h}(u)\left(\int_{\mathbb{R}^n} \xi(r)\overline{\eta}_\lambda(r+\frac{b}{\lambda})e^{-2\pi i a\cdot r}dr\right)e^{2\pi i a\cdot x}\eta_\lambda(x+\frac{b}{\lambda})\frac{dadb}{|\lambda|^n} \\
&= \int_{\mathbb{R}^n} \hat{h}^2(-,b) * (\xi\overline{\eta}_{\lambda,b})(x)\eta_\lambda(x+\frac{b}{\lambda})\frac{db}{|\lambda|^n} \\
&\quad (\text{where } \eta_{\lambda,b}(s) := \eta_\lambda(s+\frac{b}{\lambda}), s\in\mathbb{R}^n) \\
&= \int_{\mathbb{R}^{2n}}\int_{\mathbb{R}^n} \hat{h}^2(x-s,b)\xi(s)\overline{\eta}_\lambda(s+\frac{b}{\lambda})\eta_\lambda(x+\frac{b}{\lambda})\frac{db}{|\lambda|^n}ds \\
&= \int_{\mathbb{R}^{2n}}\int_{\mathbb{R}^n} \hat{h}^2(x-s,|\lambda|^{1/2}b)\xi(s) \\
&\quad \overline{\eta}(|\lambda|^{1/2}s+\text{sign}\lambda\cdot b)\eta(|\lambda|^{1/2}x+\text{sign}\lambda\cdot b)dbds.
\end{aligned}$$

The kernel function $h_\lambda(x,s)$ of $\nu_\lambda(h)$ is in $\mathcal{S}(\mathbb{R}^{2n})$ if $h\in\mathcal{S}(\mathbb{R}^{2n})$. In particular, $\nu_\lambda(h)$ is a compact operator and we have the following estimate for the Hilbert-Schmidt norm, $\|\cdot\|_{H.S}$, of $\nu_\lambda(h)$:

$$\begin{aligned}
\|\nu_\lambda(h)\|^2_{H.S} &= \int_{\mathbb{R}^n\times\mathbb{R}^n} |\int_{\mathbb{R}^n} \hat{h}^2(x-s,b)\overline{\eta}_\lambda(s+\frac{b}{\lambda})\eta_\lambda(x+\frac{b}{\lambda})\frac{db}{|\lambda|^n}|^2 dsdx \\
&\le \int_{\mathbb{R}^{3n}} |\hat{h}^2(x-s,\lambda(b-x))|^2|\eta_\lambda(s-x+b)|^2 dbdxds \\
&= \int_{\mathbb{R}^{3n}} |\hat{h}^2(x,\lambda(b+s))|^2|\eta_\lambda(b)|^2 dbdxds \\
&= \int_{\mathbb{R}^{2n}} |\hat{h}^2(x,\lambda s)|^2 dxds < \infty.
\end{aligned}$$

Let us show that $\|\nu_\lambda(h)\|_{\text{op}} \le \|\hat{h}\|_\infty$. Indeed

$$\begin{aligned}
\|\nu_\lambda(h)\xi\|^2_2 &= \int_{\mathbb{R}^n} |\int_{\mathbb{R}^n} \hat{h}^2(-,b) * (\xi\overline{\eta}_{\lambda,b})(x)\eta_\lambda(x+\frac{b}{\lambda})\frac{db}{|\lambda|^n}|^2 dx \\
&\le \frac{1}{|\lambda|^{2n}}\int_{\mathbb{R}^n}\int_{\mathbb{R}^n} |\hat{h}^2(-,b) * (\xi\overline{\eta}_{\lambda,b})(x)|^2 dbdx \\
&\le \frac{\|\hat{h}\|^2_\infty}{|\lambda|^n}\int_{\mathbb{R}^n} \|\xi\eta_{\lambda,b}\|^2_2 db \\
&= \frac{\|\hat{h}\|^2_\infty}{|\lambda|^n}\int_{\mathbb{R}^n}\int_{\mathbb{R}^n} |\xi(x)\eta_\lambda(x+\frac{b}{\lambda})|^2 dxdb \\
&= \|\hat{h}\|^2_\infty\|\xi\|^2_2.
\end{aligned}$$

Let $h\in S(\mathbb{R}^{2n})$. Then $\overline{\hat{h}} = \hat{h^*}$. This gives

$$\begin{aligned}
\nu_\lambda(h)^* = (\int_{\mathbb{R}^{2n}} \hat{h}(u)P_{\eta(\lambda,u)}\frac{du}{|\lambda|^n})^* &= \int_{\mathbb{R}^{2n}} \overline{\hat{h}(u)}P_{\eta(\lambda,u)}\frac{du}{|\lambda|^n} \\
&= \int_{\mathbb{R}^{2n}} \hat{h^*}(u)P_{\eta(\lambda,u)}\frac{du}{|\lambda|^n} = \nu_\lambda(h^*).
\end{aligned}$$

Theorem 6.5.8 *Let $a \in C^*(H_n)$ and let A be the operator field $A = F(a)$, i.e.*

$$A(\lambda) = \pi_\lambda(a), \lambda \in \mathbb{R}^*, A(0) = \rho(a) \in C^*(\mathbb{R}^{2n}).$$

Then

$$\lim_{\lambda \to 0} \|A(\lambda) - \nu_\lambda(A(0))\|_{\text{op}} = 0.$$

Proof 6.25 *Let $f \in \mathcal{S}(H_n)$, $\xi \in L^2(\mathbb{R}^n)$, $\eta \in \mathcal{S}(\mathbb{R}^n)$, $\|\eta\|_2 = 1$. Then by (6.5.1) and (6.5.6)*

$$\begin{aligned}((\pi_\lambda(f) - \nu_\lambda(\rho(f))\xi)(x) &= \int_{\mathbb{R}^n} \hat{f}^{2,3}(x - s, -\frac{\lambda}{2}(x + s), \lambda)\xi(s)ds \\ &- \int_{\mathbb{R}^n}\int_{\mathbb{R}^n} \hat{f}^{2,3}(x - s, b, 0)\xi(s)\overline{\eta}_\lambda(s + \frac{b}{\lambda})\eta_\lambda(x + \frac{b}{\lambda})\frac{db}{|\lambda|^n}ds \\ &= \int_{\mathbb{R}^n}\int_{\mathbb{R}^n} \hat{f}^{2,3}(x - s, -\frac{\lambda}{2}(x + s), \lambda)\eta_\lambda(b)\overline{\eta}_\lambda(b)\xi(s)dbds \\ &- \int_{\mathbb{R}^n}\int_{\mathbb{R}^n} \hat{f}^{2,3}(x - s, \lambda(b - x), 0)\xi(s)\overline{\eta}_\lambda(s - x + b)\eta_\lambda(b)dbds.\end{aligned}$$

Let

$$\begin{aligned}u_\lambda(x, b) &= \int_{\mathbb{R}^n} \xi(s)\overline{\eta}_\lambda(s - x + b)(\hat{f}^{2,3}(x - s, -\frac{\lambda}{2}(x + s), \lambda) \\ &\quad -\hat{f}^{2,3}(x - s, -\frac{\lambda}{2}(x + s), 0))ds, \\ v_\lambda(x, b) &= \int_{\mathbb{R}^n} \xi(s)\overline{\eta}_\lambda(s - x + b)(\hat{f}^{2,3}(x - s, -\frac{\lambda}{2}(x + s), 0) \\ &\quad -\hat{f}^{2,3}(x - s, \lambda(b - x), 0))ds,\end{aligned}$$

and

$$w_\lambda(x) = \int_{\mathbb{R}^n}\int_{\mathbb{R}^n} \hat{f}^{2,3}(x - s, -\frac{\lambda}{2}(x + s), \lambda)\xi(s)\eta_\lambda(b)(\overline{\eta}_\lambda(b) - \overline{\eta}_\lambda(s - x + b))dbds.$$

We have

$$((\pi_\lambda(f) - \nu_\lambda(\rho(f))\xi)(x) = \int_{\mathbb{R}^n} u_\lambda(x, b)\eta_\lambda(b)db + \int_{\mathbb{R}^n} v_\lambda(x, b)\eta_\lambda(b)db + w_\lambda(x). \quad (6.5.6)$$

Thus to prove $\|\pi_\lambda(f) - \nu_\lambda(\rho(f))\|_{\text{op}} \to 0$ as $\lambda \to 0$ it is enough to show that $\|u_\lambda\|_2 \leq \delta_\lambda\|\xi\|_2$, $\|v_\lambda\|_2 \leq \omega_\lambda\|\xi\|_2$ and $\|w_\lambda\|_2 \leq \epsilon_\lambda\|\xi\|_2$, where δ_λ, ω_λ, $\epsilon_\lambda \to 0$ as $\lambda \to 0$.

We have

$$\begin{aligned}&\hat{f}^{2,3}(x - s, -\frac{\lambda}{2}(x + s)), \lambda) - \hat{f}^{2,3}(x - s, -\frac{\lambda}{2}(x + s), 0) \\ &= \lambda\int_0^1 \partial_3\hat{f}^{2,3}(x - s, -\frac{\lambda}{2}(x + s)), t\lambda)dt,\end{aligned}$$

and

$$\hat{f}^{2,3}(x-s,-\frac{\lambda}{2}(x+s)),0)-\hat{f}^{2,3}(x-s,\lambda(b-x),0)=\lambda(\frac{1}{2}(s-x)-(s-x+b))$$
$$\times\int_0^1\partial_2\hat{f}^{2,3}(x-s,\lambda(b-x)+t(\lambda(\frac{1}{2}(s-x)-(s-x+b)),0)dt.$$

Hence, since $f\in\mathcal{S}(H_n)$, there exists a constant $C>0$ such that

$$|f^{2,3}(x-s,-\frac{\lambda}{2}(x+s)),\lambda)-\hat{f}^{2,3}(x-s,-\frac{\lambda}{2}(x+s),0)|\le|\lambda|\frac{C}{(1+\|x-s\|)^{2n+1}},$$

and

$$|\hat{f}^{2,3}(x-s,-\frac{\lambda}{2}(x+s)),0)-\hat{f}^{2,3}(x-s,\lambda(b-x),0)|$$
$$\le\ |\lambda|(\|s-x+b\|+\|s-x\|)\frac{C}{(1+\|x-s\|)^{4n+1}}$$

for all $\lambda\in\mathbb{R}^*$, $x,s\in\mathbb{R}^n$. Therefore we see that

$$\begin{aligned}\|u_\lambda\|_2^2 &= \int_{\mathbb{R}^n\times\mathbb{R}^n}|u_\lambda(x,b)|^2dxdb\\ &\le \int_{\mathbb{R}^n\times\mathbb{R}^n}\left(\int_{\mathbb{R}^n}|\xi(s)\eta_\lambda(s-x+b)||\lambda|\frac{C}{(1+\|x-s\|)^{2n+1}}ds\right)^2dxdb\\ &\le |\lambda|^2C'\int_{\mathbb{R}^{3n}}\frac{|\xi(s)|^2}{(1+\|x-s\|)^2}|\eta_\lambda(s-x+b)|^2dbdxds\\ &\le C''|\lambda|^2\|\xi\|_2^2.\end{aligned}$$

Similarly

$$\begin{aligned}\|v_\lambda\|_2^2 &= \int_{\mathbb{R}^n\times\mathbb{R}^n}|v_\lambda(x,b)|^2dxdb\\ &\le \int_{\mathbb{R}^n\times\mathbb{R}^n}\left(\int_{\mathbb{R}^n}|\xi(s)\eta_\lambda(s-x+b)||\lambda|(\|s-x+b\|+\|s-x\|)\right.\\ &\quad\left.\frac{C}{(1+\|x-s\|)^{4n+1}}ds\right)^2dbdx\\ &\le C'\int_{\mathbb{R}^{3n}}|\xi(s)\eta_\lambda(s-x+b)|^2|\lambda|^2(\|s-x+b\|+\|s-x\|)^2\\ &\quad\frac{1}{(1+\|x-s\|)^{4n+1}}dsdbdx\\ &\le 2C'\int_{\mathbb{R}^{3n}}|\xi(s)|\lambda|^{n/4}\eta(|\lambda|^{1/2}(s-x+b))|^2|\lambda|^2\|s-x+b\|^2\frac{dsdbdx}{(1+\|x-s\|)^{4n+1}}\\ &+ 2C'\int_{\mathbb{R}^{3n}}|\xi(s)|\lambda|^{n/4}\eta(|\lambda|^{1/2}(s-x+b))|^2|\lambda|^2\|s-x\|^2\frac{dsdbdx}{(1+\|x-s\|)^{4n+1}}\\ &\le 2C'|\lambda|\int_{\mathbb{R}^{3n}}|\xi(s)|\lambda|^{n/4}\tilde{\eta}(|\lambda|^{1/2}(s-x+b))|^2\frac{dsdbdx}{(1+\|x-s\|)^{4n+1}}\\ &+ 2C'|\lambda|^2\int_{\mathbb{R}^{3n}}|\xi(s)|\lambda|^{n/4}\eta(|\lambda|^{1/2}(s-x+b))|^2\frac{dsdbdx}{(1+\|x-s\|)^{4n-1}}\\ &\le C''|\lambda|(\|\tilde{\eta}\|_2^2+|\lambda|\|\eta\|_2^2)\|\xi\|_2^2,\end{aligned}$$

for some constants $C', C'' > 0$, where the function $\tilde{\eta}$ is defined by $\tilde{\eta}(s) := \|s\|\eta(s)$, $s \in \mathbb{R}$.

Since $\eta \in \mathcal{S}(\mathbb{R}^n)$, we can use the same arguments to see that

$$\begin{aligned}
\|w_\lambda\|_2^2 &= \int_{\mathbb{R}^n} |w_\lambda(x)|^2 dx \\
&\leq \int_{\mathbb{R}^n} \Big(\int_{\mathbb{R}^n} \int_{\mathbb{R}^n} |\xi(s)||\eta_\lambda(b)||\lambda|^{n/4+1/2}(\|s-x\|) \frac{C}{(1+\|x-s\|)^{4n+1}} dbds \Big)^2 dx \\
&\leq C'|\lambda|^{n/2+1} \int_{\mathbb{R}^{3n}} |\xi(s)|^2 |\eta_\lambda(b)|^2 \frac{\|x-s\|^2 dsdxdb}{1+\|x-s\|^{4n+1}} \leq C''|\lambda|^{n/2+1} \|\xi\|_2^2 \|\eta\|_2^2.
\end{aligned}$$

We have proved therefore that $\|\pi_\lambda(f) - \nu_\lambda(\rho(f))\| \to 0$ as $\lambda \to 0$ for $f \in \mathcal{S}(H_n)$. Since $\mathcal{S}(H_n)$ is dense in $C^*(H_n)$, the statement holds for any $a \in C^*(H_n)$.

Definition 6.14 *For $\eta \in \mathcal{S}(\mathbb{R}^n)$ we define the linear mapping $\nu_\eta := \nu : C^*(\mathbb{R}^{2n}) \to \mathcal{F}_n$ by*

$$\nu(h)(\lambda) = \nu_\lambda(h), \lambda \in \mathbb{R}^* \text{ and } \nu(h)(0) = h.$$

Proposition 6.5.9 *The mapping $\nu : C^*(\mathbb{R}^{2n}) \to \mathcal{F}_n$ has the following properties:*

1. $\|\nu\| = 1$.
2. *For every $h, h' \in C^*(\mathbb{R}^{2n})$, we have that*

$$\lim_{\lambda \to 0} \|\nu_\lambda(h \cdot h') - \nu_\lambda(h) \circ \nu_\lambda(h')\|_{\text{op}} = 0$$

   and also

$$\lim_{\lambda \to 0} \|\nu_\lambda(h^*) - \nu_\lambda(h)^*\|_{\text{op}} = 0.$$

3. *For $(a, b) \in \mathbb{R}^{2n}$ and $h \in C^*(\mathbb{R}^{2n})$ we have that*

$$\lim_{\lambda \to 0} \|\nu(h)(\lambda)\eta(\lambda, a, b) - \hat{h}(a, b)\eta(\lambda, a, b)\|_2 = 0.$$

4. $\lim_{\lambda \to 0} \|\nu(h)(\lambda)\| = \|\hat{h}\|_\infty$.

Proof 6.26 *(1) follows from Proposition 6.5.7.*

To prove (2) we take for $h, h' \in \mathcal{S}(\mathbb{R}^{2n})$ two elements $f, f' \in \mathcal{S}(H_n)$, such that $\rho(f) = h, \rho(f') = h'$. Then $\rho(f * f') = h \cdot h'$ and

$$\begin{aligned}
\|\nu_\lambda(h \cdot h') - \nu_\lambda(h) \circ \nu_\lambda(h')\|_{\text{op}} &\leq \|\nu_\lambda(h \cdot h') - \pi_\lambda(f * f')\|_{\text{op}} \\
&+ \|\nu_\lambda(h) \circ \nu_\lambda(h') - \pi_\lambda(f) \circ \pi_\lambda(f')\|_{\text{op}} \\
&\leq \|\nu_\lambda(h \cdot h') - \pi_\lambda(f * f')\|_{\text{op}} \\
&+ \|f'\|_{C^*(H_n)} \|\nu_\lambda(h) - \pi_\lambda(f)\|_{\text{op}} \\
&+ \|h\|_{C^*(\mathbb{R}^{2n})} \|\nu_\lambda(h') - \pi_\lambda(f')\|_{\text{op}}.
\end{aligned}$$

Hence, by Theorem 6.5.8, $\lim_{\lambda\to 0} \|\nu_\lambda(f * f') - \nu_\lambda(f)\circ\nu_\lambda(f')\|_{\text{op}} = 0$. *Furthermore*

$$\begin{aligned} \|\nu_\lambda(h^*) - \nu_\lambda(h)^*\|_{\text{op}} &\le \|\nu_\lambda(h^*) - \pi_\lambda(f^*)\|_{\text{op}} + \|\nu_\lambda(h)^* - \pi_\lambda(f)^*\|_{\text{op}} \to 0. \\ &\quad \text{as } \lambda \to 0. \end{aligned}$$

We conclude by the usual approximation argument.

For assertion (3), using Propositions 6.5.2 and Theorem 6.5.8, it suffices to take for $h \in C^*(\mathbb{R}^{2n})$ *an element* $c \in C^*(H_n)$, *for which* $\rho(c) = h$.

The last statement follows from Proposition 6.5.7 and assertion (3).

Definition 6.15 *Let* $D_\nu(H_n)$ *be the subspace of the algebra* $\mathcal{F}_n$, *consisting of all the fields* $(F(\lambda))_{\lambda\in\mathbb{R}} \in \mathcal{F}_n$, *such that*

$$\lim_{\lambda\to 0} \|F(\lambda) - \nu_\lambda(F(0))\|_{\text{op}} = 0.$$

Our main theorem of this section is the following characterisation of $C^*(H_n)$.

Theorem 6.5.10 *The Heisenberg* C^**-algebra* $C^*(H_n)$ *is isomorphic to* $D_\nu(H_n)$.

Proof 6.27 *First, we show that* $D_\nu(H_n)$ *is a* $*$*-subalgebra of* $\mathcal{F}_n$. *Indeed if* $F, F' \in D_\nu(H_n)$, *then*

$$\begin{aligned} \|\nu_\lambda(F(0) + F'(0)) - \pi_\lambda(F + F')\|_{\text{op}} &\le \|\nu_\lambda(F(0)) - \pi_\lambda(F)\|_{\text{op}} \\ &+ \|\nu_\lambda(F'(0)) - \pi_\lambda(F')\|_{\text{op}} \to 0 \\ &\quad \text{as } \lambda \to 0, \end{aligned}$$

and since $\lim_{\lambda\to 0} \|\nu_\lambda(F\cdot F'(0)) - \nu_\lambda(F(0))\circ\nu(F'(0))\|_{\text{op}} = 0$ *it follows that*

$$\|\nu_\lambda(F(0)\cdot F'(0)) - \pi_\lambda(F\cdot F')\|_{\text{op}} \to 0.$$

Proposition 6.5.9 tells us that $D_\nu(H_n)$ *is also invariant under the involution* $*$.

In order to see that $D_\nu(H_n)$ *is closed, let* $F \in \mathcal{F}_n$ *be contained in the closure of* $D_\nu(H_n)$. *Let* $\varepsilon > 0$. *Choose* $F' \in D_\nu(H_n)$, *such that* $\|F - F'\|_{\mathcal{F}_n} < \varepsilon$. *In particular,* $\|F(0) - F'(0)\|_{C^*(\mathbb{R}^2)} < \varepsilon$. *Thus there exists* $\lambda_0 > 0$, *such that*

$$\|\pi_\lambda(F') - \nu_\lambda(F'(0))\|_{\text{op}} < \varepsilon$$

for all $|\lambda| < |\lambda_0|$, *whence*

$$\begin{aligned} \|\pi_\lambda(F) - \nu_\lambda(F(0))\|_{\text{op}} &= \|\pi_\lambda(F) - \pi_\lambda(F') + \pi_\lambda(F') - \nu_\lambda(F'(0)) \\ &+ \nu_\lambda(F'(0)) - \nu_\lambda(F(0))\|_{\text{op}} \\ &\le 3\varepsilon, \text{ for } |\lambda| < |\lambda_0|. \end{aligned}$$

Hence $D_\nu(H_n)$ *is a* C^**-subalgebra of* $\mathcal{F}_n$.

Let $I_0 := \{F \in \mathcal{F}_n, F(0) = 0\}$ *and let* $I_{00} = \{F \in I_0; \lim_{\lambda\to 0} \|F(\lambda)\|_{op} = 0\}$. *Then* I_0 *and* I_{00} *are closed two sided ideals of* $\mathcal{F}_n$ *and it follows from the definition of* $\mathcal{F}_n$ *that*

I_{00} is just the algebra $C_0(\mathbb{R}^*, \mathcal{K})$. It is clear that $D_\nu(H_n) \cap I_0 = I_{00}$. But $D_\nu(H_n) \cap I_0$ is the kernel in $D_\nu(H_n)$ of the homomorphism $\delta_0 : \mathcal{F}_n \to \mathbb{C}^*(\mathbb{R}^{2n}); F \mapsto F(0)$.

Since $im(\nu) \subset D_\nu(H_n)$, the canonical projection $D_\nu(H_n) \to C^*(\mathbb{R}^{2n}) : F \mapsto F(0)$ is surjective and has the ideal I_{00} as its kernel. Thus $D_\nu(H_n)/I_{00} = C^*(\mathbb{R}^{2n})$ and therefore $D_\nu(H_n)$ is an extension of I_{00} by $C^*(\mathbb{R}^{2n})$. Moreover,

$$D_\nu(H_n) \quad = \quad I_{00} + im(\nu).$$

Since for every irreducible representation π of $D_\nu(H_n)$, we have either $\pi(I_{00}) \neq 0$, and then $\pi = \pi_\lambda$ for some $\lambda \in \mathbb{R}^*$ or $\pi = 0$ on I_{00} and then π must be a character of $C^*(\mathbb{R}^{2n})$. Hence $\hat{D}_\nu(H_n) = \hat{H}_n$ as sets. That topologies of these spaces agree follows from the equality

$$\lim_{\lambda \to 0} \|\tau(h)(\lambda)\|_{\text{op}} = \|\hat{h}\|_\infty, \ \forall h \in C^*(\mathbb{R}^{2n}),$$

which is due to Proposition 6.5.9.

By Theorem 6.5.8, $F(\mathcal{S}(H_n)) \subset D_\nu(H_n)$. Hence the C^*- algebra $C^*(H_n)$ can be injected into $D_\nu(H_n)$.

Since $D_\nu(H_n)$ is a type I algebra and the dual spaces of $D_\nu(H_n)$ and of $C^*(H_n)$ are the same, we have that $F(C^*(H_n))$ is equal to $D_\nu(H_n)$ by the Stone-Weierstrass theorem (see [37]).

Remark 6.5.11 Another characterisation of the C^*-algebra $C^*(H_n)$ is given (without proof) in a short paper by Gorbachev [57]. For $n = 1$ and $\lambda \in \mathbb{R}^*$ he defines an operator-valued measure μ_λ on $\mathbb{R}^2$ given on the product of two intervals $[s,t] \times [e,d]$ by $\mu_\lambda([s,t] \times [e,d]) = P^{\frac{e}{\lambda},\frac{d}{\lambda}} F P^{s,t} F^{-1}$, where $P^{s,t}$ is the multiplication operator by the characteristic function of $[t,s]$ on $L^2(\mathbb{R})$ and F is the Fourier transform on $L^2(\mathbb{R})$. For $f \in C_0(\mathbb{R}^2)$, $\lambda \in \mathbb{R}^*$ let

$$y(f)(\lambda) = \int_{\mathbb{R}^2} f(a,b) d\mu_\lambda(a,b)$$

and $y(f)(0) = f$. Gorbachev states that $C^*(H_1)$ is isomorphic to the C^*-algebra of operator fields $B = \{B(\lambda) = y(f)(\lambda) + a, \lambda \in \mathbb{R}^*, B(0) = f, f \in C_0(\mathbb{R}^2), a \in C_0(\mathbb{R}^*, \mathcal{K})\}$.

6.5.4 Almost homomorphisms and Heisenberg property

Definition 6.16 A bounded mapping $\tau : C^*(\mathbb{R}^{2n}) \to \mathcal{F}_n$ is called an *almost homomorphism* if

$$\lim_{\lambda \to 0} \|\tau_\lambda(\alpha h + \beta f) - \alpha\tau_\lambda(h) - \beta\tau_\lambda(f)\|_{\text{op}} = 0,$$
$$\lim_{\lambda \to 0} \|\tau_\lambda(h \cdot h') - \tau_\lambda(h) \circ \tau_\lambda(h')\|_{\text{op}} = 0,$$
$$\lim_{\lambda \to 0} \|\tau_\lambda(h^*) - \tau_\lambda(h)^*\|_{\text{op}} = 0, \ \alpha, \beta \in \mathbb{C}, f, h \in C^*(\mathbb{R}^{2n}).$$

The mapping ν from the previous section is an example of such almost homomorphism.

Let τ be an arbitrary almost homomorphism such that $\tau(f)(0) = f$ for any $f \in C^*(\mathbb{R}^{2n})$. We define as before $D_\tau(H_n)$ to be the subspace of the algebra $\mathcal{F}_n$, consisting of all the fields $F = (F(\lambda))_{\lambda\in\mathbb{R}} \in \mathcal{F}_n$, such that

$$\lim_{\lambda\to 0} \|F(\lambda) - \tau_\lambda(F(0))\|_{\text{op}} = 0.$$

Using the same arguments as the one in the proof of Theorem 6.5.10 one can easily prove the following

Proposition 6.5.12 *The subspace $D_\tau(H_n)$ of the C^*-algebra $\mathcal{F}_n$ is itself a C^*-algebra. The algebra $D_\tau(H_n)$ is an extension of $C_0(\mathbb{R}^*, \mathcal{K})$ by $C^*(\mathbb{R}^{2n})$, i.e. $C_0(\mathbb{R}^*, \mathcal{K})$ is a closed $*$-ideal in $D_\tau(H_n)$ such that $D_\tau(H_n)/C_0(\mathbb{R}^*, \mathcal{K})$ is isomorphic to $C^*(\mathbb{R}^{2n})$.*

Definition 6.17 We say that an almost homomorphism $\tau : C^*(\mathbb{R}^{2n}) \to \mathcal{F}_n$ has the *Heisenberg* property, if the C^*-algebra $D_\tau(H_n)$ is isomorphic to $C^*(H_n)$.

Remark 6.5.13 As for the mapping ν we have that the dual spaces of $D_\tau(H_n)$ and of H_n coincides as sets. The necessary and the sufficient conditions for them to coincide as topological spaces is

$$\lim_{\lambda} \|\tau_\lambda(h)\|_{\text{op}} = \|\hat{h}\|_\infty, \quad h \in C^*(\mathbb{R}^n).$$

Remark 6.5.14 Using the notion of Busby invariant for a C^*-algebra extension and the pull-back algebra ([105]), one can show that any extension $\mathcal{B} \subset \mathcal{F}_n$ of $C_0(\mathbb{R}^*, \mathcal{K})$ by $C^*(\mathbb{R}^{2n})$ is isomorphic to $D_\tau(H_n)$ for some almost homomorphism τ. The Busby invariant of such extension is $b : C^*(\mathbb{R}^{2n}) \to C_b(\mathbb{R}^*, B(H))/C_0(\mathbb{R}^*, \mathcal{K})$, $b(h) = \tau(h) + C_0(\mathbb{R}^*, \mathcal{K})$.

Question. What mappings τ give us C^*-algebras $D_\tau(H_n)$, which are isomorphic to $C^*(H_n)$?

Using a procedure described in [34], one can construct families of C^*-algebras of type $D_\tau(H_n)$ which are isomorphic to $D_\nu(H_n)$ and therefore to $C^*(H_n)$.

Next example shows that there is no topological obstacle for a C^*-algebra of type $D_\tau(H_n)$ to be non-isomorphic to $C^*(H_n)$. Namely, there is a C^*-algebras $D_\tau(H_n)$ with the spectrum equal to $\widehat{H}_n$ and such that $D_\tau(H_n) \not\simeq C^*(H_n)$.

We recall first that if $\mathcal{A}, \mathcal{C}$ are C^*-algebras, then an extension of $\mathcal{C}$ by $\mathcal{A}$ is a short exact sequence

$$0 \to \mathcal{A} \overset{\alpha}{\to} \mathcal{B} \overset{\beta}{\to} \mathcal{C} \to 0 \tag{6.5.7}$$

of C^*-algebras. One says that the exact sequence splits if there is a cross-section $*$-homomorphism $s : \mathcal{C} \to \mathcal{B}$ such that $\beta \circ s = I_{\mathcal{C}}$.

It is known that the extension

$$0 \to C_0(\mathbb{R}^*, \mathcal{K}) \to C^*(H_n) \to C^*(\mathbb{R}^{2n}) \to 0 \tag{6.5.8}$$

does not split (see [97] and references therein) while there exists a large number of splitting extensions $\mathcal{B}$ and therefore non-isomorphic to $C^*(H_n)$ such that $\hat{\mathcal{B}} = \hat{H}_n$ (see [35, VII.3.4]). Here is a concrete example inspired by [34].

Example 6.2 Let $\{\xi_Z\}_{Z\in\mathbb{Z}^{2n}}$ be an orthonormal basis of the Hilbert space $L^2(\mathbb{R}^n)$. Let $P_Z, Z \in \mathbb{Z}^{2n}$, be the orthogonal projection onto the one-dimensional $\mathbb{C}\xi_Z$. We define a homomorphism ν from $C^*(\mathbb{R}^{2n})$ to $\mathcal{F}_n$ by

$$\nu(\varphi)(\lambda) := \sum_{Z\in\mathbb{Z}^{2n}} \hat{\varphi}(|\lambda|^{1/2} Z) P_Z, \lambda \in \mathbb{R}^*, \nu(\varphi)(0) := \varphi,\ \varphi \in C^*(\mathbb{R}^{2n}).$$

We note that since for each $\lambda \neq 0$ and each compact subset $K \subset \mathbb{R}^{2n}$, the set $\{Z \in \mathbb{Z}^{2n} : |\lambda|^{1/2} \in K\}$ is finite and since $\hat{\varphi} \in C_0(\mathbb{R}^{2n})$, one can easily see that $\nu(\varphi)(\lambda)$ is compact. Moreover

$$\|\nu(\varphi)(\lambda)\|_{\text{op}} = \sup_{Z\in\mathbb{Z}^{2n}} |\varphi(|\lambda|^{1/2} Z)|.$$

Since we can find for every vector $u \in \mathbb{R}^{2n}$ and $\lambda \in \mathbb{R}^*$ a vector $Z_\lambda \in \mathbb{Z}^{2n}$, such that $\lim_{\lambda\to 0} |\lambda|^{1/2} Z_\lambda = u$, we see that

$$\lim_{\lambda\to 0} \|\nu_\lambda(\varphi)\|_{\text{op}} = \|\hat{\varphi}\|_\infty = \|\varphi\|_{C^*(\mathbb{R}^{2n})}. \tag{6.5.9}$$

6.6 THE C^*-ALGEBRA OF THREAD-LIKE LIE GROUPS

This section deals with the C^*-algebra of thread-like Lie groups G_N, $N \geq 3$ introduced in Subsection 1.3.2. The group G_3 is the Heisenberg group of dimension 3, for which the associated C^*-algebra is treated in Section 6.5. The groups G_N are nilpotent Lie groups and their unitary representations can be described using the Kirillov orbit method. The topology of the dual space $\widehat{G_N}$ has been investigated in details in [2]. In particular, it was shown that like for the Heisenberg group G_3 the topology of $\widehat{G_N}$, $N \geq 3$ is not Hausdorff.

It is known that $\widehat{G_3} = \mathbb{R}^* \cup \mathbb{R}^2$ as a set with natural topology on each pieces, the limit set when $\lambda \in \mathbb{R}^*$ goes to 0 is the whole real plane $\mathbb{R}^2$. The topology of $\widehat{G_N}$, becomes more complicated with growth of the dimension N. Using a description of the limit sets of converging sequences $(\pi_k) \in \widehat{G_N}$ obtained in [1] and [2] we give a characterisation of the C^*-algebra of G_N in the spirit of one for the Heisenberg group H_n. Namely, parametrising $\widehat{G_N}$ by a set $S_N^{gen} \cup \mathbb{R}^2$, where S_N^{gen} consists of element $\ell \in \mathfrak{g}_N^*$ corresponding to non-characters (here $\mathfrak{g}_N$ is the Lie algebra of G_N), we realise $C^*(G_N)$ as a C^*-algebra of operator fields $(A = A(\ell))$ on $S_N^{gen} \cup \{0\}$, such that $A(\ell) \in \mathcal{K}$, $\ell \in S_N^{gen}$, $A(0) \in C^*(\mathbb{R}^2)$ and $(A = A(\ell))$ satisfy for each converging sequence in the dual space the generic, the character and the infinity conditions (see Definition 6.24).

6.6.1 The topology of $\widehat{G_N}$

The topology of the dual space of G_N (cf. Subsection 1.3.2.1) has been studied in detail in the papers [2] and [1] based on the methods developed in [82] and [80]. We need the following description of the convergence of sequences $(\pi_k)_k$ of representations in $\widehat{G_N}$.

Let $(\pi_k)_k$ be a sequence in $\widehat{G_N}$. It is said to be *properly convergent* if it is convergent and all cluster points are limits. It is known (see [82]) that any convergent sequence has a properly convergent subsequence.

Proposition 6.6.1 *Suppose that $(\pi_k = \pi_{\ell_k})_k, (\ell_k \in S_N^{\text{gen}}, k \in \mathbb{N})$ is a sequence in $\widehat{G_N}$ that has a cluster point. Then there exists a subsequence, (also indexed by the symbol k for simplicity), called with perfect data such that $(\pi_k)_k$ is properly converging and such that the polynomials p_k, $k \in \mathbb{N}$, associated to π_k have the following properties: The polynomials p_k have all the same degree d. Write*

$$p_k(t) := c_k \prod_{j=1}^{d} (t - a_j^k) = \hat{\ell}_k(t), t \in \mathbb{R}, \ell_k \in V.$$

There exist $0 < m \leq 2d$, real sequences $(t_i^k)_k$ and polynomials q_i of degree $d_i \leq d$, $i = 1, \cdots, m$, such that

1. $\lim_{k\to\infty} p_k(t+t_i^k) \to q_i(t)$, *$t \in \mathbb{R}$, $1 \leq i \leq m$ or equivalently* $\lim_{k\to\infty} t_i^k \cdot \ell_k \to \ell^i$, *where ℓ^i in V such that $\hat{\ell^i}(t) = q_i(t)$.*
2. $\lim_{k\to\infty} |t_i^k - t_{i'}^k| = +\infty$, *for all $i \neq i' \in \{1, \cdots, m\}$.*
3. *If $C = \{i \in \{1, \cdots, m, \}, \ell^i$ is a character $\}$ then for all $i \in C$*
   (a) $\lim_{k\to\infty} |t_i^k - a_j^k| = +\infty$ *for all $j \in \{1, \cdots, d\}$;*
   (b) *there exists an index $j(i) \in \{1, \cdots, d\}$ such that $|t_i^k - a_{j(i)}^k| \leq |t_i^k - a_j^k|$ for all $j \in \{1, \cdots, d\}$; let*
   $$\rho_i^k := |t_i^k - a_{j(i)}^k|;$$
   (c) *there exists a subset $L(i) \subset \{1, \cdots, m\}$, such that $\lim_{k\to\infty} \frac{|t_i^k - a_j^k|}{\rho_i^k}$ exists in $\mathbb{R}$ for every $j \in L(i)$ and such that $\lim_{k\to\infty} \frac{|t_i^k - a_j^k|}{\rho_i^k} = +\infty$ for $j \notin L(i)$;*
   (d) *the polynomials $(t_i^k + s\rho_i^k) \cdot p_k$ in t converge uniformly on compacta to the constants*
   $$\lim_{k\to\infty} (t_i^k + s\rho_i^k) \cdot p_k(t) = p^i(s), s \in \mathbb{R},$$
   and these constants define a real polynomial of degree $\#L(i)$ in s.
   (e) *If $i' \neq i \in C$, then $L(i) \cap L(i') = \emptyset$.*
4. *Let $D = \{1, \cdots, m\} \setminus C$ and write $\rho_i^k := 1$ for $i \in D$. For $i \in D$, let*
   $$J(i) := \{1 \leq j \leq d, \lim_{k\to\infty} |t_i^k - a_j^k| = \infty\}.$$

Suppose that $(t_k)_k$ is a real sequence, such that $\lim_{k\to\infty} t_k \cdot \ell_k \to \ell$ in $\mathfrak{g}_N^*$, then

(a) *if ℓ is a non-character, then the sequence $(|t_k - t_i^k|)_k$ is bounded for some $i \notin C$;*

(b) *if ℓ is a character, then $\lim_{k\to\infty} \left|\frac{t_k - a_j^k}{t_i^k - a_j^k}\right|$ exists for some $i \in C$ and some $j \in L(i)$ and $\ell_{|\mathfrak{b}} = q_i(s)X_{N-1}^*$ for some $s \in \mathbb{R}$.*

5. *Take any real sequence $(s_k)_k$, such that $\lim_{k\to\infty} |s_k| = +\infty$, and such that for any $i \in D$, $j \in J(i)$, $\frac{s_k}{|t_i^k - a_j^k|} \to 0$, and for $i \in C$, $j \notin L(i)$, $\frac{s_k \rho_i^k}{|a_j^k - t_i^k|} \to 0$ and $\frac{s_k}{\rho_i^k} \to 0$ as $k \to \infty$. Let*

$$S_k := (\bigcup_{i=1}^{m} [t_i^k - s_k\rho_i^k, t_i^k + s_k\rho_i^k]); \; T_k := \mathbb{R} \setminus S_k, k \in \mathbb{N}.$$

Then for any sequence $(t_k)_k$, $t_k \in T_k$, we have $t_k \cdot l_k \to \infty$. We say in this case that the sequence $(s_k)_k$ is adapted to the sequence (ℓ_k).

Proof 6.28 *We may assume that $(\pi_k)_k$ is properly convergent with limit set L. We can also assume, by passing to a subsequence, that each p_k has degree d. By [80] the number of non-characters in L is finite. Let this subset of non-characters be denoted by L^{gen}. If L^{gen} is non-empty by passing further to a subsequence we may assume the sequence $(\pi_k)_k$ converges i_σ-times to each non-character $\sigma \in L^{gen}$ (see p.34, [1] for the definition of m-convergence and p.253 [2]). Let $s = \sum_{\sigma\in L^{gen}} i_\sigma$. Then there exist non-constant polynomials $q_1, \ldots, q_s$ of degree $d_i \le d$, $i = 1, \ldots, s$, and sequences $(t_1^k)_k, \ldots, (t_s^k)_k$ such that the conditions (1) and (2) are fulfilled and for each $\sigma \in L^{gen}$ there are i_σ equal polynomials amongst $q_1, \ldots, q_s$ corresponding to σ. Then if $(t_k)_k$ is a real sequence such that $t_k \cdot \ell_k \to \ell$, ℓ is a non-character, then ℓ corresponds to some $\sigma \in L^{gen}$ and we may assume that $\hat{\ell} = q_i$ for some $i \in 1, \ldots, s$. It follows from the definition of i_σ-convergence that the sequences $(t_k \cdot \ell_k)$ and $(t_i^k \cdot \ell_k)$ are not disjoint implying $|t_k - t_i^k|$ is bounded and therefore (4a).*

If $(\pi_k)_k$ has a character as a limit point then passing if necessary to a subsequence we can find a maximal family of real sequences $(t_l^k)_k$, $l < s \le m \le d$, constant polynomials q_l, non-negative sequences $(\rho_l^k)_k$ and polynomials p_l satisfying (1)–(4) (see Definition 6.4 and the discussion before in [2]).

The condition (4b) follows from the maximality of the family of sequences $(t_l^k)_k$ and the proof of Proposition 6.2, in [2].

Suppose now that we have a sequence $(t_k)_k$ such that $t_k \in T_k$ for every k and such that some subsequence (also indexed by k for simplicity of notations) $(t_k \cdot \ell_k)_k$ converges to an $\ell \in \mathfrak{g}^*$. By condition (5) then either for some $i \in D$, the sequence $(t_k - t_i^k)_k$ is bounded, i.e. $t_k \in [t_i^k - s_k\rho_i^k, t_i^k + s_k\rho_i^k]$ for k large enough, which is impossible, or we have an $i \in C$, such that $\lim_{k\to\infty} \left|\frac{t_k - a_j^k}{t_i^k - a_j^k}\right|$ exists for some $j \in L(i)$.

But then

$$\begin{aligned}\frac{|t_k - t_i^k|}{\rho_i^k} &\leq \frac{|t_k - a_j^k|}{\rho_i^k} + \frac{|t_i^k - a_j^k|}{\rho_i^k} \\ &= \frac{|t_k - a_j^k|}{|t_i^k - a_j^k|}\frac{|t_i^k - a_j^k|}{\rho_i^k} + \frac{|t_i^k - a_j^k|}{\rho_i^k}\end{aligned}$$

and so the sequence $(\frac{|t_k-t_i^k|}{\rho_i^k})_k$ *is bounded, i.e.* $t_k \in [t_i^k - s_k\rho_i^k, t_i^k + s_k\rho_i^k] \subset S_k$ *for* k *large enough, a contradiction. Hence* $\lim_k t_k \cdot \ell_k = \infty$ *whenever* $t_k \in T_k$ *for large* k.

Example 6.3 Let us consider the Heisenberg group G_3. Then $S_3 = S_3^1 \cup \mathbb{R}^2$. Let $(\ell_k) \in S_3^1$. Then $\ell_k = \lambda_k X_1^*$, $\lambda_k \in \mathbb{R}^*$. The associated polynomials are $p_k(t) = -\lambda_k t$ ($d = 1$, $c_k = -\lambda_k$, $a_1^k = 1$). Assume that (ℓ_k) is a sequence with perfect data. Then either π_{ℓ_k} converges to π_ℓ, $\ell \in S_3^1$, or π_{ℓ_k} converges to a character and in this case $\lambda_k \to 0$ as $k \to \infty$. We shall consider now the second case. So we have $m = 1$ and $\ell^1 = X_2^*$ with the corresponding polynomial $q_1(t) = 1$ and $t_1^k = -1/\lambda_k$ and thus $\rho_1^k = 1/|\lambda_k|$. The polynomial $p^1(s)$ is the limit

$$\lim_{k\to\infty} p_k(t_1^k + s\rho_i^k + t) = \lim_{k\to\infty} (-\lambda_k)(-1/\lambda_k + s/|\lambda_k| + t) = \lim_{k\to\infty} (1 - \text{sign}\lambda_k s - \lambda_k t).$$

Since (ℓ_k) is a sequence with perfect data, the sign of λ_k is constant, implying $q_1 1(s) = 1 + \epsilon s$, where $\epsilon = \pm 1$. A real sequence (s_k) is adapted to (ℓ_k) if and only if $s_k \to \infty$ and $s_k|\lambda_k| \to 0$.

6.6.2 A C^*-condition

Let $C^*(G_N)$ be the full C^*-algebra of G_N that is the completion of the convolution algebra $L^1(G_N)$ with respect to the norm

$$\|f\|_{C^*(G_N)} = \sup_{\ell \in \mathcal{S}_N} \|\int_{G_N} f(g)\pi_\ell(g)dg\|_{\text{op}}.$$

Definition 6.18 Let $f \in L^1(G_N)$. Define the function $\hat{f}^2$ on $\mathbb{R} \times \mathfrak{b}^*$ by

$$\hat{f}^2(s, \ell) := \int_B f(s, u)e^{-2\pi i \ell(\log(u))}du, s \in \mathbb{R}, \ell \in \mathfrak{b}^*.$$

We denote by $L_c^1(G_N)$ the space of functions $f \in L^1(G_N)$, for which $\hat{f}^2$ is contained in $C_c^\infty(\mathbb{R} \times \mathfrak{b}^*)$, the space of compactly supported C^∞-functions on $\mathbb{R} \times \mathfrak{b}^*$. The subspace $L_c^1(G_N)$ is dense in $L^1(G_N)$ and hence in the full C^*-algebra $C^*(G_N)$ of G_N.

Proposition 6.6.2 *Take $f \in L^1_c(G_N)$ and let $\ell \in S_N^{gen}$. Then the operator $\pi_\ell(f)$ is a kernel operator with kernel function*

$$f_\ell(s,t) = \hat{f}^2(s-t, t\cdot\ell), s,t \in \mathbb{R}.$$

Proof 6.29 *Indeed, for $\xi \in L^2(\mathbb{R}), s \in \mathbb{R}$, we have that*

$$\begin{aligned}
\pi_\ell(f)\xi(s) &= \int_{G_N} f(g)\pi_\ell(g)\xi dg \\
&= \int_{\mathbb{R}}\int_B f(t,b)e^{-2\pi i\ell(\mathrm{Ad}(\exp((t-s)X_N)\log(b))}\xi(s-t)dbdt \ (by\ 1.3.5) \\
&= \int_{\mathbb{R}}\int_B f(s-t,b)e^{-2\pi i(\mathrm{Ad}^*(\exp(tX_N)(\ell)(\log(b))}\xi(t)dbdt \\
&= \int_{\mathbb{R}} \hat{f}^2(s-t, \mathrm{Ad}^*(\exp(tX_N(\ell))\xi(t)dt \\
&= \int_{\mathbb{R}} \hat{f}^2(s-t, t\cdot\ell)\xi(t)dt.
\end{aligned}$$

Definition 6.19 Let $\mathfrak{c} :=$ span $\{X_1,, \cdots, X_{N-2}\}$. Then $\mathfrak{c}$ is an abelian ideal of $\mathfrak{g}_N$, the algebra $\mathfrak{g}_N/\mathfrak{c}$ is abelian and isomorphic to $\mathbb{R}^2$ and $C := \exp(\mathfrak{c})$ is an abelian closed normal subgroup of G_N.

Let

$$\rho = \mathrm{ind}_C^{G_N} 1$$

be the left regular representation of G_N on the Hilbert space $L^2(G_N/C)$. Then the image $\rho(C^*(G_N))$ is the C^*-algebra of $\mathbb{R}^2$ considered as algebra of convolution operators on $L^2(\mathbb{R}^2)$ and hence $\rho(C^*(G_N))$ is isomorphic to the algebra $C_0(\mathbb{R}^2)$ of continuous functions vanishing at infinity on $\mathbb{R}^2$. As for the Heisenberg algebra we have that if $f \in L^1(G_N)$ then the Fourier transform $\widehat{\rho(f)}(a,b)$ of $\rho(f) \in C^*(\mathbb{R}^2)$ equals $\hat{f}(a,b,0,\ldots,0)$.

Our aim is to realise the C^*-algebra $C^*(G_N)$ as a C^*-algebra of operator fields.

Definition 6.20 For $a \in C^*(G_N)$ we define the Fourier transform $F(a)$ of a as operator field

$$F(a) := \{(A(\ell) := \pi_\ell(a), \ell \in S_N^{gen}, A(0) := \rho(a) \in C^*(\mathbb{R}^2)\}.$$

Remark 6.6.3 We observe that the spaces $S_N^j, j = 1, \cdots, N-2$, are Hausdorff spaces if we equip them with the topology of $\widehat{G_N}$. Indeed, let $(\ell_k)_k$ be a sequence in S_N^j, such that the sequence of representations $(\pi_{\ell_k})_k$ converges to some π_ℓ with $\ell \in S_N^j$. Then the numerical sequence $(\lambda_k := \ell_k(X_k))_k$ converges to $\lambda := \ell(X_j) \neq 0$. Suppose now that the same sequence (π_{ℓ_k}) converges to some other point $\pi_{\ell'}$. Then there exists a numerical sequence $(t_k)_k$ such that $\mathrm{Ad}^*(\exp(t_k X_N))\ell_{k|\mathfrak{b}}$ converges to $\ell'_{|\mathfrak{b}}$. In particular, $-\lambda_k t_k = \mathrm{Ad}^*(\exp(t_k X_N))\ell_k(X_{j+1}) \overset{k\to\infty}{\to} \ell'(X_{j+1})$. Hence the sequence $(t_k)_k$ converges to some $t \in \mathbb{R}$ and $\pi_{\ell'} = \pi_\ell$. Similarly, we see from (6.6.1) that for $f \in L^1_c(G_N)$, the mapping $\ell \to \pi_\ell(f)$ is norm continuous when restricted to the sets $S_N^j, j = 1, \cdots, N-2$, since for the sequence $(\pi_{\ell_k})_k$ above, the functions f_{ℓ_k} converge in the L_2-norm to f_ℓ.

Definition 6.21 Define for $t, s \in \mathbb{R}$ the self-adjoint projection operator on $L^2(\mathbb{R})$ given by

$$M_{t,s}\xi(x) := 1_{(t-s,t+s)}(x)\xi(x), x \in \mathbb{R}, \xi \in L^2(\mathbb{R}),$$

where $1_{(a,b)}, a, b \in \mathbb{R}$, denotes the characteristic function of the interval $(a, b) \subset \mathbb{R}$.

We put for $s \in \mathbb{R}$

$$M_s := M_{0,s}.$$

More generally, for a measurable subset $T \subset \mathbb{R}$, we let M_T be the multiplication operator with the characteristic function of the set T. For $r \in \mathbb{R}$, let $U(r)$ be the unitary operator on $L^2(\mathbb{R})$ defined by

$$U(r)\xi(s) := \xi(s + r), \xi \in L^2(\mathbb{R}), s \in \mathbb{R}.$$

Definition 6.22 Let $(\pi_{\ell_k})_k$ be a properly converging sequence in $\widehat{G}_N$ with perfect data $((t_i^k)_k, (\rho_i^k), (s_i^k))$. Let $i \in C$ and let $\eta \in \mathcal{D}(\mathbb{R}^n)$ such that η has L^2-norm 1. Define for $\rho_i^k, k \in \mathbb{N}, i \in C$, and $u = (a, b) \in \mathbb{R}^2$ the Schwartz function

$$\eta(i, k, u)(s) := \eta(s_k p^i\left(\frac{s}{\rho_i^k}\right) + s_k b)e^{2\pi i a\cdot s}, s \in \mathbb{R}.$$

By Example 6.3, for $N = 3$ we have

$$\eta(1, k, u) = \eta(\pm s_k|\lambda_k|s + s_k(1 + b))e^{2\pi i a\cdot s}.$$

Let $P_{i,k,u}$ be the operator of rank one defined by

$$P_{i,k,u}\xi := \langle\xi, \eta(i, k, u)\rangle\eta(i, k, u), \xi \in L^2(\mathbb{R}).$$

Definition 6.23 For an element $\varphi \in \mathcal{S}(\mathbb{R}^2)$ let

$$\nu(\varphi)(i, k) := s_k \int_{\mathbb{R}^2} \hat{\varphi}(a, -b)P_{i,k,u}dadb, k \in \mathbb{N}, i \in C.$$

Then for $\varphi \in \mathcal{S}(\mathbb{R}^2), \xi \in L^2(\mathbb{R}), s \in \mathbb{R}$, we have that

$$\begin{aligned}
\nu(\varphi)(i,k)(\xi)(s) &:= s_k \int_{\mathbb{R}^2} \hat{\varphi}(a,-b)(P_{i,k,u}\xi)(s)du \\
&= s_k \int_{\mathbb{R}^2} \hat{\varphi}(a,-b)\left(\int_{\mathbb{R}} \xi(t)\overline{\eta(s_k p^i\left(\frac{t}{\rho_i^k}\right) + s_k b)e^{-2\pi i a\cdot(t-s)}}dt\right) \\
&\quad \eta(s_k p^i\left(\frac{s}{\rho_i^k}\right) + s_k b)dbda \\
&= s_k \int_{\mathbb{R}}\int_{\mathbb{R}} \hat{\varphi}^2(s-t,-b)\xi(t)\overline{\eta(s_k p^i\left(\frac{t}{\rho_i^k}\right) + s_k b)} \\
&\quad \eta(s_k p^i\left(\frac{s}{\rho_i^k}\right) + s_k b)dtdb \\
&= \int_{\mathbb{R}}\int_{\mathbb{R}} \hat{\varphi}^2(s-t, -\frac{b}{s_k} + p^i\left(\frac{t}{\rho_i^k}\right))\overline{\eta(b)} \\
&\quad \eta(s_k\left(p^i\left(\frac{s}{\rho_i^k}\right) - p^i\left(\frac{t}{\rho_i^k}\right)\right) + b)\xi(t)dtdb.
\end{aligned}$$

Since η has L_2-norm 1, using (6.6.1) and (6.6.1) we get

$$\begin{aligned}
&(U(t_i^k)\circ\pi_{\ell_k}(f)\circ U(-t_i^k)\circ M_{s_k} - \nu(F(f)(0))(i,k)\circ M_{s_k})(\xi)(s) \\
= &\int_{-s_k}^{s_k}(\int_{\mathbb{R}} \hat{f}^2(s-t,(t+t_i^k)\cdot\ell_k)) - \hat{f}^2(s-t, -\frac{b}{s_k} + p^i\left(\frac{t}{\rho_i^k}\right), 0\ldots) \\
&\overline{\eta(b)}\eta(s_k\left(p^i(\frac{s}{\rho_i^k}) - p^i\left(\frac{t}{\rho_i^k}\right)\right) + b)db)\xi(t)dt \\
&+ \int_{-s_k}^{s_k}(\int_{\mathbb{R}} \hat{f}^2(s-t,(t+t_i^k)\cdot\ell_k))\overline{\eta(b)} \\
&(\eta(b) - \eta(s_k\left(p^i\left(\frac{s}{\rho_i^k}\right) - p^i\left(\frac{t}{\rho_i^k}\right)\right) + b))db)\xi(t)dt.
\end{aligned}$$

Proposition 6.6.4 *Let $\varphi \in C^*(\mathbb{R}^2)$, $i \in C$ and $k \in \mathbb{N}$. Then*

1. *the operator $\nu(\varphi)(i,k)$ is compact and $\|\nu(\varphi)(i,k)\|_{\text{op}} \le \|\varphi\|_{C^*(\mathbb{R}^2)}$;*
2. *we have that $\nu(\varphi)(i,k)^* = \nu(\varphi^*)(i,k)$;*
3. *furthermore*
$$\lim_{k\to\infty} \|\nu(\varphi)(i,k)\circ(\mathbb{I} - M_{s_k\rho_i^k})\|_{\text{op}} = 0$$
and hence
$$\lim_{k\to\infty} \|(\mathbb{I} - M_{s_k\rho_i^k})\circ\nu(\varphi)(i,k)\circ M_{s_k\rho_i^k}\|_{\text{op}} = 0.$$

Proof 6.30 *1.) It suffices to prove this for $\varphi \in \mathcal{D}(\mathbb{R}^2)$. We have that*

$$\begin{aligned}
\|\nu(\varphi)(i,k)\xi\|_2^2 &= \int_{\mathbb{R}} |\int_{\mathbb{R}}\int_{\mathbb{R}} \hat{\varphi}^2(s-t,-\frac{b}{s_k})\xi(t)\overline{\eta(s_k p^i(\frac{t}{\rho_i^k})+b)}dt \\
&\quad \eta(s_k p^i(\frac{s}{\rho_i^k})+b)db|^2 ds \\
&= \int_{\mathbb{R}} |\int_{\mathbb{R}} (\hat{\varphi}^2(-,-\frac{b}{s_k}) * (\xi\overline{\eta_{k,b}}))(s)\eta(s_k p^i(\frac{s}{\rho_i^k})+b)db|^2 ds \\
&\quad (\text{where } \eta_{k,b}(t) := \eta(s_k p^i(\frac{t}{\rho_i^k})+b), t \in \mathbb{R}) \\
&\le \int_{\mathbb{R}^2} |(\hat{\varphi}^2(-,-\frac{b}{s_k}) * (\xi\overline{\eta_{k,b}}))(s)|^2 dbds \\
&\le \|\varphi\|_{C^*(\mathbb{R}^2)}^2 \int_{\mathbb{R}} \|\xi\eta_{k,b}\|_2^2 db \\
&= \|\varphi\|_{C^*(\mathbb{R}^2)}^2 \int_{\mathbb{R}^2} |\xi(t)|^2 |\eta(s_k p^i(\frac{t}{\rho_i^k})+b)|^2 dbdt \\
&= \|\varphi\|_{C^*(\mathbb{R}^2)}^2 \|\xi\|_2^2.
\end{aligned}$$

Furthermore, since $\nu(\varphi)(i,k)$ is an integral of rank one operators, $\nu(\varphi)(i,k)$ must be compact. Hence for every $\varphi \in C^*(\mathbb{R}^2)$, $\nu(\varphi)(i,k)$ is a compact operator bounded by $\|\varphi\|_{C^*(\mathbb{R}^2)}$.

2.) Let $\varphi \in \mathcal{S}(\mathbb{R}^2)$. Then $\overline{\hat{\varphi}} = \hat{\varphi^*}$ and so

$$\begin{aligned}
\nu(\varphi)(i,k)^* &= (s_k \int_{\mathbb{R}^2} \hat{\varphi}(u) P_{i,k,u} du)^* = s_k \int_{\mathbb{R}^2} \overline{\hat{\varphi}(u)} P_{i,k,u} du \\
&= s_k \int_{\mathbb{R}^2} \hat{\varphi^*}(u) P_{i,k,u} du = \nu(\varphi^*)(i,k).
\end{aligned}$$

3.) Take now $\varphi \in \mathcal{S}(\mathbb{R}^2)$, such that $\hat{\varphi}$ has a compact support. We denote by $[-s_k\rho_i^k, s_k\rho_i^k]^c$ the set $\mathbb{R} \setminus [-s_k\rho_i^k, s_k\rho_i^k]$. By (6.6.1) for any $\xi \in L^2(\mathbb{R}), s \in \mathbb{R}$ we have

$$\begin{aligned}
&\nu(\varphi)(i,k) \circ (\mathbb{I} - M_{s_k\rho_i^k})(\xi)(s) = \int_{[-s_k\rho_i^k, s_k\rho_i^k]^c} \int_{\mathbb{R}} \hat{\varphi}^2(s-t, -\frac{b}{s_k} + p^i(\frac{t}{\rho_i^k}))\overline{\eta(b)} \\
&\eta(s_k(p^i(\frac{s}{\rho_i^k}) - p^i(\frac{t}{\rho_i^k})) + b)db\xi(t)dt = 0,
\end{aligned}$$

since for k large enough $\hat{\varphi}^2(s-t, -\frac{b}{s_k} + p^i(\frac{t}{\rho_i^k})) = 0$ for any $t \in [-s_k\rho_i^k, s_k\rho_i^k]^c$, $b \in supp(\eta)$, $s \in \mathbb{R}$. Hence $\nu(\varphi(i,k)) \circ (\mathbb{I} - M_{s_k\rho_i^k}) = 0$ for k large enough. Since the mapping ν is continuous, it follows that $\lim_{k\to\infty} \|\nu(\varphi)(i,k)(\mathbb{I} - M_{s_k\rho_i^k})\|_{\text{op}} = 0$ for all $\varphi \in C^*(\mathbb{R}^2)$ and every $i \in C$.

Hence also

$$\begin{aligned}
& \lim_{k\to\infty} \|(\mathbb{I} - M_{s_k\rho_i^k}) \circ \nu(\varphi)(i,k) \circ M_{s_k\rho_i^k}\|_{\text{op}} \\
&= \lim_{k\to\infty} \|(M_{s_k\rho_i^k} \circ \nu(\varphi^*)(i,k) \circ (\mathbb{I} - M_{s_k\rho_i^k})\|_{\text{op}} \\
&\le \lim_{k\to\infty} \|\nu(\varphi^*)(i,k) \circ (\mathbb{I} - M_{s_k\rho_i^k})\|_{\text{op}} = 0.
\end{aligned}$$

Definition 6.24 Let $A = (A(\ell) \in \mathcal{K}(L^2(\mathbb{R})), \ell \in S_N^{gen}, A(0) \in C^*(\mathbb{R}^2))$ be a field of bounded operators. We say that A satisfies the *generic condition* if for every properly converging sequence with perfect data $(\pi_{\ell_k})_k \subset \hat{G}_N$ and for every limit point $\pi_{\ell^i}, i \in D$, and for every adapted real sequence $(s_k)_k$

$$\lim_{k\to\infty} \|U(t_i^k) \circ A(\ell_k) \circ U(-t_i^k) \circ M_{s_k} - A(\ell^i) \circ M_{s_k}\|_{\mathrm{op}} = 0. \tag{6.6.1}$$

A satisfies the *character condition* if for every properly converging sequence with perfect data $(\pi_{\ell_k})_k$, $\ell_k \in S_N^{gen}$ and for every limit point $\pi_{\ell^i}, i \in C$, and for every adapted real sequence $(s_k)_k$

$$\lim_{k\to\infty} \|U(t_i^k) \circ A(\ell_k) \circ U(-t_i^k) \circ M_{s_k\rho_i^k} - \nu(A(0))(i,k) \circ M_{s_k\rho_i^k}\|_{\mathrm{op}} = 0.$$

A satisfies the *infinity* condition, if for any properly converging sequence (π_{ℓ_k}), $\ell_k \in S_N^{gen}$, with perfect data we have that

$$\lim_{k\to\infty} \|A(\ell_k) \circ M_{T_k}\|_{\mathrm{op}} = 0,$$

where $T_k = \mathbb{R} \setminus \left(\bigcup_{i=1}^m [t_i^k - s_k\rho_i^k, t_i^k + s_k\rho_i^k]\right)$, and that for every sequence $(\ell_k)_k \subset S_N^{gen}$, for which the sequence of orbits $G_N \cdot \ell_k$ goes to infinity we also have

$$\lim_{k\to\infty} A(\ell_k) = 0.$$

We can now define the operator field C^*-algebra D_N^*, which will be the image of the Fourier transform of $C^*(G_N)$.

Definition 6.25 Let D_N^* be the space of all bounded operator fields $A = (A(\ell)) \in \mathcal{K}(L^2(\mathbb{R})), \ell \in S_N^{gen}, A(0) \in C^*(\mathbb{R}^2)$, such that A and the adjoint field A^* satisfy the generic, the character and the infinity conditions. Let for $A \in D_N^*$

$$\|A\|_\infty := \sup\{\|A(\ell)\|_{\mathrm{op}}, \|A(0)\|_{C^*(\mathbb{R}^2)} : \ell \in S_N^{gen}\}.$$

It is clear that D_N^* is a Banach space for the norm $\|\cdot\|_\infty$, since the generic, the character and the infinity conditions are stable for the sum, for scalar multiplication and limits of sequences of operator fields.

Theorem 6.6.5 *Let $a \in C^*(G_N)$ and let A be the operator field defined by $A = F(a)$ as in Definition 6.20. Then A satisfies the generic, the character and the infinity conditions.*

Proof 6.31 *For the infinity condition, it suffices to remark that for any $f \in L_c^1(G_N)$, and k large enough, we have that $\hat{f}^2(s-t, t\cdot\ell_k) = 0$ for every $s \in \mathbb{R}$, $t \in T_k$ and so $\pi_{\ell_k}(f) \circ M_{T_k} = 0$. If $G_N \cdot \ell_k$ goes to infinity in the orbit space, then $\mathbb{R}\cdot\ell_k$ is outside any given compact subset $K \subset \mathfrak{g}_N^*$ and so $\hat{f}^2(s-t, t\cdot\ell_k) = 0, s, t \in \mathbb{R}$ and hence*

$\pi_{\ell_k}(f) = 0$ for k large enough. Using the density argument, we see that the infinity condition is satisfied for every element in the Fourier transform of $C^*(G_N)$.

For the generic condition, let $(\ell_k)_k$ be a properly converging sequence in S_N with perfect data. Take $i \in D$. Then for an adapted sequence $(s_k)_k$, $f \in L^1_c(G_N)$ and $\xi \in L^2(\mathbb{R}), s \in \mathbb{R}$, we have that

$$
\begin{aligned}
&(U(t_i^k) \circ \pi_{\ell_k}(f) \circ U(-t_i^k) \circ M_{s_k} - \pi_{\ell^i}(f) \circ M_{s_k})\xi(s) \\
= &\int_{-s_k}^{s_k} \left(\hat{f}^2(s-t,(t+t_i^k)\cdot \ell_k)) - \hat{f}^2(s-t, t\cdot \ell^i))\right) \xi(t)dt.
\end{aligned}
$$

Let p_k and q_i be the polynomials corresponding to ℓ_k and ℓ^i, respectively, i.e. $p_k(t) = \hat{\ell}_k(t)$ and $q_i(t) = \hat{\ell}^i(t)$. Since $\lim\limits_{k\to\infty} \dfrac{s_k}{|t_i^k - a_j^k|} \to 0$, $j \in J(i)$, there exists $R > 0$ such that that $(s-t,(t+t_i^k)\cdot \ell_k) = (s-t, p_k(t+t_i^k), -p_k'(t+t_i^k), \ldots)$ is out of the support of $\hat{f}^2$ if $t \in [-s_k, s_k]$ and $|t| > R$. In fact, if $t \in [-s_k, s_k]$ we have

$$
\begin{aligned}
|p_k(t+t_i^k)| &= |c_k \prod_{j=1}^{d} (t+t_i^k - a_j^k)| = |c_k \prod_{j\in J(i)} |t_i^k - a_j^k| \prod_{j\in J(i)} |\frac{t}{t_i^k - a_j^k} + 1| \\
&\quad \prod_{j\notin J(i)} |t + t_i^k - a_j^k| \\
&\geq |b_i| \prod_{j\in J(i)} |1 - \frac{s_k}{|t_i^k - a_j^k|}| \prod_{j\notin J(i)} |t+t_i^k - a_j^k|,
\end{aligned}
$$

where b_i is the leading coefficient of the polynomial q_i, giving the statement. Thus by (6.6.2)

$$
\begin{aligned}
&(U(t_i^k) \circ \pi_{\ell_k}(f) \circ U(-t_i^k) \circ M_{s_k} - \pi_{\ell^i}(f) \circ M_{s_k})\xi(s) \\
= &\int_{-R}^{R} \left(\hat{f}^2(s-t,(t+t_i^k)\cdot \ell_k)) - \hat{f}^2(s-t, t\cdot \ell^i))\right) \xi(t)dt
\end{aligned}
$$

for k large enough. It is clear now that $U(t_i^k) \circ \pi_{\ell_k}(f) \circ U(-t_i^k) \circ M_{s_k} - \pi_{\ell^i}(f) \circ M_{s_k}$ converges to 0 with respect to the Hilbert-Schmidt norm and hence in the operator norm.

Let $a \in C^*(G_N)$. Then for any $\varepsilon > 0$ there exists $f \in L^1_c(G_N)$ such that $||\pi(f) - \pi(a)||_{op} \leq ||f - a||_{C^*(G_N)} < \varepsilon$ for any representation π of $C^*(G_N)$. Thus for $A(\ell) = \pi_\ell(a)$, $\ell \in S_N^{gen}$ we have

$$
\begin{aligned}
&||U(t_i^k) \circ A(\ell_k) \circ U(-t_i^k) \circ M_{s_k} - A(\ell^i) \circ M_{s_k}||_{op} \\
&= ||U(t_i^k) \circ (A(\ell_k) - \pi_{l_k}(f)) \circ U(-t_i^k)||_{op} \\
&+||U(t_i^k) \circ \pi_{\ell_k}(f) \circ U(-t_i^k) \circ M_{s_k} - \pi_{\ell^i}(f) \circ M_{s_k}||_{op} + ||(\pi_{\ell^i}(f) - A(\ell^i))||_{op} \to 0,
\end{aligned}
$$

and hence A satisfies the generic condition.

Choose now $i \in C$. By (6.6.1), for $k \in \mathbb{N}, s \in \mathbb{R}, \xi \in L^2(\mathbb{R}), f \in L^1_c(G_N)$

$$\begin{aligned} & U(t_i^k) \circ \pi_{\ell_k}(f) \circ U(-t_i^k) \circ M_{s_k\rho_i^k} - \nu(F(f(0)))(i,k) \circ M_{s_k\rho_i^k}(\xi)(s) \\ = & \int_{-s_k\rho_i^k}^{s_k\rho_i^k} (\int_{\mathbb{R}} \hat{f}^2(s-t,(t+t_i^k)\cdot \ell_k)) - \hat{f}^2(s-t,-\frac{b}{s_k}+p^i(\frac{t}{\rho_i^k}),0\dots,0) \\ & \overline{\eta(b)}\eta(s_k(p^i(\frac{s}{\rho_i^k})-p^i(\frac{t}{\rho_i^k}))+b)db)\xi(t)dt + \\ & \int_{-s_k}^{s_k} (\int_{\mathbb{R}} \hat{f}^2(s-t,(t+t_i^k)\cdot \ell_k))\overline{\eta(b)}(\eta(b)-\eta(s_k(p^i(\frac{s}{\rho_i^k})-p^i(\frac{t}{\rho_i^k}))+b))db)\xi(t)dt. \end{aligned}$$

In order to show that

$$\|U(t_i^k) \circ \pi_{\ell_k}(f) \circ U(-t_i^k) \circ M_{s_k\rho_i^k} - \nu(F(f(0)))(i,k) \circ M_{s_k\rho_i^k}\| \to 0, \ k \to \infty, \quad (6.6.2)$$

consider

$$\begin{aligned} q(k,i)(s,b) &= \int_{-s_k\rho_i^k}^{s_k\rho_i^k} (\hat{f}^2(s-t,(t+t_i^k)\cdot \ell_k)) - \hat{f}^2(s-t,-\frac{b}{s_k}+p^i(\frac{t}{\rho_i^k}),0\dots,0)) \\ & \quad \eta(s_k(p^i(\frac{s}{\rho_i^k})-p^i(\frac{t}{\rho_i^k}))+b))\xi(t)dt = u(k,i)+v(k,i), \end{aligned}$$

where

$$\begin{aligned} u(k,i)(s,b) &= \int_{-s_k\rho_i^k}^{s_k\rho_i^k} (\hat{f}^2(s-t,p_k(t+t_i^k),-p_k'(t+t_i^k),\dots) \\ & \quad -\hat{f}^2(s-t,p_k(t+t_i^k),0,\dots))\eta(s_k(p^i(\frac{s}{\rho_i^k})-p^i(\frac{t}{\rho_i^k}))+b))\xi(t)dt, \\ v(k,i)(s,b) &= \int_{-s_k\rho_i^k}^{s_k\rho_i^k} (\hat{f}^2(s-t,p_k(t+t_i^k),0,\dots) \\ & \quad -\hat{f}^2(s-t,-\frac{b}{s_k}+p^i(\frac{t}{\rho_i^k}),0\dots))\eta(s_k(p^i(\frac{s}{\rho_i^k})-p^i(\frac{t}{\rho_i^k}))+b))\xi(t)dt. \end{aligned}$$

Let also

$$w(k,i)(s) = \int_{\mathbb{R}}\int_{-s_k}^{s_k} (\int_{\mathbb{R}} \hat{f}^2(s-t,(t+t_i^k)\cdot\ell_k))\overline{\eta(b)}(\eta(b)-\eta(s_k(p^i(\frac{s}{\rho_i^k})-p^i(\frac{t}{\rho_i^k}))+b))db)\xi(t)dt.$$

Our aim is to prove that for $p(s,b) = 1_{\mathbb{R}\times supp(\eta)}(s,b)$,

$$\|u(k,i)p\|_2 \le \omega_k||\xi||_2, \|v(k,i)p\|_2 \le \delta_k||\xi||_2 \text{ and } \|w(k,i)\|_2 \le r_k||\xi||_2 \quad (6.6.3)$$

with $\omega_k, \delta_k, r_k \to 0$ as $k \to \infty$. This will imply

$$\int_{\mathbb{R}} |\int_{\mathbb{R}} (q(k,i)(s,b))\overline{\eta(b)}db|^2 ds \le \|q(k,i)\|_2^2\|\eta\|_2 \le (\omega_k+\delta_k)^2\|\xi\|_2^2,$$

which together with $\|w(k,i)\|_2 \le r_k||\xi||_2$ will give (6.6.2).

To see this we note first that since $\frac{s_k\rho_i^k}{|a_k^j - t_i^k|} \to 0$ *if* $j \notin L(i)$, *we have that for* $|t| \leq s_k\rho_i^k$

$$\begin{aligned}|p_k(t+t_i^k)| &= |c_k\prod_{j=1}^{d}(t+t_i^k-a_j^k)| = |c_k\prod_j |t_i^k - a_j^k| \prod_{j\notin L(i)} |\frac{t}{t_i^k - a_j^k}+1| \\ &\quad \prod_{j\in L(i)} |\frac{t}{t_i^k-a_j^k}+1)| \\ &\geq \sigma \prod_{j\notin L(i)} |1 - \frac{s_k\rho_i^k}{|t_i^k - a_j^k|}| \prod_{j\in L(i)} |\frac{|t|}{\rho_i^k}\frac{\rho_i^k}{|t_i^k-a_j^k|}-1|\end{aligned}$$

for some $\sigma > 0$. *Thus for large* k *there exists* $R > 0$ *such that*

$$\hat{f}^2(s-t, p_k(t+t_i^k), -p_k'(t+t_i^k), \ldots) = 0$$

and

$$\hat{f}^2(s-t, p_k(t+t_i^k), 0, \ldots) = 0,$$

if $|t| < s_k\rho_i^k$ *and* $|t| > R\rho_i^k$. *Hence the integration over the interval* $[-s_k, s_k]$ *can be replaced by the integration over* $[-R\rho_i^k, R\rho_i^k]$ *in the expression for* $u(k,i)$, $v(k,i)p$ *and* $w(k,i)$. *Since* $f \in L_c^1(G_N)$ *we have that*

$$\begin{aligned}&|\hat{f}^2(s-t, p_k(t+t_i^k), 0, \ldots) - \hat{f}^2(s-t, p^i(\frac{t}{\rho_i^k}) - \frac{b}{s_k}, 0, \ldots)| \\ &\leq C|p_k(t+t_i^k) - p^i(\frac{t}{\rho_i^k}) + \frac{b}{s_k}|\frac{1}{1+|t-s|^m}\end{aligned}$$

for some constant $C > 0$ *and* $m \in \mathbb{N}$, $m \geq 2$. *This gives*

$$\begin{aligned}&\|v(k,i)p\|_2^2 = C\int_{\mathbb{R}^2}\left|\int_{-R\rho_i^k}^{R\rho_i^k}\eta(s_k(p^i(\frac{s}{\rho_i^k}) - p^i(\frac{t}{\rho_i^k}) + b)\right. \\ &\quad \left.(p_k(t+t_i^k) - p^i(\frac{t}{\rho_i^k}) + \frac{b}{s_k})\frac{\xi(t)}{1+|t-s|^m}dt\right|^2 dsdb \\ &\leq \frac{3C}{s_k^2}\int_{\mathbb{R}^2}\int_{-R\rho_i^k}^{R\rho_i^k}\left|\eta(s_k(p^i(\frac{s}{\rho_i^k}) - p^i(\frac{t}{\rho_i^k}) + b)\right. \\ &\quad \left.(b + s_k(p^i(\frac{s}{\rho_i^k}) - p^i(\frac{t}{\rho_i^k}))\frac{\xi(t)}{1+|t-s|^m}dt\right|^2 dsdb \\ &+3C\int_{\mathbb{R}^2}\int_{-R\rho_i^k}^{R\rho_i^k}\left|\eta(s_k(p^i(\frac{s}{\rho_i^k}) - p^i(\frac{t}{\rho_i^k})) + b)(p_k(t+t_i^k) - p^i(\frac{t}{\rho_i^k}))\frac{\xi(t)}{1+|t-s|^m}dt\right|^2 \\ &dsdb \\ &+3C\int_{\mathbb{R}^2}\int_{-R\rho_i^k}^{R\rho_i^k}\left|\xi(t)\eta(s_k(p^i(\frac{s}{\rho_i^k}) - p^i(\frac{t}{\rho_i^k})) + b)(p^i(\frac{t}{\rho_i^k}) - p^i(\frac{s}{\rho_i^k}))\frac{1}{1+|t-s|^m}dt\right|^2 \\ &dsdb\end{aligned}$$

$$\leq \frac{C_1}{s_k^2}\int_{\mathbb{R}^2}\int_{-R\rho_i^k}^{R\rho_i^k}\left|\xi(t)\tilde{\eta}(s_k(p^i(\frac{s}{\rho_i^k})-p^i(\frac{t}{\rho_i^k}))+b)\right|^2\frac{1}{1+|t-s|^m}dtdsdb$$

$$(where\ \tilde{\eta}(b)=b\eta(b))$$

$$+C_1\int_{\mathbb{R}^2}\int_{-R\rho_i^k}^{R\rho_i^k}\left|\xi(t)\eta(s_k(p^i(\frac{s}{\rho_i^k})-p^i(\frac{t}{\rho_i^k}))+b)(p_k(t+t_i^k)-p^i(\frac{t}{\rho_i^k}))\right|^2$$
$$\frac{1}{1+|t-s|^m}dtdsdb$$

$$+\int_{\mathbb{R}^2}\int_{-R\rho_i^k}^{R\rho_i^k}\left|\xi(t)\eta(s_k(p^i(\frac{s}{\rho_i^k})-p^i(\frac{t}{\rho_i^k}))+b)(p^i(\frac{t}{\rho_i^k})-p^i(\frac{s}{\rho_i^k}))\right|^2$$
$$\frac{1}{1+|t-s|^m}dtdsdb$$

$$\begin{aligned}\leq\ & \frac{C_2}{s_k^2}\|\tilde{\eta}\|_2^2\|\xi\|_2^2+C_2\|\eta\|_2^2\int_{-R}^{R}|\xi(t\rho_i^k)|^2|p_k(t\rho_i^k+t_i^k)-p^i(t)|^2\rho_i^k dt\\ +\ & C_3\|\eta\|_2^2\int_{\mathbb{R}}\int_{-R\rho_i^k}^{R\rho_i^k}|\xi(t)|^2\left|p^i(\frac{t}{\rho_i^k})-p^i(\frac{s}{\rho_i^k})\right|^2\frac{1}{1+|t-s|^m}dtds.\end{aligned}$$

As $p_k(t\rho_i^k+t_i^k)-p^i(t)$ converges to 0 uniformly on each compact,

$$\int_{-R}^{R}|\xi(t\rho_i^k)|^2|p_k(t\rho_i^k+t_i^k)-p^i(t)|^2\rho_i^k dt\leq r_k\|\xi\|_2^2$$

with $r_k\to 0$ as $k\to\infty$. Moreover, $p^i(\frac{t}{\rho_i^k})-p^i(\frac{s}{\rho_i^k})=\frac{t-s}{\rho_i^k}\sum_l\alpha_l(\frac{t}{\rho_i^k})\beta_l(\frac{t-s}{\rho_i^k})$ for some finite number of polynomials α_l, β_l which do not depend on k. Thus

$$\begin{aligned}&\int_{\mathbb{R}}\int_{-R\rho_i^k}^{R\rho_i^k}|\xi(t)|^2\left|p^i(\frac{t}{\rho_i^k})-p^i(\frac{s}{\rho_i^k})\right|^2\frac{1}{1+|t-s|^m}dtds\\ =\ &\int_{\mathbb{R}}\int_{-R\rho_i^k}^{R\rho_i^k}|\xi(t)|^2\left|\frac{t-s}{\rho_i^k}\sum_l\alpha_l(\frac{t}{\rho_i^k})\beta_l(\frac{t-s}{\rho_i^k})\right|^2\frac{1}{1+|t-s|^m}dtds\\ \leq\ &\frac{C_4}{(\rho_i^k)^2}\int_{-R\rho_i^k}^{R\rho_i^k}|\xi(t)|^2dt\leq\frac{C_4}{(\rho_i^k)^2}\|\xi\|_2^2\end{aligned}$$

for a properly chosen m. It follows now that

$$||v(k,i)p||_2\leq\delta_k\|\xi\|_2$$

for some $\delta_k\to 0$ as $k\to\infty$.

For $w(k,i)$ we have

$$\begin{aligned}\|w(k,i)\|^2 &= \int_{\mathbb{R}} \left| \int_{-R\rho_i^k}^{R\rho_i^k} \int_R \hat{f}^2(s-t,(t+t_i^k)\cdot \ell_k)\overline{\eta(b)} \right. \\ &\qquad \left. (\eta(b) - \eta(s_k(p^i(\frac{s}{\rho_i^k}) - p^i(\frac{t}{\rho_i^k})) + b))\xi(t)dbdt\right|^2 ds \\ &\le C\|\eta\|_2^2 \int_{\mathbb{R}} \left| \int_{-R\rho_i^k}^{R\rho_i^k} |s_k(p^i(\frac{s}{\rho_i^k}) - p^i(\frac{t}{\rho_i^k}))| \frac{1}{1+|t-s|^m}|\xi(t)|dt\right|^2 ds \\ &\le C\|\eta\|_2^2 \int_R \int_{-R\rho_i^k}^{R\rho_i^k} |s_k(p^i(\frac{s}{\rho_i^k}) - p^i(\frac{t}{\rho_i^k}))|^2 \frac{1}{1+|t-s|^m}|\xi(t)|^2 dtds\end{aligned}$$

for some constant C. Then using the previous arguments we get

$$\|w(k,i)\|^2 \le \frac{Ds_k^2}{(\rho_i^k)^2}\|\xi\|_2^2\|\eta\|_2^2.$$

As $\frac{s_k}{\rho_i^k} \to 0$ we get the desired inequality for $w(k,i)$.

To prove the inequality for $u(k,i)$ we have as in the previous case that

$$\begin{aligned}&|\hat{f}^2(s-t,p_k(t+t_i^k),-p_k'(t+t_i^k),\ldots) - \hat{f}^2(s-t,p_k(t+t_i^k),0,\ldots)| \\ &\le C(\sum_{n=1}^{N-2} |p_k^{(n)}(t+t_i^k)|^2)^{1/2}\frac{1}{1+|t-s|^m}\end{aligned}$$

for some constant $C>0$ and $m \in \mathbb{N}$, $m \ge 2$; here $p_k^{(n)}$ denotes the n-th derivative of p_k. For $n=1$ we have

$$|p_k'(t+t_i^k)| = |c_k \prod_j (t_i^k - a_k^j)| \sum_l \frac{1}{(t_i^k - a_k^l)} \prod_{j\ne l}(\frac{t}{t_i^k - a_k^j}+1)| \le \sigma \frac{1}{\rho_i^k}\left(\frac{|t|}{\rho_i^k}+1\right)^{d-1}$$

for some constant $\sigma>0$. Similar inequalities hold for higher order derivatives $p_k^{(n)}(t+t_i^k)$ which show that

$$\begin{aligned}&\|u(k,i)\|_2^2 = \\ &= \int_{\mathbb{R}^2} | \int_{-R\rho_i^k}^{R\rho_i^k} (\hat{f}^2(s-t,p_k(t+t_i^k),-p_k'(t+t_i^k),\ldots) - \hat{f}^2(s-t,p_k(t+t_i^k),0,\ldots)) \\ &\quad \eta(s_k(p^i(\frac{s}{\rho_i^k}) - p^i(\frac{t}{\rho_i^k})) + b))\xi(t)dt|^2 dbds \\ &\le \int_{\mathbb{R}^2} | \int_{-R\rho_i^k}^{R\rho_i^k} C(\sum_{n=1}^{N-2} |p_k^{(n)}(t+t_i^k)|^2)^{1/2} \frac{1}{1+|t-s|^m}\eta(s_k(p^i(\frac{s}{\rho_i^k}) - p^i(\frac{t}{\rho_i^k})) + b))\xi(t)dt|^2 \\ &\quad dbds \\ &\le \frac{1}{\rho_i^k}\int_{\mathbb{R}^2} | \int_{-R\rho_i^k}^{R\rho_i^k} |p\left(\frac{|t|}{\rho_i^k}\right)| \frac{1}{1+|t-s|^m}\eta(s_k(p^i(\frac{s}{\rho_i^k}) - p^i(\frac{t}{\rho_i^k})) + b))\xi(t)dt|^2 dbds \\ &\le \frac{C'}{\rho_i^k}\|\eta\|_2^2\|\xi\|_2^2\end{aligned}$$

for a polynomial p. Thus we get the required inequality for $u(k,i)$ and hence

$$\lim_{k\to\infty} \|U(t_i^k)\circ\pi_{\ell_k}(f)\circ U(-t_i^k)\circ M_{s_k\rho_i^k} - \nu(F(f(0)))(i,k)\circ M_{s_k\rho_i^k}\|_{\text{op}} = 0.$$

To show now that the character condition holds for the fields $A\in\widehat{C^*(G_N)}$, we use again the density of $L_c^1(G_N)$ in $C^*(G_N)$.

Corollary 6.6.6 *Let $(\pi_{\ell_k})_k$ be a properly converging sequence in $\widehat{G}_N$ with perfect data $((t_i^k)_k,(\rho_i^k),(s_i^k))$. Let $i\in C$. Then for every $\varphi,\psi\in C^*(\mathbb{R}^2)$ we have that*

$$\lim_{k\to\infty}\|\nu(\varphi)(i,k)\circ\nu(\psi)(i,k)-\nu(\varphi\psi)(i,k)\|_{\text{op}} = 0.$$

Proof 6.32 *Indeed, if we take first φ,ψ in $\mathcal{S}(\mathbb{R}^2)$, then we can choose $f,g\in\mathcal{S}(G_N)$, such that $\rho(f)=\varphi,\rho(g)=\psi$ and so, by Proposition 6.6.4 and Theorem 6.6.5,*

$$\begin{aligned}
&\|\nu(i,k)(\varphi)\circ\nu(i,k)(\psi)-\nu(i,k)(\varphi\psi)\|_{\text{op}}\\
\le\ &\|(\nu(i,k)(\varphi)\circ\nu(i,k)(\psi)-\nu(i,k)(\varphi\psi))\circ M_{s_k\rho_i^k}\|_{\text{op}}\\
+\ &\|(\nu(i,k)(\varphi)\circ\nu(i,k)(\psi)-\nu(i,k)(\varphi\psi))\circ(\mathbb{I}-M_{s_k\rho_i^k})\|_{\text{op}}\\
\le\ &\|(\nu(i,k)(\varphi)\circ\nu(i,k)(\psi)\\
&-(U(t_i^k)\circ\pi_{\ell_k}(f)\circ U(-t_i^k))\circ(U(t_i^k)\circ\pi_{\ell_k}(g)\circ U(-t_i^k)))\circ M_{s_k\rho_i^k}\|_{\text{op}}\\
+\ &\|(U(t_i^k)\circ\pi_{\ell_k}(f*g)\circ U(-t_i^k)-\nu(i,k)(\varphi\psi))\circ M_{s_k\rho_i^k}\|_{\text{op}}\\
+\ &\|(\nu(i,k)(\varphi)\circ\nu(i,k)(\psi)-\nu(i,k)(\varphi\psi))\circ(\mathbb{I}-M_{s_k\rho_i^k})\|_{\text{op}}\\
\le\ &\|(\nu(i,k)(\varphi)-(U(t_i^k)\circ\pi_{\ell_k}(f)\circ U(-t_i^k))\circ(\mathbb{I}-M_{s_k\rho_i^k})\circ\nu(i,k)(\psi)\circ M_{s_k\rho_i^k}\|_{\text{op}}\\
+\ &\|(\nu(i,k)(\varphi)-(U(t_i^k)\circ\pi_{\ell_k}(f)\circ U(-t_i^k))\circ M_{s_k\rho_i^k}\circ\nu(i,k)(\psi)\circ M_{s_k\rho_i^k}\|_{\text{op}}\\
+\ &\|((U(t_i^k)\circ\pi_{\ell_k}(f)\circ U(-t_i^k))\circ(\nu(i,k)(\psi)-(U(t_i^k)\circ\pi_{\ell_k}(g)\circ U(-t_i^k))\circ M_{s_k\rho_i^k}\|_{\text{op}}\\
+\ &\|(U(t_i^k)\circ\pi_{\ell_k}(f*g)\circ U(-t_i^k)-\nu(i,k)(\varphi\psi))\circ M_{s_k\rho_i^k}\|_{\text{op}}\\
+\ &\|(\nu(i,k)(\varphi)\circ\nu(i,k)(\psi)-\nu(i,k)(\varphi\psi))\circ(\mathbb{I}-M_{s_k\rho_i^k})\|_{\text{op}}\to 0 \text{ as } k\to\infty.
\end{aligned}$$

The usual density condition shows that the statement holds for all φ, $\psi\in C^*(\mathbb{R}^2)$.

Theorem 6.6.7 *The space D_N^* is a C^*-algebra, which is isomorphic with $C^*(G_N)$ for every $N\in\mathbb{N},N\ge 3$.*

Proof 6.33 *Let us first show that D_N^* is a C^*-algebra. We prove first that D_N^* is closed under multiplication. Let $A=(A(\ell),\ell\in S_N)$ and $B=(B(\ell),\ell\in S_N)$ satisfy the generic condition and let $(\pi_{\ell_k})_k\subset\hat{G}_N$ be a properly convergent sequence with perfect data such that for every limit point π_{l^i}, $i\in D$, and for every adapted real* *sequence $(s_k)_k$ the fields A, B satisfy (6.6.1). Then*

$$\begin{aligned}
&\|U(t_i^k) \circ A(\ell_k) \circ B(\ell_k) \circ U(-t_i^k) \circ M_{s_k} - A(\ell^i) \circ B(\ell^i) \circ M_{s_k}\|_{\text{op}} \\
&\leq \|(U(t_i^k) \circ A(\ell_k) \circ U(-t_i^k) \circ M_{s_k} - A(\ell^i) \circ M_{s_k}) \circ U(t_i^k) \circ B(\ell_k) \circ U(-t_i^k) \circ M_{s_k}\|_{\text{op}} \\
&+\|A(\ell^i) \circ M_{s_k} \circ (U(t_i^k) \circ B(\ell_k) \circ U(-t_i^k) \circ M_{s_k} - B(\ell^i) \circ M_{s_k})\|_{\text{op}} \\
&+\|U(t_i^k) \circ A(\ell_k) \circ U(-t_i^k) \circ (\mathbb{I} - M_{s_k}) \circ (U(t_i^k) \circ B(\ell_k) \\
&\circ U(-t_i^k) \circ M_{s_k} - B(\ell^i) \circ M_{s_k})\|_{\text{op}} \\
&+\|A(\ell^i) \circ (\mathbb{I} - M_{s_k}) \circ B(\ell^i) \circ M_{s_k}\|_{\text{op}} \\
&+\|U(t_i^k) \circ A(\ell_k) \circ U(-t_i^k) \circ (\mathbb{I} - M_{s_k}) \circ B(\ell^i) \circ M_{s_k}\|_{\text{op}}.
\end{aligned}$$

Since $B(\ell^i)$ is compact and $\mathbb{I} - M_{s_k}$ converges to 0 strongly, $\|(\mathbb{I} - M_{s_k}) \circ B(\ell^i)\|_{\text{op}} \to 0$ giving that the product $A(\ell) \circ B(\ell)$ satisfies the generic condition.

To see that the character condition is closed under multiplication we argue as before, but use $\|(\mathbb{I} - M_{s_k\rho_i^k}) \circ \nu(\varphi)(i,k) \circ M_{s_k\rho_i^k}\|_{\text{op}} \to 0$ which is due to Proposition 6.6.4.

The infinity condition is clearly closed under multiplication of fields.

By Theorem 6.6.5, the Fourier transform F maps $C^*(G_N)$ into D_N^*. Let us show that F is also onto. By the Stone-Weierstrass approximation theorem, we must only prove that the dual space of D_N^* is the same as the dual space of $C^*(G_N)$. We proceed by induction on N. If $N = 3$, then G_N is the Heisenberg group and the statement follows from Theorem 6.5.10. Let $\pi \in \widehat{D_N^*}$.

Let for $M = 3, \cdots N-1$, $R_M : D_N^* \to D_M^*$ be the restriction map, i.e. denote by $q_M : \mathfrak{g}_N \to \mathfrak{g}_N/\mathfrak{b}_{N-M} \simeq \mathfrak{g}_M$ the quotient map and by $q_M^t : \mathfrak{g}_M^* \simeq \mathfrak{b}_{N-M}^{\perp} \to \mathfrak{g}_N^*$ its transpose. Then for an operator field $A \in D_N^*$ we define the operator field $R_M(A)$ over S_M^{gen} by:

$$R_M(A)(\tilde{\ell}) := A(q_M^t(\tilde{\ell})), \tilde{\ell} \in S_M^{\text{gen}}.$$

It follows from the definition of D_N^* that the image of R_M is contained in D_M^*. Hence R_M is a homomorphism of C^*-algebras, whose kernel I_M is the ideal

$$I_M = \{A \in D_N^*, A(\ell) = 0 \text{ for all } \ell \in S_N \cap \mathfrak{b}_{N-M}^{\perp}\}.$$

Let $Q_M : C^*(G_N) \to C^*(G_M) \simeq C^*(G_N/B_{N-M})$ be the canonical projection. Then the kernel of this projection is the ideal $J_M := \{a \in C^*(G_N); \pi_\ell(a) = 0, \ell \in S_N \cap \mathfrak{b}_{N-M}^{\perp}\}$. Let us write F_M for the Fourier transform $C^*(G_M) \to D_M^*$. With these notations we have the formula

$$R_M(F_N(a)) = F_M(a \text{ modulo } J_M), a \in C^*(G_N). \tag{6.6.4}$$

Since by the induction hypothesis $\widehat{D_M^*} = F_M(\widehat{C^*(G_M)})$ for every $3 \leq M \leq N-1$ we see from (6.6.4) that $R_M(F_N(C^*(G_N))) = F_M(C^*(G_M)) = D_M^*$ and so the mapping R_M is surjective for such an M. Hence $D_N^*/I_M \simeq C^*(G_M)$. We have also

$$I_{N-1} \subseteq I_{N-2} \subseteq \ldots \subseteq I_3.$$

If $\pi(I_{N-1}) = \{0\}$ then $\pi \in \widehat{G_N/B_1} \subset \widehat{G_N}$. Suppose now that $\pi(I_{N-1}) \neq \{0\}$. Let us show that $I_{N-1} \simeq C_0(S_N^1, \mathcal{K}(L^2(\mathbb{R})))$. It is clear from the definition of D_N^* that $C_0(S_N^1, \mathcal{K}(L^2(\mathbb{R}))) \subset D_N^*$ and so is contained in I_{N-1}. It suffices to show now that $I_{N-1} \subset C_0(S_N^1, \mathcal{K}(L^2(\mathbb{R})))$. For that it is enough to see that for any element A in I_{N-1} and any sequence $(\ell_k)_k$ in S_N^1 for which either (π_{ℓ_k}) converges to infinity or to a representation π_ℓ with $\ell \notin S_N^1$, we have that $\lim_k \|A(\ell_k)\|_{\rm op} = 0$. This follows from the infinity condition in the first case. In the second case, no limit point of the sequence (π_{ℓ_k}) is in S_N^1 by Remark 6.6.3. It suffices to show then that $\lim_k \|A(\ell_k)\|_{\rm op} = 0$ for every subsequence with perfect data (also indexed by k for simplicity of notation). We have with the notations of Definition 6.24 that for $k \in \mathbb{N}$

$$A(\ell_k) \quad = \quad A(\ell_k) \circ M_{S_k} + A(\ell_k) \circ M_{T_k}.$$

where $S_k = \cup_i (t_i^k - s_k\rho_i^k, t_i^k + s_k\rho_i^k)$, $T_k = \mathbb{R} \setminus S_k$.

Since $A(\ell) = 0$ for every π_ℓ in the limit set of the sequence $(\pi_{\ell_k})_k$, the generic and the character conditions say that

$$\lim_k \|U(t_i^k) \circ A(\ell_k) \circ U(t_i^k) \circ M_{s_k\rho_i^k}\|_{\rm op} = 0.$$

Hence

$$\lim_k \|A(\ell_k) \circ M_{t_i^k, s_k\rho_i^k}\|_{\rm op} = 0,$$

and since also

$$\lim_k \|A(\ell_k) \circ M_{T_k}\|_{\rm op} = 0,$$

it follows that $\lim_k \|A(\ell_k)\|_{\rm op} = 0$. Hence $I_{N-1} \subset C_0(S_N^1, \mathcal{K}(L^2(\mathbb{R})))$ and so $I_{N-1} = C_0(S_N^1, \mathcal{K}(L^2(\mathbb{R})))$. Finally $\pi_{|I_{N-1}}$ is evaluation in some point $\ell \in S_N^1$ and so $\pi \in \widehat{G_N}$. This finishes the proof of the theorem.

6.7 ALMOST $C_0(\mathcal{K})$-C^*-ALGEBRAS

6.7.1 Introduction

Recall that a Lie group G is called *exponential* if it is a connected, simply connected solvable Lie group for which the exponential mapping $\exp : \mathfrak{g} \to G$ from the Lie algebra $\mathfrak{g}$ to its Lie group G is a diffeomorphism. The Kirillov-Bernat-Vergne-Pukanszky-Ludwig-Leptin theory shows that there is a canonical homeomorphism $K : \mathfrak{g}^*/G \to \widehat{G}$ from the space of coadjoint orbits of G in the linear dual space $\mathfrak{g}^*$ onto the unitary dual space $\widehat{G}$ of G (see [70] for details and references). Then the unitary spectrum $\widehat{C^*(G)}$ of the C*-algebra $C^*(G)$ of the locally compact group G can be identified with the unitary dual $\widehat{G}$ of G.

Since connected Lie groups are second countable, the algebra $C^*(G)$ and its dual space $\widehat{G}$ are separable topological spaces. This allows us to work in $\widehat{G}$ with sequences instead of nets. Moreover, if G is amenable, then the left regular representation $(\lambda, L^2(G))$ of $C^*(G)$, defined by $\lambda(F)\xi := F * \xi$ for $F \in L^1(G)$ and $\xi \in L^2(G)$, is injective. So we can also identify $C^*(G)$ with an algebra of convolution operators on the Hilbert space $L^2(G)$.

The method of describing group C*-algebras as algebras of operator fields defined on the dual spaces of the groups has been studied in [44] and [68]. In [71], the C*-algebra of $(ax + b)$-like groups has been characterised as the algebra of operator fields, while the C^*-algebra of the Heisenberg groups and of the thread-like groups are described in [83]. In both cases, the topologies of the orbit spaces $\mathfrak{g}^*/G$ of the groups are well understood (see for instance [2]). The isomorphism problem for C*-algebras of $ax + b$-like groups was solved in [72].

In this section, we consider the Lie group G_6, whose Lie algebra is the 6-dimensional normal j-algebra $\mathfrak{g}_6$ (which has been classified as $N_{6,28}$ by P. Turkowski in [101]); a prototype of this class of Lie algebras can be found, for example, in [61]. As a transformation group, our G_6 is the Iwasawa AN-group of $Sp(2, \mathbb{R})$. half-plane of degree 2, For a geometric background of this class of Lie groups as affine automorphisms of Siegel domains, we refer the reader, for example, to the textbook [63]. Thanks to the orbit structure of exponential groups, we can write down the dual space $\widehat{G_6}$ of our group G_6 and determine its topology. The main difference here from the previous cases, the $ax + b$-like groups, is that there are now coadjoint orbits of dimension 0, 2, 4, and 6, respectively, which decompose the orbit space into the union of a sequence $(S_i)_{i=0}^6$ of seven increasing closed subsets. On each of the sets $\Gamma_i = S_i \setminus S_{i-1}, i = 1, \ldots, 6$, the orbit space topology is Hausdorff and the main difficulty is to understand for a given $a \in C^*(G)$ the behaviour of the operators $\hat{a}(\gamma)$ for $\gamma \in \Gamma_i$, when γ approaches elements in S_{i-1}. For each of these sets Γ_i, we obtain different conditions for the C*-algebra, conditions which we shall call the *continuity*, the *infinity* and the *boundary* conditions.

Our example motivates the introduction of a special class of C^*-algebras which we call *almost* $C_0(\mathcal{K})$, where $\mathcal{K}$ is the algebra of all compact operators on a certain Hilbert space (Section 6.7.2). In Section 6.8, we describe the exponential Lie group G_6, its Lie algebra $\mathfrak{g}_6$ and its coadjoint orbits in $\mathfrak{g}_6^*$. Each of them needs a special treatment which we describe in the following sections. In Section 1.4.2, we present the topology of the dual space $\widehat{G}_6$ of G_6, i.e. we determine the boundaries of each orbit and we compute the limit sets of properly converging sequences of coadjoint orbits. In Section 6.8.1, we discover the continuity and infinity conditions and in Section 6.8.2, the most intricate one, we introduce the 6 different regions Γ_i of the dual space of G_6 according to the dimensions of the coadjoint orbits and obtain the boundary conditions for each of these regions. In the last section (Section 6.8.4), we describe the actual C*-algebra of G_6 as an algebra of operator fields and we see that this C^*-algebra has the structure of an almost $C_0(\mathcal{K})$-C^*-algebra.

6.7.2 A special class of C^*-algebras

Definition 6.26 Let Γ be a topological Hausdorff space, let $\mathcal{H}$ be a Hilbert space and denote by $\mathcal{K}$ or $\mathcal{K}(\mathcal{H})$ the algebra of compact operators on $\mathcal{H}$. Let $C_0(\Gamma, \mathcal{K})$ be the space of all continuous mappings $\varphi : \Gamma \to \mathcal{K}(\mathcal{H})$ vanishing at infinity. Equipped with the norm

$$\|\varphi\|_\infty := \sup_{\gamma \in \Gamma} \|\varphi(\gamma)\|_{\text{op}} \quad \text{for} \quad \varphi \in C_0(\Gamma, \mathcal{K}),$$

the space $C_0(\Gamma, \mathcal{K})$ is a C^*-algebra, whose spectrum is homeomorphic to the topological space Γ.

Let now A be a separable C^*-algebra and $\widehat{A}$ be the unitary dual of A.

Definition 6.27 We suppose that there exists a finite increasing family $S_0 \subset S_1 \subset \ldots \subset S_d = \widehat{A}$ of closed subsets of the spectrum $\widehat{A}$ of A such that for $i = 1, \cdots, d$, the subsets $\Gamma_0 = S_0$ and $\Gamma_i := S_i \setminus S_{i-1}$ are Hausdorff in their relative topologies. Furthermore we assume that for every $i \in \{0, \cdots, d\}$ there exists a Hilbert space $\mathcal{H}_i$ and a concrete realisation $(\pi_\gamma, \mathcal{H}_i)$ of γ on the Hilbert space $\mathcal{H}_i$ for every $\gamma \in \Gamma_i$. We also assume that the set S_0 is the collection $\mathcal{X}$ of all characters of A.

Definition 6.28 We define the Fourier transform $\mathcal{F} : A \to l^\infty(\widehat{A})$ as to be the mapping:

$$\mathcal{F}(a)(\gamma) = \hat{a}(\gamma) := \pi_\gamma(a) \quad \text{for} \quad \gamma \in \widehat{A}, a \in A.$$

For a subset $S \subset \widehat{A}$, denote by $CB(S)$ the $*$-algebra of all uniformly bounded operator fields $(\psi(\gamma) \in \mathcal{B}(\mathcal{H}_i))_{\gamma \in S \cap \Gamma_i, i=1,\cdots,d}$, which are operator norm continuous on the subsets $\Gamma_i \cap S$ for every $i \in \{1, \cdots, d\}$ for which $\Gamma_i \cap S \neq \emptyset$. We provide the algebra $CB(S)$ with the infinity-norm:

$$\|\psi\|_S := \sup_{\gamma \in S} \|\psi(\gamma)\|_{\text{op}}.$$

Definition 6.29 We say that a C^*-algebra A is "*almost* $C_0(\mathcal{K})$" if for every $a \in A$:

1. The mappings $\gamma \mapsto \mathcal{F}(a)(\gamma)$ are norm-continuous on the different sets Γ_i. (We remark that for every closed subset S of $\widehat{A}$ and every $a \in A$, the restriction $\mathcal{F}(a)_{|S}$ of the operator field $\mathcal{F}(a)$ to S is then contained in $CB(S)$).

2. For any $i = 1, \cdots, d$, we have a sequence $(\sigma_{i,k} : CB(S_{i-1}) \to CB(S_i))_k$ of linear mappings which are uniformly bounded in k such that

$$\lim_{k\to\infty} \text{dis}\Big((\sigma_{i,k}(\mathcal{F}(a)_{|S_{i-1}}) - \mathcal{F}(a)_{|\Gamma_i}), C_0(\Gamma_i, \mathcal{K}(\mathcal{H}_i))\Big) = 0$$

   and such that

$$\lim_{k\to\infty} \text{dis}\Big((\sigma_{i,k}(\mathcal{F}(a)_{|S_{i-1}})^* - \mathcal{F}(a^*)_{|\Gamma_i}), C_0(\Gamma_i, \mathcal{K}(\mathcal{H}_i))\Big) = 0.$$

   (This condition justifies the name of "almost $C_0(\mathcal{K})$").

Definition 6.30 Let $D^*(A)$ be the set of all operator fields φ defined over $\widehat{A}$ such that

1. The field φ is uniformly bounded, i.e. $\|\varphi\| := \sup_{\gamma \in \widehat{A}} \|\varphi(\gamma)\|_{\text{op}} < \infty$.
2. $\varphi_{|\Gamma_i} \in CB(\Gamma_i)$ for every $i = 0, 1, \ldots, d$.
3. For any sequence $(\gamma_k)_{k \in \mathbb{N}}$ going to infinity in $\widehat{A}$, we have $\lim_{k \to \infty} \|\varphi(\gamma_k)\|_{\text{op}} = 0$.
4. We have

$$\lim_{k \to \infty} \operatorname{dis}\Big((\sigma_{i,k}(\varphi_{|S_{i-1}}) - \varphi_{|\Gamma_i}), C_0(\Gamma_i, \mathcal{K}(\mathcal{H}_i))\Big) = 0$$

and

$$\lim_{k \to \infty} \operatorname{dis}\Big((\sigma_{i,k}(\varphi_{|S_{i-1}})^* - (\varphi_{|\Gamma_i})^*), C_0(\Gamma_i, \mathcal{K}(\mathcal{H}_i))\Big) = 0.$$

We see immediately that for every $a \in A$, the operator field $\mathcal{F}(a)$ is contained in the set $D^*(A)$. In fact, it turns out that $D^*(A)$ is a C*-subalgebra of $l^\infty(\widehat{A})$ and that A is isomorphic to $D^*(A)$.

Theorem 6.7.1 *Let A be a separable C*-algebra which is almost $C_0(\mathcal{K})$. Then the subset $D^*(A)$ of the C*-algebra $l^\infty(\widehat{A})$ is a C*-subalgebra which is isomorphic to A under the Fourier transform.*

Proof 6.34 *We remark that the conditions (1) to (4) imply that $D^*(A)$ is a closed involution-invariant subspace of $l^\infty(\widehat{A})$. For $i = 0, \cdots, d$, let D_i^* be the set of all operator fields defined over S_i satisfying conditions (1) to (4) on the sets S_j for $j = 1, \cdots, i$. Then the sets D_i^* are closed subspaces of the C*-algebra $l^\infty(S_i)$.*

Let I_C be the closed two-sided ideal in A generated by the elements of the form $ab - ba, a, b \in A$. Then the space of characters $S_0 = \mathcal{X}$ of A is the spectrum of A/I_C and D_0^* equals the algebra $C_0(S_0)$ of continuous functions on S_0 vanishing at infinity by the conditions (1) and (2). Since $\mathcal{F}(A)_{|S_0} = C_0(S_0)$ it follows that $D_0^* = \mathcal{F}(A)_{|S_0}$.

Let us assume now that for some $1 \leq i < d$, we have that $D_j^* = \mathcal{F}(A)_{|S_j}$ for $j = 0, \ldots, i-1$. We shall prove then that $D_i^* = \mathcal{F}(A)_{|S_i}$. We know already that $\mathcal{F}(A)_{|S_i}$ is a subalgebra of the closed subspace $D_i^* \subset CB(S_i)$ and it follows from its definition that the restriction of D_i^* to S_{i-1} is contained in D_{i-1}^*. Hence

$$\mathcal{F}(A)_{|S_{i-1}} \subset D^*(A)_{|S_{i-1}} \subset D_{i-1}^* = \mathcal{F}(A)_{|S_{i-1}}.$$

Therefore $D^*(A)_{|S_{i-1}} = \mathcal{F}(A)_{|S_{i-1}}$. Let $\phi_1, \phi_2 \in D_i^*$. By our assumption, there exists $a_1, a_2 \in A$ such that $\phi_{j|S_{i-1}} = \hat{a}_{j|S_{i-1}}, j = 1, 2$. The product $\phi_1 \circ \phi_2$ satisfies also the conditions (1), (2), and (3). We shall show that it also satisfies the condition (4). We now have that

$$\begin{aligned}
& \phi_1 \circ \phi_{2|\Gamma_i} - \sigma_{i,k}(\phi_1 \circ \phi_{2|S_{i-1}}) \\
= \; & \phi_{1|\Gamma_i} \circ \phi_{2|\Gamma_i} - \sigma_{i,k}(\hat{a}_{1|S_{i-1}} \circ \hat{a}_{2|S_{i-1}}) \\
= \; & \phi_{1|\Gamma_i} \circ \phi_{2|\Gamma_i} - \hat{a}_{1|\Gamma_i} \circ \hat{a}_{2|\Gamma_i} \\
& + (\hat{a}_{1|\Gamma_i} \circ \hat{a}_{2|\Gamma_i}) - \sigma_{i,k}(\hat{a}_{1|S_{i-1}} \circ \hat{a}_{2|S_{i-1}}).
\end{aligned}$$

Now

$$\begin{aligned}
& \lim_{k\to+\infty} dis((\hat{a}_{1|\Gamma_i} \circ \hat{a}_{2|\Gamma_i} - \sigma_{i,k}(\hat{a}_{1|S_{i-1}} \circ \hat{a}_{2|S_{i-1}})), C_0(\Gamma_i, \mathcal{K}(\mathcal{H}_i))) \\
= & \lim_{k\to+\infty} dis((\widehat{a_1 a_2}_{|\Gamma_i} - \sigma_{i,k}(\widehat{a_1 a_2}_{|S_{i-1}})), C_0(\Gamma_i, \mathcal{K}(\mathcal{H}_i))) \\
= & \ 0
\end{aligned}$$

by condition (2) in Definition 6.29. Furthermore,

$$\begin{aligned}
& \phi_{1|\Gamma_i} \circ \phi_{2|\Gamma_i} - \hat{a}_{1|\Gamma_i} \circ \hat{a}_{2|\Gamma_i} \\
= & \ \phi_{1|\Gamma_i} \circ (\phi_{2|\Gamma_i} - \hat{a}_{2|\Gamma_i}) + (\phi_{1|\Gamma_i} - \hat{a}_{1|\Gamma_i}) \circ \hat{a}_{2|\Gamma_i} \\
= & \ \phi_{1|\Gamma_i} \circ (\phi_{2|\Gamma_i} - \sigma_{i,k}(\phi_{2|S_{i-1}})) + \phi_{1|\Gamma_i} \circ (\sigma_{i,k}(\hat{a}_{2|S_{i-1}}) - \hat{a}_{2|\Gamma_i}) \\
& +(\phi_{1|\Gamma_i} - \sigma_{i,k}(\phi_{1|S_{i-1}})) \circ \hat{a}_{2|\Gamma_i} + (\sigma_{i,k}(\hat{a}_{1|S_{i-1}}) - \hat{a}_{1|\Gamma_i}) \circ \hat{a}_{2|\Gamma_i}.
\end{aligned}$$

Hence,

$$\lim_{k\to+\infty} dis(\phi_{1|\Gamma_i} \circ \phi_{2|\Gamma_i} - \sigma_{i,k}(\phi_1 \circ \phi_{2|S_{i-1}}), C_0(\Gamma_i, \mathcal{K}(\mathcal{H}_i))) = 0.$$

Therefore D_i^* is an algebra, i.e. a C^*-subalgebra of $CB(S_i)$. The conditions (1) and (2) for the C^*-algebra A tell us that the ideal $ker(S_{i-1}) = \{a \in A : \hat{a}_{|S_{i-1}} = 0\}$ of A is isomorphic with the algebra $C_0(\Gamma_i, \mathcal{K}(\mathcal{H}_i)) \subset \mathcal{F}(A)$ of all continuous mappings defined on Γ_i with values in $\mathcal{K}(\mathcal{H}_i)$ and vanishing at infinity. Indeed, for any $\gamma \neq \gamma' \in S_i$, there exists an $a \in A$ such that $\gamma(a) \neq 0$, but $\gamma'(a) = 0$ and also $\hat{a}_{|S_{i-1}} = 0$. Then $\hat{a}_{|\Gamma_i}$ is in $C_0(\Gamma_i, \mathcal{K}(\mathcal{H}_i))$ by the condition (1), (2), and (3) and in this way we see that $ker(S_{i-1})$ separates the points in Γ_i and that $\widehat{ker(S_{i-1})}_{|\Gamma_i} \subset C_0(\Gamma_i, \mathcal{K}(\mathcal{H}_i))$. The theorem of Stone-Weierstrass says now that $\widehat{ker(S_{i-1})}_{|\Gamma_i} = C_0(\Gamma_i, \mathcal{K}(\mathcal{H}_i))$.

Let now π be an element of the spectrum of D_i^*. Let $R_{i-1} : D_i^* \to D_{i-1}^*$ be the restriction map and let $I_{i-1} := ker(R_{i-1}) \subset D_i^*$. Suppose that $\pi(I_{i-1}) = \{0\}$. We can then consider π as a representation of the algebra D_i^*/I_{i-1}. This algebra is isomorphic with the image of R_{i-1} which is itself isomorphic with $\mathcal{F}(A)_{|S_{i-1}}$ and therefore π is an element of S_{i-1}. Suppose that π is not trivial on I_{i-1}. By the condition (4) this kernel I_{i-1} is contained in $C_0(\Gamma_i, \mathcal{K})$ and since it contains $ker(R_{i-1}) \cap \mathcal{F}(A)$, it is itself isomorphic to $C_0(\Gamma_i)$. Therefore π is an evaluation at some point in Γ_i.

We have seen that the spectrum of D_i^* coincides with the spectrum of the subalgebra $\mathcal{F}(A)_{|S_i}$. Hence by the theorem of Stone-Weierstrass (see [37]), the C^*-algebra $\mathcal{F}(A)_{|S_i}$ and D_i^* coincide.

6.8 THE C^*-ALGEBRA OF G_6

This section aims to show that the C^*-algebra of the exponential Lie group $G_6 = \exp(\mathfrak{g}_6)$ is in the class of almost $C_0(\mathcal{K})$-C^*-algebras. We shall need in our description of $C^*(G)$ the following criterion for the compactness of an operator $\pi(F), F \in A$, where π is an irreducible representation of a type I, C^*-algebra A.

Theorem 6.8.1 *The operator $\pi(F), F \in A$, is compact if and only if $\pi'(F) = 0$ for every π' in the boundary of π.*

Since the topology of the spectrum $\widehat{C^*(G)}$ of an exponential Lie group $G = \exp(\mathfrak{g})$ is that of the topology of the space of coadjoint orbits $\mathfrak{g}^*/G$ according to [70], we have the following.

Theorem 6.8.2 *If $G = exp(\mathfrak{g})$ is an exponential Lie group, ℓ is an element of $\mathfrak{g}^*$ and π_ℓ is the corresponding irreducible unitary representation, then for $F \in C^*(G)$ the operator $\pi_\ell(F)$ is compact if and only $\pi_q(F) = 0$ for any q in the boundary of the orbit Ω_ℓ of ℓ.*

We will make use therefore of the results of Subsection 1.4.2 concerning the determination of the boundary of any coadjoint orbit of our group G.

6.8.1 Continuity and infinity conditions

In this short subsection, we find the continuity and infinity conditions for $C^*(G)$.

Definition 6.31 We say that a net $(\gamma_i)_{i\in I}$ in a topological space Γ goes to infinity, if the net contains no converging subnet. In particular, a sequence of orbits $\overline{\Omega} = (\Omega_k)_{k\in\mathbb{N}} \subset \mathfrak{g}^*$ goes to infinity, if for any compact subset $K \subset \mathfrak{g}^*$ there exists an index k_0 such that $K \cap \Omega_k = \emptyset$ whenever $k \geq k_0$.

Proposition 6.8.3 (Riemann-Lebesgue Lemma) *Let A be a C^*-algebra. If a net $(\pi_k)_k \subset \widehat{A}$ goes to infinity, then $\lim_k \|\pi_k(a)\|_{\text{op}} = 0$ for all $a \in A$.*

Proof 6.35 *We know from [37, Proposition 3.3.7] that for every $c > 0$ and $a \in A$, the subset $\{\pi \in \widehat{A}; \|\pi(a)\|_{\text{op}} \geq c\}$ is quasi-compact. This shows that $\lim_k \|\pi_k(a)\|_{\text{op}} = 0$, if the net $(\pi_k)_k$ goes to infinity.*

In the following, we recall the definition of properly converging sequences and their limit sets, we also give a proposition of properly converging sequences in our group.

Definition 6.32 Let Y be a topological space. Let $\overline{y} = (y_k)_k$ be a net in Y. We denote by $L(\overline{y})$ the set of all limit points of the net $\overline{y}$. A net $\overline{y}$ is called *properly converging* if $\overline{y}$ has limit points and if every cluster point of the net is a limit point, i.e. the set of limit points of any subnet is always the same, indeed, it equals to $L(\overline{y})$.

We know that every converging net in Y admits a properly converging subnet, hence, we can work with properly converging nets in our space $\widehat{G}$.

Definition 6.33 Let $\varepsilon = \pm 1$, and

- $\mathbb{R}_{\varepsilon P^*} := \{ \operatorname{Ad}^*(G)(x^*X^* + \varepsilon P^*) = x^*X^* + \operatorname{Ad}^*(G)(\varepsilon P^*), x^* \in \mathbb{R}\}$,
- $\mathbb{R}_{\varepsilon Q^*} := \{ \operatorname{Ad}^*(G)(a^*A^* + \varepsilon Q^*) = a^*A^* + \operatorname{Ad}^*(G)(\varepsilon Q^*), a^* \in \mathbb{R}\}$,

- $\mathbb{R}_{\varepsilon R^*} := \{ \mathrm{Ad}^*(G)(b^*B^* + \varepsilon R^*) = b^*B^* + \mathrm{Ad}^*(G)(\varepsilon R^*), b^* \in \mathbb{R}\}$,
- $\mathbb{R}_{\varepsilon S^*} := \{ \mathrm{Ad}^*(G)(b^*B^* + \varepsilon S^*) = b^*B^* + \mathrm{Ad}^*(G)(\varepsilon S^*), b^* \in \mathbb{R}\}$.

Proposition 6.8.4 *Let $\underline{\Omega} = (\Omega_k)_{k\in\mathbb{N}}$ be a properly converging sequence in $\mathfrak{g}^*/G$. Then there exists a subsequence (also denoted by $\underline{\Omega} = (\Omega_k)_{k\in\mathbb{N}}$ for simplicity) such that either $\underline{\Omega}$ is a constant sequence, or if $\underline{\Omega}$ is not constant, it is contained in one of the subsets in Definition 6.33 and its limit set is the closure in $\mathfrak{g}^*/G$ of one of the points of this subset or all the elements of the subsequence are characters and the limit set is a single character.*

Proof 6.36 *The orbit space consists of four open orbits of the eight sets given in Definition 6.33, of the four orbits of $\nu P^* + \varepsilon Q^*, \varepsilon, \nu \in \{1, -1\}$, and of the set of characters $\mathcal{X}$. Hence, if the sequence $\underline{\Omega}$ contains a constant subsequence, then the limit set of the sequence itself is the closure of this constant element, since it is properly converging. If the sequence is not constant, then we can suppose that a subsequence is contained in one of the 8 sets or is made out of characters. If we deal with non-characters, then we can write $\underline{\Omega} = (\Omega + \lambda_k U^*)_{k\in\mathbb{N}}$, where $\lambda_k \in \mathbb{R}$ and U^* is one of the characters X^*, A^* or B^*, and Ω is a fixed orbit in one of sets $\mathbb{R}_{\varepsilon P^*}, \mathbb{R}_{\varepsilon Q^*}, \mathbb{R}_{\varepsilon R^*}, \mathbb{R}_{\varepsilon S^*}$. Since the sequence $\underline{\Omega}$ is properly converging, for any v in a limit orbit Ω_∞, there exists a sequence $(v_k)_k \subset \Omega$ and a real sequence $(\lambda_k)_k$ such that $\lambda_k U^* + v_k$ converges in $\mathfrak{g}^*$ to v. We can write $\mathfrak{g} = \mathbb{R}U + \mathfrak{m}$, where $\mathfrak{m}$ is an ideal of $\mathfrak{g}$ containing $[\mathfrak{g}, \mathfrak{g}]$. Then by the formulas in Subsection 1.4.1, it follows that $v_k(U) = \varphi(v_{k|\mathfrak{m}})$, where $\varphi : \mathfrak{m}^* \to \mathbb{R}$ is a continuous function. Hence the sequence $(\lambda_k + \varphi(v_{k|\mathfrak{m}}))_k$ is convergent with a limit $\lambda + v(U)$ and so the sequence $(\lambda_k)_k$ converges to some λ. Hence the limit orbit Ω_∞ is contained in $\overline{\Omega + \lambda U^*} = \lambda U^* + \overline{\Omega}$, and so the limit set of the sequence $\underline{\Omega}$ is the set $\overline{\Omega + \lambda U^*}$.*

We define now the subsets $S_j, j = 0, \ldots, 6$, which we shall need for the definition of the algebra $D^*(G)$ in the later section.

Definition 6.34 Let

- $\Gamma_0 = \mathcal{X} = \{a^*A^* + b^*B^*, a^*, b^* \in \mathbb{R}\}$,
- $\Gamma_1 := \{\mathbb{R}_{\varepsilon P^*}, \varepsilon = \pm 1\}$,
- $\Gamma_2 := \{\mathbb{R}_{\nu Q^*}, \nu = \pm 1\}$,
- $\Gamma_3 := \{\Omega_{\varepsilon P^* + \nu Q^*}, \nu, \varepsilon = \pm 1\}$,
- $\Gamma_4 := \{\mathbb{R}_{\varepsilon R^*}, \varepsilon = \pm 1\}$,
- $\Gamma_5 := \{\mathbb{R}_{\varepsilon S^*}, \varepsilon = \pm 1\}$,
- $\Gamma_6 := \{\Omega_{\varepsilon S^* + \nu Q^*}, \varepsilon, \nu = \pm 1\}$.

Let $S_i := \bigcup_{j=0}^{i} \Gamma_j \subset \mathfrak{g}^*/G$. Then S_i is closed in S_{i+1} for $i = 0, \dots, 5$ with respect to the orbit topology. The set Γ_6 is finite and open in $\widehat{G}$ while Γ_3 is finite and open in S_3. The sets $\Gamma_1, \Gamma_2, \Gamma_4$ and Γ_5 are homeomorphic to two disjoint copies of the real line $\mathbb{R}$.

Theorem 6.8.5 *The mappings $\pi \mapsto \pi(F)$, $F \in C^*(G)$, are norm-continuous on the subsets Γ_i of $\widehat{G}$ for all $i = 0, \dots, 6$.*

Proof 6.37 *This follows from the following more general result for the C^*-algebra of a second countable locally compact group G.*

Let $(\chi_k)_k$ be a sequence of unitary characters of G, which converges pointwise to a unitary character χ. Let $(\pi, \mathcal{H})$ be a unitary representation of G. Then for every $F \in L^1(G)$, we have that

$$\lim_{k\to\infty} \|\chi_k \otimes \pi(F) - \chi \otimes \pi(F)\|_{\text{op}} = 0.$$

Indeed, it suffices to remark that

$$\|\chi_k \otimes \pi(F) - \chi \otimes \pi(F)\|_{\text{op}} = \|\pi(\chi_k \cdot F) - \pi(\chi \cdot F)\|_{\text{op}} \leq \|\chi_k \cdot F - \chi \cdot F\|_1$$

and therefore by the Lebesgue's theorem of dominated convergence, $\lim_{k\to\infty} \|\chi_k \otimes \pi(F) - \chi \otimes \pi(F)\|_{\text{op}} = 0$.

6.8.2 The boundary conditions for the C^*-algebra

For each one of our subsets Γ_i defined in Definition 6.34, we must find the linear mappings $\sigma_{i,\delta}, \delta > 0$, which give the structure of the C^*-algebra of G an almost $C_0(\mathcal{K})$-algebra. These mappings $\sigma_{i,\delta}$ will be built on the representations coming from the different sets Γ_i. Hence, this forces us to make a case by case study of the different situations.

We indicate here some definitions and methods, which will be used in the proofs of this section.

Remark 6.8.6

1. For a measurable subset S of a measure space (X, μ), denote the multiplication operator with the characteristic function 1_S on $L^2(X, \mu)$ by M_S.

2. Let $H = \exp(\mathfrak{h})$ be an abelian normal subgroup of G. The subspace $L^1_c = L^1_{c,\mathfrak{h}}$ is defined to be the set of all F's in $L^1(G)$, for which the partial Fourier transform

$$\widehat{F}^{\mathfrak{h}}(s, q) := \int_H F(sh)\chi_q(h)dh, s \in G, q \in \mathfrak{h}^*$$

is a C^∞-function with compact support on $G/H \times \mathfrak{h}^*$. The vector space L^1_c is of course dense in $L^1(G)$ and hence is also dense in $C^*(G)$.

Furthermore, for every $F \in L^1_c$ there exists a function $\varphi \in C_c(G/H)$ such that

$$|\widehat{F}^{\mathfrak{h}}(s,q)| \leq \|q\| |\varphi(s)| \text{ for } s \in G, q \in \mathfrak{h}^*.$$

3. Let $F : \mathbb{R}^d \times \mathbb{R}^d \to \mathbb{C}$ be a smooth function with compact support for which there exists some continuous function $\varphi : \mathbb{R}^d \to \mathbb{R}^+$ with compact support, such that

$$|F(x,y)| \leq \varphi(x-y) \text{ for } x,y \in \mathbb{R}^d.$$

Then by the Young's inequality, the operator norm $\|\cdot\|_{\text{op}}$ of the kernel operator T_F defined on $L^2(\mathbb{R}^d)$ by

$$T_F(\xi)(x) := \int_{\mathbb{R}^d} F(x,y)\xi(y)dy \text{ for } \xi \in L^2(\mathbb{R}^d), x \in \mathbb{R}^d,$$

is bounded by the L^1-norm $\|\varphi\|_1$ of φ.

4. Let $(X,\mu),(Y,\nu)$ be two measure spaces. Let $Y \mapsto L^2(X,\mu) : y \mapsto \xi(y)$ be an integrable mapping. Then we have that

$$\| \int_Y \xi(y)d\nu(y) \|_2 \leq \int_Y \|\xi(y)\|_2 d\nu(y),$$

this is,

$$\Big(\int_X | \int_Y \xi(y)(x)d\nu(y)|^2 d\mu(x) \Big)^{1/2} \leq \int_Y \Big(\int_X |\xi(y)(x)|^2 d\mu(x) \Big)^{1/2} d\nu(y).$$

Proposition 6.8.7 *Let (X,μ) be a measure space, and $(\sigma_i)_{i\in I}$ be a family of bounded linear operators on the Hilbert space $\mathcal{H} = L^2(X,\mu)$ such that $\|\sigma_i\|_{\text{op}} \leq C$ for all $i \in I$ and some $C > 0$. Suppose furthermore that there exists families $(T_{i,j})_{i\in I}(j = 1,\ldots,N)$ and $(S_i)_{i\in I}$ of measurable subsets of X such that $T_{i,j} \cap T_{i',j} = \emptyset, j = 1,\ldots,N$, and $S_i \cap S_{i'} = \emptyset$ whenever $i \neq i'$. Then the linear operator*

$$\sigma = \sum_{j=1}^{N} \sum_{i\in I} M_{T_{i,j}} \circ \sigma_i \circ M_{S_i}$$

is bounded by NC.

Proof 6.38 *Let us write*

$$\sigma^j := \sum_{i\in I} M_{T_{i,j}} \circ \sigma_i \circ M_{S_i} \quad \textit{for} \quad j = 1,\cdots,N.$$

Then $\sigma = \sum_{j=1}^{N} \sigma^j$ *and for* $\xi \in L^2(X, \mu), j \in \{1, \ldots, N\}$ *we have*

$$\begin{aligned}
\|\sigma^j(\xi)\|_2^2 &= \int_X |\sum_{i\in I} M_{T_{i,j}} \circ \sigma_i \circ M_{S_i}(\xi)(x)|^2 d\mu(x) \\
&= \sum_{i\in I} \int_{T_{i,j}} |\sigma_i(M_{S_i}(\xi))(x)|^2 d\mu(x) \\
&\leq \sum_{i\in I} \int_X |\sigma_i(M_{S_i}(\xi))(x)|^2 d\mu(x) \\
&\leq \sum_{i\in I} C^2 \int_X |M_{S_i}(\xi)(x)|^2 d\mu(x) \\
&= C^2 (\sum_{i\in I} \int_{S_i} |\xi(x)|^2 d\mu(x)) \\
&\leq C^2 \int_X |\xi(x)|^2 d\mu(x) \\
&= C^2 \|\xi\|_2^2.
\end{aligned}$$

Hence $\|\sigma\|_{\text{op}} \leq NC$.

6.8.3 The open orbits

6.8.3.1 The orbit $\Omega_{\varepsilon,-\varepsilon} = \Omega_{\varepsilon S^* - \varepsilon Q^*}$ for $\varepsilon = \pm 1$

Let $\ell_{\varepsilon,-\varepsilon} := \varepsilon(S^* - Q^*), \varepsilon \in \{1, -1\}$. We have seen in (1.4.2) and (1.4.3) that the G-orbit of $(\ell_{\varepsilon,-\varepsilon})_{|\mathfrak{h}}$ is the subset

$$\Omega_{\varepsilon,-\varepsilon|\mathfrak{h}} = \left\{ \varepsilon \left((-e^{-b} + \frac{e^{-b}p^2}{2}) Q^* + (-e^{-\frac{a+b}{2}} p) R^* + e^{-a} S^* \right), a, b, p \in \mathbb{R} \right\}.$$

By Subsection 1.4.2.1, the boundary $\partial(\varepsilon, -\varepsilon)$ of the orbit of $(\ell_{\varepsilon,-\varepsilon})_{|\mathfrak{h}}$ is given by the union of the 5 orbits

$$\begin{aligned}
G \cdot \varepsilon S^*_{|\mathfrak{h}} &= \{E(a,b,p) \cdot \varepsilon S^* = (e^{-b} \frac{\varepsilon}{2} p^2) Q^* + (-\varepsilon p e^{-\frac{a+b}{2}}) R^* + \varepsilon e^{-a} S^*); \\
&\quad a, b, p \in \mathbb{R}\}, \\
G \cdot \varepsilon Q^*_{|\mathfrak{h}} &= \{E(a,b,p) \cdot \varepsilon Q^* = \varepsilon e^{-b} Q^*;\ a, b, p \in \mathbb{R}\}, \\
G \cdot (-\varepsilon Q^*)_{|\mathfrak{h}} &= \{-E(a,b,p) \cdot \varepsilon Q^* = -\varepsilon e^{-b} Q^*;\ a, b, p \in \mathbb{R}\}, \\
G \cdot \sqrt{2}\varepsilon R^*_{|\mathfrak{h}} &= \{E(a,b,p) \cdot (\sqrt{2}\varepsilon R^*) = \varepsilon((-e^{-b}\sqrt{2}p) Q^* + (e^{\frac{-a-b}{2}} \sqrt{2}) R^*); \\
&\quad a, b, p \in \mathbb{R}\}, \\
G \cdot (-\sqrt{2}\varepsilon R^*)_{|\mathfrak{h}} &= \{-E(a,b,p) \cdot (\sqrt{2}\varepsilon R^*) = \varepsilon((e^{-b}\sqrt{2}p) Q^* - (e^{\frac{-a-b}{2}} \sqrt{2}) R^*); \\
&\quad a, b, p \in \mathbb{R}\}
\end{aligned}$$

and $\{0\}$. Here we denote $E(a,b,p) = \exp(aA)\exp(bB)\exp(pP)$ for $(a,b,p) \in \mathbb{R}$.

We must now consider the following subsets of $\mathbb{R}^3$, which give us the points in the orbit $\Omega_{\varepsilon,-\varepsilon|\mathfrak{h}}$ "close" up to a distance δ to the corresponding boundary orbits.

Definition 6.35 *1. For $k \in \mathbb{Z}$, denote by*

$$J_{\delta,k} := \left[\log(\frac{1}{\delta^6}), +\infty\right) \times \left(-\infty, \log(\frac{1}{\delta})\right] \times I_{\delta^2,k},$$

where $I_{c,l} := [cl, cl + c) \subset \mathbb{R}$ for $l \in \mathbb{Z}, c > 0$ and let

$$L_{\delta,k} := \mathbb{R} \times \mathbb{R} \times I_{\delta^2,k}.$$

2. Let

$$K_\delta = [-\frac{1}{\delta}, \frac{1}{\delta}] \times [-\frac{1}{\delta}, \frac{1}{\delta}] \times [-\frac{1}{\delta^{1/2}}, \frac{1}{\delta^{1/2}}].$$

3. Let $S_{\delta,1} := \{(a, b, p); e^{-a} > \delta^6\}$(corresponds to $\mathbb{R}_{\varepsilon S^*}$).

4. Let $S_{\delta,2} := \{(a, b, p); e^{-a} \leq \delta^6, e^{-b} < \delta\}$(corresponds to $\mathbb{R}_{\varepsilon S^*}$).

5. Let

$$S_{\delta,3,k,+} := J_{\delta,k} \cap \{(a, b, p); \varepsilon e^{-b/2}(1 - \frac{p^2}{2}) > \delta^{1/2}, \delta \leq e^{-b} \leq \frac{1}{\delta^{5/4}}\} \text{ for } k \in \mathbb{Z},$$

(corresponds to $\mathbb{R}_{\varepsilon Q^*}$) and $S_{\delta,3,+} := \bigcup_k S_{\delta,3,k,+}$.

6. Let

$$S_{\delta,3,k,-} := J_{\delta,k} \cap \{(a, b, p); \varepsilon e^{-b/2}(1 - \frac{p^2}{2}) < -\delta^{1/2}, \delta \leq e^{-b} \leq \frac{1}{\delta^{5/4}}\} \text{ for } k \in \mathbb{Z},$$

(corresponds to $\mathbb{R}_{-\varepsilon Q^*}$) and $S_{\delta,3,-} := \bigcup_k S_{\delta,3,k,-}$.

7. Let

$$S_{\delta,4,\pm} := \{(a, b, p) \in \mathbb{R}^3; e^{-a} < \delta^6, e^{-b} \geq \delta, e^{-b/2}|\frac{p^2}{2} - 1| \leq \delta^{1/2}, \pm p \leq 0\}.$$

(corresponds to $\mathbb{R}_{\pm R^*}$).

8. Define the corresponding multiplication operators on $L^2(\mathbb{R}^3)$ by $M_{\delta,i} = M_{S_{\delta,i}}$ for $i = 1, 2$, respectively; for $k \in \mathbb{Z}$, define $M_{\delta,3,k,\pm} = M_{S_{\delta,3,k,\pm}}$,$M_{\delta,3,\pm} = M_{S_{\delta,3,\pm}}$ and $M_{\delta,4,\pm} = M_{S_{\delta,4,\pm}}$.

Remark 6.8.8

1. For any $\delta > 0$, the sets $\{S_{\delta,1}, S_{\delta,2}, S_{\delta,3,k,\pm}, S_{\delta,4,\pm}; k \in \mathbb{Z}\}$ are pairwise disjoint measurable subsets of $\mathbb{R}^3$.

2. The region

$$\{(a, b, p); e^{-b/2}|\frac{p^2}{2} - 1| > \delta^{1/2}, e^{-b} > \frac{1}{\delta^{5/4}}\}$$

does not need to be considered, since for a function $F \in L^1_c$ the kernel function $\widehat{F}^{\mathfrak{h}}(s - a, t - b, e^{\frac{a-b}{2}}(v - p), \varepsilon(-e^{-b} + \frac{e^{-b}p^2}{2})Q^* - (e^{-\frac{a+b}{2}}p)R^* + e^{-a}S^*)$ of the operator $\pi_{\varepsilon,-\varepsilon}(F)$ is 0 for δ small enough on this region.

3. Recall that for $\alpha = (a,b,p)$ and $\beta = (s,t,v) \in \mathbb{R}^3$, the multiplication of $E(s,t,v)\cdot E(a,b,p)$ is given by

$$E(s,t,v)\cdot E(a,b,p) = E(s+a, t+b, e^{\frac{-a+b}{2}}v+p).$$

For $\beta \in K_\delta$ and $\alpha \in S_{\delta,k,3,\pm}$, we have that $e^{-a} \le \delta^6, e^b < \frac{1}{\delta}, |v| < \frac{1}{\delta^{1/2}}$ and so $e^{\frac{-a+b}{2}}|v| \le \delta^2$ and

$$\begin{aligned} \beta\cdot\alpha &= (s+a, t+b, e^{\frac{-a+b}{2}}v+p) \\ &\in \mathbb{R}\times\mathbb{R}\times([-\delta^2,\delta^2]+I_{\delta^2,k}) \\ &\subset \bigcup_{i=-1}^{1} L_{\delta,k+i} =: T_{\delta,3,k}. \end{aligned}$$

We see that the set $T_{\delta,3,k}$ is the union of the 3 disjoint boxes $L_{\delta,k+i}$ for $i \in \{-1,0,1\}$.

Let $N_{\delta,3,k}$ be the multiplication operator with the characteristic function of the set $T_{\delta,3,k}$ for all $k \in \mathbb{Z}$.

4. The irreducible representation $\pi_{\varepsilon,-\varepsilon} = \mathrm{ind}_H^G \chi_{\varepsilon(S^*-Q^*)}$ acts on a $\xi \in L^2(G/H, \chi_{\varepsilon(S^*-Q^*)})$ in the following way (see (1.4.13)):

$$\begin{aligned} &\pi_{\varepsilon,-\varepsilon}(F)\xi(E(s,t,v)) \\ = &\int_{\mathbb{R}^3} e^{\frac{1}{2}(a-b)}\widehat{F}^{\mathfrak{h}}(s-a, t-b, e^{\frac{a-b}{2}}(v-p), \varepsilon(-e^{-b}(1-\frac{p^2}{2})Q^* \\ &+(-e^{-\frac{a+b}{2}}p)R^* + e^{-a}S^*))\xi(E(a,b,p))dadbdp,\ F\in L^1(G). \end{aligned}$$

5. Let $F \in L^1_c$. There exists a box $K = [-M,M]^3$ for some $M > 0$ such that

$$\widehat{F}^{\mathfrak{h}}(s,q) = 0 \text{ for } s \notin K \text{ and } q \in \mathfrak{h}^*.$$

Hence it follows from the relations (6.8.1) that for $\delta > 0$ small enough,

$$\widehat{F}^{\mathfrak{h}}(st^{-1}, t\cdot(\varepsilon(S^*-Q^*))) = 0, \text{ for } t \in S_{\delta,3,k,\pm}, s \notin T_{\delta,3,k} \text{ and } k \in \mathbb{Z}.$$

Therefore,

$$N_{\delta,3,k}\circ\pi_{\varepsilon,-\varepsilon}(F)\circ M_{\delta,3,k,\pm} = \pi_{\varepsilon,-\varepsilon}(F)\circ M_{\delta,3,k,\pm} \text{ for } k \in \mathbb{Z}. \tag{6.8.1}$$

Definition 6.36 *Let* $\sigma_l := \mathrm{ind}_H^G \chi_l$ *for* $l = \varepsilon S^*$, $\varepsilon(\pm\sqrt{2}R^* - 2Q^*)$ *and* $\varepsilon(-1+\frac{\delta^4 k^2}{2})Q^*$, *respectively. We define the linear operator* $\sigma_{\varepsilon,-\varepsilon,\delta}(F)$ *by the rule*

$$\sigma_{\varepsilon,-\varepsilon,\delta}(F) := \sum_{i=1}^{2}\sigma_{\varepsilon S^*}(F)\circ M_{\tilde{\varphi}_i}\circ M_{\delta,i} + \sigma_{\delta,3}(F) + \sigma_{\delta,4,+}(F) + \sigma_{\delta,4,-}(F),$$

where $\sigma_{\delta,4,\pm}(F) = \sigma_{\varepsilon(\pm\sqrt{2}R^*-2Q^*)}(F) \circ M_{\delta,4,\pm}$ and

$$\sigma_{\delta,3}(F) := \sum_{k\in\mathbb{Z}} \Big(N_{\delta,3,k} \circ \sigma_{\varepsilon(-1+\frac{\delta^4k^2}{2})Q^*}(F) \circ (M_{\delta,3,k,+} + M_{\delta,3,k,-})\Big),$$

where the operators $M_{\tilde{\varphi}_i}, i = 1, 2$, are multiplication operators with the functions $\tilde{\varphi}_1(a, b, p) := \varphi(-e^{-b}), \tilde{\varphi}_2 = 1$ and $\varphi : \mathbb{R} \to \mathbb{R}$ is a C^∞-function with support contained in $[-1, 1]$ such that $\varphi([-\frac{1}{2}, \frac{1}{2}]) = \{1\}$ and $\|\varphi\|_\infty = 1$.

The kernel functions of the different operators appearing in the sum above are given by

1. $\sigma_{\varepsilon S^*}(F) \circ M_{\tilde{\varphi}_i}$ gives
$$F_{\delta,i}((s,t,v),(a,b,p)) = \widehat{F}^{\mathfrak{h}}(s-a, t-b, e^{\frac{a-b}{2}}(v-p), \varepsilon((\tfrac{e^{-b}p^2}{2})Q^* + (-e^{-\frac{a+b}{2}}p)R^* + e^{-a}S^*))e^{\frac{1}{2}(a-b)}\varphi_i(a,b,p) \text{ for } i = 1, 2;$$

2. $\sigma_{\varepsilon(\pm\sqrt{2}R^*-2Q^*)}(F) \circ M_{\delta,4,\pm}$ gives
$$\begin{aligned} &F_{\delta,4,\pm}((s,t,v),(a,b,p)) \\ = \ &\widehat{F}^{\mathfrak{h}}(s-a, t-b, e^{\frac{a-b}{2}}(v-p), \varepsilon(-e^{-b}(\pm\sqrt{2}p+2))Q^* \pm \varepsilon(e^{-\frac{a+b}{2}}\sqrt{2})R^*) \\ &e^{\frac{1}{2}(a-b)}1_{S_{\delta,4,\pm}}(a,b,p); \text{ and} \end{aligned}$$

3. $\sigma_{\varepsilon(-1+\frac{\delta^4k^2}{2})Q^*}(F) \circ M_{\delta,3,k,\pm}$ gives
$$F_{\delta,3,k,\pm}((s,t,v),(a,b,p)) = \widehat{F}^{\mathfrak{h}}(s-a, t-b, e^{\frac{a-b}{2}}(v-p), \varepsilon(e^{-b}(-1+\frac{\delta^4k^2}{2})Q^*) e^{\frac{1}{2}(a-b)}1_{S_{\delta,3,k,\pm}}(a,b,p) \text{ for } k \in \mathbb{Z}.$$

Now the representation $\sigma_{\varepsilon|1-\frac{k^2\delta^4}{2}|Q^*}$ is equivalent to the representation $\sigma_{\varepsilon Q^*}$, since both linear functionals are in the same G-orbit. Therefore, for any $k \in \mathbb{Z}$ and $1 > \delta > 0$:

$$\|N_{\delta,3,k} \circ \sigma_{\varepsilon|1-\frac{\delta^4k^2}{2}|Q^*}(F) \circ M_{\delta,3,k,\pm})\|_{\text{op}} \le \|\sigma_{\varepsilon Q^*}\|_{\text{op}}(F). \tag{6.8.2}$$

Proposition 6.8.9 *For any $F \in L^1_c$ and $\delta > 0$, the operator $\sigma_{\delta,3}(F)$ is bounded in norm by $C\|\sigma_{Q^*}(F)\|_{\text{op}}$, where C is the number given by $\max_{l\in\mathbb{Z}} \#\{k \in \mathbb{Z}, T_{\delta,3,k} \cap T_{\delta,3,l} \neq \emptyset\}$.*

Proof 6.39 *Apply Proposition 6.8.7.*

Proposition 6.8.10 *For any $F \in L^1_c$ and $\delta > 0$, we have that*

$$\|\sigma_{\varepsilon,-\varepsilon,\delta}(F)\|_{\text{op}} \le 2\|\sigma_{\varepsilon S^*}(F)\|_{\text{op}} + C\|\sigma_{\varepsilon Q^*}(F)\|_{\text{op}} + 2\|\sigma_{\varepsilon R^*}(F)\|_{\text{op}}$$

and we can extend the mapping $F \mapsto \sigma_{\varepsilon,-\varepsilon,\delta}(F)$ to every $F \in C^*(G)$, then

$$\lim_{\delta\to 0} dis(\pi_{\varepsilon(S^*-Q^*)}(F) - \sigma_{\varepsilon,-\varepsilon,\delta}(F), \mathcal{K}(L^2(\mathbb{R}^3))) = 0.$$

Here $dis(a, \mathcal{K}(\mathcal{H}))$ means the distance of an operator $a \in B(\mathcal{H})$ to the algebra $\mathcal{K}(\mathcal{H})$ of compact operators on the Hilbert space $\mathcal{H}$.

Proof 6.40 *We have that:*

$$\|\sum_{i=1}^{2}\sigma_{\varepsilon S^*}(F)\circ M_{\tilde{\varphi}_i}\circ M_{\delta,i}\|_{\text{op}}\le 2\|\sigma_{\varepsilon S^*}(F)\|_{\text{op}}.$$

Similarly, we have $\|\sum_{j=\pm 1}\sigma_{j\sqrt{2}R^*-2Q^*}(F)\circ M_{\delta,4,\pm}\|_{\text{op}}\le 2\|\sigma_{\epsilon R^*}(F)\|_{\text{op}}$, *since* $\pm\sqrt{2}R^*-2Q^*$ *is contained in the G-orbit of* $\pm R^*$. *Together with Proposition 6.8.9, the inequality holds.*

The kernel function $D_{\varepsilon,-\varepsilon,\delta,1}$ *of the operator* $\pi_{\varepsilon(S^*-Q^*)}(F)-\sigma_{\varepsilon S^*}(F)\circ M_{\tilde{\varphi}_1}\circ M_{\delta,1}$ *is given by*

$$\begin{aligned}
&D_{\varepsilon,-\varepsilon,\delta,1}((s,t,v)(a,b,p))\\
=\;& e^{\frac{1}{2}(a-b)}\Big(\widehat{F}^{\mathfrak{h}}(s-a,t-b,e^{\frac{a-b}{2}}(v-p),\varepsilon(-e^{-b}+\frac{e^{-b}p^2}{2})Q^*-(e^{-\frac{a+b}{2}}p)R^*+e^{-a}S^*)\\
&-\widehat{F}^{\mathfrak{h}}(s-a,t-b,e^{\frac{a-b}{2}}(v-p),\varepsilon(e^{-b}\frac{p^2}{2})Q^*-(e^{-\frac{a+b}{2}}p)R^*+e^{-a}S^*)\varphi(-e^{-b})\\
&1_{S_{\delta,1}}(a,b,p)\Big).
\end{aligned}$$

Hence

$$\begin{aligned}
&\|(\pi_{\varepsilon,-\varepsilon}(F)-\sigma_{\varepsilon S^*}(F)\circ M_{\tilde{\varphi}_1}\circ M_{\delta,1})\circ M_{\delta,1}\|^2_{\text{H-S}}\\
=\;&\int_{\mathbb{R}^5\times\{a;e^{-a}>\delta^6\}}|(\widehat{F}^{\mathfrak{h}}(s,t,v,\varepsilon(-e^{-b}+\frac{e^{-b}p^2}{2})Q^*-(e^{-\frac{a+b}{2}}p)R^*+e^{-a}S^*)\\
&-\widehat{F}^{\mathfrak{h}}(s,t,v,\varepsilon(e^{-b}\frac{p^2}{2})Q^*-(e^{-\frac{a+b}{2}}p)R^*+e^{-a}S^*)\varphi(-e^{-b})|^2e^{a-b}dsdtdvdadbdp\\
=\;&\int_{\mathbb{R}^5\times\{a;e^{-a}>\delta^6\}}|(\widehat{F}^{\mathfrak{h}}(s,t,v,\varepsilon(-e^{-b}+\frac{e^{a}p^2}{2})Q^*-pR^*+e^{-a}S^*)\\
&-\widehat{F}^{\mathfrak{h}}(s,t,v,\varepsilon(e^{a}\frac{p^2}{2})Q^*-pR^*+e^{-a}S^*)\varphi(-e^{-b})|^2e^{\frac{3a-b}{2}}dsdtdvdadbdp\\
\le\;&\int_{\mathbb{R}^5\times\{a;e^{-a}>\delta^6\}}2|\Big(\widehat{F}^{\mathfrak{h}}(s,t,v,\varepsilon(-e^{-b}+\frac{e^{a}p^2}{2})Q^*-pR^*+e^{-a}S^*)\Big)(1-\varphi(-e^{-b})|^2\\
&\times e^{\frac{3a-b}{2}}dsdtdvdadbdp\\
&+\int_{\mathbb{R}^5\times\{a;e^{-a}>\delta^6\}}2|\Big(\widehat{F}^{\mathfrak{h}}(s,t,v,\varepsilon(-e^{-b}+\frac{e^{a}p^2}{2})Q^*-pR^*+e^{-a}S^*)\\
&-\widehat{F}^{\mathfrak{h}}(s,t,v,\varepsilon(e^{a}\frac{p^2}{2})Q^*-pR^*+e^{-a}S^*)\Big)\varphi(-e^{-b})|^2e^{\frac{3a-b}{2}}dsdtdvdadbdp\\
\le\;&\int_{\mathbb{R}^4\times\{b;e^{-b}>\frac{1}{2}\}\times\{a;e^{-a}>\delta^6\}}|\alpha(s,t,v)\beta(-e^{-b}+\frac{e^{a}p^2}{2},p,e^{-a})|^2e^{\frac{3a-b}{2}}dsdtdvdadbdp\\
&+\int_{\mathbb{R}^4\times\{b;b>-C\}\times\{a;e^{-a}>\delta^6\}}e^{-2b}|\alpha(s,t,v)\beta(\frac{e^{a}p^2}{2},p,e^{-a})|^2e^{\frac{3a-b}{2}}dsdtdvdadbdp\\
<\;&\infty
\end{aligned}$$

for two continuous functions α, β on $\mathbb{R}^3$ with compact support and some $C > 0$. This shows that the operator $(\pi_{\varepsilon(S^*-Q^*)} - \sigma_{\varepsilon S^*}(F) \circ M_{\tilde{\varphi}_1} \circ M_{\delta,1}) \circ M_{\delta,1}$ is Hilbert-Schmidt, therefore, it is compact.

Let us estimate the operator norm of the linear endomorphism

$$D_{\varepsilon,\delta,3,l,+} := (\pi_{\varepsilon,-\varepsilon}(F) - N_{\delta,3,l} \circ \sigma_{\varepsilon(-1+\frac{\delta^4 l^2}{2})Q^*}) \circ M_{\delta,3,l,+}.$$

We have seen in relation (6.8.1) that

$$\pi_{\varepsilon,-\varepsilon}(F) \circ M_{\delta,3,l,+} = N_{\delta,3,l} \circ \pi_{\varepsilon,-\varepsilon}(F) \circ M_{\delta,3,l,+}.$$

The kernel function $F_{\varepsilon,\delta,3,l,+}$ of $D_{\varepsilon,\delta,3,l,+}$ is given by

$$\begin{aligned}
&F_{\varepsilon,\delta,3,l,+}(s,t,v,a,b,p)\\
&= \Bigg(\widehat{F}^{\mathfrak{h}}(s-a,t-b,e^{\frac{a-b}{2}}(v-p),\varepsilon(e^{-b}(-1+\frac{p^2}{2})Q^* - (e^{-\frac{a+b}{2}}p)R^* + e^{-a}S^*)\\
&\quad - \widehat{F}^{\mathfrak{h}}(s-a,t-b,e^{\frac{a-b}{2}}(v-p),\varepsilon(e^{-b}(-1+\frac{\delta^4 l^2}{2})Q^*)\Bigg)\\
&\quad e^{\frac{1}{2}(a-b)} 1_{S_{\delta,3,l,+}}(a,b,p) 1_{T_{\delta,3,l}}(s,t,v).
\end{aligned}$$

On the set $S_{\delta,3,l,+}$, we have that $\frac{1}{\delta^{3/4}} \geq e^{-b} \geq \delta$, $e^{-b/2}|1-\frac{p^2}{2}| > \delta^{1/2}$ and $|p-l\delta^2| < \delta^2$. Since $F \in L^1_c$, there exists $\rho > 0$ such that $\widehat{F}^{\mathfrak{h}}(s,t,v,qQ^* + rR^* + zS^*) = 0$ whenever $|q| > \rho$ or $|r| > \frac{1}{2}\rho$.

If now $e^{-b/2} > \frac{1}{\delta^{(3/4)}} \geq \frac{\rho}{\delta^{1/2}}$ (for δ small enough), then

$$\widehat{F}^{\mathfrak{h}}(s-a,t-b,e^{\frac{a-b}{2}}(v-p),\varepsilon(e^{-b}(-1+\frac{p^2}{2})Q^* - (e^{-\frac{a+b}{2}}p)R^* + e^{-a}S^*) = 0.$$

If p is large, then $e^{-b}(\frac{p^2}{2}-1) > \rho \Rightarrow F_{\varepsilon,\delta,3,l,+}(s,t,v,a,b,p) = 0$ tells us that we can assume that $e^{-b}|p| \leq e^{-b}(\frac{p^2}{2}-1) \leq \rho$ for every p giving a non-zero value to the function $F_{\delta,3,k,+}$. For $|p|$ small, we have then $e^{-b}(|p| + |l|\delta^2) \leq \frac{C}{\delta^{1/2}}$.

Since $e^{-a} \leq \delta^6$ in the set $S_{\delta,3,l,+}$, it follows that $e^{-\frac{a+b}{2}}|p| < \delta$ whenever $F_{\delta,3,l,+}(-,e^{-\frac{a+b}{2}}p,-) \neq 0$. According to Remark 6.8.6 we can now write

$$\begin{aligned}
&|\widehat{F}^{\mathfrak{h}}(s-a,t-b,e^{\frac{a-b}{2}}(v-p),\varepsilon(-e^{-b} + \frac{e^{-b}p^2}{2})Q^* - (e^{-\frac{a+b}{2}}p)R^* + e^{-a}S^*)\\
&-\widehat{F}^{\mathfrak{h}}(s-a,t-b,e^{\frac{a-b}{2}}(v-p),\varepsilon(-e^{-b} + e^{-b}\frac{\delta^4 l^2}{2})Q^*)|\\
\leq\ & (e^{-b}|\frac{\delta^4 l^2 - p^2}{2}| + |e^{\frac{-a-b}{2}}p| + e^{-a})\psi(s-a,t-b,e^{\frac{a-b}{2}}(v-p))
\end{aligned}$$

for some non-negative C^∞-function ψ with compact support contained in $[-M,M]^3$. Hence for any $(a,b,p) \in \mathbb{R}^3$:

$$|F_{\varepsilon,\delta,3,l,+}(s,t,v,a,b,p)| \leq M\delta\psi(s-a,t-b,e^{\frac{a-b}{2}}(v-p))e^{\frac{1}{2}(a-b)},$$

for all $a, b, p, s, t, v \in \mathbb{R}$ and some $M > 0$ independent of l, F, and δ small enough. According to the Young's inequality, this tells us that

$$\|(\pi_{\varepsilon,-\varepsilon}(F) - N_{\delta,3,l} \circ \sigma_{\varepsilon(-1+\frac{\delta^4 l^2}{2})Q^*}) \circ M_{\delta,3,l,+}\|_{\mathrm{op}} \leq M\delta\|\psi\|_1,$$

for $l \in \mathbb{Z}, F \in L^1_c$ and $\delta > 0$ small enough.

We have seen in (6.8.1) that $F_{\varepsilon,\delta,3,l,+}(s,t,v,a,b,p) = 0$ whenever $(s,t,v) \notin T_{\delta,3,l}$. Consequently for any sufficiently small $1 > \delta > 0$ we see that

$$\|(\pi_{\varepsilon,-\varepsilon}(F) - \sigma_{\delta,3}(F)) \circ (M_{\delta,3,+} + M_{\delta,3,-})\|_{\mathrm{op}} \leq C\delta$$

for some constant $C > 0$.

For $i = 2$, the kernel function $D_{\delta,2}$ of the operator $(\pi_{\varepsilon,-\varepsilon}(F) - \sigma_{\varepsilon S^*}(F)) \circ M_{\delta,2}$ is given by

$$\begin{aligned} &D_{\delta,2}(s,t,v,a,b,p) = e^{\frac{1}{2}(a-b)} \\ &\Big(\widehat{F}^{\natural}(s-a, t-b, e^{\frac{a-b}{2}}(v-p)), \varepsilon(-e^{-b} + \frac{e^{-b}p^2}{2})Q^* + (-e^{-\frac{a+b}{2}}p)R^* + e^{-a}S^*) \\ &- \widehat{F}^{\natural}(s-a, t-b, e^{\frac{a-b}{2}}(v-p)), \varepsilon e^{-b}\frac{p^2}{2}Q^* - e^{-\frac{a+b}{2}}pR^* + e^{-a}S^*)\Big) 1_{S_{\delta,2}}(a,b,p). \end{aligned}$$

Then we have a similar manipulation:

$$\begin{aligned} &|D_{\delta,2}(s,t,v,a,b,p)| \\ \leq\ & e^{\frac{1}{2}(a-b)} \Big| (\widehat{F}^{\natural}(s-a, t-b, e^{\frac{a-b}{2}}(v-p), \varepsilon(-e^{-b} + \frac{e^{-b}p^2}{2})Q^* + (-e^{-\frac{a+b}{2}}p)R^* + e^{-a}S^*) \\ &- \widehat{F}^{\natural}(s-a, t-b, e^{\frac{a-b}{2}}(v-p), (\varepsilon e^{-b}\frac{p^2}{2})Q^* + (-e^{-\frac{a+b}{2}}p)R^* + e^{-a}S^*)) \Big| 1_{S_{\delta,2}}(a,b,p) \\ \leq\ & e^{-b}e^{\frac{1}{2}(a-b)}|\psi(s-a, t-b, e^{\frac{a-b}{2}}(v-p))| 1_{S_{\delta,2}}(a,b,p) \\ \leq\ & \delta e^{\frac{1}{2}(a-b)}\psi(s-a, t-b, e^{\frac{a-b}{2}}(v-p)) \end{aligned}$$

for some $\psi \in C_c(\mathbb{R}^3)$. Again we have the estimate

$$\|(\pi_{\varepsilon,-\varepsilon}(F) - \sigma_{\varepsilon S^*}(F)) \circ M_{\delta,2}\|_{\mathrm{op}} \leq \delta \int_{\mathbb{R}^3} |\psi(a,b,p)| dadbdp. \tag{6.8.3}$$

Finally for $i = 4$, the kernel function $D_{\delta,4,k,\pm}$ of the operator

$$(\pi_{\varepsilon,-\varepsilon}(F) - \sigma_{\varepsilon(\pm\sqrt{2}R^* - 2Q^*)}(F)) \circ M_{\delta,4,\pm}$$

is given by

$$\begin{aligned} &D_{\delta,4,\pm}(s,t,v,a,b,p) \\ =\ & e^{\frac{1}{2}(a-b)}(\widehat{F}^{\natural}(s-a, t-b, e^{\frac{a-b}{2}}(v-p), \varepsilon(-e^{-b} + \frac{e^{-b}p^2}{2})Q^* - \varepsilon(e^{-\frac{a+b}{2}}p)R^* + \varepsilon e^{-a}S^*) \\ &-\widehat{F}^{\natural}(s-a, t-b, e^{\frac{a-b}{2}}(v-p), -\varepsilon(e^{-b}(\pm\sqrt{2}p + 2))Q^* + \varepsilon(e^{-\frac{a+b}{2}}(\pm\sqrt{2}))R^*) \\ &1_{S_{\delta,4,\pm}}(a,b,p). \end{aligned}$$

Then we have as in the preceding case on the set $S_{\varepsilon,\delta,4,\pm}$ that

$$e^{-b/2}|\frac{p}{\sqrt{2}} \pm 1| \leq e^{-b/2}|\frac{p^2}{2} - 1| \leq \delta^{1/2},$$

and

$$e^{-b}(\frac{p \pm \sqrt{2}}{2})^2 \leq \frac{1}{2}e^{-b}(\frac{p^2}{2} - 1)^2.$$

$$\begin{aligned}
&|D_{\delta,4,k,+}(s,t,v,a,b,p)| \\
\leq\ & e^{\frac{1}{2}(a-b)} \Big| \widehat{F}^{\mathfrak{h}}(s-a, t-b, e^{\frac{a-b}{2}}(v-p), \varepsilon(-e^{-b} + \frac{e^{-b}p^2}{2})Q^* + \varepsilon(-e^{-\frac{a+b}{2}}p)R^* \\
& +e^{-a}S^*) - \widehat{F}^{\mathfrak{h}}(s-a, t-b, e^{\frac{a-b}{2}}(v-p), -\varepsilon e^{-b}(2\pm\sqrt{2}p)Q^* + \varepsilon(e^{-\frac{a+b}{2}}(\pm\sqrt{2}))R^*)) \Big| \\
& 1_{S_{\delta,4,\pm}}(a,b,p) \\
\leq\ & (|e^{-b}(-1 + \frac{p^2}{2} \pm \sqrt{2}p + 2)| + e^{-\frac{a+b}{2}}|p\pm\sqrt{2}| + e^{-a}) \\
& e^{\frac{1}{2}(a-b)}\psi(s-a, t-b, e^{\frac{a-b}{2}}(v-p))1_{S_{\delta,4,\pm}}(a,b,p) \\
\leq\ & (e^{-b}\frac{(p\pm\sqrt{2})^2}{2} + e^{-\frac{a+b}{2}}|p\pm\sqrt{2}| + e^{-a})e^{\frac{1}{2}(a-b)}\psi(s-a, t-b, e^{\frac{a-b}{2}}(v-p))1_{S_{\delta,4}}(a,b,p) \\
\leq\ & (e^{-b}(\frac{p^2}{2} - 1)^2 + e^{-\frac{a+b}{2}}\sqrt{2}|\frac{p^2}{2} - 1| + e^{-a})e^{\frac{1}{2}(a-b)}\psi(s-a, t-b, e^{\frac{a-b}{2}}(v-p)) \\
& 1_{S_{\delta,4}}(a,b,p) \\
\leq\ & (\delta + \sqrt{2}\delta^{7/2} + \delta^6)e^{\frac{1}{2}(a-b)}\psi(s-a, t-b, e^{\frac{a-b}{2}}(v-p))1_{S_{\delta,4,\pm}}(a,b,p).
\end{aligned}$$

Hence as before:

$$\|(\pi_{\varepsilon,-\varepsilon}(F) - \sigma_{-\varepsilon(\pm\sqrt{2}R^*+2Q^*)}(F)) \circ M_{\delta,4,\pm}\|_{\text{op}} \leq \delta \int_{\mathbb{R}^3} \psi(a,b,p)dadbdp. \quad (6.8.4)$$

We have seen above that $(\pi_{\varepsilon,-\varepsilon}(F) - \sigma_{\varepsilon S^*}(F) \circ M_{\tilde{\varphi}_1}) \circ M_{\delta,1}$ is compact for any $\delta > 0$. Hence it follows from the estimates above that

$$\pi_{\varepsilon,-\varepsilon}(F)(1-M_{\delta,1}) - \left(\sigma_{\varepsilon S^*}(F) \circ M_{\delta,2} + \sum_{j=\pm} \sigma_{\varepsilon(j\sqrt{2}R^*-2Q^*)}(F) \circ M_{\delta,4,j} + \sum_{j=\pm} \sigma_{\delta,3,j}(F) \right)$$

converges to 0 as δ tends to 0. Therefore,

$$\lim_{\delta\to 0} dis(\pi_{\varepsilon(S^*-Q^*)}(F) - \sigma_{\varepsilon,-\varepsilon,\delta}(F), \mathcal{K}(L^2(\mathbb{R}^3))) = 0.$$

Since L^1_c is dense in $C^*(G)$, the relation

$$\lim_{\delta\to 0} dis(\pi_{\varepsilon(S^*-Q^*)}(F) - \sigma_{\varepsilon,-\varepsilon,\delta}(F), \mathcal{K}(L^2(\mathbb{R}^3))) = 0$$

remains true for any $F \in C^*(G)$.

6.8.3.2 The orbits $\Omega_{\varepsilon,\varepsilon} = \Omega_{\varepsilon Q^*+\varepsilon S^*}, \varepsilon = \pm 1$

The boundary orbits of $\Omega_{\varepsilon,\varepsilon|\mathfrak{h}} \subset \mathfrak{h}^*$ are the orbits of εS^* and of εQ^* inside $\mathfrak{h}^*$. Therefore we must decompose $\mathbb{R}^3 \simeq G/H$ into three disjoint subsets. Since this case is similar but much easier to the preceding case, we shall omit the proofs.

Definition 6.37 *1. For $k \in \mathbb{Z}$, denote by*

$$J_{\delta,k} := \left[\log(\frac{1}{\delta^6}), +\infty\right) \times \left(-\infty, \log(\frac{1}{\delta})\right] \times I_{\delta^2,k}$$

where $I_{c,l} := [cl, cl + c] \subset \mathbb{R}$ for $l \in \mathbb{Z}, c > 0$ and let

$$L_{\delta,k} := \mathbb{R} \times \mathbb{R} \times I_{\delta^2,k}.$$

2. Let

$$K_\delta = [-\frac{1}{\delta}, \frac{1}{\delta}] \times [-\frac{1}{\delta}, \frac{1}{\delta}] \times [-\frac{1}{\delta^{1/2}}, \frac{1}{\delta^{1/2}}].$$

3. Let $S_{\delta,1} := \{(a, b, p); e^{-a} > \delta^6\}$(corresponds to $\mathbb{R}_{\varepsilon S^*}$).

4. Let $S_{\delta,2} := \{(a, b, p); e^{-a} \leq \delta^6, e^{-b} < \delta\}$(corresponds to $\mathbb{R}_{\varepsilon S^*}$).

5. Let

$$S_{\delta,3,k} := J_{\delta,k}, \ k \in \mathbb{Z}.$$

(corresponds to $\mathbb{R}_{\varepsilon Q^*}$) and $S_{\delta,3} := \bigcup_k S_{\delta,3,k}$.

6. Define the corresponding multiplication operators on $L^2(\mathbb{R}^3)$ by $M_{\delta,i} = M_{S_{\delta,i}}$ for $i = 1, 2$, respectively, and for $k \in \mathbb{Z} : M_{\delta,3,k} = M_{S_{\delta,3,k}}, M_{\delta,3} = M_{S_{\delta,3}}$.

Remark 6.8.11

1. For any $\delta > 0$, the sets $\{S_{\delta,1}, S_{\delta,2}, S_{\delta,3,k}; k \in \mathbb{Z}\}$ form a partition of $\mathbb{R}^3$.
2. Recall that for $\alpha = (a, b, p)$ and $\beta = (s, t, v) \in \mathbb{R}^3$, the multiplication of $E(s, t, v) \cdot E(a, b, p)$ is given by

$$E(s, t, v) \cdot E(a, b, p) = E(s + a, t + b, e^{\frac{-a+b}{2}} v + p).$$

For $\beta \in K_\delta, \alpha \in S_{\delta,3,k}$, we have that $e^{-a} \leq \delta^6, e^b < \frac{1}{\delta}, |v| < \frac{1}{\delta^{1/2}}$ and so $e^{\frac{-a+b}{2}} |v| \leq \delta^2$ and

$$\begin{aligned} \beta \cdot \alpha &= (s + a, t + b, e^{\frac{-a+b}{2}} v + p) \\ &\in \mathbb{R} \times \mathbb{R} \times ([-\delta^2, \delta^2] + I_{\delta,k}) \\ &\subset \bigcup_{i=-1}^{1} L_{\delta,k+i} =: T_{\delta,3,k}. \end{aligned} \tag{6.8.5}$$

Let $N_{\delta,3,k}$ be the multiplication operator with the characteristic function of the set $T_{\delta,3,k}$ for $k \in \mathbb{Z}$. We see that the set $T_{\delta,3,k}$ is contained in the union of 3 disjoint boxes $L_{\delta,k+i}$ for $i \in \{-1,0,1\}$. Hence it follows from the relations (6.8.5) that for $\delta > 0$ small enough,

$$\widehat{F}^{\mathfrak{h}}(st^{-1}, t\cdot(\varepsilon(S^*+Q^*))) = 0 \text{ for } t \in S_{\delta,3,k}, s \notin T_{\delta,3,k} \text{ and } k \in \mathbb{Z}.$$

Hence,

$$N_{\delta,3,k} \circ \pi_{\varepsilon,\varepsilon}(F) \circ M_{\delta,3,k} = \pi_{\varepsilon,\varepsilon}(F) \circ M_{\delta,3,k} \text{ for } k \in \mathbb{Z}. \tag{6.8.6}$$

The representation $\pi_{\varepsilon,\varepsilon} = \mathrm{ind}_H^G \chi_{\varepsilon(Q^*+S^*)}$ acts on $\xi \in L^2(G/H, \chi_{\varepsilon(Q^*+S^*)})$ in the following way:

$$\begin{aligned} &\pi_{\varepsilon,\varepsilon}(F)\xi(u) \\ = &\int_{\mathbb{R}^3} e^{\frac{1}{2}(a-b)} \widehat{F}^{\mathfrak{h}}(s-a, t-b, e^{\frac{a-b}{2}}(v-p), \varepsilon(e^{-b}(1+\frac{p^2}{2})Q^* + (-e^{-\frac{a+b}{2}}p)R^* \\ &+e^{-a}S^*))\xi(E(a,b,p))dadbdp. \end{aligned}$$

We define the linear operator $\sigma_{\varepsilon,\varepsilon,\delta}(F)$ by the rule

$$\sigma_{\varepsilon,\varepsilon,\delta}(F) := \sigma_{\varepsilon S^*}(F) \circ M_{\tilde{\varphi}_1} \circ M_{\delta,1} + \sigma_{\varepsilon S^*}(F) \circ M_{\delta,2} + \sigma_{\delta,3}(F), \tag{6.8.7}$$

where $\sigma_{\delta,3}(F) := \sum_{k\in\mathbb{N}} N_{\delta,k,3} \circ \sigma_{\varepsilon(1+\frac{\delta^4 k^2}{2})Q^*}(F) \circ M_{\delta,3,k}$, the operator $M_{\tilde{\varphi}_1}$ is the multiplication operator with the function $\tilde{\varphi}_1(a,b,p) := \varphi(e^{-b})$ where $\varphi : \mathbb{R} \to \mathbb{R}$ is a C^∞-function with support contained in $[-1,1]$ such that $\varphi(0) = 1$ and $\|\varphi\|_\infty = 1$. The kernel functions $F_{\delta,i}$, for $i = 1,2,3$, of these operators are given by

$$\begin{aligned} F_{\delta,1}((s,t,v),(a,b,p)) &:= \widehat{F}^{\mathfrak{h}}(s-a, t-b, e^{\frac{a-b}{2}}(v-p), \varepsilon((\frac{e^{-b}p^2}{2})Q^* + (-e^{-\frac{u+b}{2}}p) \\ &\quad R^* + e^{-a}S^*))e^{\frac{1}{2}(a-b)}\varphi(e^{-b})1_{S_{\delta,1}}(a,b,p), \\ F_{\delta,3}((s,t,v),(a,b,p)) &= \sum_{k\in\mathbb{N}} \Big(\widehat{F}^{\mathfrak{h}}(s-a, t-b, e^{\frac{a-b}{2}}(v-p), \varepsilon e^{-b}(1+\frac{\delta^4 k^2}{2})Q^*)\Big) \\ &\quad e^{\frac{1}{2}(a-b)}1_{S_{\delta,k,3}}(a,b,p), \\ F_{\delta,2}((s,t,v),(a,b,p)) &:= \widehat{F}^{\mathfrak{h}}(s-a, t-b, e^{\frac{a-b}{2}}(v-p), \varepsilon(\frac{e^{-b}p^2}{2})Q^* + (-e^{-\frac{a+b}{2}}p) \\ &\quad R^* + e^{-a}S^*))e^{\frac{1}{2}(a-b)}1_{S_{\delta,2}}(a,b,p). \end{aligned}$$

Now the representation $\sigma_{\varepsilon(1+\frac{k^2\delta^4}{2})Q^*}$ is equivalent to the representation $\sigma_{\varepsilon Q^*}$, since both linear functionals are in the same G-orbit. Therefore, for any $k \geq 0$ and $\delta > 0$,

$$\|\sigma_{\varepsilon(1+\frac{\delta^4 k^2}{2})Q^*}(F) \circ M_{\delta,3,k}\|_{\mathrm{op}} \leq \|\sigma_{\varepsilon Q^*}(F)\|_{\mathrm{op}}. \tag{6.8.8}$$

It follows from Proposition 6.8.9 that there exists $C > 0$ such that the operator $\sigma_{\delta,3}(F)$ is bounded by C for any $F \in L^1_c$.

The proof of the following proposition is similar but much easier than that of Proposition 6.8.10 and will be left to the reader.

Proposition 6.8.12 *For any $F \in L^1_c$ and $\delta > 0$, we have that*

$$\|\sigma_{\varepsilon,\varepsilon,\delta}(F)\|_{\text{op}} \leq 2\|\sigma_{\varepsilon S^*}(F)\|_{\text{op}} + C\|\sigma_{\varepsilon Q^*}(F)\|_{\text{op}}$$

and we can extend the mapping $F \mapsto \sigma_{\varepsilon,\varepsilon,\delta}(F)$ to every $F \in C^*(G)$, then

$$\lim_{\delta\to 0} dis(\pi_{\varepsilon(Q^*+S^*)}(F) - \sigma_{\varepsilon,\varepsilon,\delta}(F), \mathcal{K}(L^2(\mathbb{R}^3))) = 0.$$

6.8.3.3 *The boundary condition for* $\mathbb{R}_{\varepsilon S^*}$, $\varepsilon = \pm 1$

In this subsection, we consider the subset $\Gamma_5 = \{\mathbb{R}_{\varepsilon S^*}, \varepsilon = \pm 1\}$ of $\widehat{G}$. We use the coordinates of G according to the basis $\{X = A + B, B, P, Q, R, S\}$ of $\mathfrak{g}$. We recall that the boundary of the orbit of $b^*B^* + \varepsilon S^*$ is the orbit of the functionals $aA^* + \varepsilon Q^*, x^*X^* \pm P^*$ and $a^*A^* + b^*B^*$ for $a^*, b^*, x^* \in \mathbb{R}$. This tells us that we much find the conditions from these boundary points.

Definition 6.38 *Let*

$$\sigma_{\varepsilon S^*} := \text{ind}_H^G \chi_{\varepsilon S^*}.$$

The kernel function $\mathcal{K}_F$ of the operator $\sigma_{\varepsilon S^*}(F)$ is given by

$$\mathcal{K}_F(s,t,v;x,b,p) = \widehat{F}^{\mathfrak{h}}(s-x, t-b, e^{\frac{-b}{2}}(v-p), \varepsilon(\frac{e^{-x-b}p^2}{2})Q^*$$

$$-\varepsilon(e^{-\frac{2x+b}{2}}p)R^* + \varepsilon e^{-x}S^*)e^{-b/2}.$$

This representation is equivalent to the direct integral representation

$$\tau_{\varepsilon S^*} := \int_{\mathbb{R}}^{\oplus} \pi_{b^*B^*+\varepsilon S^*} db^*$$

acting on the Hilbert space

$$\mathcal{H}_{\tau_{\varepsilon S^*}} = \int_{\mathbb{R}}^{\oplus} L^2(G/\exp(\mathbb{R}B + \mathfrak{h}), \chi_{b^*B^*+\varepsilon S^*})db^* \simeq \int_{\mathbb{R}}^{\oplus} L^2(\mathbb{R}^2)db^*$$

with the norm

$$\|\xi\|_2^2 = \int_{\mathbb{R}} \|\xi(b^*)\|_2^2 db^* \quad \text{for} \quad \xi \in \mathcal{H}_{\tau_{\varepsilon S^*}}.$$

An intertwining operator $U_{\varepsilon S^*}$ for this equivalence is given by

$$U_{\varepsilon S^*}(\xi)(b^*)(g) := \int_{\mathbb{R}} \xi(g\exp(bB))e^{\frac{1}{4}b}e^{-2\pi i b^* b}db, \ \xi \in L^2(G/H, \chi_{\varepsilon S^*}), g \in G, b^* \in \mathbb{R}.$$

Let $C(\mathbb{R}, \mathcal{B})$ be the C^*-algebra of all continuous uniformly bounded mappings from $\mathbb{R}$ into the algebra $\mathcal{B}$ of bounded operators on the Hilbert space $L^2(\mathbb{R})$, containing the ideal $C_0(\mathbb{R}, \mathcal{K})$ of all continuous mappings from $\mathbb{R}$ into the algebra of compact operators on $L^2(\mathbb{R})$ vanishing at infinity. It follows from Section 6.8.1 that the image of $\tau_{\varepsilon S^*}$ is contained in $C(\mathbb{R}, \mathcal{B})$. On the other hand, we can consider $C(\mathbb{R}, \mathcal{B})$ as a subalgebra of $B(\mathcal{H}_{\tau_{\varepsilon S^*}})$ as for $\phi \in C(\mathbb{R}, \mathcal{B})$, let

$$\phi(\xi)(b^*) := \phi(b^*)(\xi(b^*)) \quad \text{for } \xi \in \mathcal{H}_{\tau_{\varepsilon S^*}}, b^* \in \mathbb{R}.$$

The unitary mapping $U_{\varepsilon S^*}$ induces a canonical homomorphism $\rho_{\varepsilon S^*}$ from the algebra $B(L^2(G/H, \chi_{\varepsilon S^*}))$ onto $B(\mathcal{H}_{\tau_\varepsilon S^*})$. This homomorphism is defined on $\sigma_{\varepsilon S^*}(a)$ by

$$\begin{aligned} \rho_{\varepsilon S^*}(\sigma_{\varepsilon S^*}(a)) &= U_{\varepsilon S^*} \circ \sigma_{\varepsilon S^*}(a) \circ U^*_{\varepsilon S^*} \\ &= \int_{\mathbb{R}}^{\oplus} \pi_{b^* B^* + \varepsilon S^*}(a) db^* \\ &= \tau_{\varepsilon S^*}(a) \quad \text{for} \quad a \in C^*(G). \end{aligned} \tag{6.8.9}$$

Definition 6.39 Let $\partial \mathbb{R}_{\varepsilon S^*}$ be the boundary of $\mathbb{R}_{\varepsilon S^*}$ in $\mathfrak{g}^*/G$.

It is easy to see from the relations (1.4.1) that $\partial \mathbb{R}_{\varepsilon S^*} = \{Q, R, S\}^\perp + (\mathbb{R}^{+,0}\varepsilon)Q^* \subset S^\perp$.

Definition 6.40 *1. For $k \in \mathbb{Z}$, denote by*

$$J_{\delta,k} := \left[\log(\frac{1}{\delta^6}), +\infty\right) \times \mathbb{R} \times I_{\delta^2,k},$$

where $I_{c,l} := [cl, cl + c) \subset \mathbb{R}$ for $l \in \mathbb{Z}$ and $c > 0$.

2. Let $S_{\delta,1} := \{(x, b, p); e^{-x} > \delta^6\}$.

3. Let $\delta \mapsto r_\delta \in \mathbb{R}^+$ be such that $\lim_{\delta \to 0} r_\delta = +\infty$ and $\lim_{\delta \to 0} e^{r_\delta} \delta^{1/2} = 0$.

4. For a constant $D > 0$ and $k = (k_1, k_2, k_3) \in \mathbb{Z}^3$, let

$$S_{\delta,D,3,k} := \{(x, b, p) \in \mathbb{R}^3 : e^{-x} \le \delta^6, x \in I_{r_\delta,k_1}, b \in I_{r_\delta,k_2}, p \in I_{D\delta^2 e^{r_\delta/2(k_1+k_2)},k_3}\}.$$

Proposition 6.8.13 *For every compact subset $K \subseteq \mathbb{R}^3$, we have that*

$$K S_{\delta,D,3,k} \subset \bigcup_{\substack{j \in \mathbb{Z}^3 \\ |j_i| \le 1, i=1,2,3}} S_{\delta, De^{\frac{r_\delta}{2}(-j_1-j_2)},3,k+j} =: R_{\delta,D,3,k}$$

for every $k \in \mathbb{Z}^3$ and $\delta > 0$ small enough.

Proof 6.41 *Indeed, we have an $M > 0$ such that $K \subset [-M, M]^3$ and then for $r_\delta > M$ and $(s, t, v) \in K, (x, b, p) \in S_{\delta,D,3,k}$, it follows that*

$$u := (s, t, v) \cdot (x, b, p) = (s + x, t + b, e^{b/2} v + p),$$

$(k_1+j_1)r_\delta \le s+x < (k_1+j_1+1)r_\delta$ and $(k_2+j_2)r_\delta \le t+b < (k_2+j_2+1)r_\delta$ for some $j_1, j_2 \in \{-1,0,1\}$. It follows that

$$\begin{aligned} |e^{b/2}v| &\le |e^{-x/2}v|e^{(x+b)/2} \\ &\le \delta^3|v|e^{(x+b)/2} \\ &\le \delta^3|v|e^{r_\delta/2(-j_1-j_2)}e^{r_\delta/2(k_1+k_2+j_1+j_2)}e^{r_\delta} \\ &< De^{r_\delta/2(-j_1-j_2)}\delta^2 e^{r_\delta/2(k_1+k_2+j_1+j_2)} \end{aligned}$$

for δ small enough, since $\lim_{\delta\to 0} e^{r_\delta}\delta^{1/2} = 0$. Hence,

$$\begin{aligned} p+e^b v &< (k_3+1)De^{\frac{r_\delta}{2}(-j_1-j_2)}\delta^2 e^{\frac{r_\delta}{2}(k_1+k_2+j_1+j_2)} + e^b v \\ &< (k_3+2)De^{\frac{r_\delta}{2}(-j_1-j_2)}\delta^2 e^{\frac{r_\delta}{2}(k_1+k_2+j_1+j_2)} \end{aligned}$$

and also

$$\begin{aligned} p+e^b v &\ge k_3De^{\frac{r_\delta}{2}(-j_1-j_2)}\delta^2 e^{\frac{r_\delta}{2}(k_1+k_2+j_1+j_2)} - e^b|v| \\ &\ge (k_3-1)De^{\frac{r_\delta}{2}(-j_1-j_2)}\delta^2 e^{\frac{r_\delta}{2}(k_1+k_2+j_1+j_2)}. \end{aligned}$$

Hence u is contained in $R_{\delta,D,3,k}$.

We define the multiplication operators $M_{\delta,1} = M_{S_{\delta,1}}$, respectively, $M_{\delta,D,3,k} = M_{S_{\delta,D,3,k}}$ and $P_{\delta,D,3,k} := M_{R_{\delta,D,3,k}}$ on $L^2(\mathbb{R}^3)$. Then we have the following proposition.

Proposition 6.8.14

$$\overline{M}_{\delta,1} \circ U_{\varepsilon S^*} = U_{\varepsilon S^*} \circ M_{\delta,1} \quad \text{for} \quad \delta > 0,$$

where $\overline{M}_{\delta,1} = M_{\{(x,p); e^{-x} > \delta^6\}}$ on $L^2(\mathbb{R}^2)$.

Proof 6.42 *Indeed, for $\xi \in L^2(G/H, \chi_{S^*}), b \in \mathbb{R}$ and $g \in G$, we have that*

$$\begin{aligned} U_{S^*}(M_{\delta,1}(\xi)(E(x,p)) &= \int_{\mathbb{R}} M_{\delta,1}(\xi)(E(x,p)exp(bB))e^{\frac{1}{4}b}e^{-2\pi i b^* b}db \\ &= \int_{\mathbb{R}} 1_{\{e^{-x}>\delta^6\}}\xi(E(x,p)exp(bB))e^{\frac{1}{4}b}e^{-2\pi i b^* b}db \\ &= 1_{\{e^{-x}>\delta^6\}}(E(x,p)) \int_{\mathbb{R}} \xi(E(x,p)exp(bB))e^{\frac{1}{4}b}e^{-2\pi i b^* b}db \\ &= \overline{M}_{\delta,1}(U_{S^*}\xi)(E(x,p)), \end{aligned}$$

here $E(x,p) = exp(xX)exp(pP)$ for $x,p \in \mathbb{R}$.

Definition 6.41 For $D > 0$, we define the linear operator $\sigma_{\varepsilon,\delta,D}(F)$ on $L^2(G/H, \chi_{\varepsilon S^*}) = L^2(\mathbb{R}^3)$ by the rule

$$\sigma_{\varepsilon,\delta,D}(F) := \sum_{k\in\mathbb{Z}^3} P_{\delta,D,3,k} \circ \sigma_{(k_3^2\delta^3 D^2 e^{r_\delta(k_1+k_2)})\varepsilon Q^*}(F) \circ M_{\delta,D,3,k}.$$

Now the representation $\sigma_{(k_3^2\delta^3 D^2 e^{r_\delta(k_1+k_2)})\varepsilon Q^*}$ is equivalent to the representation $\sigma_{\varepsilon Q^*}$, since both linear functionals are in the same G-orbit. Therefore for any k, δ,

$$\|\sigma_{(k_3^2\delta^3 D^2 e^{r_\delta(k_1+k_2)})\varepsilon Q^*}(F) \circ M_{\delta,D,3,k})\|_{\mathrm{op}} \leq \|\sigma_{\varepsilon Q^*}(F)\|_{\mathrm{op}}. \qquad (6.8.10)$$

Proposition 6.8.15 *For every $F \in C^*(G)$ and $\delta > 0$, the operator $\sigma_{\varepsilon,\delta,D}(F)$ is bounded in the operator norm by $3^3\|\sigma_{\varepsilon Q^*}(F)\|_{\mathrm{op}}$.*

Proof 6.43 *For $k \in \mathbb{Z}^3$, decompose the set $R_{\delta,D,3,k}$ into a disjoint union of measurable subsets $R_{k,j}, j = (j_1, j_2, j_3) \in \mathbb{Z}^3$ with $|j_i| \leq 1, i = 1,2,3$, such that $R_{k,j} \subset S_{\delta,De^{\frac{r_\delta}{2}(-j_1-j_2)},3,k+j}$ for every j. This gives us at most 3^3 such subsets $R_{k,j}$ for fixed k. These sets $R_{k,j}$ are disjoint in k for fixed j, since the sets $S_{\delta,De^{\frac{r_\delta}{2}(-j_1-j_2)},3,k+j}$ are mutually disjoint in k for fixed j. It suffices then to apply Proposition 6.8.7.*

Proposition 6.8.16 *Let $a \in C^*(G)$. Then the element*

$$\tau_{\varepsilon S^*}(a) \circ \overline{M}_{\delta,1} := \int_{\mathbb{R}}^{\oplus} \pi_{b^*B^*+\varepsilon S^*}(a) \circ \overline{M}_{\delta,1} db^*$$

is in $C_0(\mathbb{R}, \mathcal{K})$ for any $\delta > 0$.

Proof 6.44 *We must prove first that the operators $\pi_{b^*B^*+\varepsilon S^*}(a) \circ \overline{M}_{\delta,1}$ for all $b^* \in \mathbb{R}$ and $\delta > 0$ are all compact. The kernel function F_{b^*} of the operator $\pi_{b^*B^*+\varepsilon S^*}(F) \circ \overline{M}_{\delta,1}$ is given by Formula (1.4.15):*

$$\begin{aligned} F_{b^*}((t,u),(x,p)) &= 1_{\{e^{-x}>\delta^6\}} \int_{\mathbb{R}} \widehat{F^{\mathfrak{b}}}(E(t-x, u-e^{-\frac{b}{2}}p)exp(bB)) \\ &\quad \left(\varepsilon e^{-x}(\frac{1}{2}p^2Q^* - pR^* + S^*)\right) e^{ib^*b} e^{\frac{b}{4}} db. \end{aligned}$$

Hence for $F \in L^1_c$, using Remark 6.8.6:

$$\begin{aligned} \|\pi_{b^*B^*+\varepsilon S^*}(F) \circ \overline{M}_{\delta,1}\|^2_{H-S} &= \int_{\mathbb{R}^4} 1_{\{e^{-x}>\delta^6\}} |F_{b^*}(t,u,x,p)|^2 dxdtdudp \\ &= \int_{\mathbb{R}^4} 1_{\{e^{-x}>\delta^6\}} \Big| \int_{\mathbb{R}} \widehat{F^{\mathfrak{b}}}(E(t-x, u-e^{-\frac{b}{2}}p)exp(bB)) \\ &\quad \left(\varepsilon e^{-x}(\frac{1}{2}p^2Q^* - pR^* + S^*)\right) e^{ib^*b} e^{\frac{b}{4}} db\Big|^2 dxdtdudp \\ &\leq \Big(\int_{\mathbb{R}} \Big(\int_{\mathbb{R}^4} 1_{\{e^{-x}>\delta^6\}} |\widehat{F^{\mathfrak{b}}}(E(t-x, u-e^{-\frac{b}{2}}p)exp(bB)) \\ &\quad \left(\varepsilon e^{-x}(\frac{1}{2}p^2Q^* - pR^* + S^*)\right)|^2 dxdtdudp\Big)^{\frac{1}{2}} e^{\frac{b}{4}} db\Big)^2 \\ &= \Big(\int_{\mathbb{R}} \Big(\int_{\mathbb{R}^4} 1_{\{e^{-x}>\delta^6\}} |\widehat{F^{\mathfrak{b}}}(E(t,u)exp(bB)) \\ &\quad \left(\varepsilon(\frac{1}{2}e^x p^2Q^* - pR^* + e^{-x}S^*)\right)|^2 e^x dxdtdudp\Big)^{\frac{1}{2}} e^{\frac{b}{4}} db\Big)^2 \\ &< \infty, \end{aligned}$$

since the function

$$(t,u,x,p,b) \mapsto 1_{\{e^{-x}>\delta^6\}} \widehat{F^{\mathfrak{h}}}(E(t,u)exp(bB)) \left(\varepsilon(\frac{1}{2}e^x p^2 Q^* - pR^* + e^{-x}S^*)\right) e^x) e^{\frac{b}{4}}$$

has compact support.

Now we must show that $\lim_{b^*\to\infty} \|\pi_{b^*B^*+\varepsilon S^*}(F)\circ \overline{M}_{\delta,1}\|_{\mathrm{op}} = 0$. *The kernel function* F_{b^*} *of the operator* $\pi_{b^*B^*+\varepsilon S^*}(F)\circ\overline{M_{\delta,1}}$ *can be written, using partial integration, as*

$$\begin{aligned}
F_{b^*}((t,u),(x,p)) &= 1_{\{e^{-x}>\delta^6\}} \int_{\mathbb{R}} \widehat{F^{\mathfrak{h}}}(E(t-x,u-e^{-\frac{b}{2}}p)exp(bB)) \\
&\quad \left(\varepsilon e^{-x}(\frac{1}{2}p^2 Q^* - pR^* + S^*)\right) e^{ib^*b+\frac{b}{4}} db \\
&= 1_{\{e^{-x}>\delta^6\}} \int_{\mathbb{R}} \frac{1}{ib^*+\frac{1}{4}} (e^{-b/2}p\partial_2 + \partial_B)\widehat{F^{\mathfrak{h}}}(E(t-x,u-e^{-\frac{b}{2}}p) \\
&\quad exp(bB)) \left(\varepsilon e^{-x}(\frac{1}{2}p^2 Q^* - pR^* + S^*)\right) e^{ib^*b} e^{\frac{b}{4}} db \\
&= 1_{\{e^{-x}>\delta^6\}} \int_{\mathbb{R}} \frac{1}{ib^*+\frac{1}{4}} (e^{-b/2}p\partial_2 + \partial_B)\widehat{F^{\mathfrak{h}}}(E(t-x,u-e^{-\frac{b}{2}}p) \\
&\quad exp(bB)) \left(\varepsilon(\frac{1}{2}e^{-x}p^2 Q^* - e^{-x}pR^* + e^{-x}S^*)\right) e^{ib^*b} e^{\frac{b}{4}} db.
\end{aligned}$$

Therefore,

$$\begin{aligned}
&\sup_{x,p} \int_{\mathbb{R}^2} |F_{b^*}((t,u),(x,p))| dtdu \\
\le\; &\sup_{x,p} 1_{\{e^{-x}>\delta^6\}} \int_{\mathbb{R}} \frac{1}{|ib^*+\frac{1}{4}|} |(e^{-b/2}e^x(e^{-x}p)\partial_2 + \partial_B)\widehat{F^{\mathfrak{h}}}(E(t-x,u-e^{-\frac{b}{2}}p)exp(bB)) \\
&\left(\varepsilon(\frac{1}{2}e^{-x}p^2 Q^* - e^{-x}pR^* + e^{-x}S^*)\right) |e^{\frac{b}{4}} dbdtdu \\
<\; &\frac{C}{|b^*|},
\end{aligned}$$

for $|b^*| \ge 1$ *and for some constant* $C > 0$. *Furthermore,*

$$\begin{aligned}
&\sup_{t,u} \int_{\mathbb{R}^2} |F_{b^*}((u,v),(x,p))| dxdp \\
\le\; &\int_{\mathbb{R}} |\frac{1}{ib^*+\frac{1}{4}}| |1_{\{e^{-x}>\delta^6\}}(x)| (e^{-b/2}e^x(e^{-x}p)\partial_2 + \partial_B)\widehat{F^{\mathfrak{h}}}(E(t-x,u-e^{-\frac{b}{2}}p)exp(bB)) \\
&\left(\varepsilon(\frac{1}{2}e^{-x}p^2 Q^* - e^{-x}pR^* + e^{-x}S^*)\right) |e^{\frac{b}{4}} dbdxdp \\
<\; &\frac{C}{|b^*|},
\end{aligned}$$

again because the function

$$\begin{aligned}
&(t,u,x,p,b) \mapsto 1_{\{e^{-x}>\delta^6\}} \widehat{F^{\mathfrak{h}}}(E(t-x,u-e^{-\frac{b}{2}}p)exp(bB))exp(bB) \\
&\left(\varepsilon(\frac{1}{2}p^2 Q^* - pR^* + S^*)e^{-x}\right) e^{\frac{b}{4}}
\end{aligned}$$

has compact support and is C^∞ in the variables t, u, p, b. Hence, by the Young's inequality, for $F \in L^1_c$, there exists a constant $C > 0$ such that $\|\sigma_{b^*B^*+\varepsilon S^*}(F)\|_{\text{op}} \leq \frac{C}{|b^*|}$ for large enough $b^* \in \mathbb{R}^*$.

Proposition 6.8.1 *For any $F \in L^1_c$, we have that*

$$\lim_{\delta \to 0} \|\sigma_{\varepsilon S^*}(F)(1 - M_{\delta,1}) - \sigma_{\varepsilon,\delta,D}(F)\|_{\text{op}} = 0.$$

Proof 6.45 *Let $F \in L^1_c$. Then there exists $\rho > 0$ such that*

$$\hat{F}^{\natural}(s-x, t-b, e^{\frac{-b}{2}}(v-p), (\frac{e^{-x-b}p^2}{2})\varepsilon Q^* + (e^{-\frac{2x+b}{2}}p)\varepsilon R^* + e^{-x}\varepsilon S^*) = 0$$

if $e^{-x-b}p^2 > \rho$. For δ small enough, from Proposition 6.8.13 we have that $P_{\delta,D,3,k} \circ \sigma_{\varepsilon S^*}(F) \circ M_{\delta,D,3,k} = \sigma_{\varepsilon S^*}(F) \circ M_{\delta,D,3,k}$, for $k \in \mathbb{Z}$. Therefore, we must show that

$$\|P_{\delta,D,3,k} \circ \sigma_{\varepsilon S^*}(F) \circ M_{\delta,D,3,k} - P_{\delta,D,3,k} \circ \sigma_{k_3^2\delta^4D^2e^{r_\delta(k_1+k_2)}\varepsilon Q^*}(F) \circ M_{\delta,D,3,k}\|_{\text{op}} \leq C\delta,$$

for $k \in \mathbb{Z}$, and for some constant $C > 0$ independent of δ and k.

The kernel function $F_{\delta,D,3,k}$ of the operator

$$P_{\delta,D,3,k} \circ \sigma_{\varepsilon S^*}(F) \circ M_{\delta,D,3,k} - P_{\delta,D,3,k} \circ \sigma_{(\frac{k_3^2\delta^4D^2e^{r_\delta(k_1+k_2)}}{2})\varepsilon Q^*}(F) \circ M_{\delta,D,3,k}$$

is given by

$$\begin{aligned} & F_{\delta,D,3,k}((s,t,v),(x,b,p)) \\ = & \Bigg(\widehat{F}^{\natural}(s-x, t-b, e^{\frac{-b}{2}}(v-p), (\frac{e^{-x-b}p^2}{2})\varepsilon Q^* + (e^{-\frac{2x+b}{2}}p)\varepsilon R^* + e^{-x}\varepsilon S^*) \\ & - \widehat{F}^{\natural}(s-x, t-b, e^{\frac{-b}{2}}(v-p), e^{-x-b}\frac{k_3^2\delta^4D^2e^{r_\delta(k_1+k_2)}}{2}\varepsilon Q^*)\Bigg) \\ & e^{-\frac{b}{2}}1_{S_{\delta,D,3,k}}(x,b,p)1_{R_{\delta,D,3,k}}(s,t,v). \end{aligned}$$

Again this gives us the estimate

$$\begin{aligned} |F_{\delta,D,3,k}((s,t,v),(x,b,p))| \leq\ & \varphi(s-x, t-b, e^{\frac{-b}{2}}(v-p))e^{-\frac{b}{2}} \\ & (|e^{\frac{-x-b}{2}}(p - \delta^2k_3De^{r_\delta(k_1+k_2)/2})||e^{\frac{-x-b}{2}} \\ & (p + \delta^2k_3De^{r_\delta(k_1+k_2)/2})| \quad + e^{\frac{-2x-b}{2}}|p| + e^{-x}), \end{aligned}$$

where φ is as in Remark 6.8.6. On the sets $S_{\delta,D,3,k}$, if $\widehat{F}^{\natural}(-) \neq 0$ for δ small enough, we have the relations:

$$\begin{aligned} e^{-x} &\leq \delta^6, \\ e^{-x-b}p^2 &\leq \rho, \\ e^{1/2(-x-b)}|k_3|\delta^2De^{1/2r_\delta(k_1+k_2)} &\leq e^{1/2(-x-b)}(|p| + D\delta^2e^{1/2r_\delta(k_1+k_2)}) \\ &\leq \rho^{1/2} + e^{1/2(-x-b)}D\delta^2e^{1/2r_\delta(k_1+k_2)}) \\ &< \rho^{1/2} + \delta \leq 2\rho^{1/2}, \\ |e^{1/2(-x-b)}(p - k_3D\delta^2e^{1/2r_\delta(k_1+k_2)}| &\leq D\delta^2e^{(-x-b)/2}e^{1/2r_\delta(k_1+k_2)} \\ &< D\delta^{3/2}. \end{aligned}$$

Therefore,

$$\begin{aligned}|F_{\delta,D,3,k}((s,t,v),(x,b,p))| &\leq \varphi(s-x,t-b,e^{\frac{-b}{2}}(v-p))e^{-\frac{b}{2}}(2\rho^{1/2}D\delta^{3/2}+\rho^{1/2}\delta^3+\delta^6) \\ &\leq \delta\varphi(s-x,t-b,e^{\frac{-b}{2}}(v-p))e^{-\frac{b}{2}}.\end{aligned}$$

Therefore we can conclude as in Subsection 6.8.3.2.

Corollary 6.8.17 *Let $a \in C^*(G)$. Then*

$$\lim_{\delta\to 0} dis(\rho_{\varepsilon S^*}(\sigma_{\varepsilon S^*}(a)-\sigma_{\varepsilon,\delta,D}(a)), C_0(\mathbb{R},\mathcal{K})) = 0.$$

Proof 6.46 *Indeed, by Proposition 6.8.1, we know that* $\lim_{\delta\to 0}\|\sigma_{\varepsilon S^*}(a)\circ(1-M_{\delta,1})-\sigma_{\varepsilon,\delta,D}(a)\|_{\text{op}} = 0$ *for any $a \in C^*(G)$. By Proposition 6.8.16, we know that $\tau_{\varepsilon S^*}(a)\circ \overline{M}_{\delta,1} \in C_0(\mathbb{R},\mathcal{K})$. Hence, by Proposition 6.8.14 and relation* (6.8.11)*:*

$$\begin{aligned}& dis(\rho_{\varepsilon S^*}(\sigma_{S^*}(a)-\sigma_{\varepsilon,\delta,D}(a))), C_0(\mathbb{R},\mathcal{K})) \\ \leq\ & \|\rho_{\varepsilon S^*}(\sigma_{S^*}(a)-\sigma_{\varepsilon,\delta,D}(a)) - \tau_{\varepsilon S^*}(a)\circ\overline{M}_{\delta,1}\|_{\text{op}} \\ =\ & \|\rho_{\varepsilon S^*}\Big(\sigma_{S^*}(a)\circ M_{\delta,1}+\sigma_{S^*}(a)\circ(1-M_{\delta,1})-\sigma_{\varepsilon,\delta,D}(a)\Big)-\tau_{\varepsilon S^*}(a)\circ\overline{M}_{\delta,1})\|_{\text{op}} \\ =\ & \|\sigma_{\varepsilon S^*}(a)\circ(1-M_{\delta,1})-\sigma_{\varepsilon,\delta,D}(a)\|_{\text{op}} \\ \to\ & 0 \quad as \quad \delta\to 0.\end{aligned}$$

6.8.3.4 *The boundary condition for* $\mathbb{R}_{\varepsilon R^*}$

This case is similar to the preceding one (and easier). We shall omit most of the proofs.

We take the coordinates on G coming from the basis $\{X := A+B, Y := A-B, P, Q, R, S\}$. Then

$$[X,P]=0, [Y,P]=P, [X,Q]=Q, [Y,Q]=-Q, [X,R]=R, [Y,R]=0.$$

Hence for $(x,y,p) := \exp(xX)\exp(yY)\exp(pP) \in G_0$, we have

$$(x,y,p)\cdot\varepsilon R^*_{|\mathfrak{h}} = -e^{-x+y}p\varepsilon Q^* + e^{-x}\varepsilon R^*.$$

The irreducible representation $\pi_{\varepsilon R^*+b^*B^*}$ is realised as $\pi_{\varepsilon R^*+b^*B^*} := \text{ind}_L^G\chi_{\varepsilon R^*+b^*B^*}$, where $L := \exp(\text{span}\{Y,\mathfrak{h}\})$. Let $\partial\mathbb{R}_{\varepsilon R^*}$ be the boundary of $\mathbb{R}_{\varepsilon R^*}$ in $\mathfrak{g}^*/G$. It follows from the description of the coadjoint orbits (see (1.4.16)) that

$$\partial\mathbb{R}_{\varepsilon R^*} = (\{S,R\}^\perp)/G.$$

Definition 6.42 *Let*

$$\sigma_{\varepsilon R^*} := \text{ind}_H^G\chi_{\varepsilon R^*}.$$

The kernel function $\mathcal{K}_F$ of the operator $\sigma_{\varepsilon R^*}(F)$ is then given by

$$\mathcal{K}_F(s,t,v;x,y,p) = \widehat{F}^{\mathfrak{h}}(s-x,t-y,e^y(v-p),-e^{-x+y}p\varepsilon Q^*+e^{-x}\varepsilon R^*)e^y.$$

This representation is equivalent to the direct integral representation

$$\tau_{\varepsilon R^*} := \int_{\mathbb{R}}^{\oplus} \pi_{b^* B^* + \varepsilon R^*} db^*$$

acting on the Hilbert space

$$\mathcal{H}_{\tau_{\varepsilon R^*}} = \int_{\mathbb{R}}^{\oplus} L^2(G/L, \chi_{b^* B^* + \varepsilon R^*}) db^* \simeq \int_{\mathbb{R}}^{\oplus} L^2(\mathbb{R}^2) db^*$$

with the norm

$$\|\xi\|_2^2 = \int_{\mathbb{R}} \|\xi(b^*)\|_2^2 db^* \quad \text{for} \quad \xi \in \mathcal{H}_{\tau_{\varepsilon R^*}}.$$

An intertwining operator $U_{\varepsilon R^*}$ for this equivalence is given by

$$U_{\varepsilon R^*}(\xi)(b^*)(g) := \int_{\mathbb{R}} \xi(g\exp(yY)) e^{-\frac{1}{2}b} e^{-2\pi i b^* y} dy,\ \xi \in L^2(G/H, \chi_{\varepsilon R^*}), g \in G, b^* \in \mathbb{R}.$$

Similar to Subsection 6.8.3.3, we can again consider the C*-algebra $C(\mathbb{R}, \mathcal{B})$ and see that the image of $\tau_{\varepsilon R^*}$ is contained in $C(\mathbb{R}, \mathcal{B})$. Similarly, we can also consider $C(\mathbb{R}, \mathcal{B})$ as a subalgebra of $B(\mathcal{H}_{\tau_{\varepsilon R^*}})$: for $\phi \in C(\mathbb{R}, \mathcal{B})$, let

$$\phi(\xi)(b^*) := \phi(b^*)(\xi(b^*)) \text{ for } \xi \in \mathcal{H}_{\tau_{\varepsilon R^*}}, b^* \in \mathbb{R}.$$

The unitary mapping $U_{\varepsilon R^*}$ induces a canonical homomorphism $\rho_{\varepsilon R^*}$ from the algebra $B(L^2(G/H, \chi_{\varepsilon R^*}))$ onto $B(\mathcal{H}_{\tau_\varepsilon R^*})$. This homomorphism is defined on $\sigma_{\varepsilon R^*}(a)$ by

$$\begin{aligned} \rho_{\varepsilon R^*}(\sigma_{\varepsilon R^*}(a)) &= U_{\varepsilon R^*} \circ \sigma_{\varepsilon R^*}(a) \circ U_{\varepsilon R^*}^* \\ &= \int_{\mathbb{R}}^{\oplus} \pi_{b^* B^* + \varepsilon R^*}(a) db^* \\ &= \tau_{\varepsilon R^*}(a) \quad \text{for} \quad a \in C^*(G). \end{aligned}$$

Definition 6.43 1. Let $S_{\delta,1} := \{(x, y, p); e^{-x} > \delta^6\}$.

2. Let $\delta \mapsto r_\delta \in \mathbb{R}^+$ be such that $\lim_{\delta \to 0} r_\delta = +\infty$ and $\lim_{\delta \to 0} e^{r_\delta} \delta^{1/2} = 0$.

3. For a constant $D > 0$ and $k = (k_1, k_2, k_3) \in \mathbb{Z}^3$, let

$$S_{\delta,D,2,k,} := \{(x, y, p) \in \mathbb{R}^3; e^{-x} \leq \delta^6, x \in I_{r_\delta,k_1}, y \in I_{r_\delta,k_2}, p \in I_{D\delta^2 e^{r_\delta (k_1 - k_2)},k_3}\}.$$

Proposition 6.8.18 *For every compact subset $K \subseteq \mathbb{R}^3$, we have that*

$$KS_{\delta,D,2,k} \subset \bigcup_{\substack{j \in \mathbb{Z}^3 \\ |j_i| \leq 1, i=1,2,3}} S_{\delta, De^{r_\delta(-j_1 - j_2)},2,k+j} =: R_{\delta,D,2,k}.$$

for every $k \in \mathbb{Z}^3$ and $\delta > 0$ small enough.

Proof 6.47 *Indeed, we have an $M > 0$ such that $K \subset [-M, M]^3$ and then for $r_\delta > M$, $(s,t,v) \in K$ and $(x,y,p) \in S_{\delta,D,2,k}$. It follows that*

$$u := (s,t,v) \cdot (x,y,p) = (s+x, t+y, e^{-y}v + p),$$

$(k_1 + j_1)r_\delta \le s + x < (k_1 + j_1 + 1)r_\delta$ and $(k_1 + j_1)r_\delta \le t + y < (k_2 + j_2 + 1)r_\delta$ for some $k = (k_1, k_2) \in \mathbb{Z}^2$ and $j_1, j_2 \in \{-1, 0, 1\}$. It follows that

$$\begin{aligned} |e^{-y}v| &\le |e^{-x}v|e^{x-y} \\ &< De^{r_\delta(-j_1+j_2)}\delta^2 e^{r_\delta(k_1-k_2+j_1-j_2)}, \end{aligned}$$

since for δ small enough $M\delta^6 e^{2r_\delta} < D\delta^2$. Hence:

$$\begin{aligned} p + e^{-y}v &< (k_3+1)De^{r_\delta(-j_1+j_2)}\delta^2 e^{r_\delta(k_1-k_2+j_1-j_2)} + e^{-y}v \\ &< (k_3+2)De^{r_\delta(-j_1+j_2)}\delta^2 e^{r_\delta(k_1-k_2+j_1-j_2)} \end{aligned}$$

and also

$$\begin{aligned} p + e^{-y}v &\ge k_3 De^{r_\delta(-j_1+j_2)}\delta^2 e^{r_\delta(k_1-k_2+j_1-j_2)} - e^{-y}|v| \\ &\ge r_\delta(k_3-1)De^{(-j_1+j_2)}\delta^2 e^{r_\delta(k_1-k_2+j_1-j_2)}. \end{aligned}$$

Hence u is contained in the set $R_{\delta,D,2,k}$.

Define the corresponding multiplication operators on $L^2(\mathbb{R}^2)$ by $M_{\delta,1} := M_{S_{\delta,1}}$, $M_{\delta,D,k,2} := M_{S_{\delta,D,k,2}}$ and $P_{\delta,D,2,k} := M_{R_{\delta,D,2,k}}$ for all $k \in \mathbb{Z}^3$.

Proposition 6.8.19 *We have that*

$$\overline{M}_{\delta,1} \circ U_{\varepsilon R^*} = U_{\varepsilon R^*} \circ M_{\delta,1} \quad \text{for} \quad \delta > 0,$$

where $\overline{M}_{\delta,1} := M_{\{(x,p);e^{-x}>\delta^6\}}$.

Proof 6.48 *Indeed, for $\xi \in L^2(G/H, \chi_{\varepsilon R^*}), y \in \mathbb{R}$ and $b \in \mathbb{R}$, we have that*

$$\begin{aligned} U_{\varepsilon R^*}(M_{\delta,1}(\xi))(E(x,p)) &= \int_{\mathbb{R}} M_{\delta,1}(\xi)(E(x,p)exp(yY))e^{-\frac{1}{2}y}e^{-2\pi i b^* y}dy \\ &= \int_{\mathbb{R}} 1_{\{e^{-x}>\delta^6\}}\xi(E(x,p)exp(yY))e^{-\frac{1}{2}y}e^{-2\pi i b^* y}dy \\ &= 1_{\{e^{-x}>\delta^6\}}(E(x,p)) \int_{\mathbb{R}} \xi(E(x,p)exp(yY))e^{-\frac{1}{2}y}e^{-2\pi i b^* y}dy \\ &= \overline{M}_{\delta,1}(U_{\varepsilon R^*}\xi)(E(x,p)). \end{aligned}$$

Definition 6.44 For $D > 0$ and $F \in L^1_c$, let

$$\sigma_{\varepsilon,\delta,D,2}(F) := \sum_{k\in\mathbb{Z}^3} P_{\delta,D,2,k} \circ \sigma_{\varepsilon\delta^2 k_3 De^{r_\delta(k_1-k_2)}Q^*}(F) \circ M_{\delta,D,2,k}.$$

The kernel functions $F_{\varepsilon,\delta,D,2}$ of this operator is given by :

$$\begin{aligned} F_{\varepsilon,\delta,D,2}((s,t,v),(x,y,p)) &= \sum_{k\in\mathbb{Z}^3} F_{\varepsilon,\delta,D,2,k}((s,t,v),(x,y,p)) \\ &:= \sum_{k\in\mathbb{Z}^3} \widehat{F}^{\mathfrak{h}}(s-x,t-y,e^{-y}(v-p),e^{-x+y}\delta^2 k_3 De^{r_\delta(k_1-k_2)} \\ &\quad \varepsilon Q^*)e^{-y} 1_{R_{\delta,D,2,k}}(s,t,v) 1_{S_{\delta,D,2,k}}(x,y,p). \end{aligned}$$

Proposition 6.8.20 *For every $F \in L^1_c$ and every small enough $\delta > 0$, there exists a constant $C > 0$ such that*

$$\|\sigma_{\varepsilon,\delta,D,2,k}(F)\|_{\text{op}} \le C\|\sigma_{\varepsilon Q^*}(F)\|_{\text{op}}.$$

Proof 6.49 *The proof is similar to the proof of Proposition 6.8.15.*

Proposition 6.8.21 *Let $a \in C^*(G)$. Then the element*

$$\tau_{\varepsilon R^*}(a) \circ \overline{M}_{\delta,1} := \int_{\mathbb{R}}^{\oplus} \pi_{b^*B^* \mid \varepsilon R^*}(a) \circ \overline{M}_{\delta,1} db^*$$

is in $C_0(\mathbb{R},\mathcal{K})$ for every $\delta > 0$.

Proof 6.50 *Let us prove that the operators $\pi_{b^*B^*+\varepsilon R^*}(a) \circ \overline{M}_{\delta,1}, \delta > 0$, are all compact. However, it is easy to see, using Remark 6.8.6, that for $F \in L^1_c$:*

$$\begin{aligned} &\|\pi_{b^*B^*+\varepsilon R^*}(F) \circ \overline{M}_{\delta,1}\|^2_{H-S} \\ = &\int_{\mathbb{R}^4} 1_{\{e^{-x}>\delta^6\}} \Big| \int_{\mathbb{R}} e^{ib^*y} e^{-\frac{y}{2}} \widehat{F}^{\mathfrak{h}}(E(t-x,u-e^y p)exp(yY)) \times \\ &(-\varepsilon pe^{-x+y}Q^* + \varepsilon e^{-x}R^*)dy\Big|^2 dxdtdudp \\ \le &\Big(\int_{\mathbb{R}}\Big(\int_{\mathbb{R}^4} 1_{\{e^{-x}>\delta^6\}} |\widehat{F}^{\mathfrak{h}}(E(t-x,u-e^y p)exp(yY) \times \\ &(-\varepsilon pe^{-x+y}Q^* + \varepsilon e^{-x}R^*)|^2 dxdtdudp\Big)^{\frac{1}{2}} e^{-y/2}dy\Big)^2 \\ = &\Big(\int_{\mathbb{R}}\Big(\int_{\mathbb{R}^4} 1_{\{e^{-x}>\delta^6\}} |e^x \widehat{F}^{\mathfrak{h}}(E(t,u)exp(yY)(-\varepsilon pQ^* + \varepsilon e^{-x}R^*)|^2 dxdtdudp\Big)^{\frac{1}{2}} e^{-y}dy\Big)^2 \\ < &\infty. \end{aligned}$$

The proof that $\lim_{b^*\to\infty} \pi_{b^*B^*+\varepsilon R^*}(a) \circ \overline{M}_{\delta,1}db^* = 0$ is similar to that of the corresponding proof of Proposition 6.8.16

Proposition 6.8.22 *For any $F \in L^1_c$, there exists a constant $K = K(D,F) > 0$ such that*

$$\|\sigma_{\varepsilon R^*}(F) \circ (1 - M_{\delta,1}) - \sigma_{\varepsilon,\delta,D,2,k}(F)\|_{\text{op}} \leq K\delta \quad \textit{for} \quad \delta > 0.$$

Proof 6.51 *Let $F \in L^1_c$. From Proposition 6.8.18, we know that for δ small enough, $P_{\delta,D,2,k} \circ \sigma_{\varepsilon R^*}(F) \circ M_{\delta,D,2,k} = \sigma_{\varepsilon R^*}(F) \circ M_{\delta,D,2,k}$ for $k \in \mathbb{Z}^3$. Hence, it suffices to show that*

$$\|P_{\delta,D,2,k} \circ (\sigma_{\varepsilon R^*}(F) \circ M_{\delta,D,2,k} - \sigma_{\varepsilon\delta^2 k_3 De^{r_\delta(k_1-k_2)}Q^*}(F) \circ M_{\delta,D,2,k})\|_{\text{op}} \leq L\delta,$$

for some constant $L > 0$ independent of δ.

The kernel function $F_{\varepsilon,\delta,D,2,k}$ of the operator

$$P_{\delta,D,2,k} \circ (\sigma_{\varepsilon R^*}(F) \circ M_{\delta,D,2,k} - \sigma_{\varepsilon\delta^2 k_3 De^{r_\delta(k_1-k_2)}Q^*}(F) \circ M_{\delta,D,2,k})$$

is given by:

$$\begin{aligned} F_{\varepsilon,\delta,D,2,k}((s,t,v),(x,y,p)) &= \Big(\widehat{F}^{\mathfrak{h}}(s-x, t-y, e^y(v-p), -e^{-x+y}p\varepsilon Q^* + e^{-x}\varepsilon R^*) \\ &\quad - \widehat{F}^{\mathfrak{h}}(s-x, t-y, e^y(v-p), e^{-x+y}\delta^2 k_3 De^{r_\delta(k_1-k_2)}\varepsilon Q^*)\Big) \\ &\quad 1_{S_{\delta,D,2,k}}(x,y,p) 1_{P_{\delta,D,2,k}}(s,t,v) e^y. \end{aligned}$$

Hence,

$$\begin{aligned} & |F_{\varepsilon,\delta,D,2,k}((s,t,v),(x,y,p))| \\ \leq\ & |\widehat{F}^{\mathfrak{h}}(s-x, t-y, v-e^y p, -e^{-x+y}p\varepsilon Q^* + e^{-x}\varepsilon R^* \\ & -\widehat{F}^{\mathfrak{h}}(s-x, t-y, v-e^y p, e^{-x+y}\delta^2 k_3 De^{r_\delta(k_1-k_2)}\varepsilon Q^*)| \\ & 1_{S_{\delta,D,2,k}}(x,y,p) 1_{P_{\delta,D,2,k}}(s,t,v) e^y \\ \leq\ & \varphi(s-x, t-y, v-e^y p) \\ & (e^{-x+y} e^{r_\delta(k_1-k_2)} D\delta^2 + e^{-x}) e^y 1_{S_{\delta,D,2,k}}(x,y,p) 1_{P_{\delta,D,2,k}}(s,t,v) \\ \leq\ & \delta\varphi(s-x, t-y, e^y(v-p)) e^y 1_{S_{\delta,D,2,k}}(x,y,p) 1_{P_{\delta,D,2,k}}(s,t,v), \end{aligned}$$

for δ small enough, where φ is as in Remark 6.8.6. We conclude as in the preceding cases.

Corollary 6.8.23 *Let $a \in C^*(G)$. Then*

$$\lim_{\delta \to 0} dis(\rho_{\varepsilon R^*}(\sigma_{\varepsilon R^*}(a) - \sigma_{\varepsilon,\delta,D,2}(a)), C_0(\mathbb{R}, \mathcal{K})) = 0.$$

Proof 6.52 *The proof is similar to that of Corollary 6.8.17.*

6.8.3.5 *The boundary condition for* $\Omega_{\varepsilon P^*+\nu Q^*}$, $\varepsilon = \pm 1, \nu = \pm 1$

The boundary of the orbit $\Omega_{\varepsilon P^*+\nu Q^*}$ is the union of the orbits $x^*X^*+\nu Q^*, y^*Y^*+\varepsilon P^*$ for all $y^*, x^* \in \mathbb{R}$, and of the set of characters (see (1.4.18)). We take the coordinates on G coming from the basis $\{Y := 2A, X := A+B, P, Q, R, S\}$. This gives us the bracket relations:

$$[X,P]=0,\ [X,Q]=Q,\ [Y,P]=P,\ [Y,Q]=0.$$

The irreducible representation $\pi_{\varepsilon P^*+\nu Q^*}$ is realised as $\pi_{\varepsilon P^*+\nu Q^*} := \mathrm{ind}_L^G \chi_{\varepsilon P^*+\nu Q^*}$, where $L := \exp(\mathfrak{l}) := \exp(\mathrm{span}\{P, \mathfrak{h}\})$.

Definition 6.45 1. *Let* $\mathfrak{m} := span\{Y, \mathfrak{l}\}$ *and* $M := exp(\mathfrak{m})$ *and let:*

$$\sigma_{\nu Q^*} := \mathrm{ind}_L^G \chi_{\nu Q^*} \simeq \tau_{\nu Q^*} := \int_{\mathbb{R}}^{\oplus} \pi_{x^*X^*+\nu Q^*} dx^*.$$

An intertwining operator $U_{\nu Q^*}$ for this equivalence is given by

$$(U_{\nu Q^*}(\xi)(x^*))(g) := \int_{\mathbb{R}} \xi(g\exp(xY))e^{-2\pi i x^* x} dx$$

mappings from $\mathbb{R}$ into the algebra $\mathcal{B}$ of bounded operators on the Hilbert space $L^2(\mathbb{R})$ containing the ideal $C_0(\mathbb{R},\mathcal{K})$ of all continuous mappings from $\mathbb{R}$ into the algebra $\mathcal{K}$ of compact operators on $L^2(\mathbb{R})$ vanishing at infinity. For $\xi \in L^2(G/L, \chi_{\nu Q^*}), g \in G$ and $x^* \in \mathbb{R}$, the unitary mapping $U_{\nu Q^*}$ induces an isomorphism $\rho_{\nu Q^*}$ from the algebra $B(L^2(G/L, \chi_{\nu Q^*}))$ onto

$$B(\int_{\mathbb{R}}^{\oplus} B(L^2(G/M, \chi_{x^*X^*+\nu Q^*}))dx^*).$$

This homomorphism is defined by

$$\begin{aligned}\rho_{\nu Q^*}(\sigma_{\nu Q^*}(a)) &= U_{\nu Q^*} \circ \sigma_{\nu Q^*}(a) \circ U^*_{\nu Q^*} \\ &= \int_{\mathbb{R}}^{\oplus} \pi_{x^*X^*+\nu Q^*}(a) dx^* \ \text{ for } a \in C^*(G).\end{aligned}$$

2. Let $\mathfrak{k} := \mathrm{span}\{X, \mathfrak{l}\}$ and $K := \exp(\mathfrak{k})$ and let:

$$\sigma_{\varepsilon P^*} := \mathrm{ind}_L^G \chi_{\varepsilon P^*} \simeq \tau_{\varepsilon P^*} := \int_{\mathbb{R}}^{\oplus} \pi_{y^*Y^*+\varepsilon P^*} dy^*.$$

An intertwining operator $U_{\varepsilon P^*}$ for this equivalence is given by

$$(U_{\varepsilon P^*}(\xi)(y^*))(g) := \int_{\mathbb{R}} \xi(g\exp(yX))e^{-2\pi i y^* y} dy$$

for $\xi \in L^2(G/L, \chi_{\varepsilon P^*}), g \in G$ and $x^* \in \mathbb{R}$. The unitary mapping $U_{\varepsilon P^*}$ induces an isomorphism $\rho_{\varepsilon P^*}$ from the algebra $B(L^2(G/L, \chi_{\varepsilon P^*}))$ onto

$$B(\int_{\mathbb{R}}^{\oplus} B(L^2(G/K, \chi_{y^*Y^*+\varepsilon P^*}))dy^*).$$

This homomorphism is defined by

$$\begin{aligned}\rho_{\varepsilon P^*}(\sigma_{\varepsilon P^*}(a)) &= U_{\varepsilon P^*} \circ \sigma_{\varepsilon P^*}(a) \circ U^*_{\varepsilon P^*} \\ &= \int_{\mathbb{R}}^{\oplus} \pi_{y^*Y^*+\varepsilon P^*}(a)dy^* \text{ for } a \in C^*(G).\end{aligned}$$

3. Let $\sigma_0 := \mathrm{ind}_L^G \chi_0$.

Definition 6.46 *1. Let* $S_{\delta,1} := \{(x,y); e^{-x} > \delta^6, e^{-y} > \delta^6\}$,

2. $S_{\delta,2} := \{(x,y); e^{-x} > \delta^6, e^{-y} \leq \delta^6\}$,

3. $S_{\delta,3} := \{(x,y); e^{-x} \leq \delta^6, e^{-y} > \delta^6\}$,

4. $S_{\delta,4} := \{(x,y); e^{-x} \leq \delta^6, e^{-y} \leq \delta^6\}$,

and as usual we denote by $M_{\delta,i}$ *the multiplication operator on* $L^2(\mathbb{R}^2)$ *with the function* $1_{S_{\delta,i}}$ *for* $1 \leq i \leq 4$.

Proposition 6.8.24 *For every* $a \in C^*(G)$*, the operator* $\pi_{\varepsilon P^*+\nu Q^*}(a) \circ M_{\delta,1}$ *is compact.*

Proof 6.53 *Let* $F \in L^1_c := \{F \in L^1(G), \widehat{F}^L \in C_c(G/L, \mathfrak{l}^*)\}$*. The kernel function* $F_{\varepsilon,\nu}$ *of the operator* $\pi_{\varepsilon P^*+\nu Q^*}(F) \circ M_{\delta,1}$ *is given by*

$$F_{\varepsilon,\nu}((s,t),(x,y)) = (\widehat{F}^L(s-x,t-y), e^{-x}P^* + e^{-y}Q^*)1_{S_{\delta,1}}(x,y).$$

Since $F \in L^1_c$ *is of compact support in all variables and therefore* $\pi_{\varepsilon P^*+\nu Q^*}(F) \circ M_{\delta,1}$ *is Hilbert-Schmidt. Since* L^1_c *is dense in* $C^*(G)$*, we have that* $\pi_{\varepsilon P^*+\nu Q^*}(a) \circ M_{\delta,1}$ *is compact for every* $a \in C^*(G)$.

Definition 6.47 For $F \in L^1_c$ and $\delta > 0$, let

$$\sigma_{\varepsilon,\nu,\delta}(F) := \sigma_{\nu Q^*}(F) \circ M_{\delta,2} + \sigma_{\varepsilon P^*}(F) \circ M_{\delta,3} + \sigma_0(F) \circ M_{\delta,4} \in \mathcal{B}(L^2(\mathbb{R}^2)).$$

The kernel function $F_{\varepsilon,\nu,\delta}$ of this operator is given by

$$\begin{aligned}F_{\varepsilon,\nu,\delta}((s,t),(x,y)) &= \widehat{F}^L((s-x,t-y), e^{-x}P^*)1_{S_{\delta,2}}(x,y) \\ &\quad + \widehat{F}^L((s-x,t-y), e^{-y}Q^*)1_{S_{\delta,3}}(x,y) \\ &\quad + \widehat{F}^L((s-x,t-y), 0)1_{S_{\delta,4}}(x,y).\end{aligned}$$

Similar to previous cases, we have the following propositions and corollary.

Proposition 6.8.25 *For every* $F \in L^1_c$ *and every* $\delta > 0$,

$$\|\sigma_{\varepsilon,\nu,\delta}(F)\|_{\mathrm{op}} \leq \|\sigma_{\varepsilon P^*}(F)\|_{\mathrm{op}} + \|\sigma_{\nu Q^*}(F)\|_{\mathrm{op}} + \|\sigma_0(F)\|_{\mathrm{op}}.$$

Proposition 6.8.26 *For any $F \in L^1_c$, there exists a constant $K = K(F) > 0$ such that*

$$\|\pi_{\varepsilon P^*+\nu Q^*}(F) \circ (1 - M_{\delta,1}) - \sigma_{\varepsilon,\nu,\delta}(F)\|_{\text{op}} \leq K\delta \quad \textit{for} \quad \delta > 0.$$

Proof 6.54 *Let $F \in L^1_c$. The kernel function $F_{\varepsilon,\nu,\delta,0}$ of the operator $\pi_{\varepsilon P^*+\nu Q^*}(F) \circ (1 - M_{\delta,1}) - \sigma_{\varepsilon,\nu,\delta}(F)$ is given by:*

$$\begin{aligned} & F_{\varepsilon,\nu,\delta,0}((s,t),(x,y)) \\ = \ & (\widehat{F}^L((s-x,t-y), e^{-x}P^* + e^{-y}Q^*) - \widehat{F}^L((s-x,t-y), e^{-x}P^*))1_{S_{\delta,2}}(x,y) \\ & +(\widehat{F}^L((s-x,t-y), e^{-x}P^* + e^{-y}Q^*) - \widehat{F}^L((s-x,t-y), e^{-y}Q^*))1_{S_{\delta,3}}(x,y) \\ & +(\widehat{F}^L((s-x,t-y), e^{-x}P^* + e^{-y}Q^*) - \widehat{F}^L((s-x,t-y), 0))1_{S_{\delta,4}}(x,y). \end{aligned}$$

Since $F \in L^1_c$, there exists a function $\varphi \in C_c(\mathbb{R}^2)$ such that

$$|\widehat{F}((s,t),q)| \leq |\varphi(s,t)|\|q\| \quad \textit{for } q \in \mathfrak{l}^*, (s,t) \in \mathbb{R}^2.$$

Therefore,

$$|F_{\varepsilon,\nu,\delta,0}((s,t),(x,y))| \leq 3|\varphi(s-x,t-y)|\delta^6.$$

Corollary 6.8.27 *Let $a \in C^*(G)$. Then*

$$\lim_{\delta \to 0} dis(\pi_{\varepsilon P^*+\nu Q^*}(a) - \sigma_{\varepsilon,\nu,\delta}(a), \mathcal{K}) = 0.$$

6.8.3.6 Tho boundary conditions for $\mathbb{R}_{\varepsilon P^*}$ and $\mathbb{R}_{\nu Q^*}$

We consider only the case $\Omega_{a^*A^*+\nu Q^*}$. The other cases are similar.

We take the coordinates on G coming from the basis $\{A, B, P, Q, R, S\}$. Let $L := \exp(\mathfrak{l})$ and $\mathfrak{l} := \text{span}\{P, Q, R, S\}$. Let $\sigma_0 := \text{ind}_L^G 1$ be the left regular representation of G on $L^2(G/L)$. This representation is equivalent to

$$\tau_0 := \oint_{\mathbb{R}} \pi_{a^*A^*+b^*B^*} da^* db^*$$

acting on

$$L^2(\mathbb{R}^2) = \oint_{R^2} \mathbb{C} da^* db^*.$$

Let $U_{\nu Q^*}$ be the unitary equivalence given by

$$U_{\nu Q^*}(\xi)(a^*, b^*) := \int_{\mathbb{R}^2} \xi(\exp(aA)\exp(bB))e^{-2\pi i(aa^*+bb^*)} dadb, \ a^*, b^* \in \mathbb{R}.$$

We see that the algebra $B(L^2(G/L, \chi_{\nu Q^*}))$ is mapped into $C(\mathbb{R}^2) \subset \mathcal{B}(L^2(\mathbb{R}^2))$ by the mapping

$$\rho_{\nu Q^*}(\xi) := U_{\nu Q^*} \circ \xi \circ U^*_{\nu Q^*}.$$

Definition 6.48 For $\delta > 0$, let

1. $S_{\delta,1} := \{(a,b) \in \mathbb{R}^2, e^{-b} > \delta\}$,
2. $S_{\delta,2} := \{(a,b) \in \mathbb{R}^2, e^{-b} \leq \delta\}$,

and $M_{\delta,i}$ denotes the multiplication operator on $L^2(\mathbb{R}^2)$ with the function $1_{S_{\delta,i}}$ for $i = 1, 2$. For each $F \in C^*(G)$, we define the linear operator $\sigma_{\delta,2}(F)$ on $L^2(\mathbb{R}^2)$ by

$$\sigma_{\delta,2}(F) := \sigma_0(F) \circ M_{\delta,2}.$$

Proposition 6.8.28 *Let $a \in C^*(G)$. Then*

$$\lim_{\delta \to 0} dis((\sigma_{\nu Q^*}(a) \circ (1 - M_{\delta,1}) - \sigma_{\delta,2}(a)), C_0(\mathbb{R}, \mathcal{K})) = 0.$$

The proof is similar to that of the corresponding proof in Section 6.8.3.4.

6.8.4 The C^*-algebra of G_6

We recall that the space $\widehat{G}$ has been identified in Definition 6.34 with the disjoint union of the subsets $\Gamma_j, j = 0, \ldots, 6$, of $\mathfrak{g}^*$. We let

$$l^\infty(\widehat{G}) := \{(\phi(\gamma))_{\gamma \in \widehat{G}}; \phi(\gamma) \in B(\mathcal{H}_{\pi_\gamma}),\ \|\phi\| := \sup_{\gamma \in \widehat{G}} \|\phi(\gamma)\|_{\text{op}} < \infty\}.$$

Definition 6.49 For a subset Γ of $\widehat{G}$ and $\phi \in l^\infty(\widehat{G})$, let

$$\phi_{|\Gamma} := (\phi(\gamma))_{\gamma \in \Gamma} \in l^\infty(\Gamma).$$

Definition 6.50 For $F \in C^*(G)$ and $\gamma \in \widehat{G}$, we define the Fourier transform $\mathcal{F}_G : C^*(G) \to l^\infty(\widehat{G})$ by

$$\mathcal{F}_G(F)(\gamma) := \pi_\gamma(F) \in B(\mathcal{H}_{\pi_\gamma}).$$

Let

$$E^*(G) := \{\phi \in l^\infty(\widehat{G});\ \text{the mappings } \gamma \mapsto \phi(\gamma) \text{ are norm continuous and vanish at infinity on the sets } \Gamma_i, i = 1, 2, 4, 5\}.$$

Then the subset $E^*(G)$ is a C^*-subalgebra of $l^\infty(\widehat{G})$ containing $\mathcal{F}_G(C^*(G))$ by Section 6.8.1. We can define the following representations of $E^*(G)$ for $\varepsilon = \pm 1$:

1. $\tau_{\varepsilon S^*}(\phi) := \int_{\mathbb{R}}^{\oplus} \phi(b^*B^* + \varepsilon S^*)db^*$ on the Hilbert space $\int_{\mathbb{R}}^{\oplus} L^2(\mathbb{R}^2)db^*$. We have seen in Definition 6.42 that the restriction of this representation to $\mathcal{F}_G(C^*(G))$ is equivalent to the representations $\sigma_{\varepsilon s^* S^*} = \mathrm{ind}_H^G \chi_{\varepsilon s^* S^*}$ for $s^* \in \mathbb{R}^+$. We use now these equivalences to extend the representations $\sigma_{\varepsilon s^* S^*}$ to $E^*(G)$.

2. $\tau_{\varepsilon R^*}(\phi) := \int_{\mathbb{R}}^{\oplus} \phi(b^*B^* + \varepsilon R^*)db^*$ on the Hilbert space $\int_{\mathbb{R}}^{\oplus} L^2(\mathbb{R}^2)db^*$. This gives us representations $\sigma_{\varepsilon R^*}$ of $E^*(G)$ (which coincides with $\mathrm{ind}_H^G \chi_{\varepsilon R^*}$ on $C^*(G)$).

3. $\tau_{\varepsilon P^*}(\phi) := \int_{\mathbb{R}}^{\oplus} \phi(x^*(\frac{A^*+B^*}{2}) + \varepsilon P^*)dx^*$ on the Hilbert space $\int_{\mathbb{R}}^{\oplus} L^2(\mathbb{R})db^*$. This gives us representations $\sigma_{\varepsilon P^*}$ of $E^*(G)$ (which coincides with $\mathrm{ind}_{[G,G]}^G \chi_{\varepsilon P^*}$ on $C^*(G)$).

4. $\tau_{\varepsilon Q^*}(\phi) := \int_{\mathbb{R}}^{\oplus} \phi(a^*A^* + \varepsilon Q^*)da^*$ on the Hilbert space $\int_{\mathbb{R}}^{\oplus} L^2(\mathbb{R})da^*$. This gives us representations $\sigma_{\varepsilon Q^*}$ of $E^*(G)$, which coincides with $\mathrm{ind}_{[G,G]}^G \chi_{\varepsilon Q^*}$ on $C^*(G)$).

This allows us to define the following bounded linear mappings of $E^*(G)$, where $\varepsilon = \pm 1, \nu = \pm 1$ and $\delta > 0$:

1. (with the notations in Section 6.8.3.2)

$$\sigma_{\varepsilon S^* + \varepsilon Q^*, \delta}(\phi) := \sigma_{\varepsilon S^*}(\phi) \circ M_{\tilde{\varphi}_1} \circ M_{\delta,1} + \sigma_{\varepsilon S^*}(\phi) \circ M_{\delta,2} + \sigma_{\delta,3}(\phi),$$

   where $\sigma_{\delta,3}(\phi) := \sum_{k\in\mathbb{Z}} N_{\delta,k,3} \circ \sigma_{\varepsilon(1+\frac{\delta^4 k^2}{2})Q^*}(\phi) \circ M_{\delta,3,k}$;

2. (with the notations in Section 6.8.3.1)

$$\sigma_{\varepsilon S^* - \varepsilon Q^*, \delta}(\phi) := \sum_{i=1}^{2} \sigma_{\varepsilon S^*}(\phi) \circ M_{\tilde{\varphi}_i} \circ M_{\delta,i} + \sigma_{\pm\sqrt{2}R^* - 2Q^*}(\phi) \circ M_{\delta,4,\pm} + \sigma_{\delta,3}(\phi),$$

   where $\sigma_{\delta,3}(\phi) := \sum_{k\in\mathbb{Z}} N_{\delta,3,k} \circ \sigma_{\varepsilon(-1+\frac{\delta^4 k^2}{2})Q^*}(\phi) \circ (M_{\delta,3,k,+} + M_{\delta,3,k,-})$;

3. (with the notations in Section 6.8.3.3)

$$\sigma_{\varepsilon S^*, \delta, D}(\phi) := \sum_{k\in\mathbb{Z}^3} P_{\delta,D,3,k} \circ \sigma_{k_3^2 \delta^3 D^2 e^{r_\delta(k_1+k_2)} \varepsilon Q^*}(\phi) \circ M_{\delta,D,3,k};$$

4. (with the notations in Section 6.8.3.4)

$$\sigma_{\varepsilon R^*, \delta, D}(\phi) := \sum_{k\in\mathbb{Z}^3} P_{\delta,D,2,k} \circ \sigma_{k_3 \delta^2 D e^{r_\delta(k_1-k_2)} \varepsilon Q^*}(\phi) \circ M_{\delta,D,2,k};$$

5. (with the notations in Section 6.8.3.5)

$$\sigma_{\varepsilon P^*, \nu Q^*, \delta}(\phi) := \sigma_{\nu Q^*}(\phi) \circ M_{\delta,2} + \sigma_{\varepsilon P^*}(\phi) \circ M_{\delta,3} + \sigma_0(\phi) \circ M_{\delta,4};$$

6. similarly (with the notations of Section 6.8.3.6)

$$\sigma_{\varepsilon P^*, \delta}(\phi) = \sigma_{\varepsilon P^*} \circ M_{\delta,2},$$

   and the corresponding linear mapping

$$\sigma_{\nu Q^*, \delta}(\phi) = \sigma_{\nu Q^*} \circ M_{\delta,2}.$$

We define now the C^*-subalgebra $D^*(G)$ of $l^\infty(\widehat{G})$. It will follow from Theorem 6.7.1 that this subalgebra is the Fourier transform of $C^*(G)$.

Definition 6.51 Let $D^*(G)$ be the subset of $l^\infty(\widehat{G})$ consisting of all operator fields ϕ contained in $E^*(G)$ such that for $\varepsilon, \omega = \pm 1$.

1. $$\lim_{\delta \to 0} \operatorname{dis}(\phi(\varepsilon(S^* + Q^*)) - \sigma_{\varepsilon S^* + \varepsilon Q^*, \delta}(\phi), \mathcal{K}(L^2(\mathbb{R}^3))) = 0.$$

2. $$\lim_{\delta \to 0} \operatorname{dis}((\phi(\varepsilon(S^* - Q^*)) - \sigma_{\varepsilon S^* - \varepsilon Q^*, \delta}(\phi), \mathcal{K}(L^2(\mathbb{R}^3))) = 0.$$

3. $$\lim_{\delta \to 0} \operatorname{dis}(\rho_{\varepsilon S^*}(\sigma_{\varepsilon S^*}(\phi) - \sigma_{\varepsilon S^*, \delta, D}(\phi)), C_0(\mathbb{R}, \mathcal{K})) = 0,$$
   this is,
   $$\lim_{\delta \to 0} \operatorname{dis}(\tau_{\varepsilon S^*}(\phi) - \rho_{\varepsilon S^*}(\sigma_{\varepsilon S^*, \delta, D}(\phi)), C_0(\mathbb{R}, \mathcal{K})) = 0.$$

4. $$\lim_{\delta \to 0} \operatorname{dis}(\rho_{\varepsilon R^*}(\sigma_{\varepsilon R^*}(\phi) - \sigma_{\varepsilon R^*, \delta, D}(\phi)), C_0(\mathbb{R}, \mathcal{K})) = 0,$$
   this is,
   $$\lim_{\delta \to 0} \operatorname{dis}(\tau_{\varepsilon R^*}(\phi) - \rho_{\varepsilon R^*}(\sigma_{\varepsilon R^*, \delta, D}(\phi)), C_0(\mathbb{R}, \mathcal{K})) = 0.$$

5. $$\lim_{\delta \to 0} \operatorname{dis}((\phi(\varepsilon P^* + \nu Q^*) - \sigma_{\varepsilon P^* + \nu Q^*, \delta}(\phi)), \mathcal{K}(L^2(\mathbb{R}^3))) = 0.$$

6. $$\lim_{\delta \to 0} \operatorname{dis}(\rho_{\varepsilon P^*}(\sigma_{\varepsilon P^*}(\phi) - \sigma_{\varepsilon P^*, \delta}(\phi)), C_0(\mathbb{R}, \mathcal{K})) = 0,$$
   this is,
   $$\lim_{\delta \to 0} \operatorname{dis}(\tau_{\varepsilon P^*}(\phi) - \rho_{\varepsilon P^*}(\sigma_{\varepsilon P^*, \delta}(\phi)), C_0(\mathbb{R}, \mathcal{K})) = 0.$$

7. $$\lim_{\delta \to 0} \operatorname{dis}(\rho_{\nu Q^*}(\sigma_{\nu Q^*}(\phi) - \sigma_{\nu Q^*, \delta}(\phi)), C_0(\mathbb{R}, \mathcal{K})) = 0,$$
   this is,
   $$\lim_{\delta \to 0} \operatorname{dis}(\tau_{\nu Q^*}(\phi) - \rho_{\nu Q^*}(\sigma_{\nu Q^*, \delta}(\phi))), C_0(\mathbb{R}, \mathcal{K})) = 0.$$

8. The same conditions hold for the adjoint field ϕ^*.

We have seen in the Sections 5 and 6 that we can use Theorem 6.7.1 to show the following result.

Theorem 6.8.29 *The C^*-algebra of the group G_6 is an almost $C_0(\mathcal{K})$-C^*-algebra. In particular, the Fourier transform maps $C^*(G_6)$ onto the subalgebra $D^*(G_6) \subset l^\infty(\widehat{G_6})$.*

6.9 THE C^*-ALGEBRA OF $(AX + B)$-LIKE GROUPS

6.9.1 Introduction and Notations

In order to analyse our C^*-algebra $\mathcal{A}$ say in this setting, we use again the Fourier transform $\mathcal{F}$, which allows us to decompose $\mathcal{A}$ over its unitary dual $\widehat{\mathcal{A}}$. We consider the algebra $l^\infty(\widehat{\mathcal{A}})$ of all bounded operator fields defined over $\widehat{\mathcal{A}}$ by

$$l^\infty(\widehat{\mathcal{A}}) := \{A = (A(\pi) \in \mathcal{B}(\mathcal{H}_\pi))_{\pi \in \widehat{\mathcal{A}}}, \|A\|_\infty := \sup_\pi \|A(\pi)\|_{\text{op}} < \infty\},$$

where $\mathcal{H}_\pi$ is the Hilbert space of π. The space $l^\infty(\widehat{\mathcal{A}})$ is a (huge) C^*-algebra itself. The Fourier transform $\mathcal{F}$ defined by

$$\mathcal{F}(a) := \hat{a} := (\pi(a))_{\pi \in \widehat{\mathcal{A}}} \quad \text{for} \quad a \in \mathcal{A}$$

is then an injective, hence isometric, homomorphism from $\mathcal{A}$ into $l^\infty(\widehat{\mathcal{A}})$. The problem is now to recognise the elements of $\mathcal{F}(\mathcal{A})$ inside this big algebra $l^\infty(\widehat{\mathcal{A}})$.

One condition the elements of $\mathcal{F}(\mathcal{A})$ must fulfil is of continuity in the following sense. The dual space has a natural topology, the so-called Fell topology. A net $(\pi_k)_k \subset \widehat{\mathcal{A}}$ converges to $\pi \in \widehat{\mathcal{A}}$ if for every element ξ in the Hilbert space $\mathcal{H}_\pi$ of π, there exists a vector $\xi_k \in \mathcal{H}_{\pi_k}$ for every k such that

$$\lim_k \langle \pi_k(a)\xi_k, \xi_k\rangle = \langle \pi(a)\xi, \xi\rangle \quad \text{for} \quad a \in \mathcal{A}.$$

Therefore, if an operator field $A = (A(\pi))_{\pi \in \widehat{\mathcal{A}}}$ is contained in $\mathcal{F}(\mathcal{A})$, then $\mathcal{A}$ must satisfies the relation

$$\lim_k \langle A(\pi_k)\xi_k, \xi_k\rangle = \langle A(\pi)\xi, \xi\rangle.$$

If $\mathcal{A}$ is not abelian, then the topology of $\widehat{\mathcal{A}}$ is in general not Hausdorff. A net in $\widehat{\mathcal{A}}$ can have many limit points and simultaneously many cluster points (see [44], [34], [69], [67] for details).

In particular, if $\overline{y} = (\pi_k)_k \subset \widehat{\mathcal{A}}$ is a properly converging net, then it follows from [44] that

$$\lim_k \|\pi_k(a)\|_{\text{op}} = \sup_{\sigma \in L(\overline{y})} \|\sigma(a)\|_{\text{op}} \quad \text{for} \quad a \in \mathcal{A}.$$

Therefore for any operator field $A \in \mathcal{F}(\mathcal{A})$ we must have that

$$\lim_k \|A(\pi_k)\|_{\text{op}} = \sup_{\sigma \in L(\overline{y})} \|A(\sigma)\|_{\text{op}}. \tag{6.9.1}$$

This relation (6.9.1) is however not an algebraic condition. The main problem of determining the image of the Fourier transform inside $l^\infty(\widehat{\mathcal{A}})$ consists in finding the right algebraic conditions in $l^\infty(\widehat{\mathcal{A}})$ which imply the property (6.9.1), and forces the unitary dual of the subalgebra $D^*(\mathcal{A})$ consisting of all the operator fields in $l^\infty(\widehat{\mathcal{A}})$ which satisfy those conditions to be point evaluations, i.e. the only irreducible representations of $D^*(\mathcal{A})$ must be of the form $\delta_\pi(A) = A(\pi)$ for $A \in D^*(\mathcal{A})(\pi \in \widehat{\mathcal{A}})$.

Unfortunately for most C^*-algebras, its dual space is not known or if it is known, the topology of it is a mystery. For some classes of locally compact groups G, the unitary dual $\widehat{G}$ and its topology has been determined. This is the case of compact Lie groups and exponential solvable Lie groups. A Lie group G is called *exponential* if the exponential mapping $\exp : \mathfrak{g} \to G$ from the Lie algebra $\mathfrak{g}$ to its Lie group G is a diffeomorphism. For instance, connected simply connected nilpotent Lie groups belong to this class and many more solvable Lie groups. Let us recall the Kirillov-Bernat-Vergne-Pukanszky theory shows that there is a canonical homeomorphism $K : \mathfrak{g}^*/G \to \widehat{G}$ from the space of co-adjoint orbits of G in the linear dual space $\mathfrak{g}^*$ onto the unitary dual space $\widehat{G}$ of G (see [70] for details and references).

In this section, we consider exponential Lie groups $G(\pm)$ which are semi-direct products $G(\pm) := \mathbb{R} \ltimes V$, where V is a finite-dimensional real vector space and the reals act diagonally on the complexification $V_{\mathbb{C}}$ with eigenvalues of the form $e^{\alpha(1+i\theta)}, \alpha \in \mathbb{R}$. We assume that there is at least one eigenvalue with $\alpha < 0$ and another one with $\alpha > 0$. We also treat the case of the groups $G(+) = \mathbb{R} \times V$ where all the eigenvalues have the same sign.

The C^*-algebra in the case of real eigenvalues has been studied by Xiaolu Wang in [104] (these groups are there denoted by the symbol $G(p,q)$). Let us cite part of the review made by Palle E. T. Jorgensen in MathSciNet of this monograph :

"-*The author uses foliations in his study of the group C^*-algebras $C^*(G(p,q))^+$: He proves the existence of a short exact sequence $0 \to C^*(U,F) \to C^*(G(p,q))^+ \to C(S^1) \to 0$ with the foliation algebra $C^*(U,F)$ being again described by an exact sequence $0 \to C_0(S^{p-1} \times S^{q-1} \times \mathbf{R}_+) \otimes \mathcal{K} \to C^*(U,F) \to C(S^{p-1} \cup S^{q-1}, \mathcal{K}) \to 0$ where $\mathcal{K}$ denotes the algebra of all compact operators on $L^2(\mathbf{R})$. He obtains an alternative description using a "cross" of two short exact sequences. As a corollary, he gets an independent proof of an earlier result of Do Ngok Zep on the structure of the C^*-algebra of the "$ax+b$" group. He also uses this to get $C^*(G(p,q))^+$ as a continuous field of C^*-algebras over some dense open subset of the spectrum $\approx S^{p-1} \times S^{q-1} \times \mathbf{R}_+$. The last chapter gives a complete description of $C^*(G(p,q))$ as a subalgebra of $C^b(S^{p-1} \times S^{q-1} \times [\mathbf{R}_+, M_2(\mathcal{K})])$, and this is the basis for the explicit results mentioned above.*"

In our approach, we determine the topology of the co-adjoint orbit space $\mathfrak{g}^*(\pm)/G(\pm)$, and in particular we describe the limit sets of properly converging sequences in this space to find the correct conditions which the elements in $\mathcal{F}(C^*(G(\pm)))$ inside $l^\infty(\widehat{G(\pm)})$ must satisfy. There are four such conditions: (1) the *continuity* of the fields with respect to the operator norm on the generic part $\mathcal{D}$ and on the singular part $\mathcal{S}$ of the dual space, (2) the *infinity condition* which tells what happens at infinity to the operators, (3) the *generic condition* which takes into account the fact that properly converging sequences $(\pi_k)_k$ with limit point $\pi \notin \mathcal{D}$ have two infinite-dimensional limit points and a continuum of limit characters, and (4) the *compact condition* which takes care of the fact that for infinite-dimensional non-generic representations π of $G(\pm)$, the operators $\pi(F)$ for $F \in L^1(G(\pm))$ need not be compact and which tells us how far $\pi(F)$ is away from the space of compact operators.

The section is organised in the following way. In Section 6.9.2, we describe the groups $G(\pm)$, their Lie algebras $\mathfrak{g}(\pm)$ and the co-adjoint orbits in $\mathfrak{g}^*(\pm)$. In Section 6.9.3, we determine the limit set of a properly converging sequence of co-adjoint orbits (Theorem 6.9.3). This is the essential step to find the C^*-conditions. In Section 6.9.4, we present the irreducible representations of our group $G(\pm)$ and of its C^*-algebra. In Section 6.9.5, we describe the properties of the operators $\pi(F), F \in L^1(G(\pm))$ for infinite-dimensional representations π. In Section 6.9.6, we define our Fourier transform and discover the infinity and generic conditions. In Section 6.9.7, we describe the groups $G(+)$, which are the subgroups of $G(\pm)$ associated to eigenvalues which are ≥ 1 in module, we introduce the compact condition and determine the C^*-algebra of $G(+)$ as algebra of operator fields. In Section 6.9.8, we study a certain subalgebra of the direct sum algebra $C^*(G(+)) \oplus C^*(G(-))$. This subalgebra will appear in the proof of the main theorem. In the last section (6.9.9), we compute the range of the Fourier transform for the C^*-algebra $C^*(G(\pm))$.

We shall use the following notation: $L^p(\mathbb{R})$ denotes the space of (almost everywhere equivalence classes) p-integrable functions on $\mathbb{R}$ for $p = 1, 2$ with the norm $\|\cdot\|_p$. By $\|f\|_\infty$ we denote the supremum norm $\|f\|_\infty := \sup_{x\in\Omega} |f(x)|$ of a continuous complex-valued function f vanishing at infinity on a locally compact space Ω. We write $\mathcal{B}$ for the space $\mathcal{B}(L^2(\mathbb{R}))$ of all bounded linear operators on $L^2(\mathbb{R})$. The symbol $\mathcal{K}(\mathcal{H})$ represents the algebra of all compact operators on $\mathcal{H}$, and $\mathcal{K} = \mathcal{K}(L^2(\mathbb{R}))$.

6.9.2 The groups $G(\pm)$

Let us first consider the 2-dimensional Lie group defined by

$$\mathrm{Aff}(\mathbb{R}) = \{\begin{pmatrix} a & b \\ 0 & 1 \end{pmatrix}, \ a, b \in \mathbb{R} \text{ and } a > 0\},$$

known to be the "$ax + b$" group. Let $\mathfrak{g} = \mathbb{R}X \oplus \mathbb{R}Y$ denote its Lie algebra, where

$$X = \begin{pmatrix} 1 & 0 \\ 0 & 0 \end{pmatrix} \text{ and } Y = \begin{pmatrix} 0 & 1 \\ 0 & 0 \end{pmatrix}$$

and whose Lie bracket is given by $[X, Y] = Y$. Then $\mathfrak{g}$ is completely solvable Lie algebra said to be the Lie algebra of the affine group $\mathrm{Aff}(\mathbb{R})$ of the real line, and will be briefly denoted by "$ax + b$".

Let now V be a real finite-dimensional vector space and let $\mu \in L(V)$ be a linear endomorphism of V. We assume that μ is exponential, i.e. no eigenvalue of the complexification $\mu_\mathbb{C}$ of μ on $V_\mathbb{C} = V \otimes_\mathbb{R} \mathbb{C}$ is purely imaginary and that $\mu_\mathbb{C}$ is diagonalisable. For λ in the spectrum of $\mu_\mathbb{C}$, let $V_{\lambda,\mathbb{C}}$ denote the corresponding eigenspace of $\mu_\mathbb{C}$ and let $V_\lambda = (V_{\lambda,\mathbb{C}} + \overline{V_{\lambda,\mathbb{C}}}) \cap V$. We denote by $\sigma_+(\mu)$ the set of eigenvalues $\lambda = \alpha + i\beta$ of $\mu_\mathbb{C}$ with $\alpha > 0, \beta \geq 0$, and similarly $\sigma_-(\mu) = \{\lambda = \alpha + i\beta$ in the spectrum of $\mu_\mathbb{C}, \alpha < 0, \beta \geq 0\}$. We suppose now that $\sigma_+(\mu)$ and $\sigma_-(\mu)$ are both non-empty. Let

$$\sigma(\mu) := \sigma_+(\mu) \cup \sigma_-(\mu).$$

This gives us a decomposition

$$V = V_0 \oplus V_+ \oplus V_- = \sum_{\lambda\in\sigma(\mu)} V_\lambda$$

of V into μ-invariant subspaces. Accordingly for every $v \in V$ we have the decomposition $v = \sum_{\lambda \in \sigma(\mu)} v_\lambda$, where $v_\lambda \in V_\lambda$ for $\lambda \in \sigma(\mu)$. Here V_0 denotes the kernel of μ, and

$$V_+ = \sum_{\lambda \in \sigma_+(\mu)} V_\lambda, \quad V_- = \sum_{\lambda \in \sigma_-(\mu)} V_\lambda.$$

Write

$$V_n = V_+ \oplus V_-.$$

We define an action of the group $(\mathbb{R}, +)$ on V by

$$t \cdot v := e^{-t\mu} v \text{ for } v \in V \text{ and } t \in \mathbb{R},$$

where, $e^{t\mu} = \sum_{j=0}^{\infty} \frac{t^j \mu^j}{j} \in L(V)$. This allows to define the (Lie) group $G(\pm)$ to be the semi-direct product $\mathbb{R} \ltimes V$ with the multiplication

$$(t, v)(t', v') = (t + t', t' \cdot v + v').$$

The identity is the element $e := (0, 0)$ and

$$(t, v)^{-1} = (-t, -e^{t\mu} v).$$

The subspace V_0 is then the centre of $G(\pm)$ and the subspaces V, V_+ and V_- are closed normal abelian subgroups of $G(\pm)$.

We identify the Lie algebra $\mathfrak{g}(\pm)$ of the Lie group $G(\pm)$ with the vector space

$$\mathbb{R}X \oplus V = \mathbb{R} \times V,$$

where $X = (1, 0)$ and then

$$[xX + v, x'X + v'] = x\mu(v') - x'\mu(v).$$

This gives us the adjoint action $Ad(t, v)$ of the element (t, v) of $G(\pm)$ in the following way

$$\begin{aligned}
&\mathrm{Ad}(t, v)(t', v') \\
&= \tfrac{d}{ds} \left((t, v)(s(t', v'))(t, v)^{-1}\right)_{|s=0} \\
&= \tfrac{d}{ds} \left((st', e^{(t-st')\mu} v + e^{t\mu} s v' - e^t v)\right)_{|s=0} \\
&= (t', -t' e^{t\mu}(\mu(v)) + e^{t\mu}(v')).
\end{aligned}$$

We identify the dual vector space $\mathfrak{g}^*(\pm)$ of $\mathfrak{g}(\pm)$ with the space $\mathbb{R} \times V^*$, where V^* denotes the linear dual space of V and we write

$$V_+^* := \{f \in V^* : f(V_-) = \{0\}\}, \ V_-^* := \{f \in V^* : f(V_+) = \{0\}\}$$

and

$$V^* = V_0^* \oplus V_+^* \oplus V_-^*.$$

For $(t, v) \in G(\pm)$ and $(\rho, f) \in \mathbb{R} \times V^*$, we have

$$\mathrm{Ad}^*(t, v)(\rho, f) = (\rho + \langle f, \mu(v) \rangle, e^{-t\mu^*} f).$$

Hence, the orbit $\Omega_{(\rho,f)}$ of (ρ, f) is given by

$$\Omega_{(\rho,f)} = \{(\rho + \langle f, \mu(v)\rangle, e^{-t\mu^*} f) : t \in \mathbb{R}, v \in V\}.$$

In particular, if $f(\operatorname{Im}(\mu)) = \{0\}$, i.e. $f(V_+) = \{0\}$ and $f(V_-) = \{0\}$, then

$$\Omega_{(\rho,f)} = \{(\rho, f)\}$$

and if $f(\operatorname{Im}(\mu)) \neq 0$, then

$$\Omega_{(\rho,f)} = \mathbb{R} \times \{e^{-t\mu^*} f : t \in \mathbb{R}\}.$$

Therefore, in the case where $f \in V^*$ with $f(V_n) \neq \{0\}$ the co-adjoint orbit of (ρ, f) does not depend on ρ and hence we write

$$\Omega_{(\rho,f)} = \Omega_f.$$

For $(t, 0) \in G(\pm)$ and $f \in V^*$, let

$$t \cdot f = \mathrm{Ad}^*((t, 0)f) = e^{-tu^*} f.$$

We choose a norm $\|\cdot\|$ on V^* and define a new norm $\|\cdot\|_1$ on V^* by

$$\|f\|_1 := \sum_{\lambda \in \sigma(\mu)} \|f_\lambda\|$$

for $f = \sum_{\lambda \in \sigma(\mu)} f_\lambda \in V^*$.

The real function $t \mapsto \nu(t) = \|t \cdot f\|_1$ is strictly decreasing for $f \in V_+^* \setminus \{0\}$ and strictly increasing for $f \in V_-^* \setminus \{0\}$, since its derivative is strictly negative (resp. strictly positive) and its range is the interval $]0, \infty[$ in these cases.

We denote by $\mathfrak{g}_{gen}^*$ the open subset of $\mathfrak{g}^*(\pm)$ consisting of all the $f = (f_0, f_+, f_-)$'s for which $f_+ \neq 0, f_- \neq 0$. The subset $\mathfrak{g}_{sin}^*$ consists of all the $f = (f_0, f_+, f_-)$'s for which either $f_+ \neq 0, f_- = 0$ or $f_+ = 0, f_- \neq 0$. Both sets $\mathfrak{g}_{gen}^*$ and $\mathfrak{g}_{sin}^*$ are $G(\pm)$-invariant. Since

$$\lim_{t \to +\infty} t \cdot f_+ = 0, \quad \lim_{t \to -\infty} \|t \cdot f_+\|_1 = +\infty$$

for $f_+ \neq 0$ and similarly for $f_- \neq 0$,

$$\lim_{t \to -\infty} t \cdot f_- = 0, \quad \lim_{t \to +\infty} \|t \cdot f_-\|_1 = +\infty,$$

we see that for every $f = (f_0, f_+, f_-) \in \mathfrak{g}_{gen}^*$ there exists exactly one element $f' = (f_0, f'_+, f'_-)$ in its orbit such that $\|f'_+\|_1 = \|f'_-\|_1$.

In the same way, for $f = (f_0, f_+, 0)$(resp. $f = (f_0, 0, f_-)) \in \mathfrak{g}_{sin}^*$, there exists exactly one element $f' = (f_0, f'_+, 0)$(resp. $f' = (f_0, 0, f'_-))$ in its orbit for which $\|f'_+\|_1 = 1$ (resp. $\|f'_-\|_1 = 1$).

Then the orbit space $\mathfrak{g}^*(\pm)/G(\pm)$ can be parametrised by the following sets,

$$\mathcal{D} = \{(f_0, f_+, f_-) : \|f_+\|_1 = \|f_-\|_1 \neq 0\},$$

$\mathcal{S}_+ = \{(f_0, f_+, 0) : \|f_+\|_1 = 1\}, \mathcal{S}_- = \{(f_0, 0, f_-) : \|f_-\|_1 = 1\}, \mathcal{S} = \mathcal{S}_+ \cup \mathcal{S}_-,$
$\mathcal{X} = \mathbb{R} \times V_0^*.$

We let

$$\mathcal{O} := \mathcal{D} \cup \mathcal{S} \cup \mathcal{X} \quad \text{and} \quad \Gamma := \mathcal{D} \cup \mathcal{S} \cup \{0_n\},$$

where 0_n denotes the zero element in V_n^*.

Remark 6.9.1 We have just seen that the mapping $\mathcal{O} \to \mathfrak{g}^*(\pm)/G(\pm) : \zeta \mapsto Ad^*(G(\pm))(\zeta)$ is a bijection.

6.9.3 The topology of the orbit space

We shall use the following description of the convergence of a sequence of co-adjoint orbits (see [70]).

Proposition 6.9.2 *Let G be a Lie group, $\mathfrak{g}$ be its Lie algebra and let $\mathfrak{g}^*/G$ be the space of co-adjoint orbits of G on the linear dual $\mathfrak{g}^*$ of $\mathfrak{g}$. Then a sequence $(\Omega_k)_k$ of co-adjoint orbits converges to an orbit Ω if and only if for any $f \in \Omega$ there exists an element $f_k \in \Omega_k$ for every $k \in \mathbb{N}$ such that $\lim_k f_k = f$ in $\mathfrak{g}^*$.*

We now determine the limit set of a properly converging sequence in $\mathfrak{g}^*(\pm)/G(\pm)$.

Theorem 6.9.3 *1. A properly converging sequence $(\Omega_{f_k})_k$ with $f_k = (f_{k_0}, f_{k_+}, f_{k_-}) \in \mathcal{D}$ has either a unique limit point Ω_f for some $f \in \mathcal{D}$ and then $f = \lim_k f_k$, or $\lim_k(f_{k_+}, f_{k_-}) = 0$ and then the limit set L of the sequence is given by*

$$L = \{\Omega_{(f_0,f_+,0)}, \Omega_{(f_0,0,f_-)}, \mathbb{R} \times \{f_0\}\},$$

where $f_+ = \lim_k \frac{f_{k_+}}{\|f_{k_+}\|_1}$ and $f_- = \lim_k \frac{f_{k_-}}{\|f_{k_-}\|_1}$ and $f_0 = \lim_k f_{k_0} \in V_0^*$.

2. A properly converging sequence (Ω_{f_k}) with $f_k = (f_{k_0}, f_{k_+}, f_{k_-}) \in \mathcal{S}$ has the limit set

$$L = \{\Omega_f, \mathbb{R} \times \{f_0\}\},$$

where $f = \lim_k f_k \in \mathcal{S}$ and $f_0 = \lim_k f_{k_0} \in V_0^*$.

Proof 6.55 *Let $(f_k = (f_{k_0}, f_{k_+}, f_{k_-}))_k \subset \mathfrak{g}^*(\pm)$ be any sequence such that the sequence $(\Omega_{f_k})_k$ has limit points in the orbit space $\mathfrak{g}^*(\pm)/G(\pm)$. Then there exists an element $f_0 \in V_0^*$ such that every element in any limit orbit has the form (ρ, f) with $f = (f_0, f_n)$ for some $f_n \in V_n^*$, since V_0 is the centre of the Lie algebra $\mathfrak{g}(\pm)$ and all the co-adjoint orbits are constant on V_0.*

Let $\overline{\Omega} = (\Omega_{f_k})_k$ be a properly convergent sequence with $f_k \in \mathcal{D}$ for $k \in \mathbb{N}$. If the sequence $(f_k)_k$ is not bounded, then there is no real sequence $(s_k)_k$ such that the sequence $((\rho_k, s_k \cdot f_k))_k$ can be bounded either, which is not possible, since $\overline{\Omega}$ is convergent.

Suppose first that the sequence $(f_{k+}, f_{k-})_k$ is not converging to the zero element 0_n in V_n^*. Then, for some subsequence (f_{k_l}), we have that $\inf_l\{\|f_{k_{l+}}\|_1 = \|f_{k_{l-}}\|_1\} > 0$. Hence, passing to a new subsequence (also indexed by l for simplicity of notation) we can assume that there exists some $f \in \mathcal{D}$ such that $\lim_{l\to\infty} f_{k_l} = f$. Since our sequence is properly convergent, the orbit Ω_f is also a limit point of our sequence of *orbits. Hence, by Proposition 6.9.2, we can choose a real sequence* $(s_k)_k$ *such that* $\lim_k s_k \cdot f_k = f$. *Since* $f \in \mathcal{D}$, *it follows that* $\lim_k s_k = 1$ *and therefore* $\lim_k f_k = f$. *Let* $\Omega_{(\rho,f')}$ *be any limit point of* $\overline{\Omega}$. *By Proposition 6.9.2 again, for every* $k \in \mathbb{N}$ *we can choose* $t_k \in \mathbb{R}$ *such that the sequence* $((\rho, t_k \cdot f_k))_k$ *converges to* (ρ, f'). *Since* $\lim_k f_k = f \in \mathcal{D}$, *the sequence* $(t_k)_k$ *must be bounded and so a subsequence* $(t_{k_l})_l$ *converges to some* $t \in \mathbb{R}$ *and finally* $(\rho, f') = (\rho, t \cdot f) \in \Omega_f$.

If now the sequence $(\|f_{k_+}\|_1)_k$ *converges to 0, we can choose a real sequence* $(r_k)_k$ *such that* $\|r_k \cdot f_{k_+}\|_1 = 1$ *for each* k. *But then* r_k *goes to* $-\infty$ *and* $r_k \cdot f_{k_-}$ *goes to* 0. *Passing to a subsequence* $(r_{k_l} \cdot f_k)$, *we find a limit point of it of the form* $(f_0, f_+, 0)$ *with* $\|f_+\|_1 = 1$. *In this way, we see that the properly convergent sequence* Ω_{f_k} *has two limit points* $\Omega_{(f_0,f_+,0)}$ *and* $\Omega_{(f_0,0,f_-)}$ *with* $\|f_+\|_1 = \|f_-\|_1 = 1$. *If we find another limit point* Ω_h *for some* $h \in \mathcal{S}_+$, *then* $r'_k \cdot f_k = (r'_k - r_k) \cdot (r_k \cdot f_k)$ *converges to* h *for some sequence* $(r'_k)_k \subset \mathbb{R}$. *But then* $(r'_k - r_k)_k$ *must be bounded and so* $\Omega_h \in \{\Omega_{(f_0,f_+,0)}, \Omega_{(f_0,0,f_-)}\}$. *Since* $\lim_k \|f_{k_+}\|_1 = \lim_k \|f_{k_-}\|_1 = 0$, *we see that the one point orbits* $\{(\tau, f_0)\}$ *for* $\tau \in \mathbb{R}$ *are also contained in the limit set* L.

Take now a properly converging sequence $(\Omega_{f_k})_k$ *with* $f_k \in \mathcal{S}, k \in \mathbb{N}$. *Then a subsequence* $(f_{k_l})_l$ *converges to some* $f \in \mathcal{S}$, *since the unit sphere in* V^* *is compact. Since we have properly converging sequences, we know that* Ω_f *is a limit point of the sequence* Ω_{f_k} *and it is the only limit point contained in* $\mathcal{S}$. *We can choose any real sequence* $(r_k)_k$ *with* $\lim_k r_k = +\infty$ *if* $f \in V^*_+$ *(resp.* $\lim_k r_k = -\infty$ *if* $f \in V^*_-$*) and then* $r_k \cdot f_k$ *goes to* $f_0 \in V^*_0$. *This shows that the limit set of the sequence* $(\Omega_{f_k})_k$ *is the set* $\{\Omega_f, \{(\tau, f_0)\}, \tau \in \mathbb{R}\}$.

Corollary 6.9.4 *The orbit* Ω_f *for* $f \in \mathcal{D}$ *is closed in* $\mathfrak{g}^*(\pm)$. *The closure of the orbit* Ω_f *for* $f = (f_0, f_n) \in \mathcal{S}$ *is the set* $\{\Omega_f, \mathbb{R} \times \{f_0\}\}$.

Proof 6.56 *If* $\zeta \in \mathfrak{g}^*(\pm) \subset \overline{\Omega_f}$, *the closure of* Ω_f *for* $f \in \mathcal{D}$, *then its orbit* Ω_ζ *is a limit point of the constant sequence* $(\Omega_f)_k$, *hence by the preceding theorem,* $\Omega_\zeta = \Omega_f$.

For $f = (f_0, f_n) \in \mathcal{S}$, *the orbit* Ω_ζ *for any* $\zeta \in \overline{\Omega_f}$ *is contained in the limit set of the constant sequence* $(\Omega_f)_k$ *and the element of every limit orbit of this sequence is contained in* $\overline{\Omega_f}$. *Hence, by Theorem 6.9.3,* $\overline{\Omega_f} = \{\Omega_f, \mathbb{R} \times \{f_0\}\}$.

6.9.4 The dual space of $G(\pm)$

The Kirillov orbit theory for exponential Lie groups $G = \exp(\mathfrak{g})$ tells us that for every irreducible unitary representation π of G there exists an $\ell \in \mathfrak{g}^*$ and Pukanszky polarisation $\mathfrak{p} \subset \mathfrak{g}$ at ℓ such that π is equivalent to the induced representation $\pi_{\ell,\mathfrak{p}} = \mathrm{ind}_P^G \chi_\ell$ of the unitary character $\chi_\ell = e^{-2\pi i \ell \circ \log_{|P}}$ of $P = \exp(\mathfrak{p})$. Furthermore for two linear functionals ℓ, ℓ' on $\mathfrak{g}$ and Pukanszky polarisations $\mathfrak{p}$ at ℓ (resp. $\mathfrak{p}'$ at ℓ'), the representations $\pi_{\ell,\mathfrak{p}}$ and $\pi_{\ell',\mathfrak{p}'}$ are equivalent if and only if ℓ and ℓ' are contained in the same G-orbit. Furthermore the topology of $\hat{G}$ and the natural topology on the orbit space $\mathfrak{g}^*(\pm)/G(\pm)$ coincide in the following sense. Let $[\pi]$ denote the unitary equivalence class of a unitary representation π of G. The Kirillov map

$$K : \mathfrak{g}^* \to \hat{G}; \quad \ell \mapsto [\pi_{\ell,\mathfrak{p}}]$$

induces a homeomorphism of the orbit space onto $\hat{G}$ (see [70] for details).

In our case for $\zeta = (\rho, f) \in \mathbb{R} \times (\mathcal{D} \cup \mathcal{S})$, the only Pukanszky polarisation $\mathfrak{p}$ at ζ is given by the abelian ideal V of $\mathfrak{g}(\pm)$ and so we obtain the irreducible representation $\pi_f = \mathrm{ind}_V^{G(\pm)} \chi_f$, where χ_f is the unitary character given by $\chi_f(v) = e^{-2\pi i f(v)}$ for $v \in V$.

If $\zeta = (\rho, f_0)$ with $f_0 \in V_0^*$, then the corresponding irreducible representation of G is one-dimensional, i.e. a unitary character, which we denote by χ_ζ.

Hence the unitary dual can be parametrised by the set $\mathcal{O}$. We have the generic set

$$\hat{\mathcal{D}} := \{\pi_f : f \in \mathcal{D}\}.$$

Since $\mathcal{D}$ is open, dense and $G(\pm)$-invariant in $\mathfrak{g}^*$, it follows that $\hat{\mathcal{D}}$ is open and dense in $\widehat{G(\pm)}$. Let

$$\hat{\mathcal{S}} := \{\pi_f : f \in \mathcal{S}\} \text{ and } \hat{\mathcal{X}} := \{\chi_\zeta : \zeta \in \mathcal{X}\}.$$

Then $\hat{\mathcal{X}}$ is the closed subset of $\widehat{G(\pm)}$ consisting of the one-dimensional representations of $G(\pm)$ and $\hat{\mathcal{S}}$ represents infinite-dimensional representations of $G(\pm)$ which are contained in the boundary of $\hat{\mathcal{D}}$. Applying the fact that the spaces $\mathfrak{g}^*(\pm)/G(\pm)$ and $\widehat{G(\pm)}$ are homeomorphic, we obtain the following property of the topology of $\widehat{G(\pm)}$.

Theorem 6.9.5 *A properly converging sequence $(\pi_{f_k})_k \subset \hat{\mathcal{D}}$ has either a unique limit point π_f with $f \in \mathcal{D}$, or the limit set given by*

$$L = \{\pi_{(f_0, f_+, 0)}, \pi_{(f_0, 0, f_-)}, \chi_{(\rho, f_0)} : \rho \in \mathbb{R}\}$$

for some $f_0 \in V_0^*$, $f_+ \in V_+^* \setminus \{0\}$ and $f_- \in V_-^* \setminus \{0\}$.

A properly converging sequence $(\pi_{f_k})_k$ with $f_k \in \mathcal{S}$ has the limit set

$$L = \{\pi_f, \chi_{(\rho, f_0)} : \rho \in \mathbb{R}\}$$

for some $f = (f_0, f_n) \in \mathcal{S}$ and $f_0 \in V_0^*$.

Corollary 6.9.6 *For every $a \in C^*(G(\pm))$, the operator $\pi_f(a)$ is compact for $f \in \mathcal{D}$, and if $f = (f_0, f_n) \in \mathcal{S}$ then π_f is compact if and only if $\chi_{(\rho, f_0)}(a) = 0$ for every $\rho \in \mathbb{R}$.*

Proof 6.57 *Since the Kirillov bijection is a homeomorphism, for $\zeta \in \mathfrak{g}^*$ the co-adjoint orbit Ω_ζ of an exponential Lie group G is closed if and only if $ker(\pi_\zeta)$, the kernel of the representation π_ζ in $C^*(G)$, is maximal. Since $\pi_\zeta(C^*(G))$ always contains non-trivial operators, the fact that the orbit is closed is equivalent to the fact that for every $a \in C^*(G)$ the operator $\pi_\zeta(a)$ is compact.*

If the orbit Ω_ζ is not closed, then the two-sided closed ideal K_ζ of $C^*(G)$ consisting of all a's in $C^*(G)$ such that $\pi_\zeta(a)$ is compact has the property that it is minimal modulo $ker(\pi_\zeta)$. Hence $K_\zeta = \bigcap_{\pi' \in \overline{\pi_\zeta} \setminus \pi} ker(\pi')$. Therefore in our case, by Corollary 6.9.4, for $f \in \mathcal{D}$, we have that $\pi_f(a)$ is always compact. And for $f = (f_0, f_n) \in \mathcal{S}$, $\pi_f(a)$ is compact if and only if $\chi_{(\rho, f_0)}(a) = 0$ for all $\rho \in \mathbb{R}$.

6.9.5 Properties of the representations $\pi_f, f \in \mathcal{D} \cup \mathcal{S}$.

Definition 6.52 We denote by L_c^1 the subspace of $L^1(G(\pm))$ consisting of all F's in $L^1(G(\pm))$ for which $\widehat{F}^2 \in C_c^\infty(\mathbb{R} \times V^*)$, where $\hat{F}^2$ is the partial Fourier transform of F in the second variable.

This space L_c^1 is dense in $L^1(G(\pm))$ and hence is also dense in $C^*(G(\pm))$. Let $f \in \mathcal{D} \cup \mathcal{S}$, then $\pi_f = \operatorname{ind}_V^{G(\pm)} \chi_f$ acts on the Hilbert space $L^2(\mathbb{R})$. Let us compute the operators $\pi_f(F)$ for $F \in C^*(G(\pm))$ explicitly.

Let $\xi \in L^2(\mathbb{R})$, $(t, v) \in G(\pm)$ and $f \in \mathcal{D} \cup \mathcal{S}$, then

$$\begin{aligned}
\pi_f(t,v)\xi(s) &= \xi((t,v)^{-1}(s,0)) \\
&= \xi((-t,-e^t v)(s,0)) \\
&= \xi((s-t,0)\cdot(0,-e^{-s\mu+t\mu}v)) \\
&= e^{-2i\pi f(e^{(-s+t)\mu}v)} \cdot \xi(s-t).
\end{aligned}$$

Hence

$$\begin{aligned}
\pi_f(F)\xi(s) &= \int_G F(g)\pi_f(g)\xi(s)dg \\
&= \int_{\mathbb{R}}\int_V F(t,v)\pi_f(t,v)\xi(s)dtdv \\
&= \int_{\mathbb{R}}\int_V F(t,v)e^{-2i\pi f(e^{(-s+t)\mu}v)}\xi(s-t)dtdv \\
&= \int_{\mathbb{R}} \hat{F}^2(t+s, e^{t\mu}f)\xi(-t)dt \\
&= \int_{\mathbb{R}} \hat{F}^2(s-t, e^{-t\mu}f)\xi(t)dt \\
&= \int_{\mathbb{R}} \hat{F}^2(s-t, t\cdot f)\xi(t)dt,
\end{aligned}$$

where $t \cdot f := e^{-t\mu}f$. That is,

$$\pi_f(F)\xi(s) = \int_{\mathbb{R}} F_f(s,t)\xi(t)dt, \tag{6.9.2}$$

where $F_f(s,t) := \hat{F}^2(s-t, t\cdot f)$.

Proposition 6.9.7 *Let $a \in C^*(G(\pm))$. The mapping $f \mapsto \pi_f(a)$ is norm-continuous on $\mathcal{D} \cup \mathcal{S}$ and $\lim_{f\to\infty} \|\pi_f(a)\|_{\mathrm{op}} = 0$ on $\mathcal{D}$.*

Proof 6.58 *For $f \in \mathcal{D}$, since $\lim_{t\to\infty} \|t\cdot f\|_1 = +\infty$, by (6.9.2), $\pi_f(F)$ is a Hilbert-Schmidt operator if $F \in L_c^1$. Therefore $\pi_f(C^*(G(\pm)))$ is compact on $L^2(\mathbb{R})$. Moreover,* *for f and $f' \in \mathcal{D}$, we have that*

$$\begin{aligned}
\|\pi_f(F) - \pi_{f'}(F)\|_{H.S}^2 &= \int |F_f(s,t) - F_{f'}(s,t)|^2 ds dt \\
&= \int |\hat{F}^2(s-t, e^{-t\mu} f) - \hat{F}^2(s-t, e^{-t\mu} f')|^2 ds dt \\
&= \int |\hat{F}^2(s, e^{-t\mu} f) - \hat{F}^2(s, e^{-t\mu} f')|^2 ds dt \\
&\to 0 \quad \text{if} \quad f \to f' \text{ in } \mathcal{D}.
\end{aligned}$$

This shows in particular that the mapping $f \mapsto \pi_f(a)$ is continuous on $\mathcal{D}$ for $a \in C^*(G(\pm))$.

Furthermore, for $F \in L_c^1(G(\pm))$ the function

$$f \mapsto \|\pi_f(F)\|_{H.S}^2 = \int |\hat{F}^2(s-t, e^{-t\mu} f)|^2 ds dt$$

is of compact support on $\mathcal{D}$. Therefore it follows that $\lim_{f\to\infty} \|\pi_f(a)\|_{\text{op}} = 0$ for every $a \in C^*(G(\pm))$.

If now a sequence $(f_k)_k \subset \mathcal{S}$ converges to some $f \in \mathcal{S}_+$, we see by Young's inequality that for $F \in L_c^1$,

$$\begin{aligned}
\|\pi_{f_k}(F) - \pi_f(F)\|_{\text{op}} &\leq \max\left(\sup_{s\in\mathbb{R}}\Big(\int_{\mathbb{R}} |\hat{F}^2(s-t, t\cdot f_k) - \hat{F}^2(s-t, t\cdot f)| dt\Big),\right. \\
&\qquad \left.\sup_{t\in\mathbb{R}}\Big(\int_{\mathbb{R}} |\hat{F}^2(s-t, t\cdot f_k) - \hat{F}^2(s-t, t\cdot f)| ds\Big)\right) \\
&\leq \max\left(\sup_{s\in\mathbb{R}}\Big(\int_{-C}^{\infty} |\hat{F}^2(s-t, t\cdot f_k) - \hat{F}^2(s-t, t\cdot f)| dt\Big),\right. \\
&\qquad \left.\sup_{t\in\mathbb{R}}\Big(\int_{-C}^{+C} |\hat{F}^2(s, t\cdot f_k) - \hat{F}^2(s, t\cdot f)| ds\Big)\right) \\
&\leq \max\left(\sup_{s\in\mathbb{R}}\Big(\int_{-C}^{\infty} C\|t\cdot f_k - t\cdot f\|\alpha(s-t) dt\Big),\right. \\
&\qquad \left.\sup_{t\in\mathbb{R}}\Big(\int_{-C}^{+C} \|t\cdot f_k - t\cdot f\| ds\Big)\right) \\
&\to 0 \quad \text{as} \quad k\to\infty,
\end{aligned}$$

since $|\hat{F}^2(s-t, t\cdot f_k) - \hat{F}^2(s-t, t\cdot f)| \leq C\|t\cdot f_k - t\cdot f\|\alpha(s-t)$ for some $0 \leq \alpha \in C_c(\mathbb{R})$ and some constant $C > 0$. Therefore we have that

$$\lim_{k\to\infty} \|\pi_{f_k}(a) - \pi_f(a)\|_{\text{op}} = 0 \tag{6.9.3}$$

for every $a \in C^*(G(\pm))$.

6.9.6 The Fourier Transform

Definition 6.53 For any $a \in C^*(G(\pm))$, let $\Phi(a)$ be the element of $C^*(\mathbb{R} \times V_0)$ defined by $\widehat{\Phi(a)}(\zeta) := \langle \chi_\zeta, a \rangle$ for $\zeta \in \mathbb{R} \times V_0^*$. The mapping $\Phi : C^*(G(\pm)) \to C^*(\mathbb{R} \times V_0)$ is a surjective homomorphism. Let the kernel of Φ be denoted by $I_{\mathcal{X}}$, then $C^*(G(\pm))/I_{\mathcal{X}} \simeq C^*(\mathbb{R} \times V_0)$.

For $\eta \in C_c(G(\pm))$, the element $\Phi(\eta) \in C^*(\mathbb{R} \times V_0)$ is the C_c-function

$$\Phi(\eta)(t, v_0) = \int_{V_n} \eta(t, v_0, v)dv \quad \text{for} \quad (t, v_0) \in \mathbb{R} \times V_0.$$

Choose $q \in C_c(V_n)$ with $q \geq 0$ and

$$\int_{V_n} q(v)dv = 1,$$

define the mapping $\beta : C_c(\mathbb{R} \times V_0) \to C_c(G(\pm)) \subset C^*(G(\pm))$ by

$$\beta(\varphi)(s, v_0, v_n) \quad = \quad \varphi(s, v_0)q(v_n) \quad \text{for } \varphi \in C_c(\mathbb{R} \times V_0) \text{ and } v_n \in V_n.$$

We see from (6.9.2) that for $f \in \mathcal{D} \cup \mathcal{S}$,

$$\pi_f(\beta(\varphi))\xi(s) \quad = \quad \int_{\mathbb{R}} \varphi(s-t)\hat{q}(t \cdot f)\xi(t)dt \quad \text{for } \xi \in L^2(\mathbb{R}) \text{ and } s \in \mathbb{R}.$$

Since $\|\hat{q}\|_\infty = 1$, we get

$$\|\pi_f(\beta(\varphi))\xi\|_2 \leq \|\hat{\varphi}\|_\infty \sqrt{\int_{\mathbb{R}} |\hat{q}(t \cdot f)\xi(t)|^2 dt} \leq \|\hat{\varphi}\|_\infty \|\xi\|_2.$$

Therefore we can extend the linear mapping β to the Banach space $C^*(\mathbb{R} \times V_0)$ (we denote this extension also by β) and then $\beta : C^*(\mathbb{R} \times V_0) \to C^*(G(\pm))$ is bounded.

Furthermore, for a unitary character $\chi_{(\rho, f_0)}$ of $G(\pm)$ we have that

$$\begin{aligned}
\langle \chi_{(\rho,f_0)}, \beta(\varphi) \rangle &= \int_{\mathbb{R} \times V_0 \times V_n} e^{-2\pi i(\rho t + \langle f_0, v_0 \rangle)} \varphi(t, v_0) q(v_n) dv_n dt dv_0 \\
&= \int_{\mathbb{R} \times V_0} e^{-2\pi i(\rho t + \langle f_0, v_0 \rangle)} \varphi(t, v_0) dt dv_0 \\
&= \hat{\varphi}(\rho, f_0) \quad \text{for} \quad \varphi \in C_c(\mathbb{R} \times V_0).
\end{aligned}$$

Lemma 6.9.8 *For every $\varphi \in C^*(\mathbb{R} \times V_0)$, we have that $\Phi(\beta(\varphi)) = \varphi$.*

Proof 6.59 *Indeed, by (6.9.4), $\widehat{\Phi(\beta(\varphi))} = \hat{\varphi}$, hence $\Phi(\beta(\varphi)) = \varphi$ by the definition of the mapping Φ in Definition 6.53.*

Definition 6.54 The Fourier transform $\hat{a} = \mathcal{F}(a)$ of an element $a \in C^*(G(\pm))$ is defined as the field of bounded linear operators defined over the set $\Gamma = \mathcal{D} \cup \mathcal{S} \cup \{0_n\}$ by

$$\begin{aligned} \hat{a}(f) = \mathcal{F}(a)(f) &:= \pi_f(a) \in \mathcal{K} \quad \text{for } f \in \mathcal{D}; \\ \hat{a}(f) = \mathcal{F}(a)(f) &:= \pi_f(a) \in \mathcal{B}(L^2(\mathbb{R})) \quad \text{for } f \in \mathcal{S}; \\ \hat{a}(0_n) = \mathcal{F}(a)(0_n) &:= \Phi(a) \in C^*(\mathbb{R} \times V_0), \end{aligned}$$

where Φ is defined in Definition 6.53.

For $s < t \in \mathbb{R} \cup \{\pm\infty\}$, denote by $M_{(s,t)}$ the multiplication operator on $L^2(\mathbb{R})$ with the characteristic function of the interval (s, t).

Definition 6.55 Let $A = (A(\gamma), \gamma \in \Gamma)$ be an operator field.

1. We say that A satisfies the *continuity condition* if the mapping $f \mapsto A(f)$ is norm continuous on $\mathcal{D} \cup \mathcal{S}$.

2. We say that A satisfies the *infinity condition* if $\|A(f)\|_{\text{op}}$ converges to 0 if f goes to ∞ in $\mathcal{D}$, and if for every pair of real sequences $(R_k)_k, (S_k)_k$ such that $R_k > S_k \to +\infty$, $\lim_k(R_k - S_k) = \infty$ we have that

$$\lim_k \sup_{f \in \mathcal{D} \cup \mathcal{S}} \|M_{(-\infty, -R_k)} \circ A(f) \circ M_{(-S_k, \infty)}\|_{\text{op}} = 0$$

and

$$\lim_k \sup_{f \in \mathcal{D} \cup \mathcal{S}} \|M_{(R_k, \infty)} \circ A(f) \circ M_{(-\infty, S_k)}\|_{\text{op}} = 0,$$

and the same is true for the field A^*.

Definition 6.56 *Let $F^*(G(\pm))$ be the space of all uniformly bounded operator fields A defined over the set Γ having the properties that*

1. *$A(f) \in \mathcal{K}$ for $f \in \mathcal{D}$,*

2. *$A(f) \in \mathcal{B}$ for $f \in \mathcal{S}$,*

3. *$A(0_n) \in C^*(\mathbb{R} \times V_0)$, and*

4. *A satisfies the continuity and infinity conditions.*

We define a norm $\|\cdot\|_\infty$ on $F^*(G(\pm))$ by

$$\|A\|_\infty := \sup_{\gamma \in \Gamma} \|A(\gamma)\|_{\text{op}}.$$

Proposition 6.9.9 *The vector space $F^*(G(\pm))$ is a C^*-algebra which contains the algebra $\mathcal{F}(C^*(G(\pm)))$.*

Proof 6.60 *In order to show that $F^*(G(\pm))$ is an algebra, it suffices to prove that for A and $B \in F^*(G(\pm))$ the operator field $A \circ B$ also satisfies the infinity condition. It is clear that $A(f) \circ B(f)$ goes to 0 in norm if f tends to ∞ in $\mathcal{D}$. For the second condition in the infinity condition, choose a sequence $(R'_k)_k$ such that $R_k > R'_k > S_k$ for all $k \in \mathbb{N}$ and $\lim_k(R_k - R'_k) = \infty, \lim_k(R'_k - S_k) = \infty$. Then*

$$\begin{aligned} & M_{(-\infty,-R_k)} \circ A(f) \circ B(f) \circ M_{(-S_k,\infty)} \\ = \; & M_{(-\infty,-R_k)} \circ A(f) \circ (M_{(-\infty,-R'_k)} + M_{(-R'_k,\infty)}) \circ B(f) \circ M_{(-S_k,\infty)} \\ = \; & M_{(-\infty,-R_k)} \circ A(f) \circ (M_{(-\infty,-R'_k)} \circ B(f) \circ M_{(-S_k,\infty)}) \\ & +(M_{(-\infty,-R_k)} \circ A(f) \circ M_{(-R'_k,\infty)}) \circ B(f) \circ M_{(-S_k,\infty)} \end{aligned}$$

and hence

$$\begin{aligned} & \|M_{(-\infty,-R_k)} \circ A(f) \circ B(f) \circ M_{(-S_k,\infty)}\|_{\mathrm{op}} \\ \leq \; & \|M_{(-\infty,-R_k)} \circ A(f) \circ (M_{(-\infty,-R'_k)} \circ B(f) \circ M_{(-S_k,\infty)})\|_{\mathrm{op}} \\ & +\|(M_{(-\infty,-R_k)} \circ A(f) \circ M_{(-R'_k,\infty)}) \circ B(f) \circ M_{(-S_k,\infty)}\|_{\mathrm{op}} \\ \leq \; & \|M_{(-\infty,-R'_k)} \circ B(f) \circ M_{(-S_k,\infty)}\|_{\mathrm{op}}\|A\|_\infty \\ & +\|M_{(-\infty,-R_k)} \circ A(f) \circ M_{(-R'_k,\infty)}\|_{\mathrm{op}}\|B\|_\infty. \end{aligned}$$

It is clear that $F^*(G(\pm))$ is complete and hence is a C^*-algebra. On the other hand, by Proposition 6.9.7 we know that for $a \in C^*(G(\pm))$, $\mathcal{F}(a)(f)$ goes to 0 if f tends to ∞ in $\mathcal{D}$. Moreover, for $F \in L^1_c$ we have the identity

$$\begin{aligned} M_{(-\infty,-R_k)} \circ \pi_f(F) \circ M_{(-S_k,\infty)}(t)\xi(s) \;=\; & \int_{-\infty}^{\infty} 1_{(-\infty,-R_k)}(s)1_{(-S_k,\infty)}(t) \\ & \hat{F}^2(s-t, t\cdot f)\xi(t)dt \end{aligned}$$

for $\xi \in L^2(\mathbb{R})$, $s, t \in \mathbb{R}$ and $f \in V^*$. Since $\lim_k(R_k - S_k) = +\infty$, it follows that

$$1_{(-\infty,-R_k)}(s)1_{(-S_k,\infty)}(t)\hat{F}^2(s-t, t\cdot f) = 0, s, t \in \mathbb{R},$$

for k large enough. Hence $M_{(-\infty,-R_k)} \circ \pi_f(F) \circ M_{(-S_k,\infty)} = 0$ for every $f \in \mathcal{D} \cup \mathcal{S}$ and for k large enough. Similarly for the third approximation in the infinity condition. Therefore $\mathcal{F}(L^1_c) \subset F^*(G(\pm))$, and by the density of L^1_c in $C^*(G(\pm))$ we have $\mathcal{F}(C^*(G(\pm))) \subset F^*(G(\pm))$.

Definition 6.57 Let $(\Omega_{f_k})_k$ ($f_k = (f_{k_0}, f_{k_+}, f_{k_-}) \in \mathcal{D}$ for all k) be a properly converging sequence in $\widehat{G(\pm)}$, whose limit set contains the orbits $\Omega_{(f_0,f_+,0)}$ and $\Omega_{(f_0,0,f_-)}$. Let $r_k, s_k \in \mathbb{R}$ be such that $\|r_k \cdot f_{k_+}\|_1 = 1$ and $\|s_k \cdot f_{k_-}\|_1 = 1$ for $k \in \mathbb{N}$. Then $\lim_k r_k = -\infty$ and $\lim_k s_k = +\infty$. Choose two positive sequences $(\rho_k)_k, (\sigma_k)_k$ such that $\rho_k > s_k, -r_k < \sigma_k$ for all $k \in \mathbb{N}$, $\lim_{k\to\infty} \rho_k - s_k = \infty, \lim_{k\to\infty} \sigma_k + r_k = \infty$ and $\lim_{k\to\infty} \frac{\rho_k - s_k}{r_k} = 0, \lim_{k\to\infty} \frac{\sigma_k + r_k}{s_k} = 0$. We say that the sequences $(\rho_k, \sigma_k)_k$ are *adapted* to the sequence $(f_k)_k$.

Remark 6.9.10 The reason for the condition $\lim_{k\to\infty}\frac{\rho_k-s_k}{r_k}=0$ is that then $\|e^{(\rho_k-s_k)\mu^*}f_{k_+}\|_1$ goes to 0 as k goes to ∞, since $\|e^{(\rho_k-s_k)\mu^*}f_{k_+}\|_1=\|e^{(\frac{\rho_k-s_k}{r_k})r_k\mu^*}f_{k_+}\|_1$ and for any sequence $(\epsilon_k)_k$ tending to 0, we have that $\lim_{k\to\infty}\|e^{\epsilon_k r_k\mu^*}f_{k_+}\|_1=0$. Similarly, it follows from the condition $\lim_{k\to\infty}\frac{\sigma_k+r_k}{s_k}=0$ that $\lim_{k\to\infty}\|e^{(\sigma_k+r_k)\mu^*}f_{k_-}\|_1=0$.

Definition 6.58 For $r\in\mathbb{R}$, let $U(r)$ be the unitary operator on $L^2(\mathbb{R})$ defined by

$$U(r)\xi(s)\ :=\ \xi(s+r),\quad \text{for all } \xi\in L^2(\mathbb{R}) \text{ and } s\in\mathbb{R}.$$

Definition 6.59 Let $A=(A(f)\in\mathcal{B}, f\in\Gamma)$ be a field of bounded operators. We say that A satisfies the *generic condition* if for every properly converging sequence $(\pi_{f_k})_k\subset\widehat{G(\pm)}$ with $f_k\in\mathcal{D}$ for every $k\in\mathbb{N}$, which admits limit points $\pi_{(f_0,0,f_-)},\pi_{(f_0,f_+,0)}$ and for every pair of sequences $(\rho_k,\sigma_k)_k$ adapted to the sequence $(f_k)_k$ we have that

1. $$\lim_{k\to\infty}\|U(r_k)\circ A(f_k)\circ U(-r_k)\circ M_{(-\infty,\sigma_k)}-A(f_0,f_+,0)\circ M_{(-\infty,\sigma_k)}\|_{\text{op}}\ =\ 0,$$

2. $$\lim_{k\to\infty}\|U(s_k)\circ A(f_k)\circ U(-s_k)\circ M_{(-\rho_k,\infty)}-A(f_0,0,f_-)\circ M_{(-\rho_k,\infty)}\|_{\text{op}}\ =\ 0.$$

Proposition 6.9.11 *Let A and B be two operator fields in $F^*(G(\pm))$ satisfying the generic and infinity conditions. Then $A\circ B$ satisfies the generic condition.*

Proof 6.61 *Let $f_k=(f_{k_0},f_{k_+},f_{k_-})\in\mathcal{D}$ for all $k\in\mathbb{N}$. Then*

$$\begin{aligned}
& U(s_k)\circ A(f_k)\circ B(f_k)\circ U(-s_k)\circ M_{(-\rho_k,\infty)}-A(f_0,0,f_-)\circ B(f_0,0,f_-)\circ M_{(-\rho_k,\infty)}\\
=\ & U(s_k)\circ A(f_k)\circ U(-s_k)\circ B(f_0,0,f_-)\circ M_{(-\rho_k,\infty)}-A(f_0,0,f_-)\circ B(f_0,0,f_-)\\
& \circ M_{(-\rho_k,\infty)}+U(s_k)\circ A(f_k)\circ U(-s_k)\circ(U(s_k)\circ B(f_k)\circ U(-s_k)\circ M_{(-\rho_k,\infty)}\\
& -B(f_0,0,f_-)\circ M_{(-\rho_k,\infty)}).
\end{aligned}$$

Hence

$$\begin{aligned}
& \lim_k\|U(s_k)\circ A(f_k)\circ B(f_k)\circ U(-s_k)\circ M_{(-\rho_k,\infty)}-A(f_0,0,f_-)\\
& \circ B(f_0,0,f_-)\circ M_{(-\rho_k,\infty)}\|_{\text{op}}\\
\leq\ & \lim_k\|U(s_k)\circ A(f_k)\circ U(-s_k)\circ B(f_0,0,f_-)\circ M_{(-\rho_k,\infty)}-A(f_0,0,f_-)\circ B(f_0,0,f_-)\\
& \circ M_{(-\rho_k,\infty)}\|_{\text{op}}+\lim_k\|U(s_k)\circ A(f_k)\circ U(-s_k)\circ\\
& \Big(U(s_k)\circ B(f_k)\circ U(-s_k)\circ M_{(-\rho_k,\infty)}-B(f_0,0,f_-)\circ M_{(-\rho_k,\infty)}\Big)\|_{\text{op}}\\
\leq\ & \lim_k\|U(s_k)\circ A(f_k)\circ U(-s_k)\circ B(f_0,0,f_-)\circ M_{(-\rho_k,\infty)}-A(f_0,0,f_-)\\
& \circ B(f_0,0,f_-)\circ M_{(-\rho_k,\infty)}\|_{\text{op}}+\delta_k,
\end{aligned}$$

where

$$\delta_k := \|U(s_k)\circ A(f_k)\circ U(-s_k)\circ\Big(U(s_k)\circ B(f_k)\circ U(-s_k)\circ M_{(-\rho_k,\infty)} - B(f_0,0,f_-)\circ M_{(-\rho_k,\infty)}\Big)\|_{\text{op}}$$

tends to 0 as k goes to infinity by the generic condition on the field B. Furthermore, choosing an adapted sequence $(\rho'_k)_k$ *such that* $\rho'_k > \rho_k > \sigma_k$ *for* $k \in \mathbb{N}$ *and* $\lim_k(\rho'_k - \rho_k) = +\infty$, *we obtain for*

$$\eta_k := U(s_k)\circ A(f_k)\circ U(-s_k)\circ B(f_0,0,f_-)\circ M_{(-\rho_k,\infty)} - A(f_0,0,f_-)\circ B(f_0,0,f_-)\circ M_{(-\rho_k,\infty)}.$$

We have:

$$\begin{aligned}
&\|\eta_k\|_{\text{op}}\\
\leq\ &\|U(s_k)\circ A(f_k)\circ U(-s_k)\circ M_{(-\rho'_k,\infty)}\circ B(f_0,0,f_-)\circ M_{(-\rho_k,\infty)}\\
&\quad -A(f_0,0,f_-)\circ M_{(-\rho'_k,\infty)}\circ B(f_0,0,f_-)\circ M_{(-\rho_k,\infty)}\|_{\text{op}}\\
&+\|U(s_k)\circ A(f_k)\circ U(-s_k)\circ M_{(-\infty,-\rho'_k)}\circ B(f_0,0,f_-)\circ M_{(-\rho_k,\infty)}\\
&\quad -A(f_0,0,f_-)\circ M_{(-\infty,-\rho'_k)}\circ B(f_0,0,f_-)\circ M_{(-\rho_k,\infty)}\|_{\text{op}}\\
\leq\ &\|U(s_k)\circ A(f_k)\circ U(-s_k)\circ M_{(-\rho'_k,\infty)}\\
&-A(f_0,0,f_-)\circ M_{(-\rho'_k,\infty)}\|_{\text{op}}\|B(f_0,0,f_-)\circ M_{(-\rho_k,\infty)}\|_{\text{op}}\\
&+\|U(s_k)\circ A(f_k)\circ U(-s_k) - A(f_0,0,f_-)\|_{\text{op}}\|M_{(-\infty,-\rho'_k)}\circ B(f_0,0,f_-)\\
&\circ M_{(-\rho_k,\infty)}\|_{\text{op}}.
\end{aligned}$$

Hence by the generic condition for the field A and the infinity condition for the field B, $\lim_k \|\eta_k\|_{\text{op}} = 0$. *Similar estimations hold for the second generic approximation. Hence the field* $A\circ B$ *also satisfies the generic condition.*

Proposition 6.9.12 *For every* $a \in C^*(G(\pm))$, *the operator field* $\mathcal{F}(a)$ *satisfies the generic condition.*

Proof 6.62 *We use the notations in Definition 6.59. Take any* $F \in L^1_c$, *for* $\xi \in L^2(\mathbb{R})$ *and* $s \in \mathbb{R}$, *we have that*

$$U(s_k)\circ\hat{F}(f_k)\circ U(-s_k)\circ M_{(-\rho_k,\infty)}(\xi)(s) = \int_{-\rho_k}^{\infty}\hat{F}^2(s-t,(t+s_k)\cdot f_k)\xi(t)dt.$$

Hence, by Young's inequality,

$$\begin{aligned}
&\|U(s_k)\circ\hat{F}(f_k)\circ U(-s_k)\circ M_{(-\rho_k,\infty)} - \hat{F}(f_0,0,f_-)\circ M_{(-\rho_k,\infty)}\|_{\text{op}}\\
\leq\ &\max\Bigg(\sup_{s\in\mathbb{R}}\Big|\int_{-\rho_k}^{\infty}\hat{F}^2(s-t,(t+s_k)\cdot f_k) - \hat{F}^2(s-t,t\cdot(f_0,0,f_-))dt\Big|\\
&\sup_{t\in\mathbb{R}}\Big|\int_{-\rho_k}^{\infty}\hat{F}^2(s-t,(t+s_k)\cdot f_k) - \hat{F}^2(s-t,t\cdot(f_0,0,f_-))ds\Big|\Bigg).
\end{aligned}$$

Since $F \in L^1_c$, it follows that there exists a function $0 \le g \in C_c(\mathbb{R})$ such that

$$\begin{aligned} & |\hat{F}^2(s-t,(t+s_k)\cdot f_k) - \hat{F}^2(s-t,t\cdot(f_0,0,f_-))| \\ = \ & |\hat{F}^2(s-t,f_0,e^{(-t-s_k)\mu^*}f_{k_+},e^{(-t-s_k)\mu^*}f_{k_-}) - \hat{F}^2(s-t,f_0,0,e^{-t\mu^*}f_-)| \\ \le \ & (\|e^{(-t-s_k)\mu^*}f_{k_+}\|_1 + \|e^{-t\mu^*}f_- - e^{(-t-s_k)\mu^*}f_{k_-}\|_1)g(s-t) \text{ for all } s,t\in\mathbb{R}, \end{aligned}$$

and $k \in \mathbb{N}$. There exists therefore $M > 0$ such that

$$\hat{F}^2(s-t,f_0,e^{(-t-s_k)\mu^*}f_{k_+},e^{(-t-s_k)\mu^*}f_{k_-}) - \hat{F}^2(s-t,f_0,0,e^{-t\mu^*}f_-) = 0$$

for $t > M$ and $s \in \mathbb{R}$, and $\hat{F}^2(u,h) = 0$ for any $h \in V$ and $|u| > M$. Furthermore, we have that $\lim_k e^{-s_k\mu^*}f_{k_-} = f_-$ by Proposition 6.9.3. Hence

$$\begin{aligned} & \sup_{s\in\mathbb{R}} \left| \int_{-\rho_k}^{\infty} \hat{F}^2(s-t,f_0,e^{(-t-s_k)\mu^*}f_{k_+},e^{(-t-s_k)\mu^*}f_{k_-}) - \hat{F}^2(s-t,f_0,0,e^{-t\mu^*}f_-)dt \right| \\ \le \ & \sup_{s\in\mathbb{R}} \int_{-\rho_k}^{M} (\|e^{(\rho_k-s_k)\mu^*}f_{k_+}\|_1 + \|e^{-t\mu^*}f_- - e^{(-t-s_k)\mu^*}f_{k_-}\|_1)g(s-t)dt \to 0 \end{aligned}$$

as $k \to \infty$, and

$$\begin{aligned} & \sup_{t\in\mathbb{R}} \left| \int_{-\rho_k}^{\infty} \hat{F}^2(s-t,(t+s_k)\cdot f_k) - \hat{F}^2(s-t,t\cdot(f_0,0,f_-))ds \right| \\ = \ & \sup_{t\in[-\rho_k-M,M]} \left| \int_{-\rho_k}^{\infty} \hat{F}^2(s-t,f_0,e^{(-t-s_k)\mu^*}f_{k_+},e^{(-t-s_k)\mu^*}f_{k_-}) \right. \\ & \left. -\hat{F}^2(s-t,f_0,0,e^{-t\mu^*}f_-)ds \right| \\ \le \ & \sup_{t\in[-\rho_k-M,M]} \left| \int_{-\rho_k}^{\infty} (\|e^{(\rho_k+M-s_k)\mu^*}f_{k_+}\|_1 + \|e^{-t\mu^*}f_- - e^{(-t-s_k)\mu^*}f_{k_-}\|_1)g(s-t)ds \right| \\ \le \ & \sup_{t\in[-\rho_k-M,M]} (\|e^{(\rho_k+M-s_k)\mu^*}f_{k_+}\|_1 + \|e^{-t\mu^*}f_- - e^{(-t-s_k)\mu^*}f_{k_-}\|_1)\int_{-\infty}^{\infty} g(s)ds \\ \to \ & 0 \quad \text{as} \quad k\to\infty. \end{aligned}$$

This shows that the second approximation in the generic condition holds for $F \in L^1_c$. Similarly for the first approximation in the generic condition. Since L^1_c is dense in $C^*(G(\pm))$, the proof follows.

6.9.7 The C^*-algebras of the groups $G(+)$ and $G(-)$

Definition 6.60 The group $G(+) := \mathbb{R} \ltimes (V_0 \times V_+)$ is defined to be the subgroup of $G(\pm)$ and similarly $G(-) := \mathbb{R} \ltimes (V_0 \times V_-)$.

We consider rather both groups as quotient groups of $G(\pm)$, this is,

$$G(+) = G(\pm)/V_- \quad \text{and} \quad G(-) = G(\pm)/V_+.$$

The quotient mapping

$$p_+ : G(\pm) \to G(+); (t,v_0,v_+,v_-) \mapsto (t,v_0,v_+) \simeq (t,v_0,v_+,v_-) \text{ modulo } V_-$$

maps $G(\pm)$ onto $G(+)$ and similarly for the corresponding mapping $p_- : G(\pm) \to G(-)$.

We obtain then surjective homomorphisms $P_+ : C^*(G(\pm)) \to C^*(G(+))$ (resp. $P_- : C^*(G(\pm)) \to C^*(G(-))$), where for $\eta \in C_c(G(\pm))$ the element $P_+(\eta)$ (resp. $P_-(\eta)$) is a C_c-function defined by

$$P_+(\eta)(t, v_0, v_+) \quad := \quad \int_{V_-} \eta(t, v_0, v_+, v)dv, v_0 \in V_0,\ v_+ \in V_+,\ t \in \mathbb{R}$$

(resp.

$$P_-(\eta)(t, v_0, v_-) \quad := \quad \int_{V_+} \eta(t, v_0, v, v_-)dv, v_0 \in V_0,\ v_- \in V_-,\ t \in \mathbb{R}).$$

According to Kirillov's theory, the dual space $\widehat{G(+)}$ (resp. $\widehat{G(-)}$) consists of the irreducible infinite-dimensional representations $\pi_f = \mathrm{ind}_V^G \chi_f$ for $f = (f_0, f_+) \in V_0^* \times V_+^*$ with $\|f_+\|_1 = 1$ i.e. $f \in \mathcal{S}_+$ (resp. $f \in \mathcal{S}_-$) and the unitary characters χ_ζ for $\zeta \in \mathbb{R} \times V_0^*$.

Similar to Definition 6.53, for any $a \in C^*(G(+))$ ($b \in C^*(G(-))$) we define the surjective homomorphism $\Phi_+ : C^*(G(+)) \to C^*(\mathbb{R} \times V_0)$ (resp. $\Phi_- : C^*(G(-)) \to C^*(\mathbb{R} \times V_0)$) by

$$\hat{\Phi}_+(a)(\zeta) := \langle \chi_\zeta, a \rangle \quad (\text{resp. } \hat{\Phi}_-(b)(\zeta) := \langle \chi_\zeta, b \rangle).$$

Definition 6.61 Let $A = (A(f) \in \mathcal{B}, f \in \Gamma)$ be a field of bounded operators. We say that A satisfies the *compact condition* if $A(f) - \pi_f(\beta(A(0_n))) \in \mathcal{K}$ for every $f \in \mathcal{S}$, where $\beta : C^*(\mathbb{R} \times V_0) \to C^*(G(\pm))$ is given in Definition 6.53.

Remark 6.9.13 Then $\beta_+ : C^*(\mathbb{R} \times V_0) \to C^*(G(+))$ (resp. $\beta_- : C^*(\mathbb{R} \times V_0) \to C^*(G(-))$) defined by $\beta_+ := P_+ \circ \beta$ (resp. $\beta_- := P_- \circ \beta$) is a bounded linear mapping such that $\Phi_+(\beta_+(\varphi)) = \varphi$ (resp. $\Phi_-(\beta_-(\varphi)) = \varphi$) for $\varphi \in C^*(\mathbb{R} \times V_0)$.

Note that for $f = (f_0, f_+, 0)$ (resp. $f = (f_0, 0, f_-)$) in $\mathcal{S}$ and $a \in C^*(G(\pm))$, the operator $\pi_f(a)$ is equal to $\pi_{(f_0,f_+)}(P_+(a))$ (resp. $\pi_{(f_0,f_-)}(P_-(a))$) and $\pi_f(\beta(a(0_n))) = \pi_{(f_0,f_+)}(\beta_+(P_+(a)(0_n)))$ (resp. $\pi_f(\beta(a(0_n))) = \pi_{(f_0,f_-)}(\beta_-(P_-(a)(0_n))))$.

Definition 6.62 We denote by $D^*(G(+))$ (resp. $D^*(G(-))$) the space of all operator fields A defined over $\mathcal{S}_+ \cup \{0_n\}$ (resp. $\mathcal{S}_- \cup \{0_n\}$) which satisfy the following conditions

1. $A(f) \in \mathcal{B}(L^2(\mathbb{R}))$ for $f \in \mathcal{S}_+$ (resp. $f \in \mathcal{S}_-$);
2. The mapping $f \mapsto A(f)$ is norm continuous on $\mathcal{S}_+$ (resp. on $\mathcal{S}_-$);
3. A and A^* satisfy the compact condition, which means that $A(f) - \pi_f(\beta_+(A(0_n))) \in \mathcal{K}$ for $f \in \mathcal{S}_+$ (resp. $A(f) - \pi_f(\beta_-(A(0_n))) \in \mathcal{K}$ for $f \in \mathcal{S}_-$) and similarly for A^*.

We provide the space $D^*(G(+))$ and $D^*(G(-))$ with the norms $\|A\| := \sup_{f\in\mathcal{S}_+\cup\{0_n\}} \|A(f)\|_{\text{op}}$ and $\|A\| := \sup_{f\in\mathcal{S}_-\cup\{0_n\}} \|A(f)\|_{\text{op}}$, respectively.

Definition 6.63 We define the Fourier transform $\mathcal{F} : C^*(G(+)) \to D^*(G(+))$ by

$$\mathcal{F}(a)(\gamma) := \hat{a}(\gamma) := \begin{cases} \pi_f(a) \in \mathcal{B}(L^2(\mathbb{R})) & \text{for } \gamma = f \in \mathcal{S}_+, \\ \Phi_+(a) \in C^*(\mathbb{R}\times V_0) & \text{for } \gamma = 0_n. \end{cases}$$

Proposition 6.9.14 *For any $a \in C^*(G(+))$ we have that $\hat{a} \in D^*(G(+))$.*

Proof 6.63 *Let $F \in C_c(G(+))$. We have seen in Proposition 6.9.7 that the mapping $\mathcal{S} \to \mathcal{B}(L^2(\mathbb{R})) : f \mapsto \pi_f(F)$ is norm continuous. Furthermore, for the element $H := F - \beta(\Phi_+(F)) \in C_c(G(+))$ we have that*

$$\begin{aligned} \langle \chi_{(\rho,f_0)}, H\rangle &= \int_{\mathbb{R}\times V_0\times V_n} \Big(F(t,v_0,v_n) - (\int_{V_n} F(t,v_0,w_n)dw_n)q(v_n)\Big)\, e^{-2\pi i(\rho t+\langle f_0,v_0\rangle)} \\ &\quad dv_0 dv_n dt \\ &= \int_{\mathbb{R}\times V_0\times V_n} F(t,v_0,v_n)e^{-2\pi i(\rho t+\langle f_0,v_0\rangle)} dv_0 dv_n dt \\ &\quad - \int_{\mathbb{R}\times V_0} (\int_{V_n} F(t,v_0,w_n)dw_n)e^{-2\pi i(\rho t+\langle f_0,v_0\rangle)} dv_0 dt \int_{V_n} q(v_n)dv_n \\ &= \int_{\mathbb{R}\times V_0\times V_n} (F(t,v_0,w_n) - F(t,v_0,w_n))e^{-2\pi i(\rho t+\langle f_0,v_0\rangle)} dv_0 dw_n dt \\ &= 0 \quad \text{for } \rho\in\mathbb{R} \text{ and } f_0 \in V_0^*. \end{aligned}$$

By Corollary 6.9.6, $\pi_f(F - \beta(\Phi_+(F)))$ is compact for each $f \in \mathcal{S}$. Since $C_c(G(+))$ is dense in $C^*(G(+))$, the mapping β is bounded and $\Phi_+(F) = F(0_n)$ for $F \in C_c(G(+))$, every operator field $\hat{a}$ for $a \in C^*(G(+))$ satisfies the continuity and the compact condition.

Proposition 6.9.15 *The space $D^*(G(+))$ is a C^*-algebra containing $\mathcal{F}(C^*(G(+)))$ as a C^*-subalgebra.*

Proof 6.64 *In order to see that the space $D^*(G(+))$ is complete, let $(A_k)_k \subset D^*(G(+))$ be a Cauchy sequence. Then for every $\gamma \in \mathcal{S}_+ \cup \{0_n\}$ the sequence $(A_k(\gamma))_k$ is a Cauchy sequence in $\mathcal{B}(L^2(\mathbb{R}))$ (resp. in $C^*(\mathbb{R}\times V_0)$). Hence we have an operator field $A(f) := \lim_k A_k(f) \in \mathcal{B}(L^2(\mathbb{R}))$ for $f \in \mathcal{S}_+$ and $A(0_n) := \lim_k A_k(0_n) \in C^*(\mathbb{R}\times V_0)$. Since A is a uniform limit, it follows that the map $f \mapsto A(f)$ is norm continuous on $\mathcal{S}_+$. Furthermore, for every $f \in \mathcal{S}_+$, we have that $A(f) - \pi_f(\beta(A(0_n))) = \lim_k A_k(f) - \pi_f(\beta(A_k(0_n))) \in \mathcal{K}$. Hence A is an element of $D^*(G(+))$.*

To see that $D^*(G(+))$ is a $*$-algebra, let A, B be two elements of $D^*(G(+))$. It is clear that $A + B, A^*$ are also contained in $D^*(G(+))$ and that the mapping $f \mapsto A(f)\circ B(f)$ is continuous on $\mathcal{S}_+$. Since

$$\begin{aligned} & A(f)\circ B(f) - \pi_f(\beta(A(0_n)))\circ\pi_f(\beta(B(0_n))) \\ = {} & A(f)\circ B(f) - \pi_f(\beta(A(0_n)))\circ B(f) \\ & +\pi_f(\beta(A(0_n)))\circ B(f) - \pi_f(\beta(A(0_n)))\circ\pi_f(\beta(B(0_n))) \\ \in {} & \mathcal{K}, \end{aligned}$$

it suffices to see that $\pi_f(\beta(A(0_n))) \circ \pi_f(\beta(B(0_n))) - \pi_f(\beta((A \cdot B)(0_n))) \in \mathcal{K}$ *for* $f \in \mathcal{S}_+$.

Choose $a, b \in C^*(G(+))$ *such that* $\Phi_+(a) = \hat{a}(0_n) = A(0_n)$ *and* $\Phi_+(b) = \hat{b}(0_n) = B(0_n)$. *Since* $C^*(G(+))$ *is an algebra, we have that*

$$\begin{aligned} & \pi_f(\beta(A(0_n))) \circ \pi_f(\beta(B(0_n))) - \pi_f(\beta((A \cdot B)(0_n))) \\ = \; & \pi_f(\beta(\hat{a}(0_n))) \circ \pi_f(\beta(\hat{b}(0_n))) - \pi_f(\beta((a \cdot b)\widehat{\ }(0_n))) \\ = \; & \pi_f(\beta(\hat{a}(0_n))) \circ \pi_f(\beta(\hat{b}(0_n))) - \hat{a}(f) \circ \hat{b}(f) \\ & + (a \cdot b)\widehat{\ }(f) - \pi_f(\beta((a \cdot b)\widehat{\ }(0_n))) \\ \in \; & \mathcal{K}. \end{aligned}$$

The proof is therefore finished.

Theorem 6.9.16 *The Fourier transform* $\mathcal{F}$ *maps* $C^*(G(+))$ *onto* $D^*(G(+))$.

Proof 6.65 *We know from Proposition 6.9.15 that* $\mathcal{F}(C^*(G(+)))$ *is contained in* $D^*(G(+))$. *Let us show that the* C^**-algebra* $D^*(G(+))$ *is of type I and that its dual space coincides with* $\widehat{G(+)}$. *Let* $C(\mathcal{S}_+, \mathcal{K})$ *be the* C^**-algebra of all continuous maps* $A : \mathcal{S}_+ \to \mathcal{K}$ *from the compact space* $\mathcal{S}_+$ *to the algebra* $\mathcal{K}$ *of all compact operators on the Hilbert space* $L^2(\mathbb{R})$. *We can identify* $C(\mathcal{S}_+, \mathcal{K})$ *with the subalgebra of* $D^*(G(+))$ *of all operator fields* A *for which* $A(f) \in \mathcal{K}$ *for* $f \in \mathcal{S}_+$ *in the following sense.*

Let $R_{\mathcal{X}} : D^*(G(+)) \to C^*(\mathbb{R} \times V_0)$ *be the restriction map defined by* $R_{\mathcal{X}}(A) := A(0_n)$ *for* $A \in D^*(G(+))$. *The kernel* $I_{\mathcal{X}}$ *of this homomorphism is given by*

$$\begin{aligned} I_{\mathcal{X}} &:= \{A \in D^*(G(+)) : A(0_n) = 0\} \\ &= \{A \in D^*(G(+)) : A(f) \in \mathcal{K} \text{ for all } f \in \mathcal{S}_+\} \\ & \quad (\text{by the compact condition of all operator fields in } D^*(G(+))). \end{aligned}$$

Hence, we can identify in this way the ideal $I_{\mathcal{X}}$ *with the algebra* $C(\mathcal{S}_+, \mathcal{K})$.

For every $f \in \mathcal{S}_+$, *we have an irreducible representation* δ_f *of* $D^*(G(+))$ *defined by* $\delta_f(A) := A(f)$ *for every* $A \in D^*(G(+))$. *Furthermore the character* χ_ζ *of* $D^*(G(+))$ *is defined by* $\langle \chi_\zeta, A \rangle := \langle \chi_\zeta, A(0_n) \rangle$ *for* $A \in D^*(G(+))$ *and* $\zeta \in \mathbb{R} \times V_0^*$.

Let $\pi \in \widehat{G(+)}$. *If* π *is not 0 on* $I_{\mathcal{X}}$, *then* π *determines an irreducible representation of the algebra* $C(\mathcal{S}_+, \mathcal{K})$ *and hence there exists an* $f \in \mathcal{S}_+$ *such that* $\pi(A) = A(f)$ *for* $A \in I_{\mathcal{X}}$. *But then* $\pi = \delta_f$, *since* $I_{\mathcal{X}}$ *is a two-sided ideal.*

If $\pi(I_{\mathcal{X}}) = \{0\}$, *then we can identify* π *with an irreducible representation of the algebra* $D^*(G(+))/I_{\mathcal{X}} \simeq range(R_{\mathcal{X}}) = C^*(\mathbb{R} \times V_0)$. *Hence* π *is a character of* $C^*(\mathbb{R} \times V_0)$ *and therefore* $\pi = \chi_\zeta$ *for some* $\zeta \in \mathbb{R} \times V_0^*$.

We can now apply the theorem of Stone-Weierstrass, which tells us that $\mathcal{F}(C^*(G(+))) = D^*(G(+))$.

6.9.8 The algebra $D^*_\pm$

Definition 6.64 *We define the space* $D^*_\pm$ *as the subspace of* $D^*(G(+)) \oplus D^*(G(-))$ *consisting of all pairs* $(A, B) \in D^*(G(+)) \oplus D^*(G(-))$ *such that* $A(0_n) = B(0_n) \in C^*(\mathbb{R} \times V_0)$, *where* $D^*(G(+))$ *and* $D^*(G(-))$ *are defined in Definition 6.62.*

Proposition 6.9.17 *The vector space $D^*_\pm$ is a C^*-subalgebra of $D^*(G(+)) \oplus D^*(G(-))$.*

Proof 6.66 *It is known from Proposition 6.9.15 that $D^*(G(+))$ (resp. $D^*(G(-))$) is a C^*-algebra, and $D^*_\pm$ is obviously a closed $*$-subalgebra of $D^*(G(+)) \oplus D^*(G(-))$.*

Definition 6.65 Let $C^*(G_\pm)$ be the C^*-subalgebra of $C^*(G(+)) \oplus C^*(G(-))$ consisting of all pairs $(a, b) \in C^*(G(+)) \oplus C^*(G(-))$ such that $\Phi_+(a) = \Phi_-(b)$, where $\Phi_+ : C^*(G(+)) \to C^*(\mathbb{R} \times V_0)$ and $\Phi_- : C^*(G(-)) \to C^*(\mathbb{R} \times V_0)$ are the surjective homomorphisms defined by $\hat{\Phi}_+(a)(\zeta) = \langle \chi_\zeta, a\rangle$ and $\hat{\Phi}_-(b)(\zeta) = \langle \chi_\zeta, b\rangle$ in Section 6.9.7.

Proposition 6.9.18 *The algebras $D^*_\pm$ and $C^*(G_\pm)$ are isomorphic.*

Proof 6.67 *We know from Section 6.9.7 that the algebra $C^*(G(+))$ is isomorphic to the algebra $D^*(G(+))$ of operator fields. Hence we have a homomorphism J from $C^*(G_\pm)$ into $D^*_\pm$ given by*

$$J(a, b) := (\hat{a}, \hat{b}) \text{ on } \mathcal{S} \quad \text{and} \quad J(a, b)(0_n) := a(0_n)(= b(0_n)).$$

This mapping J is clearly bijective.

Definition 6.66 Define a linear mapping $P : C_c(G(\pm)) \to C_c(G_+) \oplus C_c(G_-)$ by

$$P(\eta) \quad := \quad (\eta_+, \eta_-) = (P_+(\eta), P_-(\eta)) \text{ for } \eta \in C_c(G(\pm)),$$

where P_+ and P_- are defined in Definition 6.60.

We consider the canonical homomorphisms $\Phi_+ : C_c(G(+)) \to C_c(\mathbb{R} \times V_0)$ and $\Phi_- : C_c(G(-)) \to C_c(\mathbb{R} \times V_0)$ as defined in Definition 6.53 and Section 6.9.7 by

$$\begin{aligned} \Phi_+(\eta_+)(t, v_0) &:= \int_{V_+} \eta_+(t, v_0, v)dv, \ \eta_+ \in C_c(G(+)), \\ \Phi_-(\eta_-)(t, v_0) &:= \int_{V_-} \eta_-(t, v_0, v)dv, \ \eta_- \in C_c(G(-)). \end{aligned}$$

Let $C_c(G_\pm)$ be the subspace of $C_c(G(+)) \oplus C_c(G(-))$ consisting of the pairs (η_+, η_-) for which $\Phi_+(\eta_+) = \Phi_-(\eta_-)$.

Proposition 6.9.19 *For every $\eta \in C_c(G(\pm))$, the function $P(\eta)$ is contained in $C_c(G_\pm)$ and the mapping $P : C_c(G(\pm)) \to C_c(G_\pm)$ is surjective.*

Proof 6.68 *Indeed, we have for $\eta \in C_c(G(\pm))$ that*

$$\begin{aligned} \Phi_+(\eta_+)(t, v_0) &= \int_{V_+} \int_{V_-} \eta(t, v_0, v_+, v_-)dv \\ &= \int_{V_-} \int_{V_+} \eta(t, v_0, v_+, v_-)dv = \Phi_-(\eta_-)(t, v_0), \end{aligned}$$

for all $t \in \mathbb{R}$ and $v_0 \in V_0$. Hence $P(\eta) \in C_c(G_\pm)$.

Take $(\eta_+, \eta_-) \in C_c(G_\pm)$. Since the mapping $P_+ : C_c(G(\pm)) \to C_c(G(+))$ by $\eta \mapsto \eta_+$ is surjective, there exists an element $\sigma \in C_c(G(\pm))$ such that $\sigma_+ = \eta_+$. Choose any function $q \in C_c(V_+)$ such that $\int_{V_+} q(v)dv = 1$ and let $\gamma_- := \eta_- - \sigma_-$. Then

$$\begin{aligned} \int_{V_-} \gamma_-(t, v_0, v_-)dv &= \Phi_-(\eta_-)(t, v_0) - \int_{V_-}\int_{V_+} \sigma(t, v_0, v_+, v_-)dv \\ &= \Phi_+(\eta_+)(t, v_0) - \Phi_+(\sigma_+)(t, v_0) = 0, \end{aligned}$$

for all $t \in \mathbb{R}$ and $v_0 \in V_0$. Define the function $\psi \in C_c(G(\pm))$ by

$$\psi(t, v_0, v_+, v_-) := \sigma(t, v_0, v_+, v_-) + \gamma_-(t, v_0, v_-)q(v_+) \quad \text{for all} \quad (t, v_0, v_+, v_-) \in G(\pm).$$

Then $\psi \in C_c(G(\pm))$, and

$$\begin{aligned} \psi_+(t, v_0, v_+) &= \sigma_+(t, v_0, v_+) + q(v_+)\int_{V_-} \gamma_-(t, v_0, v_-)dv = \sigma_+(t, v_0, v_+) + 0 \\ &= \eta_+(t, v_0, v_+) \quad \text{for } t \in \mathbb{R},\ v_0 \in V_0 \text{ and } v_+ \in V_+, \\ \psi_-(t, v_0, v_-) &= \sigma_-(t, v_0, v_-) + \gamma_-(t, v_0, v_-)\int_{V_+} q(v_+)dv \\ &= \sigma_-(t, v_0, v_-) + \eta_-(t, v_0, v_-) - \sigma_-(t, v_0, v_-) \\ &= \eta_-(t, v_0, v_-) \quad \text{for } t \in \mathbb{R},\ v_0 \in V_0 \text{ and } v_- \in V_-. \end{aligned}$$

This shows that the mapping P is surjective.

Definition 6.67 For $a \in C^*(G(\pm))$, define the homomorphism $R : C^*(G(\pm)) \to D^*_\pm$ by

$$\begin{aligned} Ra(f) &:= \hat{a}(f) \quad \text{for } f \in \mathcal{S}, \\ Ra(0_n) &:= \hat{a}(0_n) = \Phi(a). \end{aligned}$$

Proposition 6.9.20 *The canonical homomorphism $R : C^*(G(\pm)) \to D^*_\pm$ is surjective.*

Proof 6.69 *It is known in Proposition 6.9.18 that $C^*(G_\pm)$ is isomorphic to $D^*_\pm$. On the other hand, the surjective map $P : C_c(G(\pm)) \to C_c(G_\pm)$ can be extended to a surjective map from $C^*(G(\pm))$ onto $C^*(G_\pm)$ in the following sense. Let*

$$P_+ : C_c(G(\pm)) \to C_c(G(+)) \ (\textit{resp. } P_- : C_c(G(\pm)) \to C_c(G(-)))$$

be defined by sending $\eta \mapsto \eta_+$ (reps. $\eta \mapsto \eta_-$). Then P_+ (resp. P_-) extends to a surjective homomorphism from $C^*(G(\pm))$ onto $C^*(G(+))$ (resp. onto $C^*(G(-))$. Let $(a, b) \in C^*(G_\pm)$ and $\varepsilon > 0$. Since $C_c(G(+))$ (resp. $C_c(G(-))$) is dense in $C^*(G(+))$ (resp. $C^*(G(-))$), there are $\eta_+ \in C_c(G(+))$ and $\eta_- \in C_c(G(-))$ such that $\|a - \eta_+\| < \varepsilon/2$ and $\|b - \eta_-\| < \varepsilon/2$. Since $(a, b) \in C^*(G_\pm)$, we have that $\Phi_+(a) = \Phi_-(b)$, and since $\Phi_+(\eta_+), \Phi_-(\eta_-) \in C_c(\mathbb{R} \times V_0)$, $\|\Phi_+(a) - \Phi_+(\eta_+)\| < \varepsilon/2$ and $\|\Phi_-(b) -$

$\Phi_-(\eta_-)\| < \varepsilon/2$. Hence, $\|\Phi_+(\eta_+) - \Phi_-(\eta_-)\| < \varepsilon$. Let $\varphi := \Phi_-(\eta_-) - \Phi_+(\eta_+) \in C_c(\mathbb{R} \times V_0)$. Since $C^*(G(-))$ is $*$-regular (see [25]), we know that the kernel of Φ_- in $C_c(G(-))$ is dense in the kernel of Φ_- in $C^*(G(-))$ and therefore there exists an η'_- in $C_c(G(-))$ with $\|\eta'_-\|_{C^*} < \varepsilon$ such that $\Phi_-(\eta'_-) = \varphi$. Then we have $\Phi_-(\eta_- - \eta'_-) = \Phi_+(\eta_+)$, hence $(\eta_+, \eta_- - \eta'_-) \in C_c(G_\pm)$. Moreover $\|(\eta_+, \eta_- - \eta'_-) - (a, b)\| < 2\varepsilon$, and by Proposition 6.9.19 there is an $\eta \in C_c(G(\pm))$ such that $P(\eta) = (\eta_+, \eta_- - \eta'_-)$. This shows that P maps $C^*(G(\pm))$ onto $C^*(G_\pm)$. Then the proof follows from Proposition 6.9.18 and the fact that $R = J \circ P$ is now a surjective homomorphism.

6.9.9 The isomorphism

Definition 6.68 Let $D^*(G(\pm))$ be the subspace of $F^*(G(\pm))$ consisting of all operator fields A defined over the set Γ such that $A(f) \in \mathcal{K}$ for $f \in \mathcal{D}$, $A(f) \in \mathcal{B}$ for $f \in \mathcal{S}$ and $A(0_n) \in C^*(\mathbb{R} \times V_0)$ such that:

1. the mapping $f \mapsto A(f)$ is continuous on $\mathcal{D} \cup \mathcal{S}$,
2. A satisfies the infinity condition defined in Definition 6.55,
3. A satisfies the generic property defined in Definition 6.59, and
4. A satisfies the compact condition defined in Definition 6.61.

Proposition 6.9.21 *The vector space $D^*(G(\pm))$ is a C^*-algebra.*

Proof 6.70 *It clear that $D^*(G(\pm))$ is a $*$-algebra by Propositions 6.9.9 and 6.9.11. Let $A \in F^*(G(\pm))$ and $(A_n)_n$ be a sequence in $D^*(G(\pm))$ converging to A in $F^*(G(\pm))$. It is clear that A satisfies the infinity condition. Using the notations of the generic condition in Definition 6.59, we have*

$$\begin{aligned}
&\lim_{k\to\infty} \|U(r_k) \circ A(f_k) \circ U(-r_k) \circ M_{(-\rho_k,\infty)} - A(f) \circ M_{(-\rho_k,\infty)}\|_{\text{op}} \\
&\le \lim_{k\to\infty} \|U(r_k) \circ A_n(f_k) \circ U(-r_k) \circ M_{(-\rho_k,\infty)} - A_n(f) \circ M_{(-\rho_k,\infty)}\|_{\text{op}} \\
&\quad + \lim_{k\to\infty} \|U(r_k) \circ (A_n - A)(f_k) \circ U(-r_k) \circ M_{(-\rho_k,\infty)} - (A_n - A)(f) \circ M_{(-\rho_k,\infty)}\|_{\text{op}} \\
&\le \lim_{k\to\infty} \|U(r_k) \circ A_n(f_k) \circ U(-r_k) \circ M_{(-\rho_k,\infty)} - A_n(f) \circ M_{(-\rho_k,\infty)}\|_{\text{op}} \\
&\quad +2\|A_n - A\|_{\text{op}},
\end{aligned}$$

where $f_k \in \mathcal{D}$ and $f \in \mathcal{S}_-$. Similarly for $f_k \in \mathcal{D}$ and $f \in \mathcal{S}_+$, the generic approximation (1) in Definition 6.59 holds. That is, A satisfies the generic condition. Furthermore, for every $f \in \mathcal{S}$, we have that $A(f) - \pi_f(\beta(A(0_n))) = \lim_k A_k(f) - \pi_f(\beta(A_k(0_n))) \in \mathcal{K}$. Hence A is an element of $D^*(G(\pm))$.

It is proved in Proposition 6.9.9 that $F^*(G(\pm))$ is a C^*-algebra and we have just seen that $D^*(G(\pm))$ is a closed $*$-subalgebra of $F^*(G(\pm))$, hence it is a C^*-algebra.

Theorem 6.9.22 *The Fourier transform $\mathcal{F}$ maps $C^*(G(\pm))$ onto $D^*(G(\pm))$. In other words, $C^*(G(\pm))$ is isomorphic to $D^*(G(\pm))$.*

Proof 6.71 *We know from Proposition 6.9.9 that* $\mathcal{F}(C^*(G(\pm))) \subset F^*(G(\pm))$*. By Proposition 6.9.12 the fields* $\mathcal{F}(a)$ *for* $a \in C^*(G(\pm))$ *satisfy the generic condition. Since for* $(f_0, f_+, 0), (f_0, 0, f_-)$ *in* $\mathcal{S}$ *and* $a \in C^*(G(\pm))$ *the operator* $\pi_{(f_0,f_+,0)}(a)$ *is equal to the operator* $\pi_{(f_0,f_+)}(P_+(a))$*, it follows from Theorem 6.9.16 that* $\pi_{(f_0,f_+,0)}(a) - \pi_{(f_0,f_+,0)}(\beta(a(0_n))) = \pi_{(f_0,f_+)}(P_+(a)) - \pi_{(f_0,f_+)}(\beta_+(P_+(a)(0_n))) \in \mathcal{K}$*. Similarly for* $(f_0, 0, f_-) \in \mathcal{S}$*. This shows that* $\mathcal{F}(a)$ *for* $a \in C^*(G(\pm))$ *satisfies the compact condition. Hence* $\mathcal{F}(C^*(G(\pm))) \subset D^*(G(\pm))$*.*

In order to finish the proof, we have to show that $\mathcal{F}(C^*(G(\pm)))$ *is rich in* $D^*(G(\pm))$ *and that* $D^*(G(\pm))$ *is of type I. Let us determine the dual space of* $D^*(G(\pm))$*. Let* $\pi \in \widehat{D^*(G(\pm))}$*. Let* $I_{\mathcal{S}}$ *be the ideal of* $D^*(G(\pm))$ *of all operator fields in* $D^*(G(\pm))$ *vanishing on* $\mathcal{S}$*. We observe that the algebra* $C_0(\mathcal{D}, \mathcal{K})$ *can be injected into* $I_{\mathcal{S}}$*. Indeed, if* $\psi \in C_0(\mathcal{D}, \mathcal{K})$*, then we identify* ψ *with the operator field* $\Psi(\ell) = \psi(\ell)$ *for* $\ell \in \mathcal{D}$ *and* $\Psi(\ell) = 0$ *for* $\ell \in \Gamma \setminus \mathcal{D}$*. It is easy to see that the operator field* Ψ *satisfies the generic condition, since* $\lim_k \|\psi(f_k)\|_{\rm op} = 0$ *if* f_k *tends to* 0 *as* k *goes to infinity, and it satisfies also trivially the compact condition. Hence we can consider* $C_0(\mathcal{D}, \mathcal{K})$ *to be an ideal of* $D^*(G(\pm))$ *contained in* $I_{\mathcal{S}}$*. Take now* $A \in I_{\mathcal{S}}$ *and a sequence* $(f_k)_k \subset \mathcal{D}$ *tending to* 0 *as* k *goes to infinity. By the generic condition we have (with the notations in Definition 6.59) that*

$$\lim_{k\to\infty} \|U(r_k) \circ A(f_k) \circ U(-r_k) \circ M_{(-\infty,\sigma_k)}\|_{\rm op} = 0$$

and

$$\lim_{k\to\infty} \|U(s_k) \circ A(f_k) \circ U(-s_k) \circ M_{(-\rho_k,\infty)}\|_{\rm op} = 0.$$

Hence

$$\lim_{k\to\infty} \|A(f_k) \circ M_{(-\infty,\sigma_k+r_k)}\|_{\rm op} = 0$$

and

$$\lim_{k\to\infty} \|A(f_k) \circ M_{(-\rho_k+r_k,\infty)}\|_{\rm op} = 0.$$

Therefore

$$\lim_{k\to\infty} \|A(f_k)\|_{\rm op} = 0.$$

This shows that $C_0(\mathcal{D}, \mathcal{K}) = I_{\mathcal{S}}$*.*

If $\pi(I_{\mathcal{S}}) \neq \{0\}$*, then the representation* π *determines an element of* $\widehat{C_0(\mathcal{D}, \mathcal{K})}$ *and is so an evaluation at some point* $\ell \in \mathcal{D}$*. If* $\pi(I_{\mathcal{S}}) = \{0\}$*, then we can consider* π *as a representation of the quotient algebra* $D^*(G(\pm))/I_{\mathcal{S}}$*. Since* $I_{\mathcal{S}}$ *is the kernel of the homomorphism*

$$R' : D^*(G(\pm)) \to D^*_\pm \text{ by } A \mapsto A_{|\mathcal{S}\cup\{0_n\}}$$

and since $R = R' \circ \mathcal{F}$*, we know from Proposition 6.9.20 that* $D^*(G(\pm))/I_{\mathcal{S}}$ *is isomorphic to* $D^*_\pm$*. Hence we can identify* π *with an element of* $\widehat{D^*_\pm}$*. Let* I_0 *be the ideal of all* *the operator fields $A \in D^*_{\pm}$ vanishing at 0_n. Using the compact condition we see that I_0 is the algebra $C_0(\mathcal{S}, \mathcal{K})$. Hence, if π is not vanishing on I_0, then π is just an evaluation at some $\ell \in \mathcal{S}$. If $\pi(I_0) = \{0\}$, then π is a representation of $D^*_{\pm}/I_0$. Now I_0 is just the kernel of the homomorphism $A \mapsto A(0_n)$ for $A \in D^*_{\pm}$. Since $D^*_{\pm} = R(C^*(G(\pm)))$ by Proposition 6.9.20, we see that $D^*_{\pm}/I_0$ is isomorphic to $C^*(\mathbb{R} \times V_0)$. Hence π determines a character χ_ς of $C^*(\mathbb{R} \times V_0)$ and so $\pi(A) = \chi_\varsigma(A(0_n))$ for $A \in D^*(G(\pm))$. Having determined the dual space of $D^*(G(\pm))$, we know that this algebra is of type I and that $\mathcal{F}(C^*(G(\pm)))$ is rich in $D^*(G(\pm))$. Hence by the theorem of Stone-Weierstrass (see [37]), the two C^*-algebras $\mathcal{F}(C^*(G(\pm)))$ and $D^*(G(\pm))$ coincide.*

Bibliography

[1] R.J. Archbold, E. Kaniuth, J. Ludwig, G. Schlichting, and D.W.B. Somerset. Strength of convergence in duals of c^*-algebras and nilpotent Lie groups. *Adv. Math*, 158(1):26–65, 2001.

[2] R.J. Archbold, J. Ludwig, and G. Schlichting. Limit sets and strengths of convergence for sequences in the duals of thread-like Lie groups. *Math. Z.*, 255(2):245–282, 2007.

[3] D. Arnal, H. Fujiwara, and J. Ludwig. Opérateurs d'entrelacement pour les groupes de Lie exponentiels. *Amer. J. Math*, 118:839–878, 1996.

[4] W. B. Arveson. Subalgebras of C^*-algebras. *Acta Mathematica*, 123:141–224, 1969.

[5] A. Baklouti. Harmonic analysis on differential operators on nilpotent homogeneous spaces. *Russian J. Math. Physics*, 6:125–136, 1999.

[6] A. Baklouti. *Deformation Theory of Discontinuous Groups.* De Gruyter Expositions in Mathematics. De Gruyter, 2022.

[7] A. Baklouti, S. Dhieb, and D. Manchon. Orbites coadjointes et variétés caractéristiques. *J. Geom. Phys.*, 54:1–41, 2005.

[8] A. Baklouti and H. Fujiwara. The closedness of the product set of two Pukanszky polarizations of exponential Lie groups. *J. Math. Soc. Japan.* to appear.

[9] A. Baklouti and H. Fujiwara. A solution to Duflo's polynomial problem for nilpotent Lie groups monomial representations. *Int. J. Math.* to appear.

[10] A. Baklouti and H. Fujiwara. Commutativité des opérateurs différentiels sur l'espace des représentations restreintes d'un groupe de Lie nilpotent. *J. Math. Pures Appl.*, 83:137–161, 2004.

[11] A. Baklouti and H. Fujiwara. Discrete type restricted representations of exponential groups and differential operators. *Int. J. Math.*, 36(3), 2025.

[12] A. Baklouti, H. Fujiwara, and J. Ludwig. The polynomial conjecture for discrete type monomial representations of exponential groups. *Kyoto J. Math.* to appear.

[13] A. Baklouti, H. Fujiwara, and J. Ludwig. A variant of the frobenius reciprocity for restricted representations on nilpotent Lie groups. In *Infinite Dimensional Harmonic Analysis IV*, pages 13–31. World Scientific, 2008.

[14] A. Baklouti, H. Fujiwara, and J. Ludwig. Intertwining operators of irreducible representations for exponential solvable Lie groups. *Forum Math.*, 27:2231–2258, 2015.

[15] A. Baklouti, H. Fujiwara, and J. Ludwig. The polynomial conjecture for restrictions of some nilpotent groups representations. *J. Lie Theory*, 29:311–341, 2019.

[16] A. Baklouti, H. Fujiwara, and J. Ludwig. Monomial representations of discrete type of an exponential solvable Lie group. In *Springer Proceedings in Mathematics and Statistics*, volume 290, pages 1–55. Springer, 2020.

[17] A. Baklouti, H. Fujiwara, and J. Ludwig. *Representation Theory of Solvable Lie Groups and Related Topics.* Springer, 2021.

[18] A. Baklouti, H. Fujiwara, and J. Ludwig. A proof of the polynomial conjecture for restrictions of nilpotent Lie groups representations. *Represent. Theory. A.M.S*, 26:616–634, 2022.

[19] A. Baklouti, H. Fujiwara, and J. Ludwig. A proof of the polynomial conjecture for nilpotent Lie groups monomial representations. *Trans. Amer. Math. Soc.*, 376:6015–6032, 2023.

[20] A. Baklouti and H. Hamrouni. The multiplicity function of mixed representations on completely solvable Lie groups. *Tokyo J. Math.*, 30:41–55, 2007.

[21] A. Baklouti and J. Ludwig. Invariant differential operators on certain nilpotent homogeneous spaces. *Monatsh. Math.*, 134:19–37, 2001.

[22] Y. Benoist. Analyse harmonique sur les espaces symétriques nilpotents. *J. Funct. Anal.*, 59:211–253, 1984.

[23] P. Bernat et al. *Représentations des groupes de Lie résolubles.* Dunod, Paris, 1972.

[24] D. P. Blecher and Ch. Le Merdy. *Operator algebras and their modules — an operator space approach*, volume 30 of *London Mathematical Society Monographs, New Series.* Oxford Science Publications, 2004.

[25] J. Boidol. $*$-regularity of exponential Lie groups. *Invent. Math.*, 56(3):231–238, 1980.

[26] N. Bourbaki. *Topologie générale, Chap. 1–4.* Springer, 2007.

[27] R. C. Busby. Double centralizers and extensions of C^*-algebras. *Trans. Amer. Math. Soc.*, 132:79–99, 1968.

[28] M. D. Choi and E. G. Effros. The completely positive lifting problem for C^*-algebras. *Ann. of Math. (2)*, 104(3):585–609, 1976.

[29] M. D. Choi and E. G. Effros. Nuclear C^*-algebras and injectivity: the general case. *Indiana Univ. Math. J.*, 26(3):443–446, 1977.

[30] L. Corwin and F. P. Greenleaf. Spectrum and multiplicities for restrictions of unitary representations in nilpotent Lie groups. *Pacific J. Math.*, 135:233–267, 1988.

[31] L. Corwin and F. P. Greenleaf. *Representations of nilpotent Lie groups and their applications, Part I: Basic theory and examples.* Cambridge University Press, 1990.

[32] L. Corwin and F. P. Greenleaf. Commutativity of invariant differential operators on nilpotent homogeneous spaces with finite multiplicity. *Comm. Pure Appl. Math.*, 45:681–748, 1992.

[33] L. Corwin and F. P. Greenleaf. Spectral decomposition of invariant differential operators on certain nilpotent homogeneous spaces. *J. Funct. Anal.*, 108:374–426, 1992.

[34] C. Delaroche. *Extensions des C^*-algèbres*, volume 29 of *Bull. Soc. Math. France Mém.* Société Mathématique de France, Paris, 1972. Supplément au Bull. Soc. Math. France, Tome 100, 142 pp.

[35] J. Dixmier. Points séparés dans le spectre d'une C^*-algèbre. *Acta Sci. Math. Szeged*, 22:115–128, 1961.

[36] J. Dixmier. Représentations irréductibles des algèbres de Lie résolubles. *J. Math. Pures Appl.*, 45:1–66, 1966.

[37] J. Dixmier. *C^*-algebras*, volume 15 of *North-Holland Mathematical Library.* North-Holland Publishing Co., Amsterdam, New York, Oxford, 1977.

[38] J. Dixmier and P. Malliavin. Factorisation de fonctions et de vecteurs indéfiniment différentiables. *Bull. Sci. Math.*, 102:305–330, 1978.

[39] M. Duflo. Opérateurs différentiels bi-invariants sur un groupe de Lie. *Ann. Éc. Norm. Sup.*, 10:265–288, 1977.

[40] M. Duflo. Open problems in representation theory of Lie groups. In T. Oshima, editor, *Katata in Japan*, pages 1–5. 1986.

[41] A. Dynin. Inversion problem for singular integral operators: C^*-approach. *Proc. Natl. Acad. Sci. U.S.A.*, 75(10):4668–4670, 1978.

[42] S. Echterhoff. *Crossed products with continuous trace*, volume 123 of *Mem. Amer. Math. Soc.* 1996.

[43] J. M. G. Fell. The dual spaces of C^*-algebras. *Trans. Amer. Math. Soc.*, 94:365–403, 1960.

[44] J. M. G. Fell. The structure of algebras of operator fields. *Acta Math.*, 106:233–280, 1961.

[45] J. M. G. Fell. A Hausdorff topology for the closed subsets of a locally compact non-Hausdorff space. *Proc. Amer. Math. Soc.*, 13:472–476, 1962.

[46] H. Fujiwara. Certains opérateurs d'entrelacement pour des groupes de Lie résolubles exponentiels et leurs applications. *Mem. Fac. Sci. Kyushu Univ. Ser. A*, 36:13–72, 1982.

[47] H. Fujiwara. Représentations monomiales des groupes de Lie nilpotents. *Pacific J. Math.*, 127:329–351, 1987.

[48] H. Fujiwara. Représentations monomiales des groupes de Lie résolubles exponentiels. In M. Duflo, N. V. Pedersen, and M. Vergne, editors, *The orbit method in representation theory*, volume 82 of *Progress in Math.*, pages 61–84. Birkhäuser, Boston, 1990.

[49] H. Fujiwara. Sur les restrictions des représentations unitaires des groupes de Lie résolubles exponentiels. *Invent. Math.*, 104:647–654, 1991.

[50] H. Fujiwara. Sur la conjecture de Corwin-Greenleaf. *J. Lie Theory*, 7:121–146, 1997.

[51] H. Fujiwara. Analyse harmonique pour certaines représentations induites d'un groupe de Lie nilpotent. *J. Math. Soc. Japan*, 50:753–766, 1998.

[52] H. Fujiwara. Une réciprocité de frobenius. In H. Heyer, editor, *Infinite Dimensional Harmonic Analysis*, pages 17–35. World Scientific, 2005.

[53] H. Fujiwara, G. Lion, and B. Magneron. Algèbres de fonctions associées aux représentations monomiales des groupes de Lie nilpotents. Prépub. Math. Univ. Paris 13, 2002.

[54] H. Fujiwara, G. Lion, B. Magneron, and S. Mehdi. Commutativity criterion for certain algebras of invariant differential operators on nilpotent homogeneous spaces. *Math. Ann.*, 327:513–544, 2003.

[55] H. Fujiwara, G. Lion, and S. Mehdi. On the commutativity of the algebra of invariant differential operators on certain nilpotent homogeneous spaces. *Trans. Amer. Math. Soc.*, 353:4203–4217, 2001.

[56] H. Fujiwara and J. Ludwig. *Harmonic analysis on exponential solvable Lie groups*. Springer Japan, 2014.

[57] N. V. Gorbachev. C^*-algebra of the Heisenberg group. *Uspekhi Mat. Nauk*, 35(6(216)):157–158, 1980.

[58] G. Grélaud. *Désintégration des représentations induites des groupes de Lie résolubles exponentiels.* Thèse de 3ᵉ cycle, Université de Poitiers, 1973.

[59] D. Handelman and H. S. Yin. Toeplitz algebras and rotational automorphisms associated to polydiscs. *Amer. J. Math.*, 110(5):887–920, 1988.

[60] P. Hochs, Y. Song, and S. Yu. A geometric formula for multiplicities of k-types of tempered representations. *Trans. Amer. Math. Soc.*, 372:8553–8586, 2019.

[61] J. Inoue. Fourier transforms for affine automorphism groups on siegel domain. *J. Math. Soc. Japan*, 43:247–259, 1991.

[62] A. Jawhar and J. Ludwig. Disintegrating tensor representations of nilpotent Lie groups. *Trans. Amer. Math. Soc*, 361(2):819–848, 2009.

[63] S. Kaneyuki. *Homogeneous bounded domains and Siegel domains*, volume 241 of *Lecture Notes in Mathematics.* Springer, Berlin, 1971.

[64] F. Kassel and T. Kobayashi. Invariant differential operators on spherical homogeneous spaces with overgroups. *J. Lie Theory*, 29(3):663–754, 2019.

[65] A. A. Kirillov. Unitary representations of nilpotent Lie groups. *Russian Math. Surveys*, 17:57–110, 1962.

[66] E. Kurniadi and H. Ishi. Harmonic analysis for 4-dimensional real frobenius Lie algebras. In *Springer Proceedings in Mathematics and Statistics*, volume 290, pages 95–109. Springer, 2019.

[67] R. Y. Lee. Full algebras of operator fields trivial except at one point. *Indiana Univ. Math. J.*, 25(4):303–314, 1976.

[68] R. Y. Lee. On the C^*-algebras of operator fields. *Indiana Univ. Math. J.*, 25(4):303–314, 1976.

[69] R. Y. Lee. On the C^*-algebras of operator fields. *Indiana Univ. Math. J.*, 26(2):351–372, 1977.

[70] H. Leptin and J. Ludwig. *Unitary representation theory of exponential Lie groups.* De Gruyter, 1994.

[71] Y. F. Lin and J. Ludwig. The C^*-algebras of ax+b-like groups. *J. Funct. Anal.*, 259:104–130, 2010.

[72] Y. F. Lin and J. Ludwig. An isomorphism between group C^*-algebras of $ax+b$-like groups. *Bull. London Math. Soc.*, 45(2):257–267, 2013.

[73] G. Lion. Intégrale d'entrelacement sur des groupes de Lie nilpotents et indice de Maslov. In J. Carmona and M. Vergne, editors, *Non Commutative Harmonic Analysis*, volume 587 of *Lecture Notes in Math.*, pages 160–176. Springer-Verlag, New York, 1977.

[74] G. Lion. Indice de Maslov et représentation de Weil. *Pub. Math. Univ. Paris VII*, 2:45–79, 1978.

[75] G. Lion. Groupes de Lie résolubles et indices de Maslov. *C. R. Acad. Sci. Paris, Ser. A*, 291:7–10, 1980.

[76] R. L. Lipsman. Orbital parameters for induced and restricted representations. *Trans. Amer. Math. Soc.*, 313:433–473, 1989.

[77] R. L. Lipsman. The Penney-Fujiwara Plancherel formula for homogeneous spaces. In M. Duflo, N. V. Pedersen, and M. Vergne, editors, *The orbit method in representation theory*, volume 82 of *Progress in Math.*, pages 135–145. Birkhäuser, Boston, 1990.

[78] R. L. Lipsman. The Penney-Fujiwara Plancherel formula for homogeneous spaces. In *Fuji-Kawaguchiko Conference on Representation Theory of Lie Groups and Lie Algebras*, pages 120–139. World Scientific, 1992.

[79] R. L. Lipsman and J. Rosenberg. The behavior of Fourier transforms for nilpotent Lie groups. *Trans. Amer. Math. Soc.*, 348(3):1031–1050, 1996.

[80] J. Ludwig. On the behaviour of sequences in the dual of a nilpotent Lie group. *Math. Ann.*, 287:239–257, 1990.

[81] J. Ludwig and H. Regeiba. C^*-algebras with norm controlled dual limits and nilpotent Lie groups. *J. Lie Theory*, 25(3):613–655, 2015.

[82] J. Ludwig, G. Rosenbaum, and J. Samuel. The elements of bounded trace in the C^*-algebra of a nilpotent Lie group. *Invent. Math.*, 83(1):167–190, 1985.

[83] J. Ludwig and L. Turowska. The C^*-algebras of the Heisenberg group and of thread-like Lie groups. *Math. Z.*, 268(3-4):897–930, 2011.

[84] H. Moscovici. An infinitesimal characterization of square integrable representations of nilpotent Lie groups. *Rev. Roumaine Math. Pures Appl.*, 21:529–530, 1976.

[85] H. Moscovici and A. Verona. Coherent states and square integrable representations. *Ann. Inst. H. Poincaré Sect. A*, 29:139–156, 1978.

[86] A. I. Ooms. On Lie algebras with primitive envelopes, supplements. *Proc. Amer. Math. Soc.*, 58:67–72, 1976.

[87] A. I. Ooms. On Frobenius Lie algebras. *Comm. Algebra*, 8:13–52, 1980.

[88] N. V. Pedersen. On the infinitesimal kernel of irreducible representations of nilpotent Lie groups. *Bull. Soc. Math. France*, 112:423–467, 1984.

[89] N. V. Pedersen. On the symplectic structure of coadjoint orbits of (solvable) Lie groups and applications, part i. *Math. Ann.*, 281:633–669, 1988.

[90] N. V. Pedersen. Geometric quantization and the universal enveloping algebra of a nilpotent Lie group. *Trans. Amer. Math. Soc.*, 315(2):511–563, 1989.

[91] N. V. Pedersen. Orbits and primitive ideals of solvable Lie algebras. *Math. Ann.*, 298(2):275–326, 1994.

[92] R. Penney. Abstract Plancherel theorem and a Frobenius reciprocity theorem. *J. Func. Anal.*, 18:177–190, 1975.

[93] N. C. Phillips. *Equivariant K-theory and freeness of group actions on C^*-algebras*, volume 1274 of *Lecture Notes in Mathematics*. Springer-Verlag, Berlin, 1987.

[94] M. Raïs. Solutions élémentaires des opérateurs différentiels biinvariants sur un groupe de Lie nilpotent. *C. R. Acad. Sci. Paris, Série A*, 273:495–498, 1971.

[95] H. Regeiba and J. Ludwig. The C^*-algebra of some 6-dimensional nilpotent Lie groups. *Adv. Pure Appl. Math.*, 5(3):173–200, 2014.

[96] H. Regeiba and J. Ludwig. The C^*-algebra of some 6-dimensional nilpotent Lie groups. *Adv. Pure Appl. Math.*, 5(3):173–200, 2014.

[97] J. Rosenberg. Homological invariants of extensions of C^*-algebras. In *Operator algebras and applications, Part 1 (Kingston, Ont., 1980)*, volume 38 of *Proc. Sympos. Pure Math.*, pages 35–75. Amer. Math. Soc., Providence, RI, 1982.

[98] O. Takenouchi. Sur le facteur-représentation des groupes de Lie de type (e). *Math. J. Okayama Univ.*, 7:151–161, 1957.

[99] M. Takesaki. *Theory of Operator Algebras III*, volume 127 of *Encyclopaedia of Mathematical Sciences*. Springer-Verlag, Berlin, 2003. Operator Algebras and Non-commutative Geometry, Vol. 8.

[100] Y. Tanimura. An algebraic approach to Duflo's polynomial conjecture in the nilpotent case. *J. Lie Theory*, 29:839–879, 2019.

[101] P. Turkowski. Solvable Lie algebras of dimension six. *J. Math. Phys.*, 31:1344–1350, 1990.

[102] J. Vargas and B. Orsted. Restriction of square integrable representations: Discrete spectrum. *Duke Math. J.*, 123(3):609–633, 2004.

[103] D. Voiculescu. Remarks on the singular extension in the C^*-algebra of the heisenberg group. *J. Operator Theory*, 5(2):147–170, 1981.

[104] X. Wang. *The C^*-algebras of a class of solvable Lie groups*, volume 199 of *Pitman Research Notes in Mathematics Series*. Longman Scientific & Technical, Harlow, 1989.

[105] N. E. Wegge-Olsen. *K-theory and C^*-algebras: A friendly approach.* Oxford Science Publications, New York, 1993.

Index

For Product Safety Concerns and Information please contact our EU representative GPSR@taylorandfrancis.com
Taylor & Francis Verlag GmbH, Kaufingerstraße 24, 80331 München, Germany

www.ingramcontent.com/pod-product-compliance
Ingram Content Group UK Ltd.
Pitfield, Milton Keynes, MK11 3LW, UK
UKHW052219190926
475702UK00006B/1098

* 9 7 8 1 0 4 1 1 6 2 2 3 0 *